Benoit Mandelbrot

A Life in Many Dimensions

Fractals and Dynamics in Mathematics, Science, and the Arts: Theory and Applications

ISSN: 2382-6320

Vol. 1 Benoit Mandelbrot: A Life in Many Dimensions
edited by Michael Frame (Yale), Nathan Cohen (Fractal Antenna Systems, Inc., USA).

Fractals and Dynamics in Mathematics, Science and the Arts
Vol 1
Theory and Applications

Benoit Mandelbrot

A Life in Many Dimensions

editors

Michael Frame
Yale University

Nathan Cohen
Fractal Antenna Systems, Inc.

World Scientific

Published by

World Scientific Publishing Co. Pte. Ltd.
5 Toh Tuck Link, Singapore 596224
USA office: 27 Warren Street, Suite 401-402, Hackensack, NJ 07601
UK office: 57 Shelton Street, Covent Garden, London WC2H 9HE

Library of Congress Cataloging-in-Publication Data
Benoit Mandelbrot : a life in many dimensions / edited by Michael Frame (Yale), Nathan Cohen (Fractal Antenna Systems, Inc., USA).
pages cm. -- (Fractals and dynamics in mathematics, science, and the arts : theory and applications ; volume 1)
Includes bibliographical references and index.
ISBN 978-9814366069 (hardcover : alk. paper) -- ISBN 978-9814366076 (electronic : alk. paper)
1. Mandelbrot, Benoit B. 2. Mathematicians--France--Biography. 3. Mathematicians--United States--Biography. 4. Fractals. I. Frame, Michael, editor. II. Cohen, Nathan (Nathan L.), editor.
QA29.M34B46 2015
510.92--dc23
[B]

2015001994

British Library Cataloguing-in-Publication Data
A catalogue record for this book is available from the British Library.

Typeset by Stallion Press
Email: enquiries@stallionpress.com

Contents

Contributors

Patrice Abry
CNRS, Laboratoire de Physique, Ecole Normale Supérieure de Lyon
46, allée d'Italie
69364 Lyon cedex 7, France

Michael F. Barnsley
Mathematical Sciences Institute
Australian National University
Canberra ACT 0200, Australia

Julien Barral
Département de Mathématiques
LAGA (UMR 7539), Institut Galilée
Université Paris-13
Sorbonne Paris Cité, 99 Avenue Jean-Baptiste Clément
93430 Villetaneuse, France

Javier Barrallo
The University of the Basque Country UPV/EHU
E.T.S. Arquitectura, Plaza Oñati, 2
20018 Donostia-San Sebastián, Spain

Harlan J. Brothers
Director of Research & Development
Brothers Technology, LLC
1204 Main Street, P.O. Box 1016
Branford, CT 06405-8016, USA

Nathan Cohen
Fractal Antenna Systems, Inc.
213 Burlington Rd
Bedford, MA 01730, USA

Kate E. Ellis
Department of Mathematics and Statistics
California State Polytechnic University, Pomona
3801 West Temple Avenue
Pomona, CA 91768, USA

Kenneth Falconer
Mathematical Institute, University of St Andrews
North Haugh, St Andrews
Fife, KY16 9SS, UK

Michael Frame
Department of Mathematics, Yale University
PO Box 208283
New Haven, CT 06520-8283, USA

Brendan Harding
Mathematical Sciences Institute
Australian National University
Canberra ACT 0200, Australia

Richard L. Hudson
CEO & Editor, Science|Business
Science Business Publishing Ltd
166 Northwood Way, Northwood
Middlesex, HA6 1RB, UK

Stéphane Jaffard
LAMA, Université Paris-Est — Créteil Val-de-Marne
61 avenue du Général de Gaulle, Bâtiment P2
94010 Créteil, France

Michel L. Lapidus
Department of Mathematics
University of California, Riverside
900 University Avenue
Riverside, CA 92521-0135, USA

Nigel Lesmoir-Gordon
The Broadcasting, Entertainment, Cinematograph
and Theatre Union (BECTU)
373-377 Clapham Road
London, SW9 9BT, UK

Michael C. Mackenzie
Department of Mathematics
University of Connecticut
196 Auditorium Road U-3009
Storrs, CT 06269-3009, USA

Demetri Martin
Comedian and Writer
Santa Monica, California

William Martino
Senior Science Advisor, Quantitative Analytics Unit
Securities and Exchange Commission (SEC)
New York Regional Office
Brookfield Place, 200 Vesey Street, Suite 400
New York, NY 10281-1022, USA

Nial Neger
Mathematics Department Chairperson (retired)
Trumbull High School
Trumbull, CT 06611, USA

Emer O'Daly
Love & Robots,
(also affiliated with O'Daly Architects)
59 Dawson St.
Dublin 2, Ireland

Jacques Peyrière
Université Paris-Sud, Mathématique bât. 425
CNRS UMR 8628
91405 Orsay Cedex, France

Vladas Pipiras
Department of Statistics and Operations Research
University of North Carolina, Chapel Hill
Hanes Hall, CB#3260
Chapel Hill, NC 27599, USA

Peter Renz, Retired.
Formerly variously serving as
editor for W. H. Freeman and Company,
associate director of the Mathematical Association of America, editor for Birkhauser, Boston, and Academic Press, and
professor at Reed, Wellesley, and Bard Colleges

John A. Rock
Department of Mathematics and Statistics
California State Polytechnic University, Pomona
3801 West Temple Avenue
Pomona, CA 91768, USA

Michael Shlesinger
Office of Naval Research, Code 30
875 N. Randolph Street
Arlington, VA 22203, USA

Murad S. Taqqu
Department of Mathematics and Statistics
Boston University
111 Cummington Street
Boston, MA 02215, USA

Christian Walter
Ethics and Finance Chair
Collège d'études mondiales
Fondation Maison des sciences de l'homme
190-196, avenue de France, CS 71345
F-75648 Paris Cedex 13, France

Ewald R. Weibel
Institute of Anatomy, University of Bern
Baltzerstrasse 2
CH-3000 Bern 9, Switzerland

Herwig Wendt
CNRS, IRIT, University of Toulouse
2 rue Camichel, BP 7122
31071 Toulouse cedex 7, France

Bruce J. West
Information Sciences Directorate
US Army Research Office
Research Triangle Park, NC 27709, USA

Damien West
Physics Department
Renesselaer Polytechnique Institute
Troy, NY 12108, USA

Charles Wuorinen
Composer, New York City
Member, American Academy of Arts and Letters
Member, American Academy of Arts and Sciences

Preface

Benoit Mandelbrot died on October 14, 2010. He had been diagnosed with pancreatic cancer about six weeks earlier, but he had not looked well when my wife Jean and I visited Benoit and his wife Aliette in Cambridge over the summer. Benoit and I had started writing two papers and a book a couple of years earlier, but perhaps from intuition, he put aside these and other projects to work full-time on his memoirs. This book was very important to him; it was his chance to tell the story of his complicated life. Benoit did not live to finish his memoirs, but aided by his IBM colleague Merry Morse, Aliette completed the book, *The Fractalist. Memoirs of a Scientific Maverick*, Pantheon, 2012.

The Fractalist is Benoit's narrative of his life, and a little of his work. Certainly, details of his work are more familiar, so focusing on the who, what, where, and when of the path he took from child in Warsaw to Sterling Professor of Mathematics at Yale is the story that more needed to be told. Indeed, Benoit was very much more than his work, but his work was so central to his life that a full picture of the man cannot be disentangled from the work. Benoit had planned four more chapters for his memoirs: complex iteration, dimension and roughness, finance, and physics. We had discussed these shortly after Benoit started writing his memoirs, but we left them for the future. There would be time enough, there always would be time enough. But it turned out that there wasn't.

After his diagnosis, we had a careful discussion of these chapters. Who could write them? I offered to try, but Benoit knew that my understanding of his work is too superficial. Still, I believe the offer pleased him, loyalty between friends. We thought of different people for each chapter, but didn't get very far with this conversation before our conversations ended. I had one other idea, to cobble together these chapters by cutting and pasting bits from Benoit's *Selecta*. Complex iteration seemed the simplest, so I wrote a short chapter, all Benoit's words, taken from *Fractals and Chaos. The Mandelbrot Set and Beyond*, Springer Verlag, 2004, volume C of his *Selecta*. Probably wisely, the publisher of *The Fractalist* declined. I'd got the words right, but not the tune.

Doubtless there will be historical analyses of Benoit's work, though the breadth and depth of his interests will make this a challenge. Lacking that, I suggested a memorial volume. World Scientific, publisher of the journal *Fractals*, agreed. I invited people who had worked with Benoit, or whose work Benoit knew and appreciated. Some chose other ways to celebrate Benoit's life, but those who contributed to this project constitute a good cross-section of Benoit's interests. Mathematicians, physicists, and financiers, to be sure, and also teachers, artists, a composer, a musician, an architect, a film maker, a publisher, and a comdeian. Aliette wrote a delightful sketch of watching Benoit at work. Also included are some of the stories told at the memorial meeting held at Yale for Benoit in April, 2011.

The delays in producing this volume are due entirely to my health challenges. I have tested the patience of the volume's contributors, and especially of Joy Quek of World Scientific Publishers, beyond all reasonable bounds. That Joy never showed anything other than concern and understanding is a gift I treasure.

Since I was a kid in 1950s West Virginia, I wanted to be a scientist, suprising my definitely non-scientist parents. Early on, I discovered that I lacked the intellect to make any real contribution to science. Accepting this disappointment took some effort, but it did have a surprising consequence. Because I'm confused almost all the time about almost everything, I'm pretty patient at crafting explanations for my students. Benoit noticed this and brought me into his world. Against all expectations, I wound up seeing the growth of new science from near its center. This book is part of my thank you to Benoit. It is a sketch of the size of his ideas. They are very big ideas indeed.

Michael Frame
New Haven, Connecticut
24 July 2014

Preface

What Made Mandelbrot Great

> *"This book is... nontraditional and its style and organization are unconventional"*
>
> — *Benoit Mandelbrot*

Benoit Mandelbrot died in October, 2010, mere days after the announcement of winners of the 2010 economics Nobel Prize (a misnomer; it's actually the "Sveriges Roksbank Prize in memory of Alfred Nobel"). He was not so named. Thus evaporated the last chance to honor the world's best- known mathematician for creating a solution (some may contend 'unconventional') for understanding financial markets. Like many things in Benoit's life, this snub was one of those outrageous ironies — based on fractals and the principles of 'roughness', Mandelbrot had long warned of the economic depression in which the world then stood deeply. Markets aren't pure Gaussian noise, but envelopes of fascinating complexity best described by Benoit's roughness tools that filled, in Benoit's words, "the unwelcome gap". Like the falling particles that make perfect Gaussians in an hour glass, some people still had their heads in the sand.[1]

Fortunately the rest of the world did not. Benoit's passing was news on a worldwide basis, a sad nadir to more than a generation of heady times created by Benoit's marvel of fractals and the tools of roughness.

Of course, in addition to being an economist (indeed, Mandelbrot had a brief offer to be professor of economics at the University of Chicago, rescinded because of fears he would further stray beyond that realm: he did), Mandelbrot was the undisputed founder of fractal geometry; he was a first-rate pure and applied mathematician. He was also a military man, a psychology researcher, a physicist, an historian, a geophysicist, a computer scientist (who never touched a computer), a geographer, a linguist, and a cosmologist. To be fair, the list should probably be dramatically longer, if one includes those fields in which Benoit cultivated disciples and colleagues, those he made the roughness connection for, and those who made it by education and inspiration. Benoit bore a remarkable number of what he called "intellectual fruit".

Benoit was a Renaissance man who created a renaissance.

Few centuries can claim true polymaths but ours can comfortably claim Benoit Mandelbrot. Because such remarkable individuals are rare, it behooves us to use our experience to understand that gift. Benoit was not beyond exploring some of these self-aspects with true humility and honesty.[2] This volume contains additional important examples, in tribute by Benoit's friends, colleagues, and family, that detail fields born and facilitated by Benoit's fertile mind. It also provides examples of Benoit the man or, more appropriately in Yiddish, the 'mensch'.

Here, anticipating the question these examples suggest, I shall try to briefly detail the unique combination of talents of Benoit, and the situation in which he found himself, that nurtured such broad expertise and influence.

To wit: What made Mandelbrot great?

First, Mandelbrot had a commanding memory. If pressed, he would admit to memorizing book pages, and a recall of old conversations. I discovered this by picking up a conversation with Benoit from some six months earlier, seemingly left mid- sentence prior. He seamlessly continued as if the delay had been but a second. Of course, such memories can have

[1]Mandelbrot's former student Gene Fama won the prize in 2013.
[2]Mandelbrot, B., The Fractalist, Pantheon Book, New York 2012.

profound downsides, as perceived dismissals and tough times from decades past — and Benoit saw many — seem fresh and with renewed sting. The memory of his youth during World War II likely held both gratitude (he was hidden as a Gentile peasant boy in Tulle, France) as well as terror and heartbreak, which the military side of Mandelbrot likely meted and hid with great alacrity. Benoit could be, and was, a 'tough guy' as needed. He often needed to be. He never forgot.

Armed with such a memory, Benoit put it to work. Benoit was at home in the library stacks worldwide, whether in Paris, New Haven, or Cambridge (both of them), and sucked in a vast realm of information on countless subjects. No pedant, Benoit was a walking but more erudite extension of an encyclopedia. This allowed him to see connections across very disparate fields and ideas, at a time when the intellectual flavor of the century was to suck the narrowness of reductionism until it was bone dry. Benoit's favorite word for intellectual extensions via reductionism: "trivial".

With relish, Mandelbrot was a visual person. By that I mean he not only likely saw deeper detail with his sight–perhaps by training–but also could project solutions visually in his mind's eye. As such, Mandelbrot could 'see' answers to important mathematical questions, and visually pull together those disparate details.[3] This allowed him to take a passel of long-ignored 'pathological curves' and find commonality and utility by defining an entire field of geometry–fractal geometry — of which they represented but small examples. Then, just to thrust home the point, he commanded — *he saw* — that fractals are *the* geometry of nature. And he saw this with great clarity. Note the absence of 'a' or 'one of'. He left no doubt to that interpretation.

We drank up this revelation giddily, feeling its inherent truth. Yet we are left with interesting and long-term questions on the fractality of nature: what processes make things fractal, why do they manifest over finite iterations, why do they often skip iterations of scaling, leaving 'scaling gaps'? We are children in the tapping of this rich insight of nature posed by Mandelbrot.

Mandelbrot was, first and foremost, a great mathematician and the formalism and analytics of math were second nature to him. He was notably

[3]Mandelbrot 'visualized' and constructed many mathematical conjectures, and attributed the awarding of 3 Fields Medals to colleagues for proving three of them.

of a deductive mindset–when needed. Yet Mandelbrot was one of the first to show in modern times that *inductive* math, math through experiments, was not only alive but essential. The exquisitely beautiful Mandelbrot set, naively a map of the relation and realm of *all* numbers, came alive not through mere equations (actually one simple one), but the experimental application of computer graphics. "Let's try this". The squashed bugs and ink splats of the early graphics of the Mandelbrot set told of new worlds that otherwise would have been approximated away for the sake of analytical simplicity. In many ways, Mandelbrot's efforts in the 1970's lent permission to explore and solve interesting problems, big problems, with computer graphics and iterative simulation.

Oddly, Mandelbrot often described himself as a modern day Kepler. I am convinced this is a bit of self-deprecation for those who unwisely wished to mistake Benoit's genius as vanity or in-your-face ego. Not so. Benoit never had any need to exaggerate. And he was no Kepler: Kepler was, in comparison, minor league. Kepler achieved his aspirations via perspiration and confabulation. His talent lay in fortitude and a maniacal unwillingness to stop chasing the windmill posed by Mars' orbit. It may be appropriate to say that Kepler achieved success through induction, and lots of it. He tried to solve Mars' orbit with every shape literally under the Sun, and posed the first true challenge to the assumption that the universe was made of perfect Euclidean shapes. But that is where the similarity between Mandelbrot and Kepler ends. Goodness!

The root of Genius Mandelbrot lay in talent, but it was enabled by circumstance. For decades Mandelbrot chased a sense of place but took decades to find an academic home. Not for lack of trying — name the top universities in the United States: each looked like they were ready to offer Benoit a permanent professorship at some point. Each time he was penalized for his scholarship in intellectual variety. Luckily, he had one great non-university home, at IBM. His open-ended job held versatility and places to explore. IBM gave Mandelbrot permission to wander. Later, Yale understood this as well, finally rewarding him with a professorial position that had eluded for nearly fifty years. In this game, Harvard lost. We all won.

Despite the grumblings of a few academic competitors, Mandelbrot was a kind and giving person. This volume exists because of the gracious and patient efforts of its authors, and most significantly to head editor Michael Frame. We all knew Benoit's generosity well. There is no sense of obligation to reciprocate, it is our pleasure. In particular, Michael, a renowned fractalist and teacher in his own right, worked tirelessely, under challenging personal circumstances, to bring this volume to reality, and to make its contents mathematically and scientifically valuable beyond a mere tribute. In addition, as Benoit's closest colleague, much of what the world knows and owes Mandelbrot from his later years comes from direct or indirect efforts of his work with Michael. Theirs was a collaboration and friendship rare in scientific circles.

Lastly, what made Mandelbrot great was all of us: the family that supported and loved him; the friends and colleagues that were enthralled by his erudition, insight, kindness, humor, and collaboration; and the students and readers who lit up to seeing the mathematical and real world in a new and exciting way, rich with insight, alive with possibilities. Generations will be humbled and grateful to Benoit Mandelbrot as we begin, with this book, the celebration of a life lived rich in many dimensions.

Nathan Cohen
Bedford, Massachusetts
19 August 2014

Chapter 1

Watching Benoit at work

Aliette Mandelbrot

I remember Benoit (Fig. 1.1) always sitting at his desk (Fig. 1.2), years of sitting for many long hours, writing, crossing out, rewriting, cutting, pasting. While writing, he would often check one of his many dictionaries. It was important for him to use words with their exact meanings.

He would then give a long draft to his secretary for retyping. When he got back the clean copy, he repeated the exact same process on the clean copy. In no time, the copy was barely readable. This process went back and forth many times. Benoit was highly demanding of himself. He felt that the writing improved with each passage back from the secretary, and he did so until he was satisfied that his text was written as well as it could be.

He may have been one of the last rare scientists not using the computer ever, but writing by hand, and having assistants and programmers doing the computer rendering of his algorithms under his direction. He understood the possibilities of the computer very well, but said he was bad at and unwilling to follow precise instructions. He also had an uncanny eye for seeing results in pictures that no one else saw.

Figure 1.1 Benoit.

Figure 1.2 Benoit's desk.

He always said that it was so hard for him to write articles or books, because he did not know who he was writing for. Mathematicians, physicists, astronomers, hydrologists, economists, or others. He did not know the language of each field, and could not use one language for all.

When he did not write, but still sitting at his desk, he would read. Many journals, many requests, colleagues' books, papers, numerous letters from all over the world, including letters from fans. He tried to answer them all.

He also used the telephone a great deal to communicate with people, still sitting at the same desk. The telephone bills were from many countries in the world, and for many hours of conversation. He loved to have information on science, on people, on politics. However, he always said that he was not a politician for his own affairs, and that fact did cost him a lot of good will.

To conclude, I will say that Benoit loved Science above everything else, was immensely happy when he thought of an interesting new fact, and was frustrated that he could not get results faster on a number of problems. He always felt that there are still many more questions waiting to be tackled.

Chapter 2

Benoit Mandelbrot: nor does lightning travel in a straight line

Michael Frame

"There were in those days giants in the earth."

Genesis vi:4

Benoit and I had one discussion about mortality (well, two, but the first, soon after we met, was humorous), though we never discussed religion. So why do I start my chapter with a Genesis quote? I want to try to sketch a picture of the brilliant, complex, and in an important way, very playful person who, for reasons I'll never understand, included me in his world, and who was for twenty years — a long time, but not nearly long enough — a dear friend. But about the Genesis quote I'll say this: the reference is to Rölvaag's novel *Giants in the Earth*, about pioneers in the Dakotas. I believe Benoit is better described as "pioneer" than as "explorer." Yes, he explored many new ideas, clearly saw vistas of whole worlds unnoticed by others. But he did not simply mark the territory and move on. In each new area he stayed a while, looked carefully all over the place. He built things, surveyed the land for hordes of others, including the contributors to this volume. So pioneer it is.

Every life is intricate, layers upon layers, dreams, fears, quiet places where the mind goes alone in the emptiness of the night. What a preposterous claim that we can know anyone; maybe important aspects of even our own lives are unknowable to each of us. Walled off by a guarded nature, some people are hard to know. A few others show the opposite problem: the reach of their interests is so vast that a complete picture is far too complex. Benoit was of the latter kind, more than anyone else I've known, or even imagined. I saw only the tiniest fraction of Benoit's remarkable mind. But what I did see, I am happy to try to share.

The most immediate expression of the breadth of Benoit's interests is the table of contents of the memorial in your hands now. Other memorials will cast different nets, but what you have here shows the perspectives of people in architecture, art, biology, chemical engineering, comedy, economics, education, electrical engineering, film, finance, journalism, mathematics, music, physics, and publishing, and this is only a tiny shadow of the fields Benoit influenced. Some contributions are quite technical, others are stories told by people recounting their work with Benoit or their work inspired by Benoit. As an editor of this collection, I decided that a straight line through Benoit's scientific work would not capture enough of his infectious curiosity. Rather, we take a more complicated path. The range of topics presented here, and the range of levels of personal interactions recounted, hint at the kinds of work Benoit did, at the ways in which he worked.

There were giants in the earth, still are some giants in the earth. One has fallen now. Come, look at the size of his footprint.

2.1 Work

I won't try to sketch the details of Benoit's life; his memoirs [1], and the memorials of the National Academy of Sciences [2] and of the American Mathematical Society [3], are good sources for this information. However, the first two sentences of the epilogue of [1],

> You have now heard my story. Does not the distribution of my personal experiences remind one of the central topic of my scientific work, namely, extreme fractal unevenness?

hint at the zigs and zags of his life. For some, complicated lives do not much perturb the simple arc of their work. Benoit was not such a person. His life was complex; his work was complex, squared. Even a very rough list of topics he introduced or extended is dizzying in its breadth: word frequencies and thermodynamics of language; fractional Brownian motion; Lévy flights; self-similar and self-affine sets; approximating dimensions of physical objects by scaling; multifractals and turbulence; computer-generated images of Julia sets; the first pictures of, the connectivity, boundary dimension =2, and hyperbolicity conjectures for, the Mandelbrot set; fast generation of circle inversion and Kleinian group limit sets; statistics of large DLA clusters; Brownian bridges and the 4/3 conjecture; multifractal finance cartoons, the trading time theorem; lacunarity; and negative dimensions. And these are just the main points of his work that Benoit and I discussed. Benoit did so much more.

I'll comment a bit more extensively on a few areas. In addition to many, many papers, main references are his books [4–9]. More were planned, some were started. Without him, I doubt any will be finished, at least as he had planned. But the areas he explored are so very fertile. The work he started will continue to grow, twisting and turning and branching into new directions, a limitless supply of surprises. Benoit is gone, he will do nothing else direct in the world. But the ideas he started and nurtured through many dry years until finally the whole field of fractals exploded, will continue long after we, too, are gone. Our understanding of the world is so much more nuanced for his having passed through, staying a while, and asking really, really good questions.

2.2 Scaling in us and around us

Often Benoit said that scaling in nature has been known for centuries to children and artists. For children, moss is pasture for toy livestock, pebbles are boulders, twigs are logs. Janson [10] describes Alexander Cozens' method for assisting invention of landscape paintings. Starting from da Vinci's observation that stains on walls could stimulate the imagination, Cozens suggested crumpling a sheet of paper, smoothing it, and blotting ink on it. Crumpling produces folds on many levels, mimicking the appearance of weathering on mountains. Two centuries later, in his invention of frottages, Max Ernst was inspired by the grain of an often-scrubbed floor. Passeron [11] reports Ernst, "placed some sheets of paper on this rough surface and rubbed them with a lead pencil.... Of course this is a process known to all children, but in terms of the creative and quasi-hallucinatory attitude of the painter, Ernst had discovered an unusually ambiguous poetic technique."

That trees and ferns, the respiratory, circulatory, and nervous systems, river networks, mountain ranges, clouds, and diffusion tracks in rocks exhibit scaling is recognized by everyone who looks and thinks for even a moment. Scaling is in us and around us. Two natural fractals are shown on the left of Fig. 2.1.

Mathematical examples of scaling begin with Bolzano's 1830 nowhere differentiable function and continue through the late 19th and early 20th centuries with the Cantor set, the Koch curve, and the Sierpinski gasket and carpet. Let's think about nowhere differentiable curves. For the functions to which it applies, calculus works because as we look ever more closely at a point on the graph of the function, the graph looks more and more like its tangent line at that point. One way to make this fail is to build the function so each little bit of the graph resembles the whole graph. Zigs and zags on all scales preclude tangent lines.

Bolzano showed his continuous function was non-differentiable on a countable, dense set of points; now we know it is nowhere differentiable everywhere. Bolzano's is not the best-known of the nowhere differentiable functions because this work went unpublished and was made widely available first in 1921 by M. Jasek [12]. Nevertheless, this is the example we use. Start with an *initiator*, here we take the line segment between $(0,0)$ and $(1,1)$. The *generator* is the broken line segment with vertices $(0,0)$, $(1/3,2/3)$, $(1/2,1/2)$, $(5/6,7/6)$, and $(1,1)$. Then iterate this process: replace each line segment with a scaled copy of the generator, reflected vertically if the left endpoint of the segment is higher than the right. The right side of Fig. 2.1 shows the sixth iterate. The unit square is included to provide reference for size.

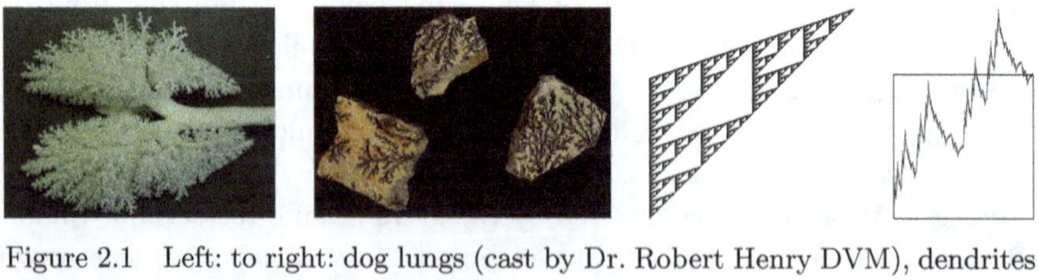

Figure 2.1 Left: to right: dog lungs (cast by Dr. Robert Henry DVM), dendrites in limestone, a Sierpinski gasket, Bolzano's function.

We see scaling is a natural ingredient for counterexamples to conjectures in analysis. Nevertheless, when these examples were created, many mathematicians regarded them as monsters. Benoit recognized that the scaling observed in nature could be understood by adapting the mathematics of scaling in these monsters, that in addition to the familiar symmetries under translation, reflection, and rotation, symmetry under magnification expresses many of nature's processes, both for growth and for erosion.

2.3 Random fractals and fractals in finance

Deterministic fractals, for example, the two right images of Fig. 2.1, work just fine for mimicking all but two aspects of the scaling symmetry of complex natural shapes. The first exception is that mathematical fractals, the limits of iterative geometric constructions, exhibit the same structure down to arbitrarily small scales. No physical object can do this: scaling can be manifested for only a finite range, the *scaling range*, over which the forces responsible for that structure are dominant: the mechanism by which ocean waves sculpt coastlines is the same over distances from tens of kilometers down to meters. Any claim of physical fractality must be accompanied by an estimate of the scaling range, and the larger that range, the more important fractality is in describing that object. In the youth of the field, some zealous investigators ignored this; others claimed fractality with only about a decade of data. These actions contributed to the difficulty some had with accepting that fractal geometry had anything useful to say about science.

The other exception is that the deterministic fractals of Fig. 2.1 have exactly the same structure on all levels. In contrast, the growth of any natural fractal is influenced by complicated environmental perturbations, so the fine detail will vary under magnification. Mimicking this feature can be achieved with a little more work.

For example, consider the Bolzano function. The generator consists of two short segments of negative slope and two longer segments of positive slope, arranged in the order long, short, long, short. Any generator arrangement of these two short and two long segments will start at $(0, 0)$ and end at $(1, 1)$, so the Bolzano function can be randomized by replacing each line segment with any combination of two short and two long segments, selected randomly. The left image of Fig. 2.2 shows an example. A variation on this randomized Bolzano function gives Benoit's finance cartoons, with which he famously mislead seasoned traders. "Two of these time series are

Figure 2.2 Left: to right: randomized Bolzano function, randomized product of Cantor sets, DLA cluster.

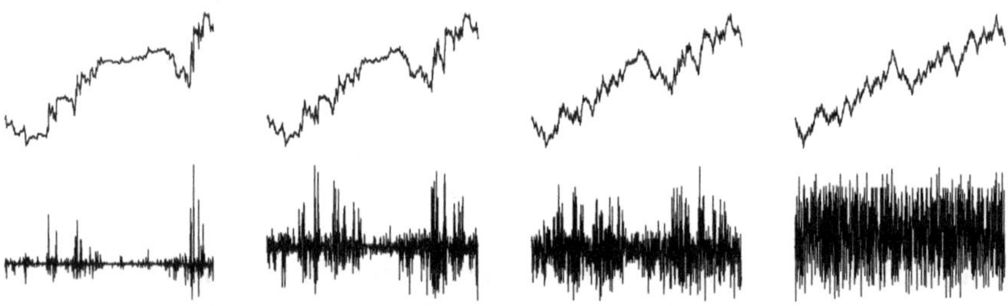

Figure 2.3 Top: finance cartoons. Bottom: difference graphs, with vertical scales taken to occupy the same space. If the same scale were used for all, the jumps on the left would shrink to near invisibility all the jumps on the right.

the differences in daily closing prices of stocks, three are generated by my cartoons. Can you tell which is which?" To Benoit's great amusement, the replies were wrong.

For instance, the top rows of Fig. 2.3 are randomized iterates of cartoons where the generator has endpoints $(0,0)$, $(a,2/3)$, $(5/9,1/3)$ and $(1,1)$, where from left to right $a = 1/9, 2/9, 3/9$, and $4/9$. The right-most of these is Brownian motion. From the difference graphs, plots of differences of successive values measured at equal time intervals, we see that as a decreases we observe larger spikes, suggesting Lévy flights, and also longer clumps of contiguous large jumps and of contiguous small jumps, suggesting fractional Brownian motion. So very simple: move one generator point and induce all kinds of interesting statistics. This is typical of Benoit's approach. Applied cleverly, simple constructions can yield remarkable complexity.

On to other kinds of random fractals. The middle image in Fig. 2.2 is a randomized version of a product of two Cantor sets. Start with the filled-in square with vertices $(0,0)$, $(1,0)$, $(1,1)$, and $(0,1)$, then replace this with four smaller squares, the lower left with $(0,0)$ as a vertex, the lower right with $(1,0)$ as a vertex, the upper right with $(1,1)$, and the upper left with $(0,1)$. The side lengths of the squares are selected independently and uniformly randomly from $[1/4, 1/2]$. Different choices of side lengths will produce fractals distinct in fine structure but having similar general properties.

These examples illustrate ways to incorporate randomness into mathematical fractals, and so begin to suggest constructions of fractals that are better forgeries of natural processes. Much of Benoit's work focused on other approaches to random fractals.

First, two early examples, precursors to the graphs of Fig. 2.3. Used by Louis Bachelier to model the French stock market around 1900, Brownian motion exhibits scaling and rightly can be called a random fractal. However, Brownian motion possesses two properties making it inappropriate for Bachelier's purpose. The increments of Brownian motion are independent of one another, and are normally distributed. Two of Benoit's early models, fractional Brownian motion and Lévy flights, address these issues separately. Increments of fractional Brownian motion are dependent but still normally distributed. Increments of Lévy flights are not normally distributed — increment size decays as a power law, not as an exponential — but are independent. In contrast, increments of Benoit's finance cartoons are dependent and not normally distributed. Though not suited for finance, fractional Brownian motion and Lévy flights are used successfully to model natural phenomena.

Diffusion-limited aggregation is another random fractal Benoit studied. The right image of Fig. 2.2 is an example. Introduced in 1981 by Witten and Sander [13], the process is simple. Particles diffuse until they encounter a nucleus to which they stick. Other particles stick to those already stuck, and the cluster grows. Branches are unstable, so tend to grow side branches exhibiting the same structure, inducing similar structures across a range of scales. Thickets of side branches, of side-side branches, of side-side-side branches, and so on, shield the central regions of the cluster from additional growth, so fractal features eventually remain (mostly) unchanged. These properties were known from early on. Even at the start of his investigations, Benoit's intuition guided him to want ever larger clusters, challenging the memory limitations of his computers and the cleverness of his programmers. Indeed, he did find interesting features in large clusters [14, 15].

In a teacher training workshop, once Benoit described a conference talk he gave on DLA. Benoit reported that he had gotten a question he was unable to answer. This was interesting, because while Benoit didn't know the answer to every question, almost always he had good intuitions about the direction of the answer. So what question had flummoxed him? At the end of his one hour lecture on DLA, someone in the front row said he'd followed everything except one point: didn't Benoit mean DNA instead of DLA? Benoit looked at us helplessly and said, "What could I say? I went on to the next question."

A final comment on Brownian motion. Benoit was bothered by an asymmetry of planar Brownian motion: the regions near the starting point and near the ending point appeared unlike the other regions. Say $\vec{B}(t) = \langle x(t), y(t)\rangle$ denotes the time t location of a point undergoing Brownian motion, for $0 \leq t \leq T$. To make a more symetrical shape, Benoit asked his programmer to plot the path

$$\vec{C}(t) = \vec{B}(t) + \frac{t}{T}(\vec{B}(0) - \vec{B}(T))$$

Note that $\vec{C}(0) = \vec{C}(T)$, so this path returns to its starting point. The picture still was too complicated, so Benoit's next request was to shade in the region bounded by the curve $\vec{C}$. With the intricate interior structure obscured, this was an island. Numerical experiments supported what Benoit expected from analyzing maps of real islands, that the boundary of the island has dimension close to 4/3. Careful estimates of the regions inside and outside the boundary led Benoit to conjecture that the dimension of the boundary is exactly 4/3. This became the 4/3-conjecture. In 2001 Lawler, Schramm, and Werner [16] applied stochastic Loewner evolution to prove the 4/3 conjecture, now the 4/3 theorem. For this work, in 2006 Werner received a Fields Medal at the 25th International Congress of Mathematicians.

2.4 Dimensions physical and mathematical

What do we mean by dimension =4/3 in the last section? Thinking of dimension as counting the number of independent directions in which we can move, a dimension of 4/3 is a bit hard to understand. Dimension can mean other things, including how the amount of material M making up an object scales with the size r of the object. If the object exhibits a power law scaling, $M = k \cdot r^d$, then the exponent d is called the *mass dimension* of the object. Measured by this d, scaling of the distribution of the galaxies has cosmological implications.

Moving in another direction, count the number $N(\epsilon)$ of boxes in an efficient cover (a cover using the minimum number of boxes of that size) of an object by boxes of side length ϵ. Smaller boxes are sensitive to more detail, so we expect $N(\epsilon)$ to increase as ϵ decreases. If present, a power law scaling $N(\epsilon) = k \cdot (1/\epsilon)^d$ gives the box-counting dimension.

For Euclidean shapes, mass and box-counting dimensions agree with our intuitions about dimension, but these can be calculated for shapes too complex to have an obvious Euclidean dimension. After he had been working with dimension for some time, Benoit realized that dimension is a way — in fact, the first way — to quantify roughness. Physical applications abound.

One of the earliest examples built on L. F. Richardson's observations when measuring the lengths of coastlines at different scales. Fix a length ϵ and walk along a coastline taking steps of length ϵ. Say $N(\epsilon)$ steps are needed. Then the length, measured at scale ϵ, is

$$L(\epsilon) = N(\epsilon)\epsilon. \tag{2.1}$$

For smooth curves, as $\epsilon \to 0$, $L(\epsilon)$ approaches a limit, the length of the curve. Richardson found that for coastlines, smaller ϵ detect more intricate features and so $L(\epsilon)$ increases, with no bound in sight, as $\epsilon \to 0$. For example, using $10\,m \leq \epsilon \leq 10^3\,m$, for the west coast of Britain Richardson observed $L(\epsilon) \approx (1/\epsilon)^{.25}$. Benoit [17] interpreted Richardson's data differently, by hypothesizing a power law relation, $N(\epsilon) \approx (1/\epsilon)^d$. Substituting into Eq. (2.1), we obtain

$$d \approx 1 - \log(L(\epsilon))/\log(\epsilon)$$

with the approximation improving for smaller ϵ. Because the value depends significantly on the measurement scale, length makes no sense for coastlines. Instead Benoit proposed that the relevant quantity for a coastline is d, sometimes called the *divider dimension* because on a map lengths would be approximated by walking dividers along the coastline. Interpreted this way, Richardson's measurements show the west coast of Britain has dimension about 1.25, while the coast of South Africa has dimension about 1.02. The craggy west coast of Britain is much rougher than the smooth coast of South Africa; an early tie between dimension and roughness.

Before coming to Yale, one of Benoit's appointments was at Harvard, where he taught the first fractal geometry course. One the the exam questions Benoit gave was "How long is the coastline of Britain?" Expecting the answer "It depends on the scale at which the length is maesured," Benoit was surprised to receive variations of "Sir, I did not realize this was a geography course!" I doubt that these answers contributed to Benoit's leaving Harvard and coming to Yale, but still.

Back in mathematics world, after Euclidean the most common dimension is Hausdorff. Because the definition is fairly lengthy and readily available in many familiar sources, including Wikipedia, we omit the general definition and focus on special cases. Suppose a fractal A can be decomposed into N smaller copies of itself, $A = A_1 \cup \cdots \cup A_N$, with each A_i similar to A, scaled by a factor of r. Then A is self-similar and the *similarity dimension* is given by

$$d = \frac{\log(N)}{\log(1/r)} \tag{2.2}$$

So for the Sierpinski gasket, the left image of Fig. 2.4 we see $N = 3$ and $r = 1/2$, giving $d = \log_2(3)$. If the pieces are scaled by different amounts r_i, then Eq. (2.2) generalizes to the *Moran equation*

$$r_1^d + \cdots + r_N^d = 1 \tag{2.3}$$

For the second shape of Fig. 2.4, $r_1 = r_2 = r_3 = 1/2$ and $r_4 = 1/4$. Taking $x = (1/2)^d$, the Moran equation (2.3) becomes $3x + x^2 = 1$. The positive solution is $x = (-3 + \sqrt{13})/2$, giving $d = \log_2((-3 + \sqrt{13})/2)$.

By a theorem of Hutchinson [18], if the transformations $T_1, \ldots, T_N$, generating the fractal satisfy the open set condition, then the similarity dimension equals the Hausdorff dimension. The T_i satisfy the *open set condition* if there is an open set O for which $O \supset T_1(O) \cup \cdots \cup T_N(O)$, with the $T_i(O)$ pairwise disjoint. For the Sierpinski gasket rules, the open unit square works for O.

The third shape of Fig. 2.4 is generated by four transformations, $T_i(x, y) = (x/2, y/2) + (a_i, b_i)$, with $(a_i, b_i) = (0, 0)$, $(1/2, 0)$, $(0, 1/2)$, and

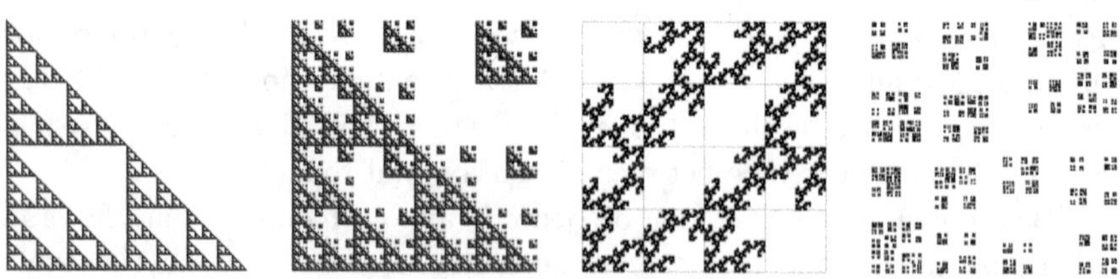

Figure 2.4 Subjects of our dimension computations.

$(1/2, 1/2)$ for $i = 1, 2, 3, 4$. The IFS using these transformations generates the filled-in unit square. However, if some compositions are forbidden, more complicated shapes can be produced. The fractal being discussed is made by forbidding $T_1 \circ T_4$, $T_2 \circ T_2$, $T_3 \circ T_3$, and $T_4 \circ T_1$. Denote by $[m_{ij}]$ the transition matrix,

$$m_{ij} = \begin{cases} 1 & \text{if } T_i \circ T_j \text{ is allowed} \\ 0 & \text{if } T_i \circ T_j \text{ is forbidden} \end{cases}$$

and the Moran equation (2.3) becomes

$$\rho\Big(\Big[m_{ij} r_i^d\Big]\Big) = 1 \tag{2.4}$$

where $\rho(M)$ is the spectral radius on M, that is, the maximum of the magnitudes of the eigenvalues of M. For this example, the eigenvalues of the transition matrix are -1, -1, 1, and 3, and each $r_i = 1/2$, giving $d = \log(3)/\log(2)$. Noting that this shape doesn't look much like the gasket points out that dimension does not characterize fractals. But then, how could it? One single number characterizing every possible fractal? No one should expect that.

For the right image of Fig. 2.4, a random fractal, the Moran equation becomes

$$E(r_1^d + r_2^d + r_3^d + r_4^d) = 1 \tag{2.5}$$

where $E(X)$ denotes the expected value of X. Recalling that for each piece of this random fractal, r_i is uniformly distributed in the range $[1/4, 1/2]$, and using the calculation

$$E(r^d) = \int_{1/4}^{1/2} 4t^d dt = \frac{4}{d+1}\left(\left(\frac{1}{2}\right)^{d+1} - \left(\frac{1}{4}\right)^{d+1}\right)$$

we find $d \approx 1.42481$.

Of course, there are more complicated types of fractals, nonlinearly self-similar, self-affine, and others even more exotic, BPI (Big Piece of Itself) spaces [19], for example.

Dimensions of nonlinear fractals sometimes can be computed by using the eigenvalues of the derivative matrix to estimate the contraction factors of the pieces of the fractal. Adapting methods from statistical mechanics, the thermodynamic formalism, the Moran equation now corresponds to an expression involving the topological pressure. Chapter 5 of [20] is an excellent reference.

Self-affine fractals are those exhibiting different scalings in diferent directions. Some examples are easy: the Hausdorff dimension of the cartesian product of Cantor sets is the sum of the Hausdorff dimensions of the Cantor sets. This is simple and sensible, and might make us expect a straightforward variation of the Moran equation involving the scaling factors in each direction. In general, the situation is much more complicated. Early on, Benoit [21] pointed out that even for simple self-affine curves the situation is complicated: their box-counting, mass, and divider dimensions can differ.

Even slight variations on the product of two Cantor sets can be much more complicated. The first major result was obtained by McMullen [22]. Divide the unit square into a collection of $m \times n$ congruent rectangles. Discard some of these rectangles. Subdivide each remaining rectangle into an $m \times n$ collection of still smaller rectangles and discard some of these in the same pattern as the original. Two examples are given in Fig. 2.5. McMullen proved that for these fractals the Hausdorff dimension d_h and the box-counting dimension d_b are given by

$$d_h = \log_m \left(\sum_{j=1}^{m} t_j^{\log_n(m)} \right) \qquad d_b = \log_m(r) + \log_n(N/r)$$

where t_j is the number of occupied rectangles in row j, r is the number of non-empty rows, and N is the total number of rectangles occupied in the first step of the construction.

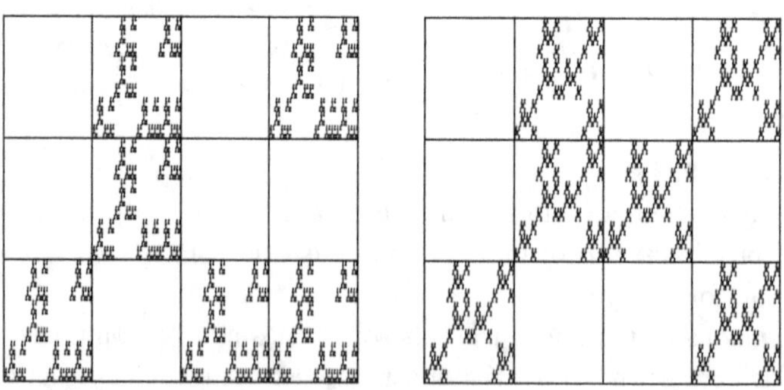

Figure 2.5 Two examples of McMullen's construction. The 3×4 subdivision grid is shown for reference.

For the left fractal of Fig. 2.5, $t_1 = 3$, $t_2 = 1$, $t_3 = 2$, $r = 3$, and $N = 6$, so

$$d_h = \log_3\left(3^{\log_4(3)} + 1^{\log_4(3)} + 2^{\log_4(3)}\right) \approx 1.48664$$
$$d_b = \log_3(3) + \log_4(6/3) = 1.5$$

an example with $d_h < d_b$. Interestingly, note that for the right fractal of Fig. 2.5, $t_1 = t_2 = t_3 = 2$ giving

$$d_h = \log_3\left(3 \cdot 2^{\log_4(3)}\right) = 1.5$$

Both fractals consist of the same number of pieces, 6, so in contrast to self-similar fractals satisfying the open set condition (and note these self-affine fractals satisfy the open set condition with O the open unit square) for which d_h is determined by the number and size of the pieces, not their relative positions, here d_h can depend on the locations of the pieces.

In addition to McMullen, Bedford, Falconer, Gatzouras, Kenyon, Lalley, and Peres, among others, have obtained results on dimensions of self-affine fractals. The world of self-affine fractals is, as Benoit observed long ago, a very complicated place.

2.5 The Mandelbrot set

Following the suggestion of his uncle Szolem, Benoit was familiar with the work of Fatou and Julia on the dynamics of complex iteration. From this, Benoit knew that the boundary of the set of complex numbers z_0 for which all iterates $z_{n+1} = z_n^2 + c$ remain bounded, the Julia set for the function z^2+c, has one of two topological types, connected and totally disconnected. Moreover, a Julia set is connected if and only if the iterates of the critical point $z = 0$ remain bounded. By hand, these calculations are impossibly tedious, and can be done abstractly in only a few cases: the Julia set for $c = 0$ is the unit circle, the Julia set for $c = -2$ is the interval $[-2, 2]$ on the real axis. Once computers could be used to generate (approximate) pictures of Julia sets, a natural project was to plot the set of c for which the Julia set is connected; that is, the set of c for which the iterates of $z = 0$ remain bounded. This is what came to be called the Mandelbrot set. The first crude pictures were produced at Harvard, where Benoit was visiting. The general shape appeared, surrounded by smudges and blotches. Not trusting Harvard's primitve computer equipment, because he suspected something very interesting was going to appear, Benoit drove to Yorktown Heights, to run the program on the IBM mainframe. Some of the dust around the Mandelbrot set disappeared. Those bits that remained were symmetrically arranged, unlikely behavior for real dirt. Zooming in, these specks of dust turned out to be small copies of the Mandelbrot set. Though it may not have been perceptible at that moment, the mathematical universe rotated when this picture came off the printer.

The pictures of the Mandelbrot set are familiar across the world. No need to add a comments here. Fig. 2.6 shows the whole Mandelbrot set (upper left) and some magnifications. Googling Mandelbrot set images in July, 2014, returned over 150,000 results. But what about the mathematics that came from that picture and those that followed? I'll mention five conjectures — four conjectures and a question, really — Benoit made about the Mandelbrot set.

For each disc attached to the big cardioid of the Mandelbrot set, the iterates of $z = 0$ converge to a cycle. Quick inspection suggests that the higher the cycle number, the smaller the disc. More careful measurement refines this: the radius of the n-cycle disc is about $1/n^2$. When Benoit presented this conjecture at a conference, two people, John Guckenheimer

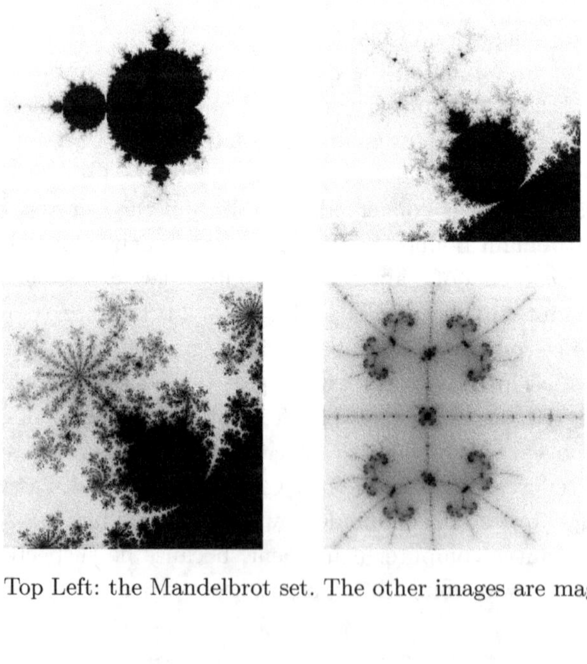

Figure 2.6 Top Left: the Mandelbrot set. The other images are magnifications.

and Richard McGehee, began writing furiously. By the end of the talk, both had proofs [23] of the conjecture.

Every picture of the Mandelbrot set suggests that the small copies of the Mandelbrot set are islands, separate from the main body of the set. Comparing pictures of the Mandelbrot set with those of connected Julia sets made Benoit doubt these copies are islands. This time he wasn't quite sure enough to make a conjecture, so he asked a question. Is the Mandelbrot set connected? In 1984, Adrien Douady and John Hubbard [24] proved the Mandelbrot set is connected.

As Benoit continued to flip between pictures of Julia sets and of the Mandelbrot set, he noticed that the Mandelbrot set "has an uncanny 'hieroglyphic' character. It includes within itself a whole deformed collection of miniature versions of all the Julia sets." This was proved at the Misiurewicz points, dense in the boundary of the Mandelbrot set, by Tan Lei [25] in 1990.

Counting the branching structure of the decorations off the Mandelbrot set, and noting how densely they appear to accumulate, Benoit conjectured that the boundary of the Mandelbrot set has dimension 2. This was proved by Shishikura [26] in 1998.

The hyperbolicity conjecture, that every open set in the Mandelbrot set consists of c for which the iterates of $z = 0$ converge to a cycle, is a special case of Douady's and Hubbard's conjecture that the Mandelbrot set is locally connected (MLC). MLC remains unproven despite substantial efforts of very talented mathematicians, including Fields medalists J.-C. Yoccoz and C. McMullen.

2.6 Kleinian group limit sets

The right isosceles Sierpinski gasket G is the only compact subset of the plane satisfying

$$G = T_1(G) \cup T_2(G) \cup T_3(G)$$

where $T_i(x, y) = (x/2, y/2) + (a_i, b_i)$ with $(a_1, b_1) = (0, 0)$, $(a_2, b_2) = (1/2, 0)$, and $(a_3, b_3) = (0, 1/2)$. One familiar algorithm for generating G is to start with any point (x_0, y_0) in the plane and generate a sequence

$$O = \{(x_0, y_0), (x_1, y_1), (x_2, y_2), \dots\}$$

where $(x_i, y_i) = T_{j_i}(x_{i-1}, y_{i-1})$ and the T_{j_i} are chosen randomly from $\{T_1, T_2, T_3\}$. Mathematically, G is the set of limit points of O;

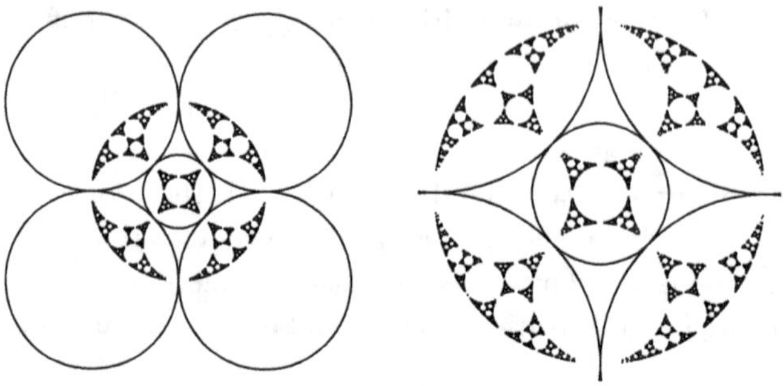

Figure 2.7 (Most of) a circle inversion limit set, generated point by point.

computationally, ignore the first 20 or 30 points and plot the rest. The points dance across the gasket and fill it very quickly, using contemporary computers.

This approach can be applied to other functions. If the functions are nonlinear, inversion in circles or complex linear fractional transformations, for instance, even very long dances can leave regions of the limit set unvisited. On the left of Fig. 2.7 we see five circles, four large and a smaller circle in the middle, together with several hundred thousand points of the limit set of inversion in these circles. On the right is a magnification of the limit set. We see points fill in very slowly near tangencies of the inverting circles. The reason is clear: near the circle, the derivative of the inverting map is very close to the identity, so successive inversions in two tangent circles do not move the iterates much closer to the point of tangency.

Benoit often said a fractal can be understood as well by what has been removed as by what remains. For some families of inverting circles, Benoit found a very quick algorithm [27] for generating limit sets. The inverting circles of Fig. 2.7 give a simple example. First, Benoit argued the limit set is contained in the circle orthogonal to the four large inverting circles. Begin with that circle filled in. Next, the limit set contains no points in the four grey discs on the left of Fig. 2.8. In the next picture of that figure, these grey discs are inverted in the five circles (recall inversion in a circle preserves a disc intersecting the circle orthogonally), and the region outside the black disc is inverted to the small grey disc inside the smallest inverting circle. Continue inverting these grey discs; their complement in the black disc approximates the limit set. The resut of only a few more iterations is shown on the right of Fig. 2.8. To appreciate this method, compare this with the right side of Fig. 2.7.

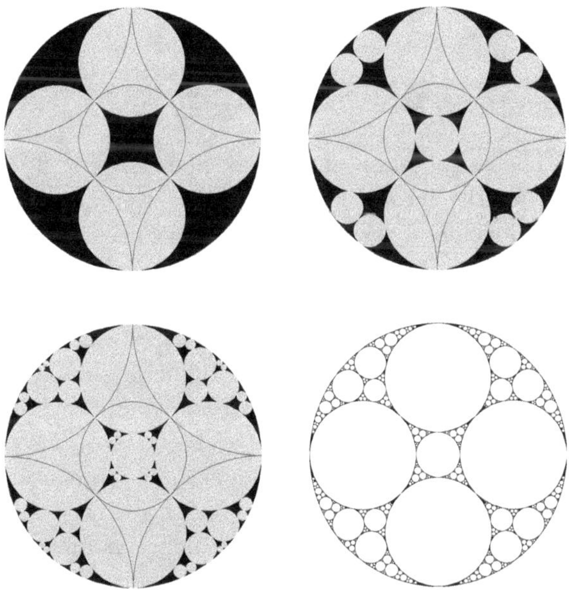

Figure 2.8 The first few stages of constructing the limit set of Fig. 2.7 by removing discs disjoint from the limit set. The right picture is a later stage of the construction, with the removed discs painted white instead of grey.

With minor modifications, this method can generate limit sets of Kleinian group actions, an important constructon in William Thurston's geometrization program for 3-manifolds. An early talk at Harvard on nonlinear iteration led to an interesting interchange between Benoit and David Mumford, reported in [1].

> My talk made David wonder whether I could also look at Kleinian limit sets. Delighted, I responded that — at least for a very important special case — I had indeed discovered a construction and showed him the draft of a paper underway.

In addition to important insights in topology, Kleinian group limit sets are the subject of a stunningly beautiful book [28].

2.7 Multifractals

In his 1974 study of turbulence [29], Benoit introduced the notion of multifractals. These involve a measure μ with support A and at each point

$x \in A$ the *local Hölder exponent*

$$\alpha(x) = \lim_{r \to 0} \frac{\log(\mu(D_r(x)))}{\log(r)} \tag{2.6}$$

where $D_r(x)$ denotes the disc of radius r and center x. If the limit does not exist, we use the limsup and liminf to obtain the upper and lower local Hölder exponents. Typically, the α values occur in a limited range, $\alpha_{\min} \leq \alpha \leq \alpha_{\max}$. For each α in this range, take $E_\alpha = \{x \in A : \alpha(x) = \alpha\}$. That is, stratify A into subsets of constant α. Finally, $f(\alpha) = d_h(E_\alpha)$. The plot of $f(\alpha)$ as a function of α quantifies how the dimension of a set changes with its roughness, measured by the local Hölder exponent.

These $f(\alpha)$ curves can be calculated from experimental data in several ways, from histograms or using wavelets (cf. Chap. 3), for example. Here we present a simple mathematical approach, for measures generated by multiplicative cascades.

Suppose the set A is defined by iterating the transformations $T_1, \ldots, T_N$, each T_i a contraction with contraction factor r_i and applied with probability p_i. Then the region $A_i = T_i(A)$ can be assigned a measure $\mu(A_i) = p_i$, the region $A_{ij} = T_i(T_j(A))$ the measure $\mu(A_{ij}) = p_i p_j$, and so on. Taking the probabilities into account, the Moran equation (2.3) becomes

$$\sum_{i=1}^{N} p_i^q r_i^{\beta(q)} = 1 \tag{2.7}$$

Incorporating the exponent q for p_i forces the exponent of r_i to be a function of q. With some work, we can show

$$\alpha = -\frac{d\beta}{da} \quad \text{and} \quad f(\alpha) = q\alpha + \beta$$

Points on the $f(\alpha)$ curve can be plotted parametrically, $(\alpha(q), f(\alpha(q)))$, say for $-20 \leq q \leq 20$. Negative q correspond to the right side of the $f(\alpha)$ curve. In early work, Benoit and others avoided this side of the curve because it corresponds to small measures, and small measures raised to large negative powers are sensitive to even tiny error. In some cases, the calculation of the $f(\alpha)$ curve can be carried out exactly. See Appendix B of [30], for example.

This approach can be generalized to every setting in which a measure can be assigned to a set whose dimension can be computed. For instance, if only some compositions of the T_i are allowed, then Eq. (2.7) is replaced by

$$\rho\left(\left[p_{ij}^q r_i^{\beta(q)}\right]\right) = 1 \tag{2.8}$$

where p_{ij} denotes the probability of applying T_i immediately after T_j. The comparison with Eq. (2.4) is apparent.

Many, many more examples are in [6]. Multifractals are applied in fields ranging from fluid mechanics through hyperbolic topology to finance (The finance cartoons mentioned in Sec. 2.3 are multifractal.) and art. When he introduced this area of study in the 1970s, I doubt Benoit had any guess of how many fields multifractals would touch.

2.8 Fractals in the classroom

In the spring semester of 1993 I introduced the first fractal geometry course at Yale. Benoit had invited me to visit for a year, and because David Peak and I had taught a similar course at Union College, I offered to teach fractal geometry here. The Yale math department provided as much support as anyone could wish, but well-meaning colleagues said I should not expect many students. Neither the course nor I had any reputation with the students; a class of 20 would be a great success. Benoit made no comment at all, beyond supporting my initial course proposal. The first class meeting saw about 80, and grew quickly to 180, where it stayed for the rest of the semester. Still, Benoit said nothing about this to me, but he smiled a lot. After the success of the class became clear, Aliette told me that Benoit had thought his work would be taught in school, but did not expect to live to see this. In this expectation he was mistaken: he saw fractals taught in many, many classes. Even at the end of his life, he was charmed by emails and letters from students.

After two decades, the Yale course remains popular, 130 students took it in the fall semester of 2014. The power of fractal images, together with the accessibility of so much of the underlying mathematics, spread these ideas to other courses, other schools. Benoit was an energetic supporter of all applications of fractals in the classroom. One concrete manifestation of his interest was the NSF-funded summer workshops for high school and college teachers that Benoit, Nial Neger, Harlan Brothers, and I ran for six years. Chapter 16 contains a detailed description of this project. Here I will mention one point: in many discussions with these teachers we realized that the right place to put fractals in the high school curriculum is not as stand-alone courses, but as effective ways to introduce engaging explorations of topics already in the curriculum. Benoit participated in all these workshops, to the considerable pleasure of the teachers, but he did so much more to support the use of fractals in education. Yet only occasionally did Benoit teach a course. "I am a preacher, not a teacher," he said often, and indeed Benoit was very good at preaching. The actual teaching fell to hundreds of the rest of us, each using these amazing visual and geometric tools in our own ways. The last twenty-five years have been an exciting time to be a mathematics teacher, elementary school to graduate school, due in part to the excitement generated by fractals.

2.9 Fractals in the arts

Benoit often spoke about the importance of the eye in science, and that artists use some of the same perceptiveness in unpacking aspects of the world. Of the several common aspects of fractals and art, one is paramount: nature. That fractals occur in many natural objects is undeniable, so these objects are more successfully imaged in representational art informed by scaling symmetry. Benoit saw this clearly in the work of Hokusai, especially in *Fuji from the Seashore*, *Fuji at Sea*, and *Fuji in a Thunderstorm*, all from his *One Hundred Views of Mt. Fuji.* Notice the seaweed on the beach in *Seashore*, the substructure of breaking waves (and anticipating Escher, the transition from wave foam to bird flock) in *Sea*, and the trees and border of the snowfield of *Thunderstorm.* Hokusai understood, and could reproduce, scaling patterns in nature.

Another of Benoit's favorites was Augusto Giacometti. For his paintings inspired by fields of wildflowers pushing through Alpine snow, Giacometti applied thick blobs of paint to the canvas, patches of color on a white background. Why were these so interesting to Benoit? By intuition or careful observation, Giacometti produced a power law distribution of blob sizes. He did not paint a careful representation of a single flower, but he understood, or intuited, the distribution of flower sizes. This same scaling occurs in his more stylized painting of the Zurich Stadtpllizei entrance hall ceiling.

Fractals arise through the physical process of applying paint to a trowel, pressing it against the canvas, and pulling the trowel away. Air invades the viscous paint, producing patterns branching on many levels. This is *decalcomania*, used in the early 20th century by Oscar Dominguez, Boris

Margo, Hans Bellmer, Max Ernst, Marcel Jean, Enrico Donati, Andre Masson, and Max Bucaille. Ernst's *Europe after the Rain* is a good example. Though he appreciated the process, Benoit was not particularly fond of decalcomania as art.

Still on the topic of physical methods, Jackson Pollock's drip paintings are familiar. While not accepted by everyone, Richard Taylor's analysis of the box-counting dimension [31, 32] of the drip paintings did convince Benoit.

Larry Poons took a different approach, preparing the canvas by affixing bits of fiber and foam, suspending the canvas vertically, and flinging buckets of paint at the canvas. One critic called the effect "geological, muddy, alluvial." Daniel Robbins [33] analyzed the fractal character of Poons's work.

Rhonda Roland Shearer [34] has written about revolutions in art catalyzed by revolutions in geometry — perspective geometry and Renaissance art, non-Euclidean geometry and modern art — and speculates that fractal geometry may signal a new revolution in art. Recognizing a revolution from inside can be tricky, so we'll need to wait a while to see if this prediction comes true.

Some kolams, mandalas, and African art [35] have fractal characteristics. These examples fascinated Benoit. He had a high opinion of Ron Eglash's book, and he and I spend some enjoyable hours looking through and discussing books about mandalas and kolams from the Yale library. This was Benoit at play, lightning quick, excited, happy.

There are many, many other instances of fractal art; I'll mention one more. Benoit was fascinated by Salvador Dali's *Visage of War*, the artist's depiction of the infinite horrors of the Spanish civil war. The largest feature of this painting is a face, with each eye and mouth filled with a face, with each eye and mouth filled with a face, with each eye and mouth filled with a face. An intentionally distrubing image, but a clear example of self-similarity. So in 1940, Dali intuited that repeating the same pattern over scales generates a sense of the infinite. Wanting to learn a bit more about the painting, I consulted [36], where I found a preliminary study of the painting. In the study, the mouth is filled with a face, with each eye and mouth filled with a face. However, one eye of the original face is filled with a honeycomb and the other eye with a cut tree trunk showing concentric rings. Also a disturbing image, but lacking the impact of the final version. When I showed this to Benoit, he stared at the study for a long time (long for him; really just a few moments), then looked at me and said, "Well, well. You have earned your pay today."

Benoit knew how to work hard; he worked hard every day I saw him. But when looking for new examples of fractals, he was playful, childlike, even. Everyone who saw this saw Benoit's intellect at its liveliest. This was a joy to watch.

2.10 Sense of wonder, sense of humor

In fact, my happiest moments of working with Benoit — and indeed some of the happiest moments of my life — were when a new idea or question occurred to one of us. To be sure, most often to him, but every now and again, to me. If we were in my office, I'd turn to Google for a first pass at gathering information. These were pretty specific questions, so usually the search got good results. Somehow Benoit interpreted this as my having a talent for Google searches, when really it was his talent for asking focused questions. I protested that I'm not talented, it's just luck. Quoting Pasteur, Benoit replied that chance favors the prepared mind. After a few of these successes, from IBM or from Cambridge, Benoit began telephoning me to do Google searches for him. Always, these were fascinating processes. I'd read off some entries, Benoit would talk about a few, and off we'd go. These were fast, exciting views of Benoit's curiosity going at full speed, at least as fast as he could go while hobbled by the constraints of language. On the rare occasions I knew something he didn't, Benoit was delighted.

Using a new geometry, whose details we learn as we go along, much of the world appears fresh. We talked about how this must be a shadow of how a child learns to move in the world. So much is new, filled with surprises. Serious, grumpy adults don't get this. For me, and I believe for Benoit, this was a return to the guileless enthusiasm of childhood. This was Benoit's gift, unexpected, and undeserved, to me. Even without direct contact with him, the fascination of these ideas is open to all who look and think. When the trials of the world sadden me, I reflect that humaitiy's great strength is our curiosity. This will save us, if anything will. Benoit has given us another powerful way to exercise our curiosity.

This has gotten too serious for a descripton of events that were fun. Here's an explicit excursion into humor.

Occasionally Benoit and I discussed precursors of fractals. Certainly, versions of the Sierpinski gasket appeared as floor tilings of European cathedrals hundreds and hundreds of years ago. Many, many other examples were familiar. The semester before I moved permanently to Yale, one of our conversations wandered to Leibnitz's *The Monadology*. Seemed plausible, as well as either of us could remember. But my volume of Leibnitz was in one of several hundred boxes of books I'd already packed, and Benoit was preparing for a long trip that summer, so we filed this away for later investigation.

I moved to New Haven, Benoit went on his trip. Every morning before coming to school, I unpacked a box or two of books. One morning, several weeks before Benoit was supposed to return, I found my volume of Leibnitz [37]. I took it to my office, read George Montgomery's translation of *The Monadology* that morning, and found the relevant paragraphs, numbered by Leibnitz.

> 67. Every portion of matter may be conceived as like a garden full of plants and like a pond full of fish. But each branch of each plant, every member of an animal, and every drop of fluid within it, is also such a garden or pond.
>
> 68. And although the ground and air which lies between the plants of the garden, and the water which is between the fish of the pond, are not themselves plants or fish, yet they nevertheless contain these, usually so small however as to be imperceptible to us.

Seems pretty clear. I got back to work, writing webpages for the fall semester courses. Later that morning, Benoit showed up in my office. I cut my trip short. I've written an essay on fractals for the Wolf Prize. In a few minutes I've got a meeting, but I'll be back in about an hour. Could you have a look at the essay?

The essay included a mention of Leibnitz, but listed the wrong paragraphs. When Benoit returned, I made some comments about the essay, then pointed out the error in the Leibnitz references. Benoit said he didn't find the references himself, but described what he wanted to a friend familiar with the work of Leibnitz. Your friend didn't understand your question, I replied, because, and here I pulled the volume of Leibnitz from my bookbag, these are the paragraphs we want. Benoit looked around my office, one desk, one laptop, two chairs, empty bookcase. You didn't expect me back for a couple of weeks. How in the world did you know to bring Leibnitz today?

It was just a coincidence, a simple coincidence, a bit of random luck. I could have said that, but instead, I said you know, Benoit, time is more complicated than even you can imagine. He stared at me for a moment, frowned, muttered something I didn't catch. For the rest of the day, he looked at me as if he wanted to ask something, but he didn't. And I didn't say anything else about this to him that day, or, as it turns out, any day.

Did he know I was kidding? Certainly. Did he know precisely how? I'll never know, so who really played the joke on whom? Benoit loved jokes and humor, intellectual play. He didn't often display his sense of humor, so when he did, it was surprising, all the more enjoyable because it was seen so rarely.

2.11 What did Benoit teach me?

Specific technical points are too numerous to recount. On a larger scale, clear lessons include the worth of curiosity and intellectual play, that science is based on evidence rather than personal preference, that looking and thinking are an unbeatable combination, and that you may be in for a long lonely ride if your intuition tells you to pursue an unfashionable direction. More subtle things we didn't discuss, but I am beginning to infer. Most I'll save for another day, but I'll end by mentioning one, probably the most important lesson for living well in the world.

> By actions clearer than any words, Benoit, and my parents, taught me that our responsibility to help one another is not a burden, but is a precious gift.

When I started working with Benoit, I expected to learn tons of mathematics. Indeed, I did, but oh my, I learned so very much more.

Bibliography

[1] B. Mandelbrot, *The Fractalist. Memoir of a Scientific Maverick* (Pantheon, New York, 2012).

[2] M. Frame, *Nat. Acad. Sci.*, http://www.nasonline.org/publications/biographical-memoirs/memoir-pdfs/mandelbrot-benoit.pdf

[3] M. Barnsley, M. Frame, *Notices Amer. Math. Soc.* **59** (2012), 1056, 1208.

[4] B. Mandelbrot, *The Fractal Geometry of Nature* (W. H. Freeman, San Francisco, 1982).

[5] B. Mandelbrot, *Fractals and Scaling in Finance. Discontinuity, Concentration, Risk* (Springer-Verlag, New York, 1997).

[6] B. Mandelbrot, *Multifractals and 1/f Noise. Wild Self-Affinity in Physics* (Springer-Verlag, New York, 1999).

[7] B. Mandelbrot, *Gaussian Self-Affinity and Fractals* (Springer-Verlag, New York, 2002).

[8] B. Mandelbrot, *Fractals and Chaos. The Mandelbrot Set and Beyond* (Springer-Verlag, New York, 2004).
[9] B. Mandelbrot, R. Hudson, *The (Mis)Behavior of Markets. A Fractal View of Risk, Ruin, and Reward* (Basic Books, New York, 2004).
[10] H. Janson, *History of Art*, 4th ed. (Abrams, New York, 1991).
[11] R. Passeron, *Phaidon Encyclopedia of Surrealism* (Dutton, New York, 1978).
[12] M. Jasek, *Véstník královské české společnosti nauk*, Třída mat. přírodověd (1921) 1–32.
[13] T. Witten, L. Sander, *Phys. Rev. Lett.* **47** (1981) 1400.
[14] B. Mandelbrot, B. Kol, A. Aharony, *Phys. Rev. Lett.* **88** (2002) 055501.
[15] B. Mandelbrot, B. Kol, A. Aharony, arXiv:cond-mat/0109426.
[16] G. Lawler, O. Schramm, W. Werner, *Math. Res. Lett.* **8** (2001) 401.
[17] B. Mandelbrot, *Science* **15** (1967) 636.
[18] J. Hutchinson, *Indiana Univ. Math. J.* **30** (1981) 713.
[19] S. Semmes, *Some Novel Types of Fractal Geometry* (Oxford Univ. Pr., Oxford, 2001).
[20] K. Falconer, *Techniques in Fractal Geometry* (Wiley, Chichester, 1997).
[21] B. Mandelbrot, *Physica Scripta* **32** (1985) 257.
[22] C. McMullen, *Nagoya Math. J.* **96** (1984) 1.
[23] J. Guckenheimer, R. McGehee, *A proof of the Mandelbrot n^2 conjecture*, Institute Mittag-Leffler, Report 15, 1984.
[24] A. Douady, J. Hubbard, *Publ. Math. D'Orsay*, 1er partie 84-02, 2me partie 85-04.
[25] Tan Lei, *Commun. Math. Phys.* **134** (1990) 587.
[26] M. Shishikura, *Ann. Math.* **147** (1998) 225.
[27] B. Mandelbrot, *Math. Intelligencer* **5** (1983) 9.
[28] D. Mumford, C. Series, D. Wright, *Indra's Pearls. The Vision of Felix Klein* (Cambridge Univ. Pr., Cambridge, 2002).
[29] B. Mandelbrot, *J. Fluid. Mech.* **62** (1974), 331.
[30] H.-O. Peitgen, H. Jürgens, D. Saupe, *Chaos and Fractals. New Frontiers in Science*, 1st edn. (Springer-Verlag, Berlin, 1992).
[31] R. Taylor, A. Micolich, D. Jonas, *Nature* **399** (1999) 422.
[32] R. Taylor, *Sci. Am.* (December 2002) 116.
[33] D. Robbins, in *Larry Poons: Paintings 1963-1990* (Salander-O'Reilly Galleries, New York, 1990), pp. 9–24.
[34] R. R. Shearer, *Leonardo* **25** (1992) 143.
[35] R. Eglash, *African Fractals* (Rutgers Univ. Pr., 1999).
[36] R. Descharnes, *Dali* (Abrams, New York, 1985).
[37] G. Leibnitz. *Basic Writings. Discourse on Metaphysics, Correspondence with Arnauld, Monadology* (Open Court Publ. Co., La Salle, 1968).

Chapter 3

Irregularities and scaling in signal and image processing: multifractal analysis

Patrice Abry, Stéphane Jaffard, Herwig Wendt

Abstract. B. Mandelbrot gave a new birth to the notions of scale invariance, self-similarity and non-integer dimensions, gathering them as the founding corner-stones used to build up *fractal geometry*. The first purpose of the present contribution is to review and relate together these key notions, explore their interplay and show that they are different facets of a single intuition. Second, we will explain how these notions lead to the derivation of the mathematical tools underlying multifractal analysis. Third, we will reformulate these theoretical tools into a wavelet framework, hence enabling their better theoretical understanding as well as their efficient practical implementation. B. Mandelbrot used his concept of fractal geometry to analyze real-world applications of very different natures. As a tribute to his work, applications of various origins, and where multifractal analysis proved fruitful, are revisited to illustrate the theoretical developments proposed here.

3.1 Introduction

As many students, before and after us, we first met Benoit Mandelbrot through his books [1, 2]; indeed, they were scientific bestsellers, and could be found in evidence in most scientific bookshops. A quick look was sufficient to realize that Benoit was an unorthodox scientist; these books radically differed from what we were used to, and were violating several taboos of the time. First, they could not be straightforwardly associated with any of the standardly labelled fields of science, which had been taught to us as separate subjects. This could already be inferred from their titles: *The fractal geometry of nature* [2] constitutes a statement that mathematics and natural sciences are intrinsically mixed, and that the purpose will not be to disentangle them artificially, but rather to explore their interactions. The contents of these books confirmed this first feeling: though they contained a large proportion of very serious mathematics, they did not follow the old *definition-theorem-proof* articulation we were used to. The focus was directly on the observation of figures and on their pertinence for modeling; the mathematics were there to help and comfort the deep geometric intuition that Benoit had developed. This does not mean that he considered mathematics as an ancillary discipline. To the contrary, many testimonies confirm how enthusiastic he was when one of his mathematical intuitions was proven correct, and how he immediately incorporated the new mathematical methods in his own toolbox, which then helped him to sharpen his intuition and go further.

Another revolutionary point was that these books reversed the classical focusses. While one did find familiar mathematical objects (the nowhere differentiable Weierstrass functions, or the *devil's staircase* had been encountered before), the role he made them play was radically different. So far, they had only been counterexamples, or pathological objects pedagogically used to shed light on possible pitfalls. Instead of this restrictive role, Benoit used them as the key examples and cornerstones to found a new geometry based on roughness and self-similarity. These books were opening a subject, and drawing tracks through a new territory; their purpose was not to give the usual polished, final description of a well-understood area, but to open windows on how science advances, and they had a strong influence on our vocations to research.

Later, personal meetings played a decisive role; let us only mention one such occasion. At the end of his PhD, one of us (S. J.) met Benoit who was visiting Ecole Polytechnique. Benoit had heard of wavelets, which, at the time, were a new tool still in its infancy, and immediately envisaged possible interactions between wavelets and fractal analysis. A key feature of fractals is scaling invariance, and, because wavelets can be deduced one from the others by translations and dilations of a given function, their particular algorithmic structure should reveal the scaling invariance properties of the analyzed objects. Driven by his insatiable curiosity, Benoit wanted to learn everything about wavelets (which was actually not so much at the time!) and then, with his characteristic generosity, he shared with this student he barely knew his intuitions on the subject. Thus, the present contribution can be read as a tribute to some of the scientific lines Benoit had dreamed about at the time. It will show how and why wavelet techniques yield powerful tools for analyzing fractal properties of signals and images, and thus supply new collections of tools for their classification and modeling.

3.1.1 Scale invariance and self-similarity

Exact (or deterministic) self similarity. Fractal analysis can roughly be thought of as a way to characterize and investigate the properties of irregular sets. Note that *irregularity* is a negatively defined word; the description of the scientific domain it covers is therefore not clearly feasible, and had not been tried before B. Mandelbrot developed the concepts of fractal geometry. One of his leading ideas was a notion of *regularity in irregularity or self-similarity*: the object is irregular, but there is some invariance in its behavior across scales. For some sets, the notion of *self-similarity* replaces the notion of smoothness. For instance, the triadic Cantor set is self-similar, in the sense that it is made of two pieces which are the same as the whole set is shrunk by a factor 3. The purpose of multifractal analysis is to transpose this idea from the context of geometrical sets to that of functions. Sometimes, this transposition can be directly achieved, when the geometrical set consisting of the graph of the function also displays some self-similarity. Consider the example of the *Weierstrass-Mandelbrot functions* (a slight variant of the famous Weierstrass functions, see [2] pp. 388–390)

$$\mathcal{W}_{a,H}(x) = \sum_{n=-\infty}^{+\infty} \frac{\sin(a^n x)}{a^{Hn}} \quad \text{for } a > 1 \text{ and } H \in (0,1) \tag{3.1}$$

(note that the series converges when $n \to +\infty$ because $a > 1$ and when $n \to -\infty$ because $|\sin(a^n x)| \le |a^n x|$ and $H < 1$). These functions clearly satisfy the exact self-similarity relationship

$$\forall x \in \mathbb{R}, \quad \mathcal{W}_{a,H}(ax) = a^H \mathcal{W}_{a,H}(x). \tag{3.2}$$

Note that self-similar functions associated with a self-similarity exponent $H > 1$ can be obtained using slight variants. For instance, the function

$$\sum_{n=-\infty}^{+\infty} \frac{\cos(a^n x) - 1}{a^{Hn}} \quad \text{for } a > 1 \text{ and } H \in (0,2)$$

is self-similar with a self-similarity exponent H, which can go up to 2; and the function

$$\sum_{n=-\infty}^{+\infty} \frac{\sin(a^n x) - a^n x}{a^{Hn}} \quad \text{for } a > 1 \text{ and } H \in (1,3)$$

is self-similar with a self-similarity exponent H which can go up to 3. Note that this "renormalization technique" can be pushed further to deal with higher and higher values of H, yielding smoother and smoother functions. One thus obtains functions that are everywhere $C^{[H]}$ and nowhere $C^{[H]+1}$, where $[H]$ denotes the integer part of H.

Random (or statistical) self similarity. Functions satisfying Eq. (3.2) are very rare; therefore the notion of exact self-similarity can be considered too restrictive for real world data modeling. Fruitful generalizations of this concept can be developed into several directions. A first possibility lies in weakening the exact deterministic relationship of Eq. (3.2) to a probabilistic one. The (random) functions $a^H f(ax)$ do not coincide sample path by sample path, but share the same statistical *laws*. A stochastic process X_t is said to be *self-similar*, with self-similarity exponent H if and only if

$$\forall a > 0, \; \{X_{at}, \; t \in \mathbb{R}\} \stackrel{\mathcal{L}}{=} \{a^H X_t, \; t \in \mathbb{R}\} \tag{3.3}$$

(see Definition 3.4.1 where the precise definition of equality in law of processes is recalled).

The first example of such self-similar processes is fractional Brownian motion (hereafter referred to as fBm), introduced by Kolmogorov [3]. Its importance for the modeling of scale invariance and fractal properties in data was made explicit and clear by Mandelbrot and Van Ness in [4]. This article is characteristic of Mandelbrot's genius, which could perceive similarities between very distant disciplines. Motivations simultaneously arose from hydrology (cumulated water flow), economic time series and fluctuations in solids. Notably, Mandelbrot explained that fBm (with $0.5 < H < 1$) was well-suited to model long term dependencies, or long memory effects, in data, a property he liked to refer to as the *Joseph effect* in a colorful and self-describing metaphoric comparison to the character of the Bible (cf. e.g., [5]). Self-similar processes have since been the subject of numerous exhaustive studies, see for instance the books [6–9]. Note also that Mandelbrot put self-similarity (which he actually more accurately called *self-affinity*) as a central notion in the study of stochastic processes in views of applications, see [10] for a panorama of his works in this direction.

3.1.2 Self-similarity and wavelets

Let us now briefly come back to Mandelbrot's intuition and see how wavelet analysis can reveal relationships such as Eq. (3.2) or (3.3), as well as other important probabilistic properties of stochastic processes. In its simplest form, an orthonormal wavelet basis of $L^2(\mathbb{R})$ is defined as the collection of functions

$$2^{j/2}\psi(2^j x - k), \quad \text{where } j, k \in \mathbb{Z}, \tag{3.4}$$

and the wavelet ψ is a smooth, well localized function. Wavelet coefficients are further defined by

$$c_{j,k} = 2^j \int f(x)\, \psi(2^j x - k)\, dx.$$

A L^1-normalization is used (rather than the classical L^2-normalization), as it is better suited to express scale invariance relationships, as shown in Eq. (3.5) below.

Let f denote a function satisfying Eq. (3.2) with $a = 2$, i.e. such that

$$\forall x \in \mathbb{R}, \quad f(2x) = 2^H f(x);$$

then, clearly, its wavelet coefficients satisfy

$$\forall j, k, \quad c_{j,k} = 2^H c_{j+1,k}.$$

Similarly, in the probabilistic setting, if Eq. (3.3) holds, then the (infinite) sequences of random vectors $\mathcal{C}_j \equiv (2^{Hj} c_{j,k})_{k\in\mathbb{Z}}$ satisfy the following property:

$$\{\mathcal{C}_j\}_{j\in\mathbb{Z}} \quad \text{share the same law.} \tag{3.5}$$

Other probabilistic properties have simple wavelet counterparts; let us mention a few of them.

Let X_t be a *Markov process*, i.e. a stochastic process satisfying the property

$$\forall s > 0, \quad X_{t+s} - X_t \text{ is independent of the } (X_u)_{u \leq t};$$

typical examples of Markov processes are supplied by *Brownian motion*, or, more generally, by *Lévy processes*, i.e. processes with stationary and independent increments. Because wavelets have vanishing integrals, the Markov property implies that, if the supports of the $\psi(2^j x - k)$ are disjoint, then the corresponding coefficients $c_{j,k}$ are independent random variables.

Recall that a stochastic process X_t has *stationary increments* if and only if

$$\forall s > 0, \, X_{t+s} - X_t \text{ has the same law as } X_s - X_0 \; ;$$

typical examples are *fractional Brownian motions*, or *Lévy processes*. If X_t has stationary increments, then

$$\forall j, \quad \text{the random variables} \quad (c_{j,k})_{k \in \mathbb{Z}} \quad \text{share the same law.} \tag{3.6}$$

Note that this property holds for each given j, but implies no relationship between wavelet coefficients at different scales; however, if X_t is a self-similar process of exponent H with stationary increments, then it follows from Eqs. (3.5) and (3.6) that the random variables $2^{Hj} c_{j,k}$ all share the same law (for all j and k).

A random variable X has a *stable distribution* iff

$$\forall a, b \geq 0, \, \exists c > 0 \quad \text{and} \quad d \in \mathbb{R} \quad \text{such that } aX_1 + bX_2 \stackrel{\mathcal{L}}{=} cX + d$$

where X_1 and X_2 are independent copies of X. A random process is stable if all linear combinations of the $X(t_i)$ are stable. This property implies that the wavelet coefficients of a stable process are stable random variables. Further, it can actually be shown that any finite collection of wavelet coefficients forms a stable vector (cf. e.g., [11–14]).

Finally, if X_t is a *Gaussian process*, then any finite collection of its wavelet coefficients is a Gaussian vector. Note however a common pitfall. Unless some additional stationarity property holds, the sole Gaussianity hypothesis for the marginal distribution of the process does *not* imply that the empirical distributions of the coefficients $(c_{j,k})_{k \in \mathbb{Z}}$ are close to Gaussian. It can only be said that they will consist of *Gaussian mixtures* (which can strongly depart from a Gaussian distribution). The statistical properties of wavelet coefficients of fBm were studied in depth in the seminal works of P.

Flandrin [15, 16] that significantly contributed to the popularity of wavelet transforms in the analysis of scale invariance.

Among the results that we mentioned above, it is important to draw a difference between those that are a consequence of the sole fact that wavelet coefficients are defined as a linear mapping, and therefore would hold for any other basis (Gaussianity and stability) and those that are the consequence of the particular algorithmic nature of wavelets (selfsimilarity and stationarity). These (rather straightforward) results are either implicit or explicit in the literature concerning wavelet analysis of stochastic processes; however, in order to give a flavor of the ideas and methods involved, we give a proof of Eq. (3.5) at the end of Sec. 3.4.1.

Further, let us mention that all these wavelet properties do not constitute exact characterizations of the probabilistic properties of the corresponding processes, but are only implications, because of a deficiency of wavelet expansions that will be explained in Sec. 3.4.1.

3.1.3 Beyond self-similarity: Multifractal analysis

Beyond self-similarity. An obvious drawback of using *exact* probabilistic properties, such as self-similarity or independence of increments, is that, though they can relevantly model physical properties of the underlying data, they usually do not hold exactly for real-life signals. Two limitations of self-similarity can classically be pointed out. First, the scaling properties cannot hold for all scales (as required by the mathematical definition of self-similarity). This may be due to noise corruption or it may also result from the fact that the *physical* phenomena implying those properties intrinsically act over a finite range of scales only. Second, the appealing property of self-similarity (the sole self-similarity parameter H contains the essence of the model) may also constitute its limitation. Can one expect that the complexity of real-world data can be accurately encapsulated in one single parameter only? Let us examine such issues.

Multiplicative cascades. As mentioned above, Mandelbrot coined fBm as one of the major candidates to model scale invariance. However, combining self-similarity and stationary increments implies that all finite moments of the increments of fBm (and of any process whose definition includes these two properties) satisfy this property: for all p such that $\mathbb{E}(|X(1)|^p) < \infty$ [8, 9],

$$\forall \tau \in \mathbb{R}, \quad \mathbb{E}(|X(t+\tau) - X(t)|^p) = \mathbb{E}(|X(1)|^p)\, |\tau|^{pH}. \tag{3.7}$$

Though very powerful, this result can also be regarded as a severe limitation for applications, where empirical estimates of moments, $\int |X(t+\tau) - X(t)|^p dt$, computed on the data assuming ergodicity, may instead behave as $\int |X(t+\tau) - X(t)|^p dt = \tau^{\zeta(p)}$, for a range of τ, and where $\zeta(p)$ is, by construction, a concave function, but is not necessarily linear. This is notably the case when analyzing velocity or dissipation fields in the study of hydrodynamic turbulence where strictly concave scaling functions were obtained (cf. e.g., [17] for a review of experimental results). This was justified by a celebrated argument due to Kolmogorov and Obukhov in 1962 [18–20]. The Navier-Stokes equation, governing fluid motions, implies that the (third power of the) gradient of the velocity $u(x+r) - u(x)$ is proportional to the mean of the local energy ϵ_r dissipated into a bulk of size r:

$$\mathbb{E}(u(x+r) - u(x))^3 \propto \mathbb{E}(\epsilon_r) \cdot r.$$

If ϵ_r is constant, then any power scales as $\mathbb{E}(u(x+r) - u(x))^p \propto \epsilon_r^p r^{p/3}$. But, in turbulence, ϵ_r is not constant over space and should rather be regarded as a random variable, hence implying $\mathbb{E}(u(x+r) - u(x))^p \propto \mathbb{E}(\epsilon_r^p)\, r^{p/3}$ which naturally implies a general scaling of the form $\sim\mathbb{E}(u(x+r) - u(x))^p \propto r^{\zeta(p)}$, with $\zeta(p)$ which is likely to depart from the linear behavior pH which is implied by the self-similarity hypothesis.

To account for this form of observed scale invariance, new stochastic models — multiplicative cascades — have been proposed. Numerous declinations of cascades were proposed in the literature of hydrodynamic turbulence to model data, see e.g., [17] (on the mathematical side, see the book [21] for a detailed review, and also the contribution of J. Barral and J. Peyrière [22] in the present volume, for a lighter introduction). For the sake of simplicity, the simplest (1D) formulation is presented here; however, it contains the key ingredients present in all other more general cascade models. Starting from an interval (arbitrarily chosen as the unit interval), a cascade is constructed as the repetition of the following procedure (c being an arbitrary integer). At iteration j, each of the c^{j-1} sub-intervals $\{I_{j-1,k}, k = 0, \ldots, c^{j-1}-1\}$, obtained at iteration $j-1$, is split into c intervals of the same length c^{-j}; a mean-1 positive random variable $w_{j,k}$ (drawn independently and with identical distribution) is associated with each $\{I_{j,k}, k = 0, \ldots, c^j - 1\}$. After an infinite number of iterations, the *cascade* W is defined as the limit (in the sense of measures) of the finite products

$$W_J(x)dx = \prod_{j \leq J,\ x \in I_{j,k}} w_{j,k}\, dx.$$

Mathematically, this sequence of measures converges under mild conditions on the distribution of the multipliers w (cf. [23]). Such constructions and variations have been massively used to model the Richardson energy cascade implied by the Navier-Stokes equation [24] (cf. [17] for a review of such models). One of the major contributions of Mandelbrot to the field of hydrodynamic turbulence was proposed in [25], where he presented the themes and variations on multiplicative cascades in a single framework and studied their common properties, notably in terms of scale invariance. This celebrated contribution paved the way towards the creation of the word *multifractal*, to the notions of multifractal processes and multifractal analysis, where *multi* here refers to the fact that instead of a single scaling exponent H, a whole family of exponents $\zeta(p)$ are needed for a better description of the data. Mandelbrot also discussed the importance of the distribution of the multipliers and how multiplicative cascades may significantly depart from log-normal statistics that could be obtained from a simplistic argument:

$$\Pi w_{j,k} = \Pi e^{\log(w_{j,k})} = e^{\sum \log w_{j,k}},$$

where $\sum \log w_{j,k}$ tends to a normal distribution, hence $e^{\sum \log w_{j,k}}$ should tend to a log-normal distribution (cf. [26, 27], and also [20]). This opened many research tracks proposing alternatives to the log-normal model, the most celebrated one being the log-Poisson model [28]. This will be further discussed in Sec. 3.6.1. Much later, multiplicative cascades were connected to Compound Poisson cascades and further to Infinitely Divisible Distributions (cf. e.g., [29]). Mandelbrot himself contributed to these latter developments with the earliest vision of Compound Poisson Cascades [30] and also with his celebrated fBm in multifractal time [31], which will be considered in Sec. 3.6.2. This gave birth to multiple constructions of processes with *multifractal* properties (cf. e.g., [32–38]).

Compared with the concept of self-similarity (and in particular fBm), which is naturally related to random walks and hence to additive constructions, cascades are, by construction, tied to a multiplicative structure. This deep difference in nature explains why practitioners, for real-world applications, are often eager to distinguish between these two classes of models. This distinction is however often misleadingly confused with the opposition of mono- versus multi-fractal processes. This will be further discussed in Sec. 3.5.4.

Scaling functions and scaling exponents. The empirical observations of strictly concave scaling exponents, made on hydrodynamic turbulence data, led to the design of a new tool for analysis, *scaling functions*. Let $f : \mathbb{R}^d \to \mathbb{R}$. The *Kolmogorov scaling function* of f (see [39]) is the function $\eta_f(p)$ implicitly defined by the relationship

$$\forall p \geq 1, \quad \int |f(x+h) - f(x)|^p dx \sim |h|^{\eta_f(p)}, \tag{3.8}$$

where the symbol $\sim$ means that

$$\eta_f(p) = \lim_{|h| \to 0} \frac{\log \left(\int |f(x+h) - f(x)|^p dx \right)}{\log |h|}. \tag{3.9}$$

Note that this limit does not necessarily always exist; when it does, it reflects the fact that f shows clear scaling properties (and it must hold in order to numerically compute scaling functions). However, we are not aware of any simple and general mathematical assumption implying that it does; therefore, it is important to formulate a mathematical and slightly more general definition of the scaling function, which is well defined for any function $f \in L^p$:

$$\forall p \geq 1, \quad \eta_f(p) = \liminf_{|h| \to 0} \frac{\log \left(\int |f(x+h) - f(x)|^p dx \right)}{\log |h|}. \tag{3.10}$$

It is important to note that, if data are smooth (i.e., if one obtains that $\eta_f(p) \geq p$), then one has to use differences of order 2 or more in Eqs. (3.9) and (3.10) in order to define correctly the scaling function.

For application to real-world data, which are only available with a finite time or space resolution, Eq. (3.9) can be interpreted as a linear regression of the log of the left hand side of Eq. (3.8) vs. log($|h|$). The same remark holds for the other scaling functions that we will meet.

Multifractal analysis. In essence, multifractal analysis consists in the determination of scaling functions (variants to the original proposition of Kolmogorov η_f will be considered later). Such scaling functions can then be involved into classification or model selection procedures. Scaling functions constitute a much more flexible analysis tool than the strict relation Eq. (3.2). For many different real-world data stemming from applications of various natures, relationships such as Eq. (3.8) are found to hold, while this is usually not the case for Eq. (3.2), see Sec. 3.6 for illustrations on real world data.

An obvious key advantage of the use of the scaling function $\eta_f(p)$ is that its dependence in p can take a large variety of forms, hence providing versatility in adjustment to data. Therefore, multifractal analysis, being based on a whole function (the scaling function $\eta_f(p)$, or some of its variants introduced below), rather than on the single selfsimilarity exponent, yields much richer tools for classification or model selection. The scaling function however satisfies a few constraints, for example, it has to be a concave non-decreasing function (cf. e.g., [40, 41]).

Another fundamental difference between the self-similarity exponent and the scaling function lies in the fact that, for stochastic processes, the self-similarity exponent is, by definition, a deterministic quantity, whereas the scaling function is constructed for each sample path, and therefore is a priori defined as a random quantity. And, indeed, even if the limit in Eq. (3.10) turns out to be deterministic for many examples of stochastic processes (such as fBm, Lévy processes, or multiplicative cascades, for instance), it is not always the case, as shown, for instance, by the example of fairly general classes of Markov processes, such as the ones studied in [42]. Note that, if a stationarity assumption can be reasonably assumed, then one can define a deterministic quantity by taking an expectation instead of a space average in Eq. (3.10), which leads to a definition of the scaling function through moments of increments. Under appropriate ergodicity assumptions, time (or space) averages can be shown to coincide with ensemble averages, and, in this case, both approaches lead to the same scaling functions, see [43] for a discussion (checking such ergodicity assumptions for experimental data does, however, not seem feasible and can become an intricate issue, cf. [44]).

3.1.4 Data analysis and signal processing

In its early ages, the concept of *scale invariance* has been deeply tied to that of $1/f$-processes, that is, to second order stationary random processes whose *power spectral densities* behave as a power law with respect to frequency over a broad range of frequencies: $\Gamma_X(f) \sim C|f|^{-\beta}$, where the frequency f satisfies $f_m \leq f \leq f_M$, with $f_M/f_m \gg 1$. From that starting point, spectrum analysis was regarded as the natural tool to analyze scale invariance and estimate the corresponding scaling parameter β. In that context, the contribution of Mandelbrot and Van Ness [4] can be regarded as seminal since, elaborating on [45], it put forward fBm, and its increment process, fractional Gaussian noise (fGn), as a model that extends and encompasses $1/f$-processes: fGn is a $1/f$-process, with $\beta = 2H - 1$, fBm is a non-stationary process that shares the same scale invariance intuition. This change of paradigm immediately raised the need for new and generic estimation tools for measuring the parameter H from real world data, which were no longer based on classical spectrum estimation. Relying strongly on the intuitions beyond selfsimilarity and long memory (a concept deeply tied with self-similarity, cf. e.g. [46]), the R/S estimator (for Range-Scale Statistics) has been the first such tool proposed in the literature, to the study of which Mandelbrot also contributed [5, 47, 48]. The R/S estimator has notably been used in finance and economy, in particular by D. Strauss-Kahn and his collaborators, (see e.g., [49, 50]). Later, in the 90s, with the advent of wavelet transforms, seminal contributions studied the properties of the wavelet coefficients of fBm [15, 16, 51] and opened avenues towards accurate and robust wavelet-based estimations of the self-similarity parameter H [52–55].

To some extent, the present contribution can be read as a tribute to ideas and tracks originally addressed by Mandelbrot (and others) aiming at estimating the parameters characterizing scale invariance in real world data: The wavelet based formulations of multifractal analysis (i.e., of the estimation from data of scaling functions and multifractal spectra) devised in Secs. 3.4 and 3.5 can be read as continuations and extensions of these pioneering works towards richer models and refined variations on the characterization of scale invariance.

3.1.5 Goals and outline

In Sec. 3.2, it is explained how various paradigms pertinent in fractal geometry, such as self-similarity or fractal dimension, can be shifted to the setting of functions, thus leading to the introductions of several *scaling functions*. In Sec. 3.3, it is shown how such scaling functions can be reinterpreted in terms of function space regularity. On one hand, it allows to derive some of their properties; on other hand, it paves the way towards equivalent formulations that will be derived later. Sec. 3.4 introduces wavelet techniques, which allow to reformulate these former notions in a more efficient setting, from both a theoretical and pratical point of view. Sec. 3.5 introduces the *multifractal formalism*, which offers a new interpretation for scaling functions in terms of fractal dimensions of the sets of points where the functions considered has a given pointwise smoothness. Applications of the tools and techniques developed here will be given in Sec. 3.6, aiming at illustrating the various aspects of multifractal analysis addressed in the course of this contribution. Such applications are chosen either because Mandelbrot significantly contributed to their analysis, or because they illustrate particularly well some concepts which will be developed here. The selected applications stem from widely different backgrounds, ranging from turbulence to the internet, heart beat variability and finance (in 1 D), and from natural textures to paintings by famous masters (in 2 D), thus showing the extremely rich possibilities of these new techniques.

3.2 From fractals to multifractals

3.2.1 Scaling and fractals

Dynamics versus geometry. One of the major contributions of Mandelbrot has been to understand the benefits of using mathematical tools and objects already defined in earlier works but regarded as incidental, such as fractal dimensions, in order to explore the world of irregular sets, to classify them and to study their properties. A key example is supplied by the *von Koch curve* (cf. e.g., [56, 57] and Fig. 3.1). The important role that Mandelbrot gave to this curve is typical of his way of diverting a mathematical object from its initial purpose and of making it fit to his own views; indeed, it was initially introduced as an example of the graph of a continuous, nowhere differentiable function (see the book by G. Edgar [58], which made available many classic articles of historical importance in fractal analysis). Mandelbrot shifted the status of this curve from a peripherical example of strange set to a central element in the construction of his theory, and he baptized it the *Von Koch snowflake*, a typical example of the poetic accuracy that Mandelbrot displayed for coining new names (other examples, being the *devil's staircase*, *Lévy flights*, *Noah or Joseph effects*, ...). This example will be revisited below in Sec. 3.4.2 to validate a numerical method for multifractal analysis.

Two features — dynamics and geometry — are intimately tied into the von Koch snowflake: it is constructed from (and hence invariant under) a dynamical system consisting of the iterative repetition of a *split and reproduce* procedure (highly reminiscent of the *split and multiply* procedure entering the definition of multiplicative cascades, cf. Sec. 3.1.3 earlier); the resulting geometrical set has well defined geometrical properties that can be put into light through the determination of its *box dimension*, whose definition we now recall.

Dimensions

Definition 3.2.1 Let A be a bounded subset of $\mathbb{R}^d$; if $\varepsilon > 0$, let $N_\varepsilon(A)$ be the smallest number such that there exists a covering of A by $N_\varepsilon(A)$ balls of radius ε.

The upper and lower box dimension of A are respectively given by

$$\overline{dim_B}(A) = \limsup_{\varepsilon\to 0} \frac{\log N_\varepsilon(A)}{-\log \varepsilon}, \quad and \quad \underline{dim}_B(A) = \liminf_{\varepsilon\to 0} \frac{\log N_\varepsilon(A)}{-\log \varepsilon}. \tag{3.11}$$

When both limits coincide (as it is the case for the von Koch snowflake), they are referred to as the *box dimension* of the set A:

$$dim_B(A) = \lim_{\varepsilon\to 0} \frac{\log N_\varepsilon(A)}{-\log \varepsilon}. \tag{3.12}$$

This implies that $N_\varepsilon(A)$ displays an approximate power-law behavior with respect to scales:

$$N_\varepsilon(A) \sim \varepsilon^{-dim_B(A)}, \tag{3.13}$$

a major feature that makes the box dimension useful for practical applications, as it can be computed through a linear regression in a log-log plot.

The strong similarity (in a geometric setting) between the box dimension and the Kolmogorov scaling function (3.10) can now clearly be seen. The latter is also defined as a linear regression in log-log plots (log of the p-th moments vs. the log of the scales). Both objects hence fall under the general heading of *scale invariance*.

This relationship between geometric and analytic quantities is not only formal. On one hand, the determination of scaling functions has important consequences in the numerical determination of the fractal dimensions of graphs, and related quantities, such as the p-variation, see Sec. 3.2.2; on other hand, the *multifractal formalism*, presented in Sec. 3.5, allows to draw relationships between scaling functions and the dimensions of other types of sets, defined in a different way. They are the sets of points where the function considered has a given pointwise smoothness (see Definition 3.3.1). Note however that the notion of dimension used in the context of the multifractal formalism differs from the box dimension. One has to use the *Hausdorff dimension* which is now defined (see [59]).

Definition 3.2.2 Let A be a subset of $\mathbb{R}^d$. If $\varepsilon > 0$ and $\delta \in [0, d]$, let

$$M_\varepsilon^\delta = \inf_R \left(\sum_i |A_i|^\delta \right),$$

where R is an ε-covering of A, i.e. a covering of A by bounded sets $\{A_i\}_{i \in \mathbb{N}}$ of diameters $|A_i| \leq \varepsilon$. (The infimum is therefore taken on all ε-coverings.)

For any $\delta \in [0, d]$, the δ-dimensional Hausdorff measure of A is

$$mes_\delta(A) = \lim_{\varepsilon \to 0} M_\varepsilon^\delta.$$

One can show that there exists $\delta_0 \in [0, d]$ such that

$$\begin{aligned} &\forall \delta < \delta_0, \quad mes_\delta(A) = +\infty \\ &\forall \delta > \delta_0, \quad mes_\delta(A) = 0. \end{aligned}$$

This critical δ_0 is called the Hausdorff dimension of A, and is denoted by $\dim(A)$.

An important convention, in view of the use of these dimensions in the context supplied by the multifractal formalism, is that, if A is empty, then $\dim(\emptyset) = -\infty$.

In contradistinction with the box dimension, the Hausdorff dimension is not obtained through a regression on log-log plots of quantities that are computable on real-life data; therefore it is fitted to theoretical questions and can not be used directly in signal or image processing; however, we will see that the multifractal formalism allows to bypass this problem by supplying an indirect way to compute such dimensions.

Scaling, dimensions and dynamics. The discussions in this section aimed at illustrating the interplay between three different notions: the signal processing notion of scaling, or scale invariance, the geometrical notion of dimension and the dynamic notion of constructive split/multiply or split/reproduce procedures. Note that the relationships between fractals and dynamics have been investigated by Mandelbrot, see [60] for a panorama. The dynamic constructive procedures of geometrical sets (e.g., von Koch snowflake) or of functions (e.g., multiplicative cascades) result in geometrical properties of the constructed object. These geometrical properties can be theoretically characterized using dimensions. These dimensions can be measured practically if they additionally imply the existence of power law behaviours (or scaling) (such as those reported in Eqs. (3.13) or (3.8)). When the constructive procedure underlying the geometrical sets or functions analyzed are known and parametrized, the measure of geometrical dimension may enable to recover the corresponding dynamic parameters. However this is certainly no more true outside of a parametric model: Many different dynamic constructions may yield the same dimensions. Therefore, the inverse problem of identifying the nature of the dynamic or even its existence from the sole measure of dimensions is ill-posed and can not be achieved in general. Specifically, a concave scaling function, that hence departs from linearity in p, does not prove the existence of an underlying cascade mechanism in data.

3.2.2 From graph box-dimensions to p-variation: The oscillation scaling function

From graphs to functions. For data analysis purposes, an obvious difference between the determination of the fractal dimension of a set A and the scaling function $\eta_f(p)$ of a function f has already been pointed out: while the former reduces to a single number, the latter consists of a whole function of the variable p, hence yields a much richer tool for classification and identification. However, the gap can easily be bridged in the case where the geometrical set A consists of the graph, denoted $\mathcal{G}r(f)$, of the function f. Let us examine how a slight extension of the notion of box dimension of a graph allows to introduce a dependency on a variable p, and hence a family of exponents whose definition bears some similarity with the Kolmogorov scaling function.

Oscillations. Let $f : [0, 1]^d \to \mathbb{R}$ be a continuous function (the set $[0, 1]^d$ plays no specific role and is taken here just to fix ideas). If $A \subset [0, 1]^d$, let $Os_f(A)$ denote the *oscillation* of f on the set A, i.e.

$$Os_f(A) = \sup_A f - \inf_A f.$$

Note that this notion defines only first order oscillation; if A is a convex set, second order oscillation (which has to be used for smooth functions) is defined as

$$Os_f(A) = \sup_{x,y\in A} \left| f(x) - 2f\left(\frac{x+y}{2}\right) + f(y) \right|$$

(and so on for higher order oscillation). Let $\lambda_j(x_0)$ denote the dyadic cube of width 2^{-j} that contains x_0; Λ_j will denote the set of such dyadic cubes $\lambda \subset \mathbb{R}^d$ of width 2^{-j}. Let now p be given. The p oscillation at scale j of f is defined as

$$R_f(p,j) = 2^{-dj} \sum_{\lambda\in\Lambda_j} (Os_f(\lambda))^p. \tag{3.14}$$

This quantity allows to define the *oscillation scaling function* of f as

$$O_f(p) = \liminf_{j\to+\infty} \frac{\log\left(R_f(p,j)\right)}{\log(2^{-j})}. \tag{3.15}$$

Box dimension of graphs versus the oscillation scaling function. Since f is continuous, the graph of f is a bounded subset of $\mathbb{R}^{d+1}$. First, note that, in the definition of the upper box dimension Eq. (3.11), one can rather consider coverings by dyadic cubes of side 2^{-j}; indeed each optimal ball of width ε used in the covering is included in at most 4^{2d} dyadic cubes of width $2^{-j} = [\log_2(\varepsilon)]$; and, in the other way, a dyadic cube is contained in the ball of same center and width $\sqrt{d}\cdot 2^{-j}$. Therefore using dyadic cubes (instead of optimally positioned balls) only changes prefactors in $N_\varepsilon(\mathcal{G}r(f))$ and not scaling exponents.

Let us now consider the dyadic cubes used in the covering of the graph of f and which stand above a dyadic cube λ of dimension d and width 2^{-j} included in $[0,1]^d$. Their number $N(\lambda)$ clearly satisfies the property

$$2^j\left(\sup_\lambda f - \inf_\lambda f\right) \le N(\lambda) \le 2^j\left(\sup_\lambda f - \inf_\lambda f\right) + 2.$$

This, together with the definition of the oscillation scaling function, implies that for $p = 1$,

$$\overline{dim}_B(\mathcal{G}r(f)) = \max(d, d+1-O_f(1)).$$

However, the oscillation scaling function can be used for other (positive) values of p, and therefore yields a classification tool with similar potentialities as those offered by the Kolmogorov scaling function. We will see an example of its use in finance, in order to estimate quadratic variations, see Sec. 3.6.2. An extension of the oscillation scaling function will be considered later, defined through wavelet quantities, the *leader scaling function*, cf. Sec. 3.4.4.

Box dimension and selfsimilarity exponent. For some self-similar functions and processes, a direct relationship between the box dimension of their graphs and their self-similarity exponent can be drawn. For instance, one-dimensional fBm of selfsimilarity index H has a graph of box dimension $2 - H$, and the same result holds in a deterministic setting for the graphs of the Weierstrass-Mandelbrot functions, which have box dimension $2 - H$ (see [61, 62] where similar results can also be found for Hausdorff dimensions of graphs). Note that Mandelbrot himself had already drawn connections between self-similarity and fractal dimension (cf. e.g., [63–65]).

3.3 Geometry vs. analysis

The shift from fractal geometry to multifractal analysis essentially consists in replacing the study of selfsimilarity, and of geometric properties of sets, by the study of analytic properties of functions. This shift has several facets: first, in Sec. 3.3.1, we will explore relationships between selfsimilarity and pointwise regularity, two concepts which are often confused, and actually have subtle interplays which are far from being fully understood today. Second, in Sec. 3.3.2, we will give the interpretation of scaling functions in terms of function space regularity, which allows to use the full apparatus of functional analysis. Finally, in Sec. 3.5, the multifractal formalism is introduced: It allows to interpret the (Legendre transform of the) scaling function in terms of dimensions of sets of points where f has a given pointwise Hölder regularity.

3.3.1 Pointwise Hölder regularity

Hölder exponent. Selfsimilarity of either deterministic functions or stochastic processes is intimately related with their local regularity. Before explaining these relationships, let us start by recalling the notion of *pointwise Hölder regularity* (alternative notions of pointwise regularity, have also been considered, allowing to deal with functions that are not locally bounded, see [66–69]).

Definition 3.3.1 Let $f\colon \mathbb{R}^d \to \mathbb{R}$ be a locally bounded function, $x_0 \in \mathbb{R}^d$ and let $\gamma \geq 0$; f belongs to $C^\gamma(x_0)$ if there exist $C > 0$, $R > 0$ and a polynomial P of degree less than γ such that

$$\text{if } |x - x_0| \le R, \quad \text{then } |f(x) - P(x - x_0)| \le C|x - x_0|^\gamma.$$

The Hölder exponent of f at x_0 is

$$h_f(x_0) = \sup\{\gamma : f \text{ is } C^\gamma(x_0)\}.$$

When Hölder exponents lie between 0 and 1, the Taylor polynomial $P(x - x_0)$ boils down to $f(x_0)$ and the definition of the Hölder exponent heuristically means that,

$$|f(x) - f(x_0)| \sim |x - x_0|^{h_f(x_0)}.$$

For these cases, a relationship can be drawn with the local oscillation considered in Sec. 3.2.2. Indeed, let $\lambda_j(x_0)$ denote the dyadic cube of width 2^{-j} that contains x_0, and $3\lambda_j(x_0)$ the cube of same center and 3 times wider (it therefore contains 3^d dyadic cubes of width 2^{-j}). One easily checks that, if $0 < h_f(x_0) < 1$, then

$$h_f(x_0) = \liminf_{j\to+\infty} \frac{\log\left(Os_f(3\lambda_j(x_0))\right)}{\log(2^{-j})}. \tag{3.16}$$

This formula puts into light the fact that the Hölder exponent can theoretically be obtained through linear regressions on log-log plots of the oscillations $Os_f(3\lambda)$ versus the log of the scale. However, in practice, this formula has found little direct applications for two reasons: one is that alternative formulas based on *wavelet leaders* (cf. Sec. 3.4.4) are numerically more stable, and the other is that many types of signals display an extremely erratic Hölder exponent, so that the instability of its determination is intrinsic, no matter which method is used. We will see in Sec. 3.4.4 how to turn this deeper obstruction.

Implications of selfsimilarity on regularity and dimensions. The Hölder exponent sometimes coincides with the selfsimilarity exponent, hence resulting in a confusing identification of the two quantities. This is the case for two frequently used examples: Weierstrass-Mandelbrot functions and fBm sample paths have everywhere the same and constant Hölder exponent, $\forall x : h_f(x) = H$. The theoretical definition of such examples essentially relies on a single parameter, and therefore it comes as no surprise that, for such simple functions with a constant Hölder exponent, both their selfsimilarity and their local regularity are governed by this sole parameter (and coincide). Similarly, their scaling functions are linear in p, and therefore are governed by one parameter only, the slope of this linear function, which, for these two examples, is H.

However, one should not infer hasty conclusions from these examples: selfsimilar processes with stationary increments do not always possess a single Hölder exponent that corresponds to the selfsimilarity exponent, as shown by the example of *stable Lévy processes*: these processes are selfsimilar of exponent $H = 1/\alpha$, where α is the stability parameter. In addition, their scaling function is linear in p only for p small enough:

$$\eta_f(p) = \begin{cases} Hp & \text{if } 0 \le p \le 1/H \\ 1 & \text{if } p > 1/H \end{cases}$$

This is intimately related with the fact that their Hölder exponent is not constant: sample paths of a stable Lévy process have, for each $h \in [0, H]$, a dense set of points where their Hölder exponent takes the value h (cf. [70] for Lévy processes, and [71] for *Lévy sheets* indexed by $\mathbb{R}^d$). Thus, their Hölder exponents jump in a very erratic manner from point to point, as a consequence of the high variability of the increments: indeed, their distributions only have power-law decay, and therefore (in strong contradistinction with Gaussian variables) can take extremely large values with a relatively large probability; the importance of this phenomenon in applications was largely undervalued in the past, until B. Mandelbrot pointed out its possibly dramatic consequences, referring to them in a colorful way as *Noah's effect.*

Let us now be more explicit concerning the relationships between selfsimilarity and Hölder regularity. A first result relates selfsimilarity with uniform regularity of sample paths.

Proposition 3.3.1 *Assume that X_t is a selfsimilar process of exponent H, with stationary increments, and such that $\mathbb{E}(|X_1|^\alpha) < \infty$ for some $\alpha > 0$. Then,*

$$\forall h < H - 1/\alpha, \quad a.s.\ \exists C, \quad \forall t, s, |X_t - X_s| \le C|t-s|^h \tag{3.17}$$

The proof follows directly from the Kolmogorov-Chentsov theorem. Recall that this theorem asserts that, if X_t is a random process satisfying

$$\forall t, s, \quad \mathbb{E}(|X_t - X_s|^\alpha) \le C|t-s|^{1+\beta}$$

then,

$$\forall h < \beta/\alpha, \quad \text{a.s. } \exists C, \quad \forall t, s, |X_t - X_s| \le C|t-s|^h$$

(see [7]). But, if X is selfsimilar, then

$$\mathbb{E}(|X_t - X_s|^\alpha) = |t-s|^{H\alpha}\mathbb{E}(|X_1|^\alpha), \tag{3.18}$$

and the proposition follows.

We will reinterpret this result in Sec. 3.4.3 when the uniform Hölder exponent will be introduced. The following result goes in the opposite direction: it relates selfsimilarity with irregularity of sample paths.

Proposition 3.3.2 *Let X_t be a selfsimilar process of exponent H, with stationary increments, and which satisfies $\mathbb{P}(\{X_1 = 0\}) = 0$, i.e. the law of X at time $t = 1$ (or at any other time $t > 0$) does not contain a Dirac mass at the origin. Then*

$$a.s. \quad a.e. \quad h_X(t) \leq H. \tag{3.19}$$

Proof: We first prove the result for $t = 0$. The assumption $\mathbb{P}(\{X_1 = 0\}) = 0$ implies the existence of a decreasing sequence $(\delta_n)_{n\in\mathbb{N}}$ which converges to 0 and satisfies

$$\mathbb{P}(|X_1| \leq \delta_n) \leq \frac{1}{n^2}.$$

Let $l_n = \exp(-1/\delta_n)$. Then $X_{l_n} \stackrel{\mathcal{L}}{=} (l_n)^H X_1$. Therefore

$$\mathbb{P}\left(|X_{l_n}| \leq \frac{(l_n)^H}{\log(1/l_n)}\right) = \mathbb{P}(|X_1| \leq \delta_n) \leq \frac{1}{n^2}.$$

The Borel-Cantelli lemma implies that a.s. a finite number of these events happens, and therefore $h_X(0) \leq H$. By the stationarity of increments, this argument holds the same at each t, i.e.

$$\forall t \ \text{ a.s. } h_X(t) \leq H;$$

and Fubini's theorem allows to conclude that Eq. (3.19) holds.

In particular, assume that X_t satisfies the assumptions of Proposition 3.3.2, and also of Proposition 3.3.1 for any $\alpha > 0$. Then, both results put together show that, in this case, Eq. (3.17) is optimal. Relationships between selfsimilarity and Hölder exponents for selfsimilar processes will be further investigated in Sec. 3.5.4.

Note that, without additional assumptions, one can not expect results more precise than those given by Propositions 3.3.1 and 3.3.2, as shown by the subcase of stable, selfsimilar processes with stationary increments, see [9, 72]: Let H denote the selfsimilarity index of such a process, and α

its stability index. First, note that there is no direct relationship between α, H and Hölder regularity: they always satisfy

$$0 < \alpha < 2 \quad \text{and} \quad H \leq \max(1, 1/\alpha);$$

the case of fBm is misleadingly simple since it satisfies

$$\alpha = 2 \quad \text{and} \quad \forall x,\ h_{B_H}(x) = H \in (0,1);$$

however Lévy processes satisfy $H = 1/\alpha$ and $h_f(x)$ takes all values in $[0, H]$. Finally, Lévy fractional stable processes have nowhere locally bounded stable sample paths if $H < 1/\alpha$, and continuous sample paths if $H > 1/\alpha$ (see Sec. 3.4.3 for a more precise result). Sample paths of selfsimilar processes (with stationary increments) together with their scaling functions (and multifractal spectra) are illustrated in Fig. 3.2 in Sec. 3.5.2.

Let us now mention a few results concerning the relationship between selfsimilarity and fractional dimensions. We start with fBm which again is the simplest and best understood case. Such results are of two types:

- Dimensions of the graph of sample paths: Box and Hausdorff dimensions coincide and take the value $2 - H$, see [61, 62].
- Relationship between the Hausdorff dimension of a set A and of its image $X(A)$ (see [73] and references therein)

$$\text{a.s., for any Borel set,} \quad A \subset \mathbb{R}, \quad \dim X(A) = \min(1, \dim(A)/H). \tag{3.20}$$

We will see that this second result has important consequences for the multifractal analysis of some finance models proposed by L. Calvet, A. Fisher, and B. Mandelbrot, see Eq. (3.51). Partial results of these two types also hold for other selfsimilar processes, see [73].

Regularity versus dimension: The mass distribution principle. The example of stable Lévy processes suggests that the most fruitful connection to be drawn does not involve the notion of selfsimilarity, but rather should connect scaling functions with pointwise regularity. This is precisely what the *multifractal formalism* performs, as it establishes a relationship between scaling functions and the Hausdorff dimensions of the *isohölder sets* of f defined by

$$E_f(h) = \{x_0 : h_f(x_0) = h\}. \tag{3.21}$$

The multifractal formalism will be detailed in Sec. 3.5. A clue which shows that such relationships are natural can however be immediately given, via the *mass distribution principle* which goes back to Besicovitch (see Proposition 3.3.3 below). This principle establishes a connection between the dimension of a set A and the regularity of the measures that this set can carry. The motivation for it stems from the following remark. Mathematically, it is easy to obtain upper bounds for the Hausdorff dimension of a set A: Indeed, any well chosen sequence of particular ε-coverings (for a sequence $\varepsilon_n \to 0$) yields an upper bound. To the opposite, obtaining a lower bound by a direct application of the definition is more difficult, since it requires to consider *all possible* ε-coverings of A. The advantage of the mass distribution principle is that it allows us to obtain lower bounds of Hausdorff dimensions by just constructing one measure *carried* by the set A. Recall that a measure μ is *carried by* A if $\mu(A^c) = 0$. Note, however, that a set A satisfying such a definition is not unique; in particular, this notion should not be mistaken with the more usual one of *support* of the measure, which is uniquely defined as the intersection of all closed sets carrying μ. For example, consider the measure

$$\mu = \sum_{p/q \in [0,1], p \wedge q = 1} \frac{1}{q^3} \delta_{p/q}$$

(where δ_a denotes the Dirac mass at the point a). The support of this measure is the interval $[0, 1]$, but it is (also) carried by the (much smaller) set $A = \mathbb{Q} \cap [0, 1]$.

Proposition 3.3.3 *Let μ be a probability measure carried by A. Assume that there exists $\delta \in [0, d]$, $C > 0$ and $\varepsilon > 0$ such that, for any ball B of diameter less than ε, μ satisfies the following uniform regularity assumption*

$$\mu(B) \leq C|B|^\delta.$$

Then the δ-dimensional measure of A satisfies $mes_\delta(A) \geq 1/C$, and therefore $\dim(A) \geq \delta$.

Proof: Let $\{B_i\}_{i\in\mathbb{N}}$ be an arbitrary ε-covering of A. Then

$$1 = \mu(A) = \mu\left(\cup B_i\right) \leq \sum \mu(B_i) \leq C \sum |B_i|^\delta.$$

The result follows by passing to the limit when $\varepsilon \to 0$.

3.3.2 Scaling functions vs. function spaces

From scaling functions to function spaces. Kolmogorov scaling function $\eta_f(p)$ (as defined by Eq. (3.10)) receives a function space interpretation, which is important for several reasons. On one hand, it allows to derive a number of its mathematical properties, and on other hand, it points towards variants and extensions of this scaling function, yielding sharper information on the singularities existing in data, and therefore offering a deeper understanding of the multifractal formalism.

Function space interpretation. The function space interpretation of the Kolmogorov scaling function can be obtained through the use of the spaces $\text{Lip}(s, L^p)$ defined as follows.

Definition 3.3.2 Let $s \in (0,1)$, and $p \in [1, \infty]$; $f \in \text{Lip}(s, L^p(\mathbb{R}^d))$ if $\int |f(x)|^p dx < \infty$ and

$$\exists C > 0, \quad \forall h > 0, \quad \int |f(x+h) - f(x)|^p dx \leq C|h|^{sp}. \tag{3.22}$$

Note that larger smoothness indices s are reached by replacing in the definition the first-order difference $|f(x+h) - f(x)|$ by higher and higher order differences: $|f(x+2h) - 2f(x+h) + f(x)|, \ldots$ (which is coherent with the remark just after the introduction of Kolmogorov scaling function Eq. (3.10), where, there too, higher order differences have to be introduced in order to deal with smooth functions f).

It follows from Eqs. (3.8) and (3.22) that,

$$\forall p \geq 1, \quad \eta_f(p) = \sup\{s : f \in \text{Lip}(s/p, L^p(\mathbb{R}^d))\}. \tag{3.23}$$

In other words, the scaling function allows to determine within which spaces $\text{Lip}(s, L^p)$ data are contained, for $p \in [1, \infty]$. The reformulation supplied by Eq. (3.23) has several advantages. It will lead to a better numerical implementation, based on wavelet coefficients, and it will have the additional advantage of yielding a natural extension of these function spaces to $p \in (0,1)$; this, in turn, will lead to a scaling function also defined for $p \in (0,1)$, therefore supplying a more powerful tool for classification, see Sec. 3.4.2. Note that the reason why Kolmogorov scaling function cannot be directly extended to $p < 1$ is that L^p spaces and the spaces which are derived from them (such as $\text{Lip}(s, L^p)$) do not make sense for $p < 1$; indeed Definition 3.3.2 for $p < 1$ leads to mathematical inconsistencies (spaces of functions thus defined would not necessarily be included in spaces of distributions).

Another advantage of the function space interpretation of the scaling function is that it allows to draw relationships with areas of modeling where the a priori assumptions which are made are function space regularity assumptions. Let us mention two famous examples, which will be further used as illustrations for the wavelet methods developed in the next section: *Bounded Variation* (BV) modeling in image processing (cf. Sec. 3.4.2), and *Bounded Quadratic Variation* hypothesis in finance (cf. Secs. 3.4.4 and 3.6.2, this latter case being related with the determination of the oscillation scaling functions Eq. (3.15)).

Bounded variation and image processing. A function f belongs to the space BV, i.e., has *bounded variation*, if its gradient, taken in the sense of distributions, is a finite (signed) measure. A standard assumption in image processing is that real-world images can be modeled as the sum of a function $u \in BV$ which models the *cartoon part*, and another term v which accounts for the noise and texture parts (for instance, the first "$u+v$ model", introduced by Rudin, Osher and Fatemi in 1992 ([74]) assume that $v \in L^2$, see also [63, 64]). The BV model is motivated by the fact that if an image is composed of smooth parts separated by *contours*, which are piecewise smooth curves, then its gradient will be the sum of a smooth function (the gradient of the image inside the smooth parts) and Dirac masses along the edges, which are typical finite measures. On the opposite, characteristic functions of domains with fractal boundaries usually do not belong to BV, see Fig. 3.1 for an illustration. Therefore, a natural question in order to validate such models is to determine whether an image (or a portion an image) actually belongs to the space BV or not. We will see in Sec. 3.4.2 how this question can be given a sharp answer using a direct extension of Kolmogorov scaling function.

Bounded quadratic variations. Another motivation for function space modeling is supplied by the *finite quadratic variation* hypothesis in finance. In contradistinction with the previous image processing example, this hypothesis is not deduced from a plausible intrinsic property of financial data, but rather constitutes an ad hoc regularity hypothesis which (considering the actual state of the art in stochastic analysis) is required in order to use the tools of stochastic integration. There exist several slightly different formulations of this hypothesis. One on them, which we consider here, is that the 2-oscillation is uniformly bounded at all scales. This means that f, assumed to be defined on $[0,1]^d$, satisfies

$$\exists C, \quad \forall j \geq 0, \quad 2^{dj} R_f(2,j) = \sum_{\lambda \in \Lambda_j} (Os_f(\lambda))^2 \leq C$$

(where $R_f(p,j)$ was defined by Eq. (3.14)). The definition of the oscillation scaling function Eq. (3.15) shows that, if $O_f(2) > d$, then f has finite quadratic variation, and if $O_f(2) < d$, then this hypothesis is not valid. This will be further discussed in Sec. 3.4.2 and illustrated in Sec. 3.6.2 and Fig. 3.7.

3.4 Wavelets: A natural tool to measure global scale invariance

The general considerations developed in Sec. 3.3.2 emphasized the importance of scaling functions as a data analysis tool. To further develop this program, reformulations of scaling functions which are numerically more robust than the Kolomogorov and oscillation scaling functions introduced so far are of both practical and theoretical interests. Motivations for introducing wavelet techniques for this purpose were already mentioned in Sec. 3.1.2. We now introduce these techniques in a more detailed way, and develop and explain the benefits of rewriting scaling functions in terms of wavelet coefficients.

3.4.1 Wavelet bases

Orthonormal wavelet bases on $\mathbb{R}^d$ exist under two slightly different forms. First, one can use $2^d - 1$ functions $\psi^{(i)}$ with the following properties: The functions

$$2^{dj/2}\psi^{(i)}(2^j x - k), \quad \text{for } i = 1, \ldots, 2^d - 1, \quad j \in \mathbb{Z} \text{ and } k \in \mathbb{Z}^d \tag{3.24}$$

form an orthonormal basis of $L^2(\mathbb{R}^d)$. Therefore, $\forall f \in L^2$,

$$f(x) = \sum_{j \in \mathbb{Z}} \sum_{k \in \mathbb{Z}^d} \sum_i c^i_{j,k} \psi^{(i)}(2^j x - k), \tag{3.25}$$

where the $c^i_{j,k}$ are the wavelet coefficients of f, and are given by

$$c^i_{j,k} = 2^{dj} \int_{\mathbb{R}^d} f(x)\psi^{(i)}(2^j x - k)dx. \tag{3.26}$$

Note that, in practice, discrete wavelet transform coefficients are generally not computed through the integrals Eq. (3.26), but instead using the discrete time convolution filter bank based pyramidal and recursive algorithm (cf. [75]), referred to as the *Fast Wavelet Transform* (FWT).

However, the use of such bases gives rise to several difficulties as soon as one does not have the a priori information that $f \in L^2$. For instance, if the only assumption available is that f is a continuous and bounded function, then one can still compute wavelet coefficients of f using Eq. (3.26) (indeed, these integrals still make sense), however, the series at the right-hand side of Eq. (3.25) may converge to a function which differs from f. This is due to the fact that certain *special* functions (which do not belong to L^2) have all their wavelet coefficients that vanish, and therefore, such possible components of functions will always be missed in the reconstruction formula Eq. (3.25). These *special* functions always include polynomials, however, in the case of Daubechies compactly supported wavelets, other *fractal-type* functions also have this property, see [76]. This explains why, at the end of Sec. 3.1, we mentioned that the properties of wavelet coefficients of certain processes that we obtained were not characterizations: Indeed information on these special functions, as possible components of the processes, is systematically missed, whatever the information on the wavelet coefficients is. However, this drawback can be turned by using a slightly different basis, which we now describe.

The alternative wavelet bases that will systematically be used from now on are of the following form. There exists a function $\varphi(x)$ and $2^d - 1$ functions $\psi^{(i)}$ with the following properties. The functions

$$\varphi(x-k) \quad (\text{for } k \in \mathbb{Z}^d) \quad \text{and} \quad 2^{dj/2}\psi^{(i)}(2^j x - k) \quad (\text{for } k \in \mathbb{Z}^d, \text{ and } j \geq 0) \tag{3.27}$$

form an orthonormal basis of $L^2(\mathbb{R}^d)$. In practice, we will use Daubechies compactly supported wavelets [77], which can be chosen arbitrarily smooth. Since these functions form an orthonormal basis of L^2, it follows, as previously, that

$$\forall f \in L^2, \quad f(x) = \sum_{k \in \mathbb{Z}^d} C_k \varphi(x-k) + \sum_{j=0}^{\infty} \sum_{k \in \mathbb{Z}^d} \sum_i c^i_{j,k} \psi^{(i)}(2^j x - k); \tag{3.28}$$

the $c^i_{j,k}$ are still given by Eq. (3.26) and

$$C_k = \int_{\mathbb{R}^d} f(x)\varphi(x-k)dx. \tag{3.29}$$

This choice of basis enables to answer the following important data analysis questions (which was not permitted by the previous choice Eq. (3.24)). In real-world data, the a priori assumption that $f \in L^2$ does not need to hold (for instance, the sample paths of standard models, such as Brownian motion, do not belong to $L^2(\mathbb{R})$). As already mentioned, the sole information that the series of coefficients of a function f on an orthonormal basis is square summable, does not imply that $f \in L^2$ (consider the example of $f = 1$ and the basis Eq. (3.24)). However, for the alternative basis given by Eq. (3.27), the following property now holds. Assume that f is a function in L^1_{loc} with slow growth, i.e. satisfies

$$\exists C, A > 0 : \int_{B(0,R)} |f(x)|dx \leq C(1+R)^A, \tag{3.30}$$

where $B(x,R)$ denotes the ball of center x and radius R; then the wavelet expansion Eq. (3.28) of f converges to f almost everywhere. Additionally, if the wavelet coefficients of f are square summable, then $f \in L^2$ (in contradistinction with what happens when using the basis Eq. (3.24)). Note that the slow growth assumption is a very mild one, since it is satisfied by all standard models in signal processing, and actually is necessary, in order to remain in the mathematical setting supplied by tempered distributions.

Note that one can even go further: Eqs. (3.26) and (3.29) make sense even if f is not a function; indeed, if one uses smooth enough wavelets, these formulas can be interpreted as a duality product between smooth functions (the wavelets) and tempered distributions. This gives rise to the problem of determining if the series Eq. (3.28) converges in other functional settings, and it is indeed one of the most remarkable properties of wavelet expansions that it is very often the case. Wavelet characterizations of function spaces are detailed in [78]. Let us mention a property which is particularly useful, and shows that convergence properties of wavelet expansions are "as good as possible." Suppose that f is continuous with slow growth, i.e. that it satisfies

$$\exists C, N > 0, \ \forall x \in \mathbb{R}^d, \quad |\, f(x)\,| \leq C(1+|x|)^N; \tag{3.31}$$

then the wavelet expansion of f converges uniformly on every compact set.

However, while the differences that we pointed out between the two types of wavelet bases are important for reconstruction, they are not for the type of analysis that we will perform. Indeed, properties will be deduced from the inspection of asymptotic behaviors of wavelet coefficients at small scale, where both bases Eqs. (3.24) and (3.27) coincide.

As an example of the fruitful interplay between the algorithmic structure of wavelet bases and the concept of selfsimilarity, we now prove Eq. (3.5). We start by recalling the definition of equality in law of stochastic processes.

Definition 3.4.1 Two vectors of $\mathbb{R}^l$: $(X_1, \ldots, X_l)$ and $(Y_1, \ldots, Y_l)$ share the same law if, for any Borel set $A \subset \mathbb{R}^l$, $\mathbb{P}(\{X \in A\}) = \mathbb{P}(\{Y \in A\})$.

Two processes X_t and Y_t share the same law if, $\forall l \geq 1$, for any finite set of time points $t_1, \ldots, t_l$, the vectors of $\mathbb{R}^l$

$$(X_{t_1}, \ldots, X_{t_l}) \quad and \quad (Y_{t_1}, \ldots, Y_{t_l})$$

share the same law.

We apply this definition to the two processes X_{at} and $a^H X_t$ (such as stated in Eq. (3.3)), which are assumed to share the same law. Taking for A the domain between two parallel hyperplanes, this, in turns, implies that any finite linear combinations

$$\sum_{i=1}^{l} \alpha_i X_{t_i} \quad \text{and} \quad \sum_{i=1}^{l} \alpha_i Y_{t_i}$$

share the same law. Therefore, if X_t is a selfsimilar process of exponent H, then Eq. (3.3) means that

$$\forall t_1, \ldots, t_l, \sum_{i=1}^{l} \alpha_i X_{at_i} \quad \text{and} \quad \sum_{i=1}^{l} \alpha_i a^H X_{t_i} \text{ share the same law.} \tag{3.32}$$

Assume now that for almost every $\omega \in \Omega$, the sample path of X_t is Riemann integrable. The coefficient

$$c_{j,k} = \int 2^j X_t \psi(2^j t - k) dt$$

is a limit of the Riemann sums

$$\sum_i (t_{i+1} - t_i) 2^j X_{t_i} \psi(2^j t_i - k); \tag{3.33}$$

using Eq. (3.32), it follows that Eq. (3.33) has the same law as

$$\sum_i (t_{i+1} - t_i) 2^{-H} 2^j X_{2t_i} \psi(2^j t_i - k)$$

which, writing $u_i = 2t_i$, can be written

$$\sum_i (u_{i+1} - u_i) 2^{-H} 2^{j-1} X_{u_i} \psi(2^{j-1} u_i - k),$$

which is a Riemann sum of the integral

$$\int 2^{-H} 2^{j-1} X_u \psi(2^{j-1} u - k) du = 2^{-H} c_{j-1,k}.$$

Passing to the limit when the largest spacing between sampling points t_i (and therefore u_i) tends to 0, we obtain that

$$c_{j,k} \stackrel{\mathcal{L}}{=} 2^{-H} c_{j-1,k}.$$

The argument that we developed for one coefficient can be reproduced similarly for an arbitrary vector of coefficients, hence Eq. (3.5) holds.

3.4.2 Wavelet scaling function

We now investigate properties which arise as consequences of the wavelet characterizations of function spaces.

Notations. Compact notations are used for indexing wavelets. Instead of the three indices (i, j, k), dyadic cubes are used: $\lambda \ (= \lambda(j, k)) = \frac{k}{2^j} + \left[0, \frac{1}{2^j}\right)^d$. In order to keep notations as simple as possible, the index i is dropped in formulas and notations, with the implicit convention that sums or suprema over indices are also taken on the index i. Accordingly, $c_\lambda = c^i_{j,k}$ and $\psi_\lambda(x) = \psi^{(i)}(2^j x - k)$. The wavelet ψ_λ is localized near the cube λ. Since we use Daubechies compactly supported wavelets, $\exists C > 0$ such that $\forall i, j, k$, $supp(\psi_\lambda) \subset C \cdot \lambda$ (where $C \cdot \lambda$ denotes the cube of same center as λ and C times wider). Finally, let Λ_j denote the set of dyadic cubes λ which index a wavelet of scale j, i.e. wavelets of the form $\psi_\lambda(x) = \psi^{(i)}(2^j x - k)$ (note that Λ_j is a subset of the dyadic cubes of size 2^{j+1}), and Λ will denote the union of the Λ_j for $j \geq 0$.

Wavelet scaling function. A key property of wavelet expansions is that many function spaces have a simple characterization by conditions bearing on wavelet coefficients. This property has a direct consequence on the determination of the Kolmogorov scaling function. Let

$$S_f(p,j) = 2^{-dj} \sum_{\lambda \in \Lambda_j} |c_\lambda|^p. \tag{3.34}$$

Classical embeddings between function spaces imply that, if the wavelets are smooth enough, then the Kolmogorov scaling function can be re-expressed as (cf. [79])

$$\forall p \geq 1, \quad \eta_f(p) = \liminf_{j \to +\infty} \frac{\log\left(S_f(p,j)\right)}{\log(2^{-j})}. \tag{3.35}$$

This formulation, which, again, yields the scaling function through linear regressions in log-log plots, has several advantages when compared to the earlier version Eq. (3.10). First, it allows to extend the Kolmogorov scaling function to the range $0 < p \leq 1$; it will be shown in Sec. 3.4.4 how to go even further and define scaling functions for negative values of p (see also [80] for an extension of $\eta_f(p)$ to $p < 0$). We will call this extension of the Kolmogorov scaling function the *wavelet scaling function*, but we will keep the same notation. By construction, $\eta_f(p)$ is a concave increasing function, of derivative at most d, see [40].

Wavelet regularity. Let us say a few words about the required smoothness of wavelets. The rule of thumb is that wavelets should be smoother and have more vanishing moments than the regularity index appearing in the definition of the function space. In signal processing, one is not necessarily aware beforehand of the regularity of the data. In practice, one uses smoother and smoother wavelets, and when the results do not vary, this means that smooth enough wavelets are used. (Note that this is reminiscent of the definition of the Kolmogorov scaling function where one has to use higher and higher order differences and stop when the results are numerically stabilized.) In the following we will never specify the required smoothness, and always assume that *smooth enough wavelets* are used (the minimal regularity required being always easy to infer).

Note that, in the wavelet setting, one should also draw a difference between Eq. (3.35), which allows us to define the wavelet scaling function of any function f, and its use in data analysis, where scaling properties need to hold to enable measurements based on linear regressions in log-log plots.

Function space interpretation. Let us examine how the wavelet scaling function can be used in order to validate the function space assumptions discussed in Sec. 3.3.2. For instance, regarding the $u+v$ model, where $u \in BV$ and $v \in L^2$, the values taken by wavelet scaling function at $p = 1$ and $p = 2$ allow practitioners to determine if data belong to BV or L^2:

- If $\eta_f(1) > 1$, then $f \in BV$, and if $\eta_f(1) < 1$, then $f \notin BV$
- If $\eta_f(2) > 0$, then $f \in L^2$ and if $\eta_f(2) < 0$, then $f \notin L^2$.

Thus wavelet techniques allow to check if the function space assumptions which are made, e.g., in certain denoising algorithms relying on the $u + v$ model, are valid.

Examples of synthetic and natural images are shown in Fig. 3.1, together with the corresponding measures of $\eta_f(1)$ and $\eta_f(2)$. The image consisting of a simple discontinuity along a circle and no texture, (i.e., a typical *cartoon* part of the image in the $u + v$ decomposition) is in BV, while the image of textures or discontinuities existing on a complicated support (such as the Von Koch snowflake) are not. Y. Gousseau and J.-M. Morel were the first authors to rise the question of finding statistical tests to verify if natural images belong to BV [81]. Our results confirm that the BV requirement is seldom met; actually, images often turn out not to be even in L^2, as illustrated in Fig. 3.1 (bottom row).

3.4.3 Uniform Hölder exponent

It has been shown in the previous section that the Kolmogorov scaling function can be rewritten and extended as a wavelet coefficient based version. It is hence natural to ask whether the same can be performed for the oscillation scaling function Eq. (3.15). Because oscillations are defined only if f is locally bounded, this condition needs to be checked in practice. This test can be performed using the *uniform Hölder exponent*, which constitutes the subject of this subsection. However, before entering technicalities, the following general question concerning function space modeling needs to be addressed.

Function space modeling. The issue of determining whether experimental data, acquired by a physical device, can be modeled by a bounded function, or not, may appear meaningless at first sight. Indeed, data can be collected only with finite size and resolution, and hence consist of finite, and thus bounded, sequences. For images, a common pitfall is to conclude that images are grey-levels, so that they necessarily take values in $[0, 1]$, and, therefore, by construction, are appropriately modeled by bounded

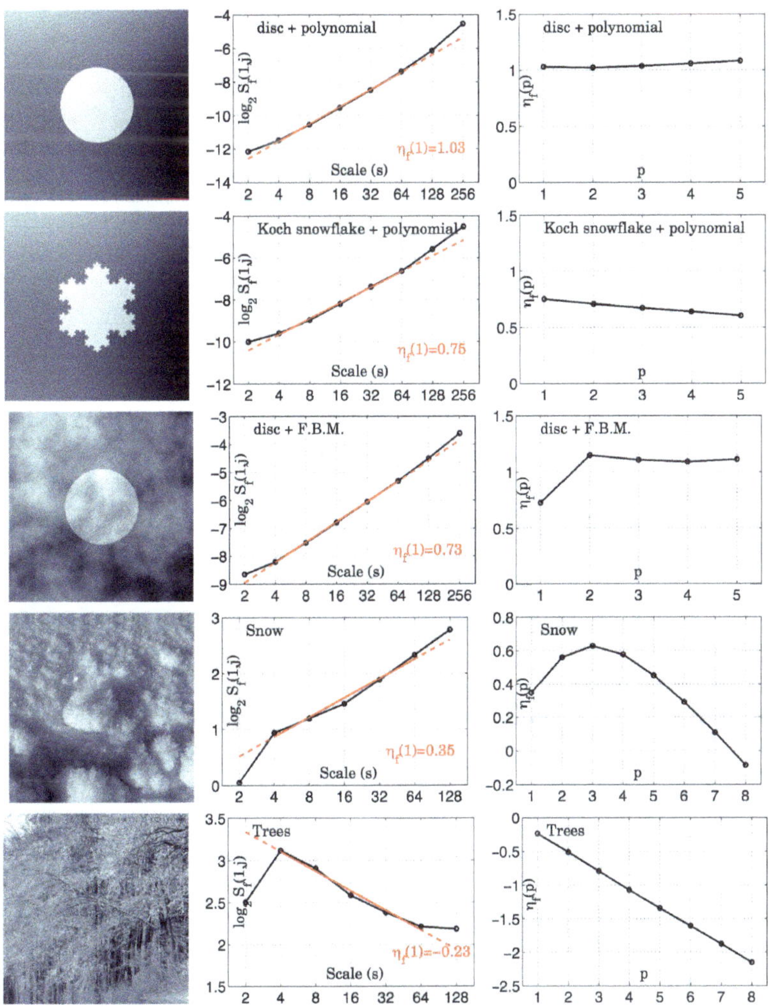

Figure 3.1 **Image processing.** Different images (synthetic, 3 top rows, and natural, 2 bottom rows) together with log-log plots of their wavelet coefficients based $S_f(1,j)$ (as defined in Eq. (3.34)) (middle column) and their wavelet scaling functions $\eta_f(p)$ (as defined in Eq. (3.35)) (right column). From the wavelet scaling function, we obtain the very accurate estimate $\hat{D} = 1 - \eta_f(1) \approx 1.25$ for the fractal dimension of the Von Koch snowflake (given by $D = \log(4)/\log(3) \approx 1.26$). Furthermore, we recover that the image of the simple disk discontinuity is just in BV and not smoother since $\eta_f(1) \gtrsim 1$, while the Von Koch snowflake and the superimposition of disk and fBm texture are not. Both natural images are not in BV, and the one at bottom row is not even in L^2.

functions. The issue is even more important regarding the whole strategy of functional space modeling. Given any finite resolution and size sequence of data practically available, they can always be extrapolated by a C^∞ function, that thus belongs to any possible function space. The resolution of this paradox requires the use of the notion of scale invariance. Let us consider again the example of image modeling and address the issue of deciding whether a digital image can be modeled by a bounded function or not. It is clear that if the image is analyzed only at its finest scale, then the answer is positive. However, following the point of view that we developed until now, the problem can be reinterpreted in terms of analysis of scaling properties, over a range of available scales, and will therefore amount to determining whether an observed scaling exponent fits with those that are compatible with bounded functions, or not. The uniform Hölder exponent, which will be considered in this section, answers this question.

More generally, numerical results, obtained from finite resolution and size data, leading to conclude that data belong to certain function spaces and not to others, should actually be understood in the following way: the function space regularity thus determined holds under the hypothesis that scaling that are observed on the range of available scales would continue to hold at finer scales if they were available.

Uniform Hölder exponent: Definition. The mathematical idea behind testing whether data are locally bounded or not is to consider boundedness as a particular case in the whole range of Lipschitz spaces, which corresponds to the regularity exponent $s = 0$. Let us be more specific.

The Lipschitz spaces $C^s_{sg}(\mathbb{R}^d)$ are defined for $0 < s < 1$ by these conditions

$$\exists C, N, \quad \forall x, y \in \mathbb{R}^d, \quad |f(x)| \le C(1 + |x|)^N$$

and

$$\exists C, N, \quad \forall x, y \in \mathbb{R}^d, \quad |f(x) - f(y)| \le C|x - y|^s(1 + |x| + |y|)^N.$$

If $s > 1$, they are then defined by recursion on $[s]$ by the condition $f \in C^s_{sg}(\mathbb{R}^d)$ if $f \in L^\infty$ and if all its partial derivatives (taken in the sense of distributions) $\partial f/\partial x_i$ (for $i = 1, \cdots d$) belong to $C^{s-1}_{sg}(\mathbb{R}^d)$.

If $s < 0$, then the C^s_{sg} spaces are composed of distributions, also defined by recursion on $[s]$ as follows: $f \in C^s_{sg}(\mathbb{R}^d)$ if f is a finite sum of partial derivatives (in the sense of distributions) of order 1 of elements of $C^{s+1}_{sg}(\mathbb{R}^d)$.

This allows to define the C^s_{sg} spaces for any $s \notin \mathbb{Z}$ (note that a consistent definition using the Zygmund classes can also be supplied for $s \in \mathbb{Z}$, see [78], however we will not need to consider these specific values in the following). We can now define the parameter that we will use for determining boundedness.

Definition 3.4.2 The *uniform Hölder exponent* of a tempered distribution f is

$$h_f^{min} = \sup\{s : \; f \in C^s_{sg}(\mathbb{R}^d)\}. \tag{3.36}$$

This definition does not make any a priori assumption: the uniform Hölder exponent is defined for any tempered distribution, and it can be positive and negative. Note that we have already met this notion: Prop. 3.3.1 can be reinterpreted as stating that, if $0 < H - 1/\alpha < 1$, then $h_X^{min} \geq H - 1/\alpha$. In particular the remark following the proof of Prop. 3.3.2 allows to recover the fact that, for fBm, $h_f^{min} = H$; additionally, it yields that the same result holds for the Rosenblatt process (see [82, 83] for the definition and the properties of the wavelet expansion of this process).

The values taken by the uniform Hölder exponent have the following interpretation:

- If $h_f^{min} > 0$, then f is a locally bounded function,
- if $h_f^{min} < 0$, then f is not a locally bounded function.

Typical examples are supplied by Lévy fractional stable processes which already appeared in Sec. 3.3.1; they satisfy $h_f^{min} = H - 1/\alpha$, and therefore (as already mentioned) they have nowhere locally bounded stable sample paths if $H < 1/\alpha$, and continuous sample paths if $H > 1/\alpha$.

The importance of this exponent lies in the fact that it does not only play the role of a prerequisite (i.e., a parameter whose value has to be positive if one wants to determine the oscillation scaling function), but also that it turns out to be a very efficient parameter for discrimination. As will be illustrated later (cf. Sec. 3.6), the classification of several types of data can actually be based on the determination of their uniform Hölder exponent.

Uniform Hölder exponent and wavelets. In practice, h_f^{min} can be derived directly from the wavelet coefficients of f through a simple regression in a log-log plot; indeed, it follows from the wavelet characterization of the spaces C^s_{sg} that

$$h_f^{min} = \liminf_{j\to+\infty} \frac{\log(\sup_{\lambda\in\Lambda_j}|c_\lambda|)}{\log(2^{-j})}. \tag{3.37}$$

This estimation procedure has been studied in more detail in [68].

The exponent h_f^{min} can also be derived from the wavelet scaling function, [68],

$$h_f^{min} = \lim_{p\to+\infty} \frac{\eta_f(p)}{p} = \lim_{p\to+\infty} \eta'_f(p).$$

Uniform Hölder exponent and applications. In Sec. 3.6, which reviews a number of real-world applications, we will see that h_f^{min} can be found either positive or negative depending on the nature of the application. For instance, velocity turbulence data (cf. 3.6.1 and Fig. 3.6, middle row, left plot) and price time series in finance (cf. Sec. 3.6.2 and Fig. 3.7, 2nd and 3rd row row, second plots and bottom row, first plot) are found to always have $h_f^{min} > 0$, while aggregated count Internet traffic time series always have $h_f^{min} < 0$ (cf. 3.6.4 and Figs. 3.10 and 3.11, top row, left plots). For biomedical applications (cf. e.g., fetal hear rate variability as in Sec. 3.6.3 and Fig. 3.8, 3rd row), as well as for image processing (cf. Sec. 3.6.5 and Fig. 3.12, 4th column), h_f^{min} can commonly be found either positive or negative. For these last two examples, h_f^{min} is found to be a highly relevant parameter for classication purposes (cf. Fig. 3.8, bottom row, and Fig. 3.12, bottom row, respectively).

Functions that are not locally bounded and fractional integration. The examples listed above indicate that it is not uncommon in real-world application to find that $h_f^{min} < 0$. However, this does not imply that further analyses which require that $h_f^{min} > 0$ (and will be described in the following) are impossible and should be discarded. Indeed, *fractional integration* offers a one to one way to associate with any distribution f (with potentially negative h_f^{min}) a function $f^{(-s)}$, whose exponent $h_{f^{(-s)}}^{min}$ is positive, and therefore classification can then be operated on this new function $f^{(-s)}$ rather than on the initial data f. We now describe this procedure, both from a theoretical and a practical point of view. Note that there exist many variants of the definition of fractional integration, which would however all lead to the same definition for the exponents used here, see [84] for a discussion of fractional integration, especially in the context of stochastic processes.

Definition 3.4.3 Let f be a tempered distribution. The fractional integral of order s of f, denoted by $f^{(-s)}$, is defined as convolution operator $(Id - \Delta)^{-s/2}$ which, in the Fourier domain, is the multiplication by the function $(1 + |\xi|^2)^{-s/2}$.

Numerically, fractional integration is intricate to perform (because of finite size and border effect issues). However, for our purpose here, these difficulties can be circumvented by instead using *pseudo-fractional integration*, defined as follows.

Definition 3.4.4 Let $s > 0$, let ψ_λ be a wavelet basis whose elements belong to C^r with $r > s + 1$; let f be a function, or a distribution, with wavelet coefficients c_λ. The *pseudo-fractional integral* (in the basis ψ_λ) of f of order s, denoted by $I^s(f)$, is the function whose wavelet coefficients (on the same wavelet basis) are

$$c_\lambda^s = 2^{-sj} c_\lambda.$$

This operation is numerically straightforward, and the following result shows that it retains the same pointwise and multifractal properties as exact fractional integration, see [40, 68, 85, 86].

Proposition 3.4.1 *Let f be a tempered distribution; then*

$$\forall s \in \mathbb{R}, \begin{cases} \forall p > 0, & \eta_{f^{(-s)}}(p) = \eta_{I^s(f)} = \eta_f(p) + sp, \\ h_{f^{-s}}^{min} = h_{I^s(f)}^{min} = h_f^{min} + s. \end{cases}$$

Additionally, the pointwise Hölder exponents satisfy

$$\forall s > -h_f^{min}, \quad \forall x \in \mathbb{R}^d, \quad h_{f^{(-s)}}(x) = h_{I^s(f)}(x).$$

This last property also holds for the leader scaling functions which will be introduced in Sec. 3.4.4 (under the same condition $s > -h_f^{min}$, so that those quantities are well defined). Therefore, in practice, for multifractal analysis, pseudo-fractional integration is preferred to exact fractional integration.

These results point toward a natural strategy in order to perform the multifractal analysis of a function which is not locally bounded (or of a distribution). First determine its exponent h_f^{min}; then, if $h_f^{min} < 0$, perform a pseudo-fractional integration of order $s > -h_f^{min}$; it follows that the uniform regularity exponent of $I^s(f)$ is positive, and therefore that it is a bounded function, to which multifractal analysis can be applied. Yet, this strategy gives rise to an important problem. There is no canonical choice for the fractional integration order, and the shape of the wavelet leader scaling function (see Definition 3.4.6 below) obtained after fractional integration may depend on this choice. In practice, when multifractal properties of a collection of data have to be studied, one follows the following strategy. One first determines the exponent h_f^{min} of each signal in the database; if some of them have negative h_f^{min}, one picks an exponent s such that $h_f^{min} + s$ is positive for all signals. Under these constraints, s should be picked as small as possible, so that data are modified by a fractional integration of order as small as possible. The rule of thumb practically used is to select s as the smallest multiple of 0.5 ensuring that $h_f^{min} + s$ is positive for all signals and to perform a pseudo-fractional integration *of the same order* s for all signals.

3.4.4 Wavelet leaders

From oscillations to wavelet leaders. Our purpose is now to obtain a wavelet reformulation of the oscillation scaling function Eq. (3.15) (in the same way as the wavelet scaling function is the wavelet reformulation of the Kolmogorov scaling function). This requires the introduction of *wavelet leaders*, which are local suprema of wavelet coefficients. Let us start with a heuristic argument showing that they are natural candidates to estimate oscillations. Recall that, if λ denotes a dyadic cube, then 3λ denotes the cube with same center and three times wider.

Definition 3.4.5 Let f be a locally bounded function satisfying *(3.31)*. The *wavelet leaders* of f are the coefficients

$$d_\lambda = \sup_{\lambda' \subset 3\lambda} |c_{\lambda'}|. \tag{3.38}$$

The fact that the supremum in Eq. (3.38) is finite is a consequence of the boundedness assumption for f. Therefore, checking that $h_f^{min} > 0$ is an absolute prerequisite for computing leaders.

Let us now sketch the reason why wavelet leaders allow to estimate the oscillations. Consider a wavelet coefficient c_λ. Since we use compactly supported wavelets,

$$c_\lambda = 2^{dj} \int_{C\lambda} f(x)\psi^{(i)}(2^j x - k)dx$$

Since wavelets have vanishing integral,

$$
\begin{aligned}
|c_\lambda| &= \left| 2^{dj} \int_{C\lambda} (f(x) - f(k2^{-j})\psi^{(i)}(2^j x - k)dx \right| \\
&\leq 2^{dj} \int_{C\lambda} |f(x) - f(k2^{-j})|\, |\psi^{(i)}(2^j x - k)|dx \\
&\leq 2^{dj} Os_f(C\lambda) \int_{C\lambda} \left|\psi^{(i)}(2^j x - k)dx\right| \\
&= C \cdot Os_f(C\lambda).
\end{aligned}
$$

Consider now a given cube K; this argument works for any wavelet whose support is included in K, so that each wavelet coefficient is bounded by $COs_f(C\lambda) \leq COs_f(K)$. This explains why wavelet leaders are bounded by the local oscillation.

Converse estimates follow from the following argument. From Eq. (3.25), we get

$$
f(x) - f(y) = \sum_{j,k} c_{j,k}^{(i)} (\psi^{(i)}(2^j x - k) - \psi^{(i)}(2^j y - k).
$$

Let j_0 be such that $2^{-j_0} \sim |x-y|$. The low frequency terms ($j \leq j_0$) are estimated using the smoothness of the wavelets, which makes the difference $\psi^{(i)}(2^j x - k) - \psi^{(i)}(2^j y - k)$ small. The terms such that $j_0 \leq j \leq Aj_0$ (where A has to be chosen correctly) are estimated using the bound of the wavelet coefficients supplied by the wavelet leader. Finally, high frequency terms ($j \geq Aj_0$) are estimated using the uniform regularity of f. This allows us to obtain the required converse estimates. Note that subtle relationships between oscillations and wavelet coefficients have recently been obtained by M. Clausel and her collaborators, see e.g. [87] and references therein.

Leader scaling function. The previous heuristic argument motivates the introduction of a new scaling function, constructed on the model of the wavelet scaling function, but replacing wavelet coefficients by wavelet leaders.

Definition 3.4.6 Let f be a locally bounded function with slow growth (i.e. satisfying Eq. (3.31)), and let

$$
T_f(p,j) = 2^{-dj} \sum_{\lambda \in \Lambda_j} |d_\lambda|^p. \tag{3.39}
$$

The *leader function* $\zeta_f(p)$ is given by

$$\forall p \in \mathbb{R}, \qquad \zeta_f(p) = \liminf_{j\to+\infty} \frac{\log(T_f(p,j))}{\log(2^{-j})}. \tag{3.40}$$

The relationship between the leader function and the oscillation scaling function is very similar to the one mentioned to exist between the wavelet scaling function and the Kolmogorov scaling function: they coincide when $p \geq 1$; therefore the former extends the later for $p < 1$, see [88].

Let us discuss two consequences of this result. First, the fact that they coincide for $p = 1$ implies that, if $h_f^{min} > 0$, then the upper box dimension of the graph of f can be derived from the leader scaling function, see [89]:

$$\overline{dim_B}(\mathcal{G}r(f)) = \max(d, d+1-\zeta_f(1)).$$

Second, when $h_f^{min} > 0$, one can determine whether f has bounded quadratic variation or not by inspecting its leader scaling function for $p = 2$, indeed:

- If $\zeta_f(2) > d$, then f has bounded quadratic variation,
- if $\zeta_f(2) < d$, then the quadratic variation of f is unbounded.

Discussion of the condition $h_f^{min} > 0$. One may wonder if the condition $h_f^{min} > 0$ is really necessary in order to obtain such results. It is actually the case, and one can not obtain estimates of the quadratic variation (or of any p-variation) using wavelet methods for functions which are locally bounded but do not have some uniform smoothness for the following reason. Recall that the Hilbert transform is defined as the convolution with $p.v.(1/x)$ (or alternatively as the Fourier multiplier given by the function $sign(\xi)$) and consider as an example the sawtooth function $x - [x]$ which is bounded and has discontinuities at the integers. It obviously has bounded p-variation for any p. However, after applying the Hilbert transform to it, discontinuities are transformed into logarithmic singularities, and therefore the property of bounded p-variation is lost for any value of p. Since the Hilbert transform does not modify the wavelet-based quantities that we have considered, such as h_f^{min}, $\zeta_f(p)$, $\eta_f(p)$, it is clear that wavelet methods can not estimate p-variations of functions that have discontinuities, and therefore some uniform regularity assumption is indeed required.

In finance, for instance, checking whether $h_f^{min} > 0$ is positive or not is of double importance. First, it suggests to reject jump models for price; second, it enables us to probe the finiteness of quadratic variations, a requested property to validate the use of stochastic integration on such data. This is further discussed in Sec. 3.6.2 and illustrated in Fig. 3.7.

The bonus of negative values of p. The leader scaling function can also be given a function space interpretation for $p > 0$ in the framework of *oscillation spaces*, studied in [90, 91]. In particular, embeddings between these spaces and other, more classical, function spaces (such as Besov spaces) allow to derive an important relationship between the two scaling functions constructed through wavelet coefficients, see [88].

Proposition 3.4.2 *Let f be a function satisfying $h_f^{min} > 0$. Then*

$$\forall p > 0, \quad \eta_f(p) \geq \zeta_f(p).$$

Let p_c be the "critical exponent" defined by the condition $\eta_f(p_c) = d$, then

$$\forall p \geq p_c, \quad \eta_f(p) = \zeta_f(p).$$

An important property of the leader scaling function is that it is *well-defined* also for $p < 0$, although it can no longer be given a function space interpretation. By well defined, it is meant that it has the following robustness properties if the wavelets belong to the Schwartz class (partial results still hold otherwise), see [40, 85]:

- $\zeta_f(p)$ is independent of the wavelet basis.
- $\zeta_f(p)$ is invariant under the addition of a C^∞ perturbation.
- $\zeta_f(p)$ is invariant under a C^∞ change of variable.

The invariance under a smooth change of variable is a mandatory property for texture classification. Indeed, it is needed that natural textures can be recognized even after the deformation produced by setting them on smooth surfaces.

The possibility of involving negative values of p in analysis can turn crucial. A celebrated example is that of hydrodynamic turbulence where two multiplicative cascade models are classically in competition. Using positive values of p only, the Kolomogorov and the wavelet scaling functions computed from experimental data fail to discriminate between the two models. However, the leader scaling function, enabling the use of negative values of p, permits to show that one model is unambiguously preferred by data. This is discussed in Sec. 3.6.1 and illustrated in Fig. 3.6 (bottom row plots).

3.5 Beyond functional analysis: Multifractal analysis

The fact that the leader scaling function extends the analysis to negative values of p shows that multifractal analysis allows to go beyond the scope of functional analysis. Indeed, the scaling function no longer has a function space interpretation if $p < 0$. Note that this extension has some canonical character, as a consequence of its independence of the chosen wavelet basis, and also because of its robustness properties. The point made in the present section is to show that even more is true by exhibiting a natural interpretation of scaling functions, which requires both positive and negative values of p. This interpretation is in terms of the pointwise singularities of the function, and will be supplied by the multifractal formalism, which we now describe.

3.5.1 Multifractal formalism for oscillation

Multifractal spectrum versus multifractal formalism. From now on, it is assumed that $h_f^{min} > 0$, which implies that f is a continuous function, and therefore that its oscillations are well defined. Recall that the notion of oscillation came up in two occurrences until now: the definition of the oscillation scaling function Eq. (3.15), and the characterization of the pointwise Hölder exponent Eq. (3.16). This points towards a possible relationship between both quantities, which can be made explicit via *multifractal formalisms*. Specifically, a multifractal formalism establishes a connection between the scaling functions and the spectrum of singularities (or multifractal spectrum) of f, which is defined as follows.

Definition 3.5.1 Let f be a locally bounded function. The *spectrum of singularities* of f is the function

$$D_f(h) = \dim(E_f(h)),$$

where we recall that dim denotes the Hausdorff dimension, and the $E_f(h)$ are the isohölder sets $E_f(h) = \{x_0 : h_f(x_0) = h\}$.

This definition justifies the name of *multifractal functions.* On the theoretical side, one typically considers functions f that have non-empty isohölder sets for h taking all values in an interval $[h_f^{min}, h_f^{max}]$, and therefore one has to deal with potentially an infinite (and even uncountable) number of fractal sets $E_f(h)$ whose Hausdorff dimensions have to be determined.

Recall that $\dim(\emptyset) = -\infty$ so that $D_f(h) = -\infty$ if h does not belong to the range of h_f. We will call the *support* of D_f the set

$$supp(D_f) = \{h : D_f(h) \geq 0\} = \{h : E_f(h) \neq \emptyset\}.$$

(Note that the support of the spectrum needs not be a closed set, in contradistinction with the usual definition of the support of a function).

The initial formulation of the multifractal formalism was due to the physicists G. Parisi and U. Frisch and was based on the Kolmogorov scaling function [92]. It was further grounded mathematically in [93], see also [94]. Here, the focus is on a formalism based on the oscillation scaling function, which we describe first since it paves the way towards a more satisfactory leader scaling function formulation.

Oscillation multifractal formalism. In this paragraph, it is assumed that $h_f(x)$ takes on values between 0 and 1, so that Eq. (3.16) is valid. Since pointwise Hölder exponents can thus be derived from the quantities $Os_f(3\lambda)$, it is natural to consider scaling functions based on these quantities, i.e., the oscillation scaling function of f defined by Eq. (3.15).

Let us now show how the spectrum of singularities can be obtained from the scaling function. The definition of the scaling function roughly means that the quantity $R_f(p,j)$ defined by Eq. (3.14) satisfies $R_f(p,j) \sim 2^{-Os_f(p)j}$. Let us estimate the contribution to $R_f(p,j)$ of the cubes λ that cover the points of $E_f(h)$; Eq. (3.16) asserts that they satisfy $Os_f(3\lambda) \sim 2^{-hj}$; since we need about $2^{-D_f(h)j}$ such cubes to cover $E_f(h)$, the corresponding contribution roughly is

$$2^{-dj}2^{D_f(h)j}2^{-hpj} = 2^{-(d-D_f(h)+hp)j}.$$

When $j \to +\infty$, the dominant contribution stems from the smallest exponent (the others bring contributions which are exponentially smaller), so that

$$Os_f(p) = \inf_h (d - D_f(h) + hp). \tag{3.41}$$

It follows from the definition of the scaling function $Os_f(p)$ that it is a concave function on $\mathbb{R}$. This is in agreement with the fact that the right-hand side of Eq. (3.41) necessarily is a concave function (as an infimum of a family of linear functions) no matter whether $D_f(h)$ is concave or not. However, if the spectrum also is a concave function, then the Legendre transform in Eq. (3.41) can be inverted (as a consequence of a general result on the duality of convex functions), which justifies the following definition.

Definition 3.5.2 Let f be a function which satisfies $h_f^{min} > 0$; f follows the multifractal formalism if its spectrum of singularities satisfies

$$D_f(h) = \inf_p (d - Os_f(p) + hp). \tag{3.42}$$

Validity of the multifractal formalism. The derivation exposed above is not a strict mathematical proof, but rather consists of a heuristic explanation of the relation between $D_f(h)$ and $Os_f(p)$. The determination of the range of validity of Eq. (3.42), or of any of its variants, is one of the main open mathematical problems in multifractal analysis. Nonetheless, let us emphasize that the cornerstone of this derivation relies heavily on Eq. (3.16), i.e., on the fact that the Hölder exponent of a function can be estimated from the set of values that its oscillation takes on the "enlarged" dyadic cubes 3λ. In particular, the variant of this multifractal formalism based on increments (i.e., on the Kolmogorov scaling function) has a narrower range of validity, see [40, 85, 95] for a discussion.

An important remark is that, in applications, one should not focus too much on the problem of the validity of the multifractal formalism. On one side, because one can not compute directly a spectrum of singularities of experimental data, the multifractal formalism can be checked only for mathematically defined functions or processes. On other side, the purpose of multifractal analysis often is to use a spectrum as a tool for classification or model selection, and, with this respect, using a scaling function (or its Legendre transform) is a perfectly valid approach, independently of the fact that the multifractal formalism holds or not. However, Sec. 3.5.4 below details an example where the multifractal formalism allows to determine effectively the multifractal spectrum: It can be the case when it is degenerate, i.e., when $h_f(x)$ is a constant function (in which case D_f is supported by only one point).

Further comments. The formulation of the multifractal formalism given by Eq. (3.42) has the following advantage. The scaling function is invariant under bi-Lipschitz deformations of the function, which is a natural requirement since the spectrum of singularities possesses this invariance property (as a consequence of the assumption that the range of h_f is included in $(0, 1)$).

To this Legendre transform based multifractal formalism, initially proposed in [92] and thoroughly theoretically grounded in [43, 88], has been opposed an alternative formulation relying on the *large deviation principle* (cf. e.g., [43] for an early introduction and [96] for a recent version involving oscillations of order 1). However, we will not follow this track for the following reasons.

- Using oscillations restricts the possible range of Hölder exponents to $h \in [0, 1]$;
- the method can not be adapted to data that are not locally bounded (which is often the case in signal and image processing).

This second drawback, as we saw, can be circumvented easily by a pseudo-fractional integration when using wavelet techniques, and this is the reason why we now turn towards these methods.

The limitation to the range $h \in [0, 1]$ implied by either the use of increments of order 1 (Kolmogorov scaling function) or oscillations of order 1 (oscillation scaling function) has long been recognized and led to the pioneering contributions of A. Arneodo and collaborators promoting the use of wavelet based methods in multifractal analysis (cf. e.g., [97–100] and references therein).

3.5.2 A fractal interpretation of the leader scaling function: The leader multifractal formalism

The introduction of wavelet leaders was motivated by the fact that they allow to estimate local oscillations, and therefore come as a natural ingredient in a wavelet reformulation of the oscillation scaling function and of pointwise Hölder regularity.

Let $x_0 \in \mathbb{R}^d$, and recall that $\lambda_j(x_0)$ denotes the dyadic cube of width 2^{-j} which contains x_0. If $h_f^{min} > 0$, then the heuristic relationship that we gave in Sec. 3.4.4 and which related oscillations and wavelet leaders actually leads to the following result (see [88] and references therein):

$$h_f(x_0) = \liminf_{j \to +\infty} \frac{\log\left(d_{\lambda_j(x_0)}\right)}{\log(2^{-j})}. \tag{3.43}$$

The heuristic arguments that were developed above for the derivation of the oscillation multifractal formalism can be reproduced in the setting of wavelet leaders. They lead to the following reformulation of the multifractal formalism.

Definition 3.5.3 Let f be a function statisfying $h_f^{min} > 0$; the *leader spectrum* of f is defined through a Legendre transform of the leader scaling function, i.e.

$$L_f(h) = \inf_{p \in \mathbb{R}} \left(d + hp - \zeta_f(p)\right).$$

The *wavelet leader multifractal formalism* holds if

$$\forall h \in \mathbb{R}, \quad D_f(h) = L_f(h). \tag{3.44}$$

This formalism holds for large classes of models used in signal and image processing, such as fBm, lacunary and random wavelet series [101], or multiplicative cascades. In this wavelet leader setting, the justification of the derivation of Eq. (3.44) relies heavily on Eq. (3.43). In particular, multifractal formalisms based on other quantities (such as the wavelet scaling function) have a narrower range of validity, see [88] for a discussion.

The multifractal formalism does not hold in all generality, nonetheless the following upper bound does, yielding a general relation between the leader scaling function and the spectrum of singularities, see [88].

Theorem 3.5.1 *If* $h_f^{min} > 0$, *then,*

$$\forall h \in \mathbb{R}, \quad d_f(h) \leq L_f(h).$$

Examples of leader multifractal spectra $L_f(h)$ computed from a single realization of various processes and fields are illustrated in Figs. 3.2, 3.3 and 3.4, and compared with the theoretical multifractal spectra $D_f(h)$. It can be observed that, for all these reference processes, the leader multifractal spectra computed from a single realization very satisfactorily match the theoretical multifractal spectra $D_f(h)$, hence suggesting that the multifractal formalism, $L_f(h) = D_f(h)$, holds in all cases.

The following relationship holds for most standard models used in multifractal analysis:

$$h_f^{min} = \inf(supp(D_f)). \tag{3.45}$$

Note however that this relationship is not true in all generality; for instance the "chirp"

$$F_{\alpha,\beta}(x) = |x|^\alpha \sin\left(\frac{1}{|x|^\beta}\right)$$

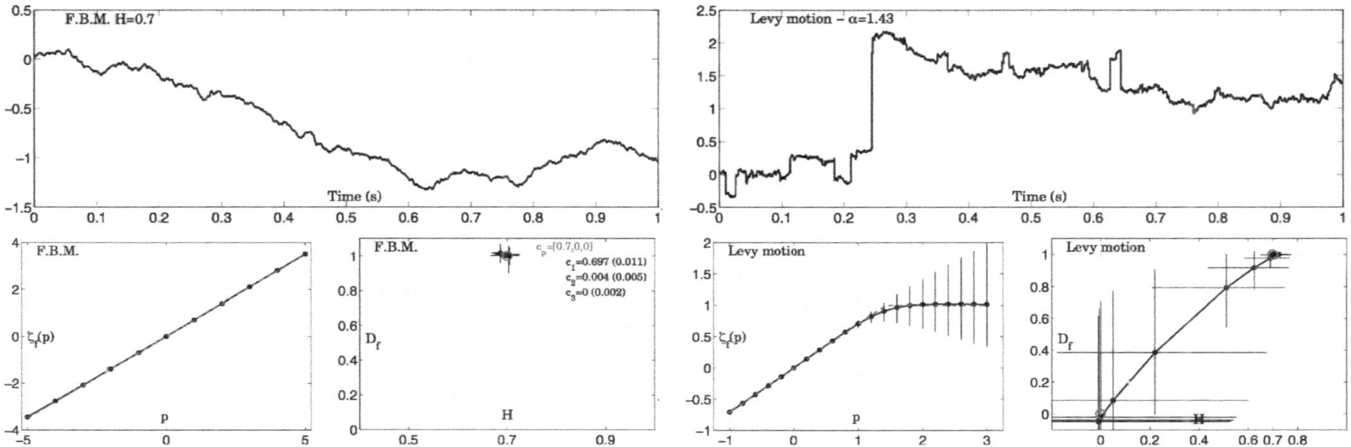

Figure 3.2 **Selfsimilar processes.** Top row: sample paths of fBm with $H = 0.7$ (left) and of stable Lévy motion with $H = 0.7$ (right). Bottom row: estimated wavelet leader scaling functions and Legendre spectra. Both processes have leader scaling functions that behave linearly in p for p close to 0: $\zeta_f(p) = pH$ for fBm, $\zeta_f(p) = Hp$ (if p is close to 0) for stable Lévy motion. These self similar processes with stationary increments are thus characterized by $c_2 = 0$. While fBm shows a linear in p leader scaling function and is mono-Hölder, stable Lévy motion is multifractal and its leader scaling function is only piecewise linear (cf. discussion in Sec. 3.5.4). Confidence intervals are constructed based on the non parametric bootstrap procedure described in [41].

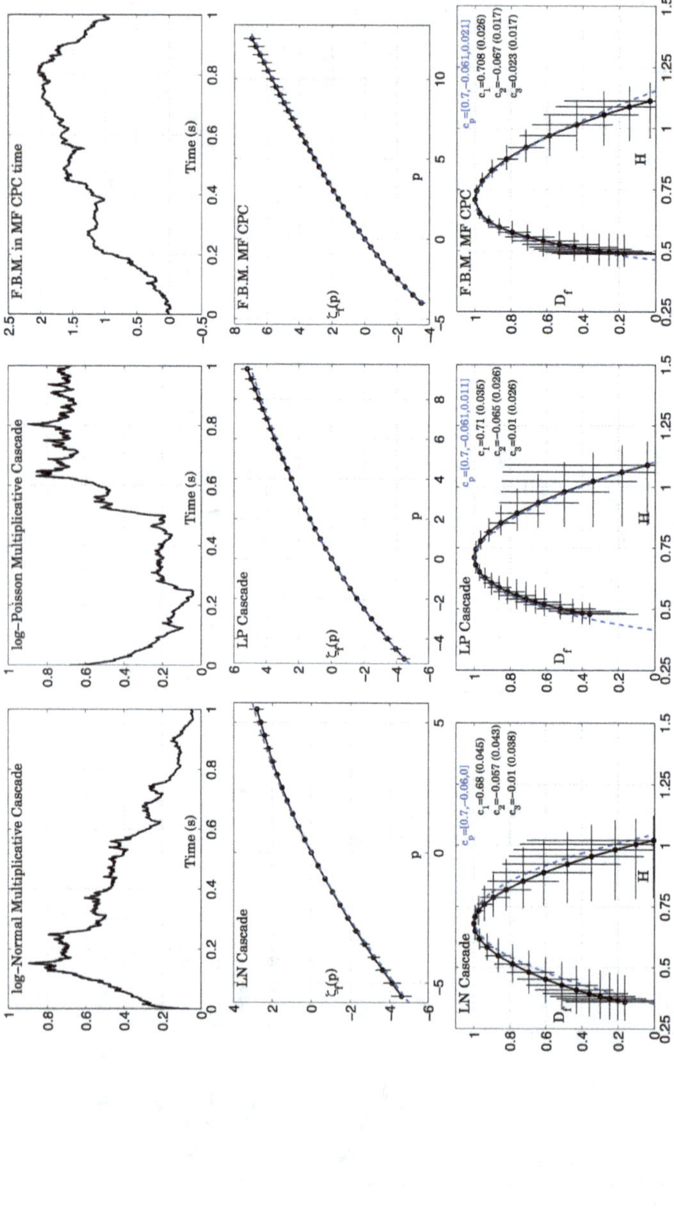

Figure 3.3 **1D multifractal processes.** Top row: sample paths of log-normal cascade (left), log-Poisson cascade (middle), and an fBm in multifractal time (as defined in Eq. (3.51)) (right). 2nd and 3rd rows show the superimposition of the theoretical (dashed) and estimated (solid) corresponding leader scaling functions and multifractal spectra. Theory and estimation are strikingly in agreement both for positive and negative *p*s and across the entire multifractal spectra (increasing and decreasing parts). Confidence intervals are obtained by a non parametric bootstrap procedure described in [41].

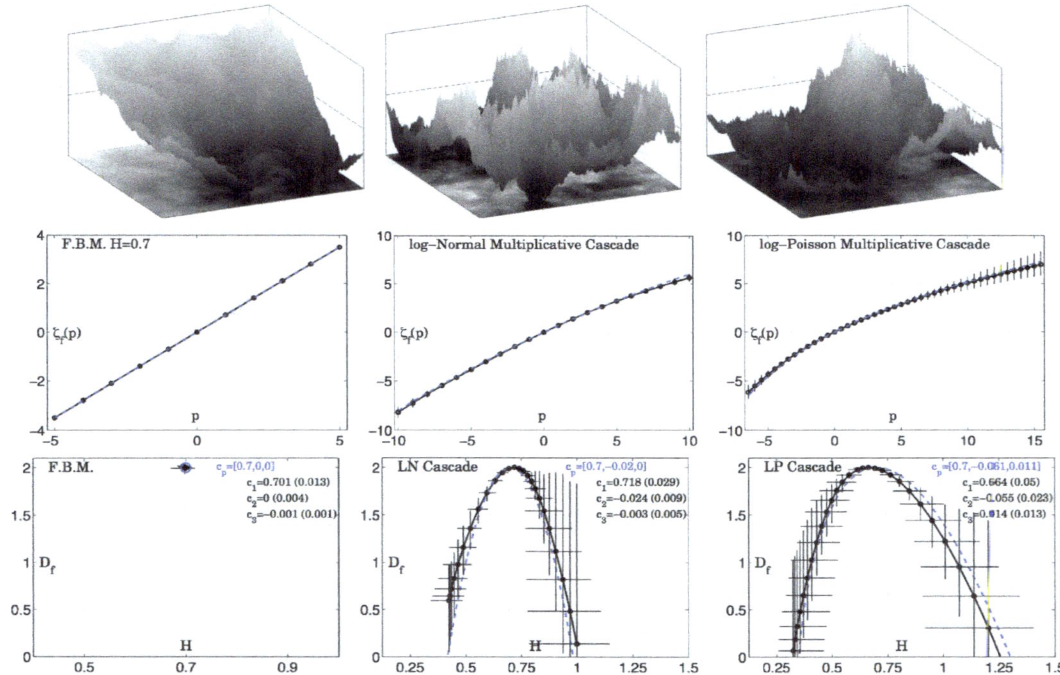

Figure 3.4 **2D multifractal processes.** Top row: 2D realizations of fBm (left), log-normal cascade (middle), log-Poisson cascade (right). 2nd and 3rd rows show the superimposition of the theoretical (dashed) and estimated (solid) corresponding leader scaling functions and multifractal spectra. As in the 1D case (cf. Figs. 3.2 and 3.3), theory and estimates are strikingly in agreement both for positive and negative *p*s and across the entire multifractal spectra (increasing and decreasing parts). Confidence interval are constructed using a non parametric bootstrap procedure described in [41].

(for α and β positive) satisfies

$$h^{min}_{F_{\alpha,\beta}} = \frac{\alpha}{\beta+1} \quad \text{and} \quad \inf\left(supp\left(D_{F_{\alpha,\beta}}\right)\right) = \alpha.$$

However, we will mention a partial result confirming Eq. (3.45) in Sec. 3.5.3.

3.5.3 Local spectrum and homogeneity

So far, little attention has been paid to the region where multifractal analysis is performed, implicitly assuming that it is conducted on the entire set of data available. However, one may wonder whether multifractal analysis would yield different results when applied to different domains. The answer is often negative: Scaling functions, and spectra of singularities, when measured on a sub-part of the data (say a non-empty open subset, in order to dismiss boundary problems) do not depend on the particular open set which is chosen. When such is the case, the function analyzed is said to have *homogeneous multifractal properties*. It is observed that large categories of experimental data have homogeneous multifractal properties: for instance it is the case for fully developed turbulence, where analyses performed on different parts of the signal always yield the same results. However, this might not systematically be the case.

A simple cause of non-homogeneity can be that data consist of a juxtaposition of different pieces, concatenated together. This situation is commonly met in image processing. Indeed, because of occlusion effects (i.e. objects in the front of the picture partly mask objects behind), images usually appear as a patchwork of homogeneous textures. In this case, it is reasonable to isolate the different components of the data, and perform multifractal analysis over each piece independently. Scaling functions computed on the whole image simply result as the minimum of the *local* scaling functions, and conversely the global multifractal spectrum consists of the maximum of the local spectra. Note that such cases can lead to simple examples of (global) non-validity of the multifractal formalism. Indeed, the scaling function of the whole image remains, by construction, a concave function, whereas the spectrum, as a maximum of concave function needs no more be concave, even if each component is. Therefore, the multifractal formalism will fail, since, by construction, the right hand side of Eq. (3.42) always yields concave functions.

However, more intricate situations may exist, where multifractal characteristics can evolve with time (or the location, for images) in a way which is more involved than piecewise constant. One can expect, for instance, that stock market data display different statistics during different market phases, such as crises or bubbles. Different multifractal characteristics could stem from such differences. One may also expect that the whole conditions of economy evolve with time, as a consequence of the ever increasing complexity of financial operations, of new financial products, and (more subtly) of mathematical models which themselves influence the way that operators react to the market; all this pleads for possible smooth (or not so smooth) modifications of multifractal characteristics with time. Such possible mechanisms may help to interpret the empirical observations that multifractal properties of market price fluctuations, such as those reported in Fig. 3.7 (bottom row plots) for the hourly Euro-USD rate, are significantly changing along time (cf. Sec. 3.6.2 for discussion).

On the mathematical side, such situations have been barely studied so far (see a contrario the results concerning heterogeneous ubiquity, by J. Barral, A. Durand and S. Seuret, where sophisticated mathematical tools have been developed to deal with the theoretical challenges of this situation [35, 36, 102, 103]). A typical example, studied in [42], is supplied by Markov processes (i.e. processes with independent increments) which do not have stationary increments (and thus are not Lévy processes). Roughly speaking, such processes have jumps of random amplitude, but each jump brings the process in a new environment, and the local multifractal properties will depend on this environment which is randomly chosen. One therefore obtains a non-homogeneous process, with a random spectrum, and random scaling functions. To the opposite, homogeneous multifractals have some additional regularity properties. For instance, the pitfall mentioned in the previous subsection does not occur: a homogeneous multifractal function satisfies Eq. (3.45). An interesting phenomenon of homogeneity breaking will be discussed in the next subsection. We will consider a homogeneous monohölder stochastic process, whose square is not homogeneous, and no longer monohölder.

Note also that a test for the constancy along time of the multifractal properties measured using the leader scaling function and relying on a non parametric bootstrap procedure was proposed in [104].

3.5.4 Mono vs. multifractality?

Disentangling issues. For practical purposes, it is often of interest to decide whether data are mono- or multifractal. However, despite apparent simplicity, as such the question is ill-posed and can be rephrased in different ways:

1. Are data characterized by a single Hölder exponent that takes the same and unique value everywhere or do there exist a variety of Hölder exponents in data?
2. Are data better described by a selfsimilar process (often implicitly meaning fBm) or by a cascade based process? In this later formulation, the important underlying question is: Is there an additive (as in random walk and hence selfsimilar models) or a multiplicative (as in cascades) structure underlying data and hence revealing significant differences in the nature of the data?
3. Are the scaling exponents (i.e., the scaling function) measured on data linear in p or not?

Relying on the fBm (Gaussian based) intuition, these three different formulations are often thought to be equivalent. This section aims at discussing some of the subtleties underlying the relations between these three questions.

Log-cumulants. Recall that, if they exist, the cumulants of a random variable X are the coefficients of the Taylor expansion of the second characteristic function of X, i.e. the coefficients c_m (if they exist) in the expansions

$$\log\left(\mathbb{E}\left(e^{pX}\right)\right) = \sum_{m=1}^{\infty} c_m \frac{p^m}{n!}. \tag{3.46}$$

Following [105–107], the cumulants of the log of the wavelet leaders are now introduced. Let $C_m(j)$ denote the m-th order cumulant of the random variables $\log(d_{j,k})$. Under stationarity and ergodicity assumptions on this sequence, $C_1(j)$ is the expectation of the $\log(d_{j,k})$, $C_2(j)$ their variance, etc. Assuming that the moments of order p of the leader $d_{j,k}$ exist, starting from the assumption $\mathbb{E}(d_{j,k}^p) = \mathbb{E}(d_{0,0}^p) \cdot 2^{-j\zeta_f(p)}$, one obtains that

$$\log(\mathbb{E}(d_{j,k}^p)) = \log(\mathbb{E}(d_{0,0}^p)) + \zeta_f(p)\log(2^{-j});$$

using Eq. (3.46), one obtains the following expansion for p close to 0:

$$\log(\mathbb{E}(d_{j,k}^p)) = \log(\mathbb{E}(e^{p\log d_{j,k}})) = \sum_{m\geq 1} C_m(j)\frac{p^m}{m!}.$$

Comparing these two expansions, we obtain that the $C_m(j)$ are necessarily of the form

$$C_m(j) = C_m^0 + c_m \log(2^{-j}), \tag{3.47}$$

and that $\zeta_f(p)$ can be expanded around 0 as

$$\zeta_f(p) = \sum_{m \geq 1} c_m \frac{p^m}{m!}. \tag{3.48}$$

The concavity of ζ_f implies that $c_2 \leq 0$. Note that, even if the moment of order p of $d_{j,k}$ is not finite, the cumulant of order m of $\log(d_{j,k})$ is likely to be finite and Eq. (3.47) above is also likely to hold. Moreover, Eq. (3.47) provides practitioners with a direct way to estimate the c_m by means of linear regressions in $C_m(j)$ versus $\log(2^{-j})$ diagrams [41]. The c_m are hence not derived from an a posteriori expansion performed from the estimates of the $\zeta_f(p)$, but instead estimated directly. For the sake of completeness, let us mention that the polynomial expansion of $\zeta_f(p)$ around 0 can be translated, via the Legendre transform, into an expansion of $L_f(h)$ around its maximum [108], on condition that $c_2 \neq 0$:

$$\begin{aligned} L_f(h) = d &+ \frac{c_2}{2!}\left(\frac{h-c_1}{c_2}\right)^2 + \frac{-c_3}{3!}\left(\frac{h-c_1}{c_2}\right)^3 \\ &+ \frac{-c_4 + 3{c_3}^2/c_2}{4!}\left(\frac{h-c_1}{c_2}\right)^4 + \cdots \end{aligned} \tag{3.49}$$

Mono vs. Multi-Hölder. The only general mathematical result concerning the validity of the multifractal formalism is supplied by Theorem 3.5.1, which, in general does not yield the spectrum. However, it does yield an interval in which the support of the spectrum is included. Indeed, let

$$h_f^{min} = \inf\{h : L_f(h) \geq 0\} \quad \text{and} \quad h_f^{max} = \sup\{h : L_f(h) \geq 0\}.$$

Then, the support of the spectrum $D_f(h)$ is included in the interval $[h_f^{min}, h_f^{max}]$. (It can indeed be checked that h_f^{min} thus defined actually coincides with the uniform Hölder exponent, defined in Sec. 3.4.3.) An important particular case stems from the situation where h_f^{min} and h_f^{max} coincide, in which case the spectrum is supported by the sole point $h_f^{min} = h_f^{max}$, so that only one single Hölder exponent exists in the data, which therefore constitute a monohölder function satisfying

$$\forall x, \quad h_f(x) = h_f^{min} = h_f^{max}.$$

This puts into light the necessity to use both positive and negative values of p in the estimation: If the leader scaling function was determined only for positive values of p, the upper bound supplied by Theorem 3.5.1 would yield an increasing function, and therefore would, at best, yield the increasing part of the spectrum. In particular, h_f^{max} would not be obtained, hence preventing from drawing any conclusion with respect to data being monohölder or not.

However, testing the theoretical equality $h_f^{min} = h_f^{max}$ is practically difficult as the estimation of these quantities turns out to be poor (cf. e.g., [41, 109]). Instead, one can estimate the parameter c_2 as defined above in Eq. (3.47). When the estimated c_2 is found strictly negative, this unambiguously indicates multi-Hölder function. As discussed below, the converse is however not necessarily true: $c_2 = 0$ does not necessarily imply monohölder function, as may be incorrectly extrapolated from the Gaussian fBm case.

Let us also note that both approaches based on multifractal analysis can be regarded as non parametric: testing for mono- or multi-Hölder can be achieved without recourse to any parametric setting, in sharp contradistinction with other existing estimators (for example, many estimators evaluate the selfsimilarity index H assuming that the data are sample paths of an fBm).

Estimated c_2 measured on real-world data are reported in Sec. 3.6: Turbulence velocity is well-known as the emblematic real-world example where a non vanishing parameter c_2 is met, (cf. Sec. 3.6.1, Fig. 3.6). This is observed to be also the case for the Euro-USD price fluctuations (cf. Sec. 3.6.2 and Fig. 3.7, bottom row, left plot) and for fetal heart rate variability (cf. Sec. 3.6.3 Fig. 3.8). However, Internet traffic aggregated count time series are found to be characterized by $c_2 = 0$ at coarse scales and weakly departing from 0 at fine scales (cf. Sec. 3.6.4 and Figs. 3.10 and 3.11, bottom row, left plots).

Selfsimilarity? Selfsimilar processes are often obtained as renormalized limits of random walks (see for instance the fBm, or stable Lévy processes). Because of the additive structure underlying their nature, selfsimilar models are appealing for describing data. From the examples shown in Sec. 3.5.2, it can even be conjectured that scaling functions often are piecewise linear functions. In particular, combining ergodicity and stationarity properties, the Kolmogorov scaling function of selfsimilar processes with stationary increments should behave linearly in p, as $\eta_f(p) = pH$ on an interval of p containing 0 (where H is the selfsimilarity parameter). Conjecturing that this property also holds for $\zeta_f(p)$, it follows that testing for selfsimilarity can be formulated as testing for $c_2 = 0$. This has been conducted in a non parametric bootstrap framework, cf. e.g., [41, 104]. Note, however, that selfsimilarity with stationary increments together with $c_2 = 0$ does not imply mono-Hölder sample paths (cf. the example of Lévy motion reported in Sec. 3.5.2 above). Furthermore, $c_2 = 0$ indicates selfsimilarity but does not necessarily imply that data follow a fBm model. This would in addition require to test for joint Gaussianity of all finite distribution functions.

Non linear point transform. Let us give some further insights into the relevance of the question of deciding whether data are mono or multi-Hölder. We aim at showing that the notion of mono-Hölder function is, in some sense, unstable since it can be altered under the action of the most simple smooth non-linear operator. To that end, let us consider fBm $B_H(t)$, which is the simplest example of a mono-Hölder stochastic process: The Hölder exponent is everywhere equal to $h = H$. Let us now further consider its square transform $Y_H(t) = (B_H(t))^2$. On one hand, at points where the sample path of fBm does not vanish, the action of the mapping $x \to x^2$ locally acts as a C^∞ diffeomorphism, and the pointwise regularity is therefore preserved. On other hand, consider now the (random) set A of points where fBm vanishes. The uniform modulus of continuity of fBm implies that a.s., for s small enough,

$$\sup_t |B_H(t+s) - B_H(t)| \leq C|s|^H \sqrt{\log(1/|s|)}.$$

therefore, if B vanishes at t, then $B_H(t+s)^2 \leq C|s|^{2H} \log(1/|s|)$, so that $h_Y(t) \geq 2H$. The converse estimate follows from the fact that, for every t,

$$\limsup_{s \to 0} \frac{|B_H(t+s) - B_H(t)|}{|s|^H} \geq 1,$$

so that, if $B_H(t) = 0$, then

$$\limsup_{s \to 0} \frac{(B_H(t+s))^2}{|s|^{2H}} \geq 1,$$

so that $h_Y(t) \leq 2H$. Thus, at vanishing points of B_H, the action of the square is to shift the Hölder exponent from $h = H$ to $h = 2H$. This set of points has been the subject of investigations by probabilists (cf. the flourishing literature on the *local time* either of Brownian motion, or of fBm); in particular, it is known to be a fractal set of dimension $1 - H$, cf. [110]. It follows that $Y_H(t)$ is a *bi-Hölder process*, whose spectrum is given by

$$D_{Y_H}(h) = \begin{cases} 1 & \text{if } h = H, \\ 1 - H & \text{if } h = 2H, \\ -\infty & \text{elsewhere.} \end{cases} \tag{3.50}$$

Because the multifractal formalism yields the Legendre transform, i.e., the concave hull, of the multifractal spectrum, the leader based estimated multifractal spectrum would suggest practitioners to (wrongly) conclude that $Y_H(t)$ is a *fully* multifractal function whose spectrum is supported by the interval $[H, 2H]$. This is illustrated in Fig. 3.5 where it can be observed that the leader based measured multifractal spectrum yields a very precise estimation of the exact concave hull of $D_{Y_H}(h)$. Again, note that this is

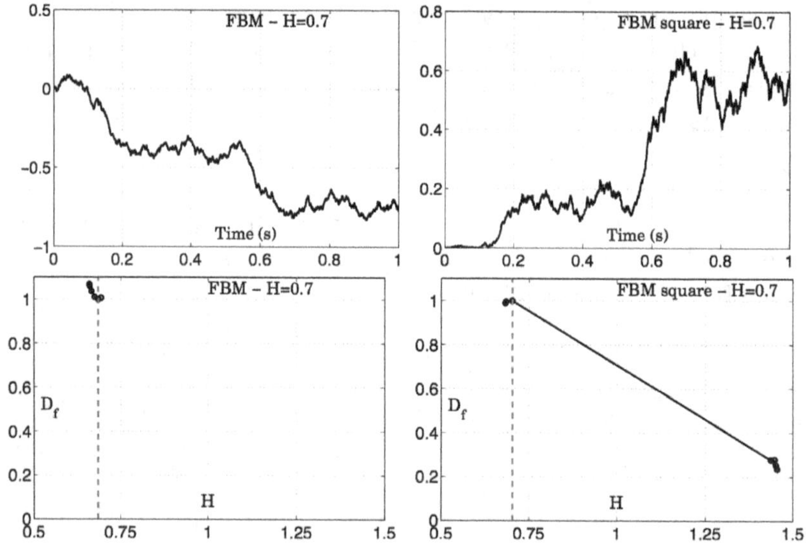

Figure 3.5 **Non linear transform.** Top row: sample path of fBm, with $H = 0.7$ (left) and its squared transform (right). Bottom row: estimated wavelet leader based Legendre spectra. fBm is a mono-Hölder process and its estimated Legendre spectrum collapses on the theoretical spectrum $D(h) = 1$ for $h = H$. Its squared non linear point transform, however, is a bi-Hölder process, with theoretical multifractal spectrum given in Eq. (3.50): The estimated Legendre spectrum perfectly reproduces its corresponding concave hull.

made possible only because negative values of p can be used in the leader framework.

Further, the analysis of the $y = x^2$ point transform of fBm raises another disturbing issue. The multifractal properties of fBm are *homogeneous* (i.e., the spectrum measured from a restriction of the sample path to any interval (a, b) on the real line is the same as that corresponding to the whole one). This is no longer true for $(B_H)^2$: the spectrum measured on a restricted interval (a, b) will vary depending on whether the interval includes or not a point where B_H vanishes. This example shows that the multifractal properties are not preserved under simple non linear transformations.

3.6 Real-world applications

Reviewing the literature dedicated to real-world applications reveals that the concepts of scaling, scale invariance, selfsimilarity, fractal geometry and multifractal analysis have been and continue to be used to analyze data in numerous contexts of very different natures, ranging from natural phenomena — physics (hydrodynamic turbulence [17, 25, 111], astrophysics and stellar plasmas [112], statistical physics [113, 114], roughness of surfaces [115]), biology (human heart rate variabilities [116–120], fMRI [121, 122], physiological signals or images), geology (fault repartition, [123]) — to human activities — computer network traffic [124], texture image analysis [115], population geographical repartition [125, 126], social behaviors or financial markets [50, 127]. B. Mandelbrot himself significantly contributed to the studies of many different applications, two of the most prominent possibly being hydrodynamic turbulence, the scientific field that actually gave birth to the essence of multifractal, (cf. e.g., [25, 26]) and finance (cf. e.g., [31, 128–130]).

As a tribute to B. Mandelbrot's outstanding research work, a small number of applications from widely different origins have been selected to be revisited. Their presentations are not intended as state-of-the-art reviews, but rather as illustrations, both of the theoretical arguments developed in the previous sections and of the benefits obtained from using multifractal analysis in these applications.

3.6.1 Hydrodynamic turbulence

In the early 20s, Richardson, combining empirical observations with the analysis of Navier-Stokes equations, that mostly govern fluid flows, formulated a heuristic description of the apparently random movements observed in hydrodynamic turbulence flows, in terms of a cascade of energy: Between a coarse scale (where energy is injected by an external forcing) and a fine scale (where energy is dissipated by viscous friction), no characteristic scale can be identified. This seminal contribution paved the road to tremendous amounts of works involving the concept of scaling to model turbulence flows (cf. [17, 111] for ancient and modern reviews, respectively). In 1941, Kolmogorov proposed a non explicit version of fBm to model velocity (which implies that $c_1 = H$, and $c_m \equiv 0$ for $m \geq 2$) [132]. However, a general consensus (a rare situation worth mentioning for the field of turbulence) amongst practitioners conducting real experiments rejected fBm as a valid model because the empirically observed scaling exponents (Kolomogorov scaling function) did not behave linearly in p, for $p \geq 0$. Framed in seminal contributions due to A. Yaglom [20] and supported by physical arguments developed by Kolmogorov (and his collaborator A. Obukhov) in 1962 [18, 19], the energy transfer from coarse to fine scales has been modeled via split/multiply iterative random procedures that account for the physical intuitions beyond the vorticity stretching mechanisms implied by the Navier-Stokes equation. In 1974, B. Mandelbrot studied these various declinations of cascades and their properties [25], paving the way for their interpretation as a collection of singularities and hence to the first formulation of the multifractal formalism by G. Parisi and U. Frisch, based on the Kolmogorov scaling function [92].

This work triggered enourmous amounts of analysis of experimental and real-world turbulence data, aiming at measuring as precisely as possible scaling exponents (i.e., the scaling function). The earliest contributions relied on the Kolmogorov scaling function (cf. e.g. [133–138]), and later work initiated by A. Arneodo, E. Bacry and J.-F. Muzy relied on the use of wavelets (cf. e.g. [41, 99, 100, 107, 139, 140]).

Figure 3.6 (top plot) shows an example of a turbulence longitudinal Eulerian velocity signal, measured with hot-wire anemometry techniques. It has been collected on a jet turbulence experiment [131], at average Reynolds number $\simeq$ 580, and has been made available to us by C. Baudet and collaborators (LEGI, Université Joseph Fourier, INPG, CNRS, Grenoble, France), who are gratefully acknowledged. Middle row plots depict its scaling property in the so-called inertial range, a priori defined from the physical characteristics of the flow conditions: Left, the wavelet leader based $\log_2 T_f(j,q)$ versus $\log_2 2^j$ (for $p = 3$), and h_f^{min} on the right. In the bottom row the measured wavelet leader scaling function and multifractal spectrum are plotted, clearly indicating multifractality.

While the generic multifractality of turbulence data (understood as a strict departure from linearity in p of the scaling function) has never been called into question by any of these studies, a major open issue remains: Which precise cascade model better describes turbulence? This question is not purely theoretical, since the precise ingredients entering the cascade that best match data may reveal physical mechanisms of importance for turbulence. Two such models are emblematic of this issue: In 1962, Obukhov and Kolmogorov [18, 19] proposed a model mostly based on a law of large numbers argument and referred to as the log-normal multifractal model. It predicts that $\zeta_f(p) = c_1 p + c_2 p^2/2$ and hence that $c_m \equiv 0$ for $m \geq 3$. In 1994, She and Lévêque [28] proposed an alternative construction based on the key ingredient that energy dissipation gradients must remain finite within turbulence flows. It is referred to as the log-Poisson model and yields a multifractal process with all non zero c_ms.

Figure 3.6 (bottom plots) compares the scaling function (left) and multifractal spectrum (right) measured from data (solid black), together with bootstrap based confidence intervals (cf. [41] for details) against those predicted by the log-normal (solid red) and log-Poisson (dashed blue) model. Together with the fact that the measured c_3 can not be distinguished from 0 ($c_3 = 0.001 \pm 0.001$), these plots clearly indicate that the log-normal model is preferred by turbulence data. This has been studied systematically and on larger data sets in [41, 139] and unambiguously confirms earlier findings reported in [107, 134].

3.6.2 Finance time series: Euro-USD rate fluctuations

Modeling stock market prices received considerable efforts and attention starting with the seminal early contribution of Bachelier in 1900 who, in his thesis, based his description of stock market on Brownian motion, i.e., on an ordinary Gaussian random walk [141]. Essentially, this model implies that daily *returns*, i.e., increments of (or differences between) prices taken along two consecutive days, $X(t) = P(t) - P(t-1)$, consist of independent, identically distributed Gaussian zero-mean random variables. This hence leaves little room to earn money from stock market investments by pure statistical model based prediction. However, it has long been observed or expected that stock market prices significantly depart from the Brownian model. First, returns may exhibit statistics that significantly depart from

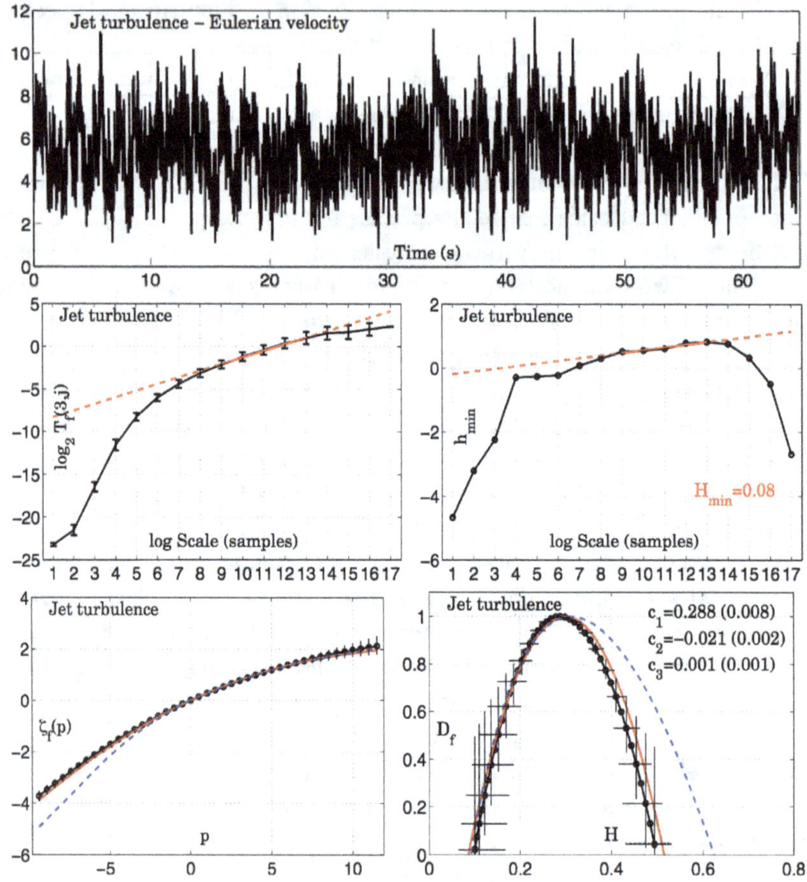

Figure 3.6 **Hydrodynamic Turbulence.** Top: longitudinal Eulerian turbulence velocity signal (hot-wire anemometry, Reynolds number: $\simeq$ 500, Courtesy C. Baudet *et al.* [131]). Middle: log-log plot for the wavelet leader based $T_f(3,j)$ (as in Eq. (3.39)) and for the wavelet coefficients based h_f^{min} (as in Eq. (3.37)). Bottom: Wavelet leader scaling function (left) and multifractal spectrum (right) measured from a single run of data (black solid lines), compared to the log-normal (gray solid lines) and log-Poisson (dashed lines) models. Data clearly select the log-normal model.

Gaussian distributions. Second, it is often observed that volatility, a quantity essentially based on the squared (or absolute) value of the returns, display some form of dependency. Notably, in their celebrated contribution, Granger and Joyeux indicated that volatility is well characterized by long memory, or long term dependence [142, 143], a model that is closely related with selfsimilarity and fBm. This motivated substantial efforts to model stock market returns with the GARCH models family or FARIMA alternatives (cf. e.g., [46, 144, 145]).

B. Mandelbrot significantly contributed to the analysis of the statistical properties of stock market returns by performing an iconoclastic analysis of commonly used finance models, emphasizing heavy tails, as opposed to Gaussian distributions (the Noah effect), and long memory, as opposed to independence (the Joseph effect), cf. [146] for a review, and several seminal contributions of B. Mandelbrot; see also e.g., [147, 148]. In particular, this called into question the Black and Scholes model strongly relying on Gaussianity, and led B. Mandelbrot and collaborators to explicitly propose that stock market prices may have multifractal properties in [149] in contradistinction with most Itô-martingale based models used so far. B. Mandelbrot's work paved the way towards numerous efforts to assess multifractality empirically in stock market prices and exploit it (cf. e.g., [50, 127, 128, 150, 151]).

Almost one century after Bachelier, B. Mandelbrot and collaborators proposed a new model for stock market prices: fBm in multifractal time [31, 130], a process that explicitly incorporates multifractal properties by a multifractal time jittering, obtained by a multiplicative cascade construction. In this model, a new generation of cascades is used, based on Compound Poisson distributions, following again an original idea of B. Mandelbrot [30, 152, 153]. Let F denote the distribution function of the multiplicative cascade μ on $[0,1]$: $F(t) = \mu([0,t))$. Then, fBm in multifractal time is defined as

$$X_t = B_H(F(t)), \tag{3.51}$$

where B_H stands for fBm with selfsimilarity parameter H (which is independent of μ). Note that, as a consequence of Eq. (3.20), the spectrum of X is deduced from the spectrum of F by a simple dilation along the h axis

$$D_X(h) = D_F(h/H);$$

see the results of R. Riedi and S. Seuret [43, 154] for additional discussions concerning *multifractal subordination*, i.e. understanding multifractal properties of composed processes. Numerous rich and interesting declinations of such constructions of cascades and multifractal processes have been proposed since (cf. e.g., [32–34, 37, 38, 127, 128, 150]). Note that the idea of *multifractal subordination* (i.e., composition by a multifractal change of time) is extremely powerful, and can be applied in many settings. Let us mention just one example: In [155], B. Mandelbrot and one of the authors performed the multifractal analysis of the Polya function (a famous example of continuous space-filling function) by simply remarking that it can be *factorized* as the composition of the repartition function of a simple dyadic cascade with a monohölder function.

As a tribute to B. Mandelbrot's contributions, though the authors have not personally contributed to the field, the multifractal properties of the Foreign exchange Euro-USD price fluctuations are now studied. Fig. 3.7 (top row) shows the hourly Euro-USD quotations (and hourly returns) since the birth of Euro. Data Courtesy Vivienne Investissement, Lyon, France www.vivienne-investment.com. The return (or increment) time series clearly shows large fluctuations that are not consistent with a constant along time Gaussian distribution. Scaling properties analyzed here are found to hold for scales ranging from the day to the month, and hence correspond to long term characteristics, as opposed to intra-day analysis, not conducted here. Multifractal spectra measured over one year long time series (second and third rows, left plots), using the wavelet leader multifractal formalism, show that the Euro-USD quotations exhibit multifractal spectra that are in agreement with the fBm in multifractal time model, yet indicate that the multifractal properties may vary along the years. To further analyze these variations, scaling attributes are measured in one year long time windows (with a 3-month sliding time shift) and reported (together with confidence intervals) in Fig. 3.7 (bottom row). Interestingly, it is found that h_f^{min} is permanently above 0: $h_f^{min} > 0$, which clearly rules out the use of jump processes to model ForEx time series. Moreover, because $h_f^{min} > 0$, the finiteness of quadratic variations can be evaluated using the leader scaling function at $p = 2$, $\zeta_f(2)$ (as in Eq. (3.39), cf. Sec. 3.4.4). Figure 3.7 (bottom row, 2nd plot) shows that $\zeta_f(2)$ fluctuates and often remains below the finite quadratic variations limit of 1: $\zeta_f(2) < 1$. This result validates the hypothesis of the finiteness of quadratic variations, underlying the use of Itô-martingale processes as models to describe the Euro-USD variations, at least when long term variations are analyzed, as is the case here; this does not exclude that shorter term (or intra-day) variations might show different properties. In Fig. 3.7 (bottom row, 3rd and 4th plots), parameters c_1, c_2 are plotted as functions of time. These plots

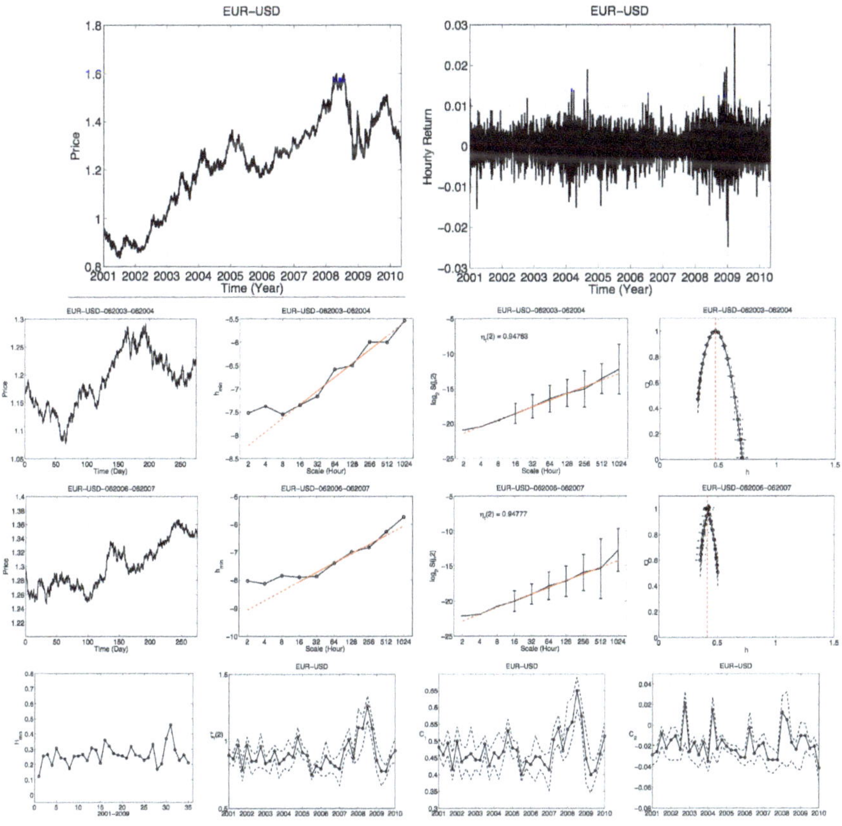

Figure 3.7 **Euro-USD.** Top: Euro-USD rate fluctuations for 10 years (starting January 1st, 2001) (left) and hourly returns (right). Data Courtesy Vivienne Investissement, Lyon, France. One year long time series (June 2003 to June 2004, 2nd row; June 2006 to June 2007, 3rd row), with log-log plots for the wavelet coefficients based h_f^{min} (cf. Eq. (3.37)), log-log plots for the wavelet leader based $T_f(2, j)$ (given by Eq. (3.39)), and multifractal spectra, measured using the wavelet leader multifractal formalism. Bottom row: Estimation in one year long time windows (with a 3-month sliding shift) of h_f^{min}, $\eta_f(2)$, c_1, c_2 (together with confidence intervals). These plots show that while $h_f^{min} > 0$, $\zeta_f(2)$ often remains below the finite quadratic variations limit of 1. Parameter c_1, c_2 variations indicate that the multifractal properties of the Euro-USD price fluctuations vary significantly along time.

furthermore indicate that though a fBm in multifractal time-like process provides a valid model to describe the price fluctuations, the multifractal properties may vary along time. Two different regimes are suggested by these plots: Before 2007, with a $c_1 \simeq 0.45 < 0.5$ and $c_2 \simeq -0.02$; and after 2008, with $c_1 \simeq 0.5$ and $c_2 \simeq -0.01$. Recalling that the ordinary Brownian motion (proposed by Bachelier) would correspond to $c_1 = 0.5$ and $c_2 = 0$, it may be concluded that the (long term) fluctuations of the Euro-USD quotations show less structure for the later period than for the former, hence making predictions for investments more difficult (see also discussion in Sec. 3.5.3 on homogeneous multifractal properties).

Assuming that the fBm in multifractal time model studied by L. Calvet, A. Fisher and B. Mandelbrot is the correct description of prices, let us now show how the parameter H of the corresponding fBm can be inferred from the observed data using the wavelet scaling function (the heuristic argument that follows can easily be turned into a mathematical proof). Let us first return to the definition of the Kolmogorov scaling function Eq. (3.8). Approximating the integral by a Riemann sum, and taking $h = 2^{-j}$, one obtains

$$\int |X_{t+h} - X_t|^p dt \sim 2^{-j} \sum_k \left| X_{\frac{k+1}{2^j}} - X_{\frac{k}{2^j}} \right|^p$$
$$= 2^{-j} \sum_k \left| B_H\left(F\left(\frac{k+1}{2^j}\right)\right) - B_H\left(F\left(\frac{k}{2^j}\right)\right) \right|^p$$

which, using the fact that $|B_H(t) - B_H(s)| \sim |t-s|^H$, is of the order of magnitude of

$$2^{-j} \sum_k \left| F\left(\frac{k+1}{2^j}\right) - F\left(\frac{k}{2^j}\right) \right|^{pH}.$$

In particular, for $p = 1/H$, using the fact that F is continuous and increasing, we obtain

$$2^{-j} \sum_k \left| F\left(\frac{k+1}{2^j}\right) - F\left(\frac{k}{2^j}\right) \right|$$
$$= 2^{-j} \sum_k \left(F\left(\frac{k+1}{2^j}\right) - F\left(\frac{k}{2^j}\right) \right) = 2^{-j}$$

Since this quantity is, by definition, of the order of magnitude of $(2^{-j})^{\eta_f(p)}$, we obtain that the exponent H of the fBm used in this model (cf. [130]) satisfies

$$\eta_f\left(\frac{1}{H}\right) = 1. \tag{3.52}$$

Using the fact that the Kolmogorov and the wavelet scaling function coincide in the range of exponents involved here, Eq. (3.52) allows to recover the selfsimilarity exponent of the fBm in this model from the wavelet scaling function. In the example considered in Fig. 3.7, one obtains that, in practice, the parameter H thus estimated is very close to c_1.

3.6.3 Fetal heart rate variability

Heart rate variability (HRV), which refers to the short time fluctuations (within the minute) of heart rate, has long been considered as a powerful tool to characterize the health status of a given subject: In a nutshell, the more variability, the healthier the subject is (cf. e.g., the seminal contribution [158]). To characterize variability in heart rate time series, spectral analysis has long been used to measure the so called LF/HF balance, i.e., the ratio of energies measured in frequency bands attached to the sympathic and parasympathic activities of the autonomous central nervous system [158]. Yet, it has long been observed that heart rate time series display *fractal* properties and that the corresponding fractal exponent could be used as a non invasive tool for non healthy status detection (cf. [158]). This opened the way to numerous analyses aiming at describing HRV data with fBm-like models and at correctly estimating the corresponding selfsimilarity parameter H; later, multifractal analysis has naturally been considered as an analysis tool potentially improving the characterization of HRV scaling properties (cf. e.g., [116–118]). Recently, multifractal analysis has been involved in the detection of per partum fetal asphyxia. Obstetricians are monitoring fetal-HRV during the delivery process to detect potential fetal asphyxia. Using a set of analysis criteria (the international FIGO rules), obstetricians measure the health status of the fetus and decide on operative delivery or not. A major difficulty they are facing is the high number of False Positive (fetuses diagnosed as unhealthy during the delivery process but a posteriori evaluated as perfectly healthy), and hence of operative delivery that might have been avoided. This constitutes an important public health issue, as operative deliveries are accompanied by a significant number of serious posterior complications for both newborns and their mothers.

Figure 3.8 (top row) shows examples of heart rate time series (in Beats-per-Minute) for a healthy subject, correctly diagnosed as such by the obstetrician (left) (True Negative), for a healthy subject, incorrectly diagnosed as suffering from per partum asphyxia (False Positive), and for a fetus actually suffering form per partum asphyixia and diagnosed as such (True Positive). Data Courtesy M. Doret, Obstetrics Dept. of the University Hospital Femme-Mère-Enfant, Lyon, France, [156, 157]. As shown in Fig. 3.8 (second and third rows), for all three classes (True Negative, True Positive, False Positive), HRV exhibits scaling properties for scales ranging from the second to the minute. The fourth row in Fig. 3.8 shows multifractal spectra, estimated from 15 minutes long heart rate time series using wavelet leaders, which suggests that the healthier the subject, the larger its variability in the sense that its c_1 parameter is smaller (multifractal spectrum shifted to the left). To confirm such an observation, multifractal parameters are estimated within 15 minutes long sliding windows (with a 50% time overlap) for the last hour before delivery. The 15 minutes window duration is chosen after the obstetrician's request that asphyxia should be detected as early as possible. Multifractal parameters are then averaged within the three classes (each containing 15 patients carefully selected as representative by obstetricians). A Wilcoxon ranksum test is conducted for each time window between the True Positive and False Positive classes. The results reported in Fig. 3.8 (bottom row) clearly indicate a significant increase of c_1 and h_f^{min} (hence a significant decrease in HRV) for the Unhealthy True Positives compared to the Healthy True Negatives and False Positives. The p-value of the Wilcoxon ranksum test is observed to be below 5% (cf. red $\diamond$ in Fig. 3.8 (bottom row)) often 30 minutes earlier than the time the decision to operate the delivery was actually taken. These results are reported in detail in [156, 157].

It is worth noting that for heart rate time series (as for many other types of human rhythms signals, cf. e.g., fMRI applications [121, 122]), h_f^{min} can be found either positive or negative depending on the subject, or its health status. This is consistent with the fact that practitioners, in applications, noticed that they often had to switch from fBm to fGn (or conversely) to model data and understood this as a puzzling inconsistency (see for instance

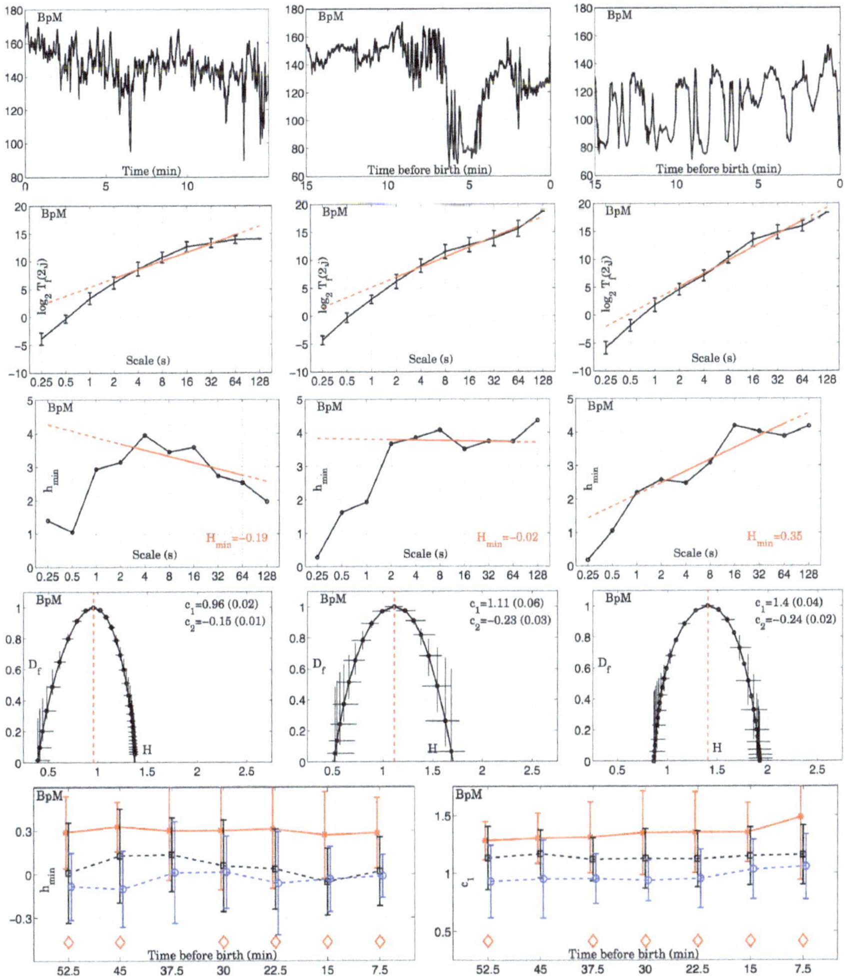

Figure 3.8 **Fetal-ECG.** Top: Examples of cardiac frequency time series (in Beats-per-Minute) for fetuses during delivery process — last 15 min before delivery, for a healthy subject, correctly diagnosed as such by the obstetrician (left) (True Negative), for a healthy subject, incorrectly diagnosed as suffering from per partum asphyxia (middle) (False Positive), for a fetus actually suffering from per partum asphyixia and diagnosed as such (True Positive). Data Courtesy M. Doret, Obstetrics Dept. of the University Hospital Femme-Mère-Enfant, Lyon, France. Second Row: log-log plots for the wavelet leader based $T_f(2,j)$ (cf. Eq. (3.39)). Third Row: log-log plots for the wavelet coefficients based h_f^{min} (as in Eq. (3.37)). Fourth Row: Multifractal spectra, measured using the wavelet leader multifractal formalism. Bottom Row: Parameters h_f^{min} (left) and c_1 (right)

Figure 3.8 (*Continued*) estimated over 15 min long sliding windows (with a 7.5 min overlap) over the last hour before delivery, values are averaged over the classes of 15 patients each: True Positive (dashed line with 'o'), False Positive (dashed line with squares), True Positive (solid line with $*$). The red $\diamond$ correspond to windows where a Wilcoxon ranksum test between True Positive and True Negative gives a $p-$value below 5%. This analysis shows that unhealthy subjects (True Positive) systematically display less HRV as compared to healthy ones (True Positive and False Positive) (cf. [156, 157]).

discussions in [122]). To some extent, the general framework developed here alleviates such inconsistency by introducing h_f^{min} as one of the various parameters that can be used to analyze scaling without a priori assumptions on the data. In the analysis of fetal HRV described above, a fractional integration of order $1/2$ (larger than the absolute value of the most negative h_f^{min} encountered in the whole database) has been applied to enable the use of the wavelet leader multifractal formalism in a consistent manner on the whole set of data.

3.6.4 Aggregated Internet traffic

Internet traffic engineering currently is a challenging task involving a variety of issues ranging from economy development planning and security assessment to statistical data characterization. The earliest versions of statistical models used for Internet traffic modeling were stemming from the experience gained on the previous large communication network: Telephone communications. B. Mandelbrot himself made very early contributions to such modeling (cf. e.g., [161]), and this certainly was a major source of inspiration for the later developments he made in the theory of fractal stochastic processes. However, such telephone communications models were all based on the assumption that there exist one or several characteristic scales of time in data, while it has soon been recognized that Internet traffic is well characterized by long memory and self similarity (cf. [53, 162–165] for seminal articles). This later raised the question of multifractality of Internet traffic (cf. e.g., [166–169]) that we briefly illustrate here.

Any activity performed on the Internet (emailing, web browsing, Video-On-Demand, data downloading, IP telephony,...) amounts to establishing a connection between hosts (server-clients, peer to peer,...), consisting of computers located all over the world, according to an agreed-on protocol,

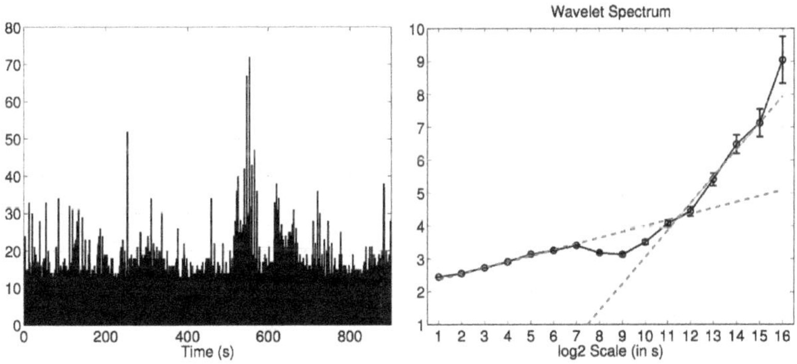

Figure 3.9 **Internet Traffic: Aggregated time series and scaling.** Left: Aggregrated time series (Courtesy MAWI research project, [159]); Right: log-log plot for the wavelet coefficient based $S_f(2, j)$ (as in Eq. (3.34)). It clearly displays two independent scaling ranges, below and above the second.

and exchanging a (very) large number of elementary quanta of information, the Internet Protocol packets (in short, IP Pkts). Collecting Internet traffic data hence naturally consists of recording the IP Pkt time stamp and header (IP source, IP destination, Port Source, Port Destination). Obviously, analyzing the marked point process resulting of the tremendous number of IP Pkts exchanged is not feasible, and it is classical to instead monitor so-called aggregated (or count) time series: Practitioners select a given aggregation-level Δ, whose choice depends on the application, and construct either the Pkt or Byte time series that counts the number of IP Pkts whose timestamp falls within $[k\Delta, (k+1)\Delta)$, or the corresponding volume in Byte.

An example of such Pkt time series is plotted in Fig. 3.9 (left). It consists of traffic from the MAWI repository, made available by the Japanese WIDE research program [159]. The MAWI repository consists of data that have been collected everyday for 15 minutes at 2 pm (Tokyo time) since January 2001 on major backbone links (100Mbps or 1Gbps) connecting Japan and the United States (cf. [160, 170, 171].

The left plot of the same figure displays the wavelet coefficient based $\log_2(S_f(2, j))$ vs. $\log_2(2^j) = j$ (as in Eq. (3.34)) and clearly shows that, for Internet traffic, there exist two ranges of scales for which scaling can be identified: Coarse scales, above 1s, and fine scales, below 1s. For each of those ranges of scales, the wavelet leader based multifractal formalism is applied.

For the coarse scales, Fig. 3.10 clearly indicates on a 15 minutes long set of data aggregated at 1ms that the leader scaling function is linear, and

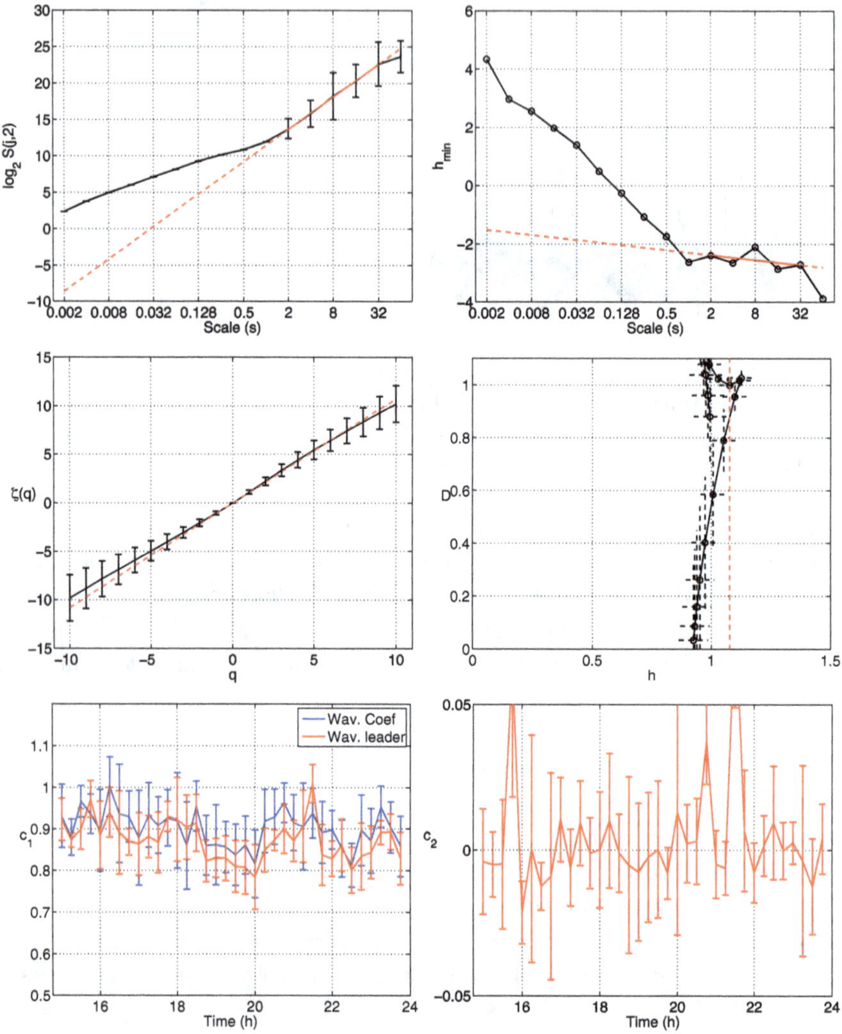

Figure 3.10 **Internet Traffic: Coarse scales.** Top: log-log plots for the wavelet leader based $T_f(2,j)$ (cf. Eq. (3.39)) (left) and for the wavelet coefficients based h_f^{min} (as in Eq. (3.37)) (right). Middle: Wavelet leader scaling function (left) and multifractal spectrum (right) measured from 15-min long data. Bottom: Evolution along time of parameters c_1 (left) and c_2 (right), measured from 15-min long sliding windows for 24 consecutive hours (9 hours shown here; cf. [160]), suggesting that scaling properties are constant along time and clearly validating that, at coarse scales, i.e., above 1s, aggregated Internet traffic are selfsimilar with long memory ($c_1 \simeq H \simeq 0.9$) and show no multifractality ($c_2 \simeq 0$).

that the multifractal spectrum collapses to a single point, hence suggesting monofractality or that multifractality is at best very weak. To further investigate this point, the parameters c_1 and c_2 are estimated on sliding 15 min long windows for a trace collected during 24 hours (on September, 22nd, 2005), for scales ranging from 1s to 1 min. Results are reported for the last 8 hours for simplicity and show, first, that the estimated parameters are remarkably constant along the day (while the changes in amplitude of the time series may be drastic) and, second, that $c_2 \simeq 0$. This indicates no multifractality at coarse scale and, because $c_1 \sim H \simeq 0.9$, a significant long memory in the data. These experimental results corroborate those reported in [168] and are consistent with the main model due to Taqqu *et al.* (cf. [172] relating selfsimilarity and heavy-tails of the web file size distribution, see also [173–175]).

For the fine scales, Fig. 3.11 shows on the same 15 minutes long set of data aggregated at 1ms that the leader scaling function is linear for p around 0, hence indicating a very weak multifractality or no multifractality at all. Parameters c_1 and c_2, estimated on sliding 15 minutes long windows for the 24 hour trace, again turn out to be constant along time, with $c_1 \simeq 0.55 \pm 0.05$ and $c_2 \simeq 0.007 \pm 0.005$. This hence validates that fine scales of aggregated traffic show at best little multifractality. Furthermore, estimations of the selfsimilarity parameters (not shown here) being close to 0.5 also indicate little correlation. These analyses tend to show that fine scales of aggregated traffic are characterized by the quasi-absence of dependence structure. Moreover, the estimated parameters c_1 and c_2 remain remarkably constant along time for the whole day and across the various days analyzed, a highly unexpected result for Internet traffic, which is often described as widely varying, and which is continuously subject to numerous occurrences of anomalies and attacks. This is essentially because data have been pre-processed using a random projection and a median procedure that robustify estimation against anomalous behaviors (cf. [160, 171]). In practice, the effect of this procedure is to eliminate abnormal components in the traffic. The results reported here can hence be associated with the properties of regular and legitimate background Internet traffic, corresponding to the usual Internet user behaviors. Furthermore, this quasi-absence of dependence structure at fine scales is consistent with the Cluster Point Process modeling of Internet traffic aggregated count time series proposed in [168], which implies that fine scales are actually not possessing strict scaling properties. However, it also shows that the statistical properties of the fine scales are satisfactorily approximated

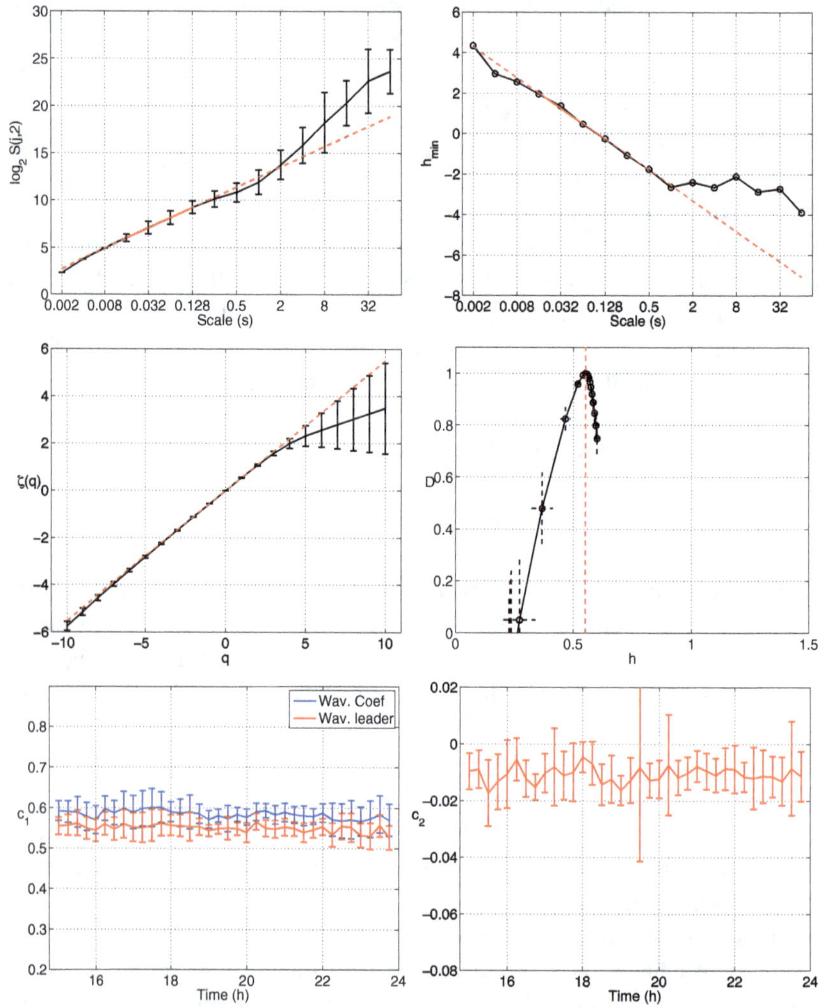

Figure 3.11 **Internet Traffic: Fine Scales.** Top: log-log plots for the wavelet leader based $T_f(2,j)$ (as in Eq. (3.39)) (left) and for the wavelet coefficients based h_f^{min} (given by Eq. (3.37)) (right). Middle: Wavelet leader scaling function (left) and multifractal spectrum (right) measured from 15-min long data. Bottom: Evolution along time of parameters c_1 (left) and c_2 (right), measured from 15-min long sliding windows for 24 consecutive hours (9 hours shown here), cf. [160], also suggest that scaling properties are constant along time and indicate that, at fine scales (i.e., below 1s), aggregated Internet traffic are characterized by $c_1 \simeq H \simeq 0.6$ and $c_2 \simeq 0.005 \pm 0.005$, revealing at best a very weak multifractality or no multifractality at all.

by scaling behaviours. Therefore, the fact that regular or legitimate traffic does not show multifractality does not imply that multifractal parameters like c_1 and c_2 can not be used for traffic monitoring, such as anomaly detection for instance. Indeed, the occurrence of anomalies is likely to change the dependence structure of the fine scale statistics (as shown in e.g., [171]) and consequently the empirically estimated parameters c_1 and c_2, which may thus prove very useful, even if scaling properties are measured that actually consist of approximations of the true statistical properties of the fine scales.

Finally, let us mention that for aggregated count time series $h_f^{min} < 0$, as shown on the top left plots of Figs. 3.10 and 3.11, hence a fractional integration of order 1 is systematically applied prior to performing the wavelet leader multifractal formalism.

3.6.5 Texture classification: Vincent Van Gogh meets Benoit Mandelbrot

Applications described so far all implied the multifractal analysis of 1D signals only. However, B. Mandelbrot always advocated that fractal analysis should be applied to images, to characterize the roughness of textures (cf. e.g., [176, 177]). The wavelet coefficient based analysis of selfsimilarity has been readily and immediately extended to 2D fields and images, thanks to the straightforward formulation of the 2D discrete wavelet transform (cf. e.g., [75]). The estimation of the corresponding H parameter has then been used to characterize textures in images, often in the context of biomedical applications (cf. [68, 178]). The extension of multifractal formalisms to images turned out to be significantly more intricate. Direct extension of the Kolmogorov scaling function were used to analyze rainfall repartitions by D. Schertzer and S. Lovejoy, see [179]. A second attempt based on the modulus maxima of the coefficients of the 2D continuous wavelet transform was proposed by A. Arneodo and his collaborators; it was employed in medical [180] or geophysical applications [115]. The wavelet leader formalism, recently proposed and based on a 2D discrete wavelet transform, enabled a significant step forward toward the use of multifractal analysis in higher dimensions. Its potential benefits have been assessed in [181], and it has recently been used to characterize texture in paintings [182].

The paintings that Van Gogh performed while he was living in France are grouped into two periods by art experts and art historians: The Paris period, ending early 1888, and the later Provence period. While art experts agree on the classification of most of Van Gogh's paintings, some are still not clearly attributed to either period. To perform such classification, art experts often make use of material or stylometric features such as the density of brush strokes, their size or scale, the thickness of contour lines, the layers, the chemicals of the colors, ...

Aiming at promoting the investigation of computer-based image processing procedures to assist art experts in their judgement, the *Image*

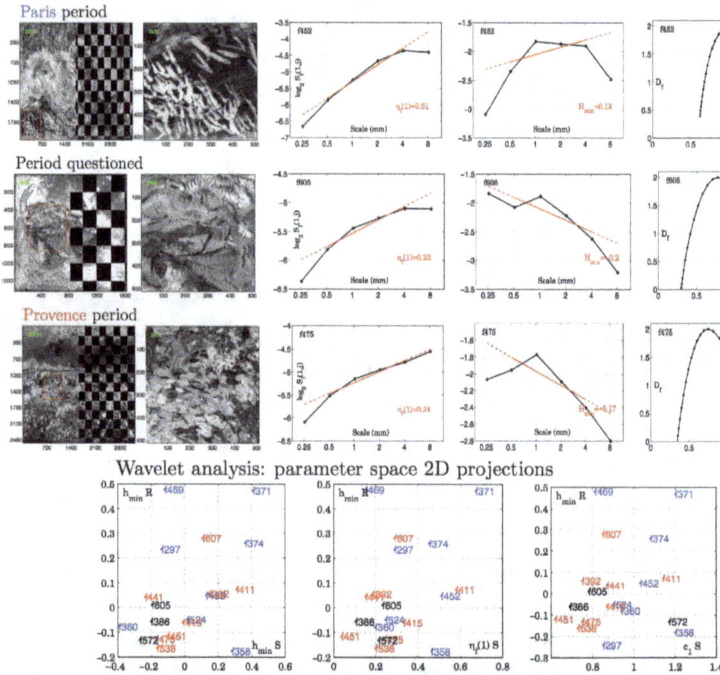

Figure 3.12 **Van Gogh's Paintings.** Left columns: Representative examples of paintings of the Paris (top), Provence (bottom) and unknown (middle) periods, with selected homogeneous patches (second column), Courtesy *Image Processing for Art Investigation* research program, cf. www.digitalpaintinganalysis.org/, and the Van Gogh and Kröller-Müller Museums (The Netherlands). Third column: log-log plots for the wavelet coefficient based $S_f(1, j)$ (as in Eq. (3.34)). Fourth column: h_f^{min}. Right column: Wavelet leader based multifractal spectra, suggesting that the questioned painting has multifractal properties close to that of the Provence period example. Bottom row: 2D projections of multifractal attributes, indicating that paintings from the Paris period (darker) are overall globally more regular than paintings from the Provence period (lighter).

Processing for Art Investigation research program (cf. www.digitalpaint inganalysis.org/), in collaboration with the Van Gogh and Kröller-Müller Museums (The Netherlands), made available to research teams two sets of (partial and low resolution digitized versions of) 8 Van Gogh's paintings each from the Paris and Provence period, and 3 paintings whose period remains undecided (nomenclature below corresponds to the Van Gogh museum catalog). The digitized copies are available through the usual RGB (Red, Blue, Green) channels and converted into the HSV (Hue, Saturation, Value) channels, hence enabling to process the gray-level light intensity information and the color information.

Three examples, corresponding to the Paris period, the Provence period and unclassified paintings, respectively, are shown in Fig. 3.12 (left column) together with examples of homogenous patches selected for analysis (second column). The 2D wavelet leader based multifractal formalism has been applied, for each selected patch, independently to the six RGB and HSV channels. Examples of obtained multifractal spectra are shown in Fig. 3.12 (right column) and suggest, for example, that $f605$ is closer to $f475$ than to $f452$ and hence is likely to belong to the Provence period. For each patch of the $2 \times 8 + 3$ paintings and each channel, multifractal attributes $h_f^{min}, \eta_f(1), c_1, \ldots$ are estimated. Fig. 3.12 (bottom row) shows 2D projections of various pairs of such attributes, which all indicate that on average the Paris paintings (blue clusters) are globally more regular than the Provence paintings (red clusters). Also, the Saturation and Red channels are found to be the most discriminant. This suggests that both a color and an intensity information (saturation and Red) are useful for discrimination. Interestingly, it has been reported by art experts that saturation is a key feature to distinguish between the Paris and Provence periods in Van Gogh's paintings. Finally, these 2D projections suggest that $f605$ and $f386$ belong to the lower left red cluster and hence to the Provence period, while conclusion is less straightforward for $f572$. Such results are left to the appreciation of art experts. These analyses are detailed in [182].

3.7 Conclusion

A key issue for the future developments of multifractal analysis is to disentangle (at least) four different acceptations that can be associated with the key word *multifractal*, and that are often misleadingly confused one with another, sometimes leading to potential misunderstandings and erroneous interpretations.

Theoretical multifractal analysis and local regularity. From the theoretical or mathematical side, multifractal analysis is deeply tied with the analysis of the fluctuations along time (or space) of the local regularity of the function or data of interest, these fluctuations being often coined as *irregularity*. Usually, this local regularity is measured in terms of Hölder exponents $h_f(x)$. The global description of the irregularity of f is gathered in the multifractal spectrum $D_f(h)$, which consists of the Hausdorf dimensions of the geometrical iso-Hölder sets. For mathematics, multifractal is hence all about Hölder exponent and (fractal, box, or Haussdorf) dimensions of geometrical sets. Though listed first here, the mathematical formulation of multifractal analysis theory can be considered as the most recent piece of the history of multifractal, as it really started in the 90s only, see [93], as a consequence of the seminal article of G. Parisi and U. Frisch [92]. Note however that precursors can be found, for instance in articles dealing with slow and fast points of Brownian motion. Recall that this specific process displays exceptional points where the modulus of continuity is slightly smaller (resp. bigger) than a.e.; the corresponding sets are random fractals. Following the seminal contribution of J.-P. Kahane, their dimensions have been precisely determined, respectively by S. Orey and S. J. Taylor for fast points, see [183], and by E. Perkins for slow points, see [184].

Multifractal formalism and scale invariance. From the more practical perspective of real-world data analysis, scaling functions, consisting of time (or space) averages of the p-th power of scale dependent quantities (such as increments, oscillations, wavelet coefficients or wavelet leaders) — also often referred to, in statistical physics, as partition or structure functions — constitute the core aspect of the notion of multifractal. Importantly, practical multifractal analysis amounts in essence to assuming that scaling functions behave as power laws with respect to the analysis scales, in the range of scales available in the data. The description of these power laws naturally leads to the notion of scaling exponents. Equivalently, these scaling exponents are termed scaling functions, with the practical and often implicit request that the limit underlying its definition actually exists (cf. Eq. (3.8) versus Eq. (3.9)).

These power law behaviors — or scale invariance, or scaling — of scale dependent quantities, and the corresponding scaling exponents (or scaling functions), were at the heart of the intuition that B. Mandelbrot developed, connecting scaling properties to (fractal) dimensions, to selfsimilarity and/or to fluctuations of local regularity, and relating scaling to self similar random walks or multiplicative cascades.

The multifractal formalism (in reference to the Thermodynamic formalism) therefore consists of the mathematical developments that relate scale invariance and scaling functions to local regularity and multifractal spectra. In some sense, the multifractal spectrum constitutes the link between formal mathematics and physics or data analysis.

Multifractal processes. Multifractal processes do not constitute a well-defined class. In essence, this notion refers to processes (or functions) that can be fruitfully used to model scale invariance properties in data. Following the seminal distinctions proposed by B. Mandelbrot, they can be thought of as falling into two major classes: Selfsimilar random walks and multiplicative cascades (or martingales).

The first one, now very classical after the seminal review of B. Mandelbrot on fBm [4], pertains to the class of additive models: A large step (or increment) can be obtained as the sum of many different small steps, yet sharing the same statistical properties. This class is hence deeply associated with selfsimilarity. Often, this class is confused with its emblematic but restrictive representative fBm, the only Gaussian selfsimilar process with stationary increments. The Gaussian nature of this process and its fully parametrized formulation lead to the formulation of various parametric estimation of its selfsimilarity exponent H. Such parametric estimation are sometimes incorrectly considered as a multifractal analysis of data. This is inaccurate since multifractal analysis aims at measuring much richer properties of data than the sole selfsimilarity and, above all, does not rely on the assumption that data are selfsimilar or Gaussian. Multifractal analysis thus has a much broader scope and can also be applied to limits of other random walks, such as Lévy processes (and, more generally, jump processes), see [21] for overviews on the mathematical side.

The second class, multiplicative cascades, is usually considered as that of archetypal multifractal processes, in the sense that their (scaling functions and) multifractal spectra can be computed theoretically and found to often consist of *continuous, smooth bell-shaped* functions. Therefore, for such processes, multifractal analysis unveils information that can not easily be obtained with other analysis tools (as opposed to the previous class of selfsimilar random walks).

Furthermore, as mentioned earlier, and again following one of B. Mandelbrot's intuitions, this particular class can be extended by combinations (or subordinations) of selfsimilar process with multiplicative cascades (for instance fBm in multifractal time mentioned above, or Lévy processes in multifractal time, as studied by J. Barral and S. Seuret [35, 36]).

It is however clear that these examples do not even give a vague idea of the tremendous variety of all possible multifractal functions. A few mathematical results actually support this idea (and allow to transform it into a precise mathematical statement). Generically, *most* functions in a given function space are multifractal and satisfy the multifractal formalism, see [185, 186] for precise statements. Therefore, multifractality is not a rare property, and is certainly not restricted to the few examples which have been investigated up to now.

The classical opposition between monofractal versus multifractal processes, often used in practical data analysis, is not well grounded and somehow irrelevant. Confusion stems from the fact that Gaussian fBm is characterized by a single Hölder parameter $h_f(x) = H$ and is hence monohölder (its spectrum $D_f(h)$ is supported by a single point). But selfsimilarity does not in general imply this monohölder property. Instead, the classification of data might be by opposing additive to multiplicative structure. Indeed, the physics (or the biology or else) underlying data production may significantly differ depending on whether it is driven by additive mechanisms, or by multiplicative ones. In that respect, multifractal analysis may help as the multifractal spectra of selfsimilar random walks generally differ from those of multiplicative constructions. However, in opposition to what is often used in applications, multifractal analysis per se, understood as the measurement of the scaling function and multifractal spectrum, is not sufficient for proving that there exist any additive or multiplicative structures underlying the data. Multifractal analysis does not aim at stating definite answers to fuzzy questions such as: Are data following a cascade model or not? but, instead, it provides quantitative answers to more restricted questions such as: Is $h_f^{min} > 0$? Is $\zeta_f(2) > 1$? In turns, these precise measurements can contribute to help practitioners to formulate hypotheses regarding data, or to classify them.

Practical multifractal analysis: Signal and image processing. As stated in the previous paragraph, practical multifractal analysis amounts to measuring from data multifractal attributes such as $\eta_f(p), h_f^{min}, \zeta_f(p), c_m, L_f(h)$. These quantities can be computed from data without assuming any a priori data model. In other words, data to which multifractal analysis is applied need not stem from any exact selfsimilar random walk or multiplicative construction. Multifractal analysis is hence, by nature, a non parametric analysis tool that can be applied to any type of data, be they (multi)fractal or not. Whether $L_f(h)$ can be related with $D_f(h)$ or not does not prevent practitioners to use these *multifractal* attributes as tools for classification, diagnosis, detection, or else. This is notably the case in image processing, where it is very unlikely that databases containing images of widely different natures can be associated with specific construction models (such as random walks or cascades).

Furthermore, in practice, one question is always central and was already raised by B. Mandelbrot: What multiresolution quantity should one start with? Increments, oscillations, wavelet coefficients, wavelet leaders? In this respect, the present contribution aims at providing clear and general answers. First, increments and oscillations only address the restricted case where $0 \leq h_f(t) < 1, \forall t$; while wavelet based quantities enable to bypass that restriction by selecting smooth enough wavelets with a large enough number of vanishing moments. Therefore, wavelet coefficients generalize increments and the wavelet leaders constitute the generalized counterpart of oscillations. Second, wavelet coefficients and wavelet leaders should not be opposed. Wavelet coefficients enable to measure information in data such as the uniform Hölder exponent and a global selfsimilarity type scaling exponent; they should hence always be applied first to data. Wavelet leaders should be applied next, when relevant ($h_f^{min} > 0$) and when seeking for additional and more refined analysis of scaling properties in data, via a full collection of scaling exponents $\zeta_f(p)$, or equivalently c_m, or equivalently $L_f(h)$.

Multifractal toolbox. The practical solutions proposed here in terms of discrete wavelet transform coefficients and leaders are believed to be one of the methods enabling the best practical achievements in terms of combining firm mathematical grounding, satisfactory estimation performance, low implementation and computational costs, robustness and versatility. It can be accompanied with non parametric bootstrap procedures, providing confidence intervals for estimates and hypothesis test procedures. Technical details are reported in [41] and *Matlab* toolboxes implementing these tools are made publicly available by the authors.

Bibliography

[1] B. Mandelbrot, *Fractals: Form, Chance and Dimension* (W. H. Freeman, San Francisco, 1977).
[2] B. Mandelbrot, *The Fractal Geometry of Nature* (W. H. Freeman, San Francisco, 1982).
[3] A. N. Kolmogorov, *Dokl. Akad. Nauk SSSR* **26** (1940) 115.
[4] B. Mandelbrot, J. van Ness, *SIAM Reviews* **10** (1968) 422.
[5] B. Mandelbrot, W. Wallis, *Water Resources Research* **4** (1968) 909.
[6] P. Abry and P. Gonçalvès, J. Lévy-Véhel, eds. *Lois d'échelle, fractale et ondelettes, Vols. 1 & 2* (Hermès Scientific Publications, 2002).
[7] S. Cohen, J. Istas, *Fractional Fields and Applications*, Springer-Verlag, Berlin, 2013.
[8] P. Embrechts, M. Maejima, *Selfsimilar Processes* (Princeton Univ. Pr., Princeton, 2001).
[9] G. Samorodnitsky, M. Taqqu, *Stable non-Gaussian Random Processes* (Chapman and Hall, New York, 1994).
[10] B. Mandelbrot, *Gaussian Self-Affinity and Fractals* (Springer-Verlag, New York, 2002).
[11] P. Abry, B. Pesquet-Popescu, M. S. Taqqu, in *Int. Conf. on Signal Proc., 16th World Computer Congress* (Beging, 2000).
[12] V. Pipiras, M. S. Taqqu, P. Abry, *Statistics & Probability Letters* **64** (2003) 147.
[13] V. Pipiras, M. S. Taqqu, P. Abry, *Bernoulli* **13** (2007) 1091.
[14] S. Stoev, V. Pipiras, M. S. Taqqu, *Signal Processing* **82** (2002) 1873.
[15] P. Flandrin, *IEEE Trans. Info. Theory* **35** (1989) 197.
[16] P. Flandrin, *IEEE Trans. Info. Theory* **38** (1992) 910.
[17] U. Frisch, *Turbulence, the Legacy of A.N. Kolmogorov* (Addison-Wesley, 1993).
[18] A. N. Kolmogorov, *J. Fluid Mech.* **13** (1962) 82.
[19] A. M. Obukhov, *J. Fluid Mech.* **13** (1962) 77.
[20] A. M. Yaglom, *Dokl. Akad. Nauk. SSR* **166** (1966) 49.
[21] J. Barral, J. Berestycki, J. Bertoin, A. Fan, B. Haas, S. Jaffard, G. Miermont, J. Peyrière, *Quelques interactions entre analyse, probabilités et fractals* (S.M.F, 2010).
[22] J. Barral, J. Peyrière, this volume.
[23] J.-P. Kahane, J. Peyrière, *Adv. Math.* **22** (1976) 131.
[24] L. F. Richardson, *Weather Prediction by Numerical Process* (Cambridge Univ. Pr., Cambridge, 1922).
[25] B. B. Mandelbrot, *J. Fluid Mech.* **62** (1974) 331.
[26] B. B. Mandelbrot, in *Frontiers of Physics, Proc. Landau Memorial Conf., Tel Aviv, 1988*, E. A. Gotsman, *et al.*, eds. (Pergamon Press, Oxford, 1990), pp. 309–340.

[27] B. B. Mandelbrot, in *Statistical Models and Turbulence, Lecture Notes in Physics*, M. Rosenblatt, C. Van Atta, eds. (Springer-Verlag, New York, 1972), pp. 333–351.

[28] Z. S. She, E. Lévêque, *Phys. Rev. Lett.* **72** (1994) 336.

[29] E. A. Novikov, *Phys. Rev. E* **50** (1994) R3303.

[30] J. Barral, B. B. Mandelbrot, *Probability Theory and Related Fields* **124** (2002) 409.

[31] B. B. Mandelbrot, *Sci. Am.* **280** (1999) 70.

[32] E. Bacry, J. Delour, J. F. Muzy, *Phys. Rev. E* **64** (2001) 026103.

[33] E. Bacry, J. F. Muzy, *Phys. Rev. E* **66** (2002) 056121.

[34] E. Bacry, J. F. Muzy, *Commun. Math. Phys.* **236** (2003) 449.

[35] J. Barral, S. Seuret, *Adv. Math.* **214** (2007) 437.

[36] J. Barral, S. Seuret, *Adv. Math.* **220** (2009) 936.

[37] P. Chainais, R. Riedi, P. Abry, *IEEE Trans. Info. Theory* **51** (2005) 1063.

[38] P. Chainais, R. Riedi, P. Abry, *Traitement du Signal* **22** (2005) 27.

[39] A. N. Kolmogorov, *Comptes Rendus De L'Academie Des Sciences De L'Urss* **30** (1941) 301.

[40] S. Jaffard, B. Lashermes, P. Abry, in *Wavelet Analysis and Applications*, T. Qian, M. I. Vai, X. Yuesheng, eds. (Birkhäuser, Basel, 2006), pp. 219–264.

[41] H. Wendt, P. Abry, S. Jaffard, *IEEE Signal Processing Mag.* **24** (2007) 38.

[42] J. Barral, N. Fournier, S. Jaffard, S. Seuret, *Ann. of Prob.* **38** (2010) 1924.

[43] R. H. Riedi, in *Theory and Applications of Long Range Dependence* P. Doukhan, G. Oppenheim, M. S. Taqqu, eds. (Birkhäuser, Basel, 2003) pp. 625–717.

[44] B. Lashermes, P. and Abry, P. Chainais, *Int. J. on Wavelets, Multiresolution and Information Processing* **2** (2004) 497.

[45] H. Hurst, *Trans. Amer. Soc. Civil Eng.* **116** (1951) 770.

[46] J. Beran, *Statistics for Long-Memory Processes*, Chapman & Hall, 1994.

[47] B. Mandelbrpot, *Multifractals and 1/f Noise*, Springer-Verlag, New York, 1999.

[48] B. Mandelbrot, H. Wallis, *Water Resources Res.* **5** (1969) 967.

[49] D. Hernad, M. Mouillard, D. Strauss-Kahn, *Rev. Stat. Appl.* **26** (1978) 61.

[50] E. Peters, *Fractal Market Analysis*, Wiley, New York, 1994.

[51] A. Tewfik, M. Kim, *IEEE Trans. Info. Theory* **2** (1992) 904.

[52] P. Abry, P. Gonçalvès, P. Flandrin, *Wavelets, Spectrum Estimation and 1/f Processes*, Springer-Verlag, New York, 1995.

[53] P. Abry, D. Veitch, *IEEE Trans. Info. Th.* **44** (1998), 2.

[54] D. Veitch, P. Abry, *IEEE Trans. Info. Th.* **45** (1999) 878.

[55] D. Veitch, P. Abry, *IEEE Trans. Signal Proces.* **49** (2001) 2325.

[56] H.-O. Peitgen, D. Saupe, *The Science of Fractal Images*, Springer-Verlag, Berlin, 1988.

[57] M. Schroeder, *Fractals, Chaos, Power Laws: Minutes from an Infinite Paradise*, Freeman, New York, 1991.
[58] G. Edgar (ed.), *Classics on Fractals*, Cambridge Univ. Pr., Cambridge, 1995.
[59] K. Falconer, *Fractal Geometry*, Wiley, Chichester, 1990.
[60] B. Mandelbrot, *Fractals and Chaos. The Mandelbrot Set and Beyond*, Springer-Verlag, New York, 2004.
[61] K. Falconer, *Fractal Geometry. Mathematical Foundations and Applications* Wiley, Chichester, 1993.
[62] K. Falconer, *Techniques in Fractal Geometry*, Wiley, 1997.
[63] B. Mandelbrot, Pages 3–15 of *Fractals in Physics*, L. Pietronero, E. Tosatti, eds., North-Holland, 1986.
[64] B. Mandelbrot, Pages 17–20 of *Fractals in Physics*, L. Pietronero, E. Tosatti, eds., North-Holland, 1986.
[65] B. Mandelbrot, *Physica Scripta* **32** (1985) 257.
[66] S. Jaffard, *Banach Center Pub. Approximation and Probability* **72** (2006) 93.
[67] S. Jaffard, *Ann. Fac. Sci. Toul.* **15** (2006) 3.
[68] S. Jaffard, P. Abry, S. Roux, *Mathematical Image Processing* **5** (2011) 1.
[69] S. Jaffard, C. Melot, *Commun. Math. Phys.* **258** (2005) 513.
[70] S. Jaffard, *Prob. Theory Related Fields* **114** (1999) 207.
[71] A. Durand, S. Jafard, *Prob. Theory Related Fields* **153** (2012) 45.
[72] N. Kôno, M. Maejima, *Tokyo J. Math.* **14** (1991) 93.
[73] N. Shieh, Y. Xiao, *Bernoulli* **16** (2010) 926.
[74] L. Rudin, S. Osher, E. Fatemi, *Physica D* **60** (1992) 259.
[75] S. Mallat, *A Wavelet Tour of Signal Processing*, Academic Pr., San Diego, 1998.
[76] P.-G. Lemarié Rieusset, *Comptes Rendus Acad. Sci. (Paris)* **318** (1994) 1083.
[77] I. Daubechies, *Comm. Pure App. Math.* **41** (1988) 909.
[78] Y. Meyer, *Ondelettes et Opérateurs*, Hermann, Paris, 1990.
[79] S. Jaffard, *SIAM J. of Math. Anal.* **28** (1997) 944.
[80] S. Jaffard, B. Lashermes, P. Abry, Pages 219–264, *Wavelet Analysis and Applications*, T. Qian, M. I. Vai, X. Yuesheng, eds., Birkhäuser Verlag, Basel, 2006.
[81] Y. Gousseau, J.-M. Morel, *SIAM J. Math. Anal.* **33** (2001) 634.
[82] V. Pipiras, P. Abry, *Sig. Proc.* **86** (2006) 2326.
[83] V. Pipiras, *J. Four. Anal. Appl.* **10** (2004) 599.
[84] V. Pipiras, M. Taqqu, Pages 165–201 of *Fractional calculus and its connexions to fractional Brownian motion*, P. Doukhan, G. Oppenheim, M. Taqqu, eds., Birkhäuser, Basel, 2003.
[85] S. Jaffard, P. Abry, S. Roux, B. Vedel, H. Wendt, Pages 51–89 of *Wavelet Methods in Mathematical Analysis and Engineering*, A. Damlamian, S. Jaffard, eds. World Scientific, Singapore, 2010.

[86] B. Lashermes, S. Jaffard, P. Abry, *IEEE Int. Conf. Acoustics, Speech, and Signal Proc.* **5** (2005) 161.

[87] M. Clausel, S. Nicolay, *J. Fourier Anal. Appl.* **18** (2012) 750.

[88] S. Jaffard, Pages 91–152 of *Fractal Geometry and Applications: A Jubilee of Benoît Mandelbrot*, M. Lapidus, M. van Frankenhuijsen, eds., Amer. Math. Soc., Providence, 2004.

[89] S. Jaffard, *Compt. Rend. Acad. Scien.* **326** (1998) 555.

[90] S. Jaffard, *Constr. Approx.* **21** (2005) 29.

[91] S. Jaffard, *J. Math. Phys.* **39** (1998) 4129.

[92] U. Frisch, G. Parisi, Pages 84–88 of *Turbulence and Predictability in Geophysical Fluid Dynamics and Climate Dynamics*, M. Ghil, R. Benzi, G. Parisi, eds., North-Holland, Amsterdam, 1985.

[93] G. Brown, G. Michon, J. Peyrière, *J. Stat. Phys.* **66** (1992) 775.

[94] T. Halsey, M. Jensen, L. Kadanoff, I. and Procaccia, B. Shraiman, *Phys. Rev. A* **33** (1986) 1141.

[95] P. Abry, S. Jaffard, S. Roux, B. Vedel, H. Wendt, Pages 1–20 of *Unexploded Ordnance Detection and Mitigation*, J. Byrnes ed. Springer, New York, 2008.

[96] J. Barral, P. Gonçalves, *J. Stat. Phys.* **144** (2011) 1256.

[97] A. Arneodo, B. Audit, N. Decoster, J.-F. Muzy, C. Vaillant, Pages 27–102 of *The Science of Disasters*, J. Kropp, H.J. Schellnhuber, eds., Springer-Verlag, New York, 2002.

[98] A. Arneodo, E. Bacry, J. Muzy, *Physica A* **213** (1995) 232.

[99] J. Muzy, E. Bacry, A. Arneodo, *Phys. Rev. Lett.* **67** (1991) 3515.

[100] J. Muzy, E. Bacry, A. Arneodo, *J. Bifurcation and Chaos* **4** (1994) 245.

[101] J. Aubry, S. Jaffard, *Commun. in Math. Phys.* **227** (2002) 483.

[102] J. Barral, S. Seuret, *Adv. Math.* **226** (2011) 3191.

[103] A. Durand, *Comm. Math. Phys.* **283** (2008) 451.

[104] H. Wendt, P. Abry, Pages 3465–3458 in *Proc. IEEE Int. Conf. Acoust., Speech, and Signal Proc.*, IEEE Las Vegas, 2008.

[105] B. Castain, *J. Phys. II* **6** (1996) 105.

[106] B. Castaing, Y. Gagne, M. Marchand, *Physica D* **68** (1993) 387.

[107] Delour, J. Muzy, A. Arneodo, *Euro. Phys. Jour. B* **23** (2001) 243.

[108] H. Wendt, *Contributions of Wavelet Leaders and Bootstrap to Multifractal Analysis: Images, Estimation Performance, Dependence Structure and Vanishing Moments. Confidence Intervals and Hypothesis Tests*, Ph.D. thesis, Ecole Normale Supérieure de Lyon, 2008.

[109] H. Wendt, S. Roux, P. Abry, S. Jaffard, *Signal Proces.* **89** (2009) 1100.

[110] D. Monrad, L. Pitt, Pages 163189 of *Progress in Probability and Statistics, Seminar on Stochastic Processes 1986*, E. Cinlar, K.L. Chung, R.K. Getoor, eds., Birkhäuser, Boston, 1986.

[111] A. Monin, A. Yaglom, *Statistical Fluid Mechanics*, MIT Press, Cambridge, 1975.

[112] J.-L. Starck, F. Murtagh, A. Bijaoui, *Image Processing and Data Analysis: The Multiscale Approach*, Cambridge Univ. Pr. Cambridge, 1998.

[113] V. Bareikis, R. Katilius, *Proceedings of the 13th International Conference on Noise in Physical Systems and 1/f Fluctuations, Palanga, Lithuania*, World Scientific, Singapore. 1995.
[114] M. Keshner, *Proc. IEEE* **70** (1982) 212.
[115] A. Arneodo, S. Roux, M. Decoster, *Experimental Chaos, AIP Conf. Proc.* **622** (2002) 80.
[116] P. Ivanov, *IEEE Eng. In Med. and Biol. Mag.* **26** (2007) 33.
[117] P. Ivanov, L. Nunes Amaral, A. Goldberger, S. Havlin, M. Rosenblum and Z. Struzik, H. Stanley, *Nature* **399** (1999) 461.
[118] K. Kotani, K. Takamasu, L. Safonov, Y. Yamamot, *Proc. SPIE* **5110** (2003) 340.
[119] S. Lowen, M. Teich, *Fractal-Based Point Processes*, Wiley, Hoboken, 2005.
[120] L. Liebovitch, A. Todorov, *J. Appl. Physiol.* **80** (1996) 1446.
[121] P. Ciuciu, P. Abry, C. Rabrait, H. Wendt, *IEEE J. Selected Topics in Signal Proces.* **2** (2009) 929.
[122] B. He, *Neuron* **66** (2010) 353.
[123] E. Foufoula-Georgiou, P. Kumar, *Wavelets in Geophysics*, Academic Press, San Diego, 1994.
[124] K. Park, W. Willinger, Pages 1–38 of *Self-Similar Network Traffic and Performance Evaluation*, K. Park, W. Willinger, eds., Wiley, New York, 2000.
[125] A. Dauphiné, *Géographie Fractale*, Hermès, Lavoisier, Paris, 2011.
[126] P. Frankhauser, *Population* **4** (1997) 1005.
[127] L. Calvet, A. Fisher, *Multifractal Volatility: Theory, Forecasting and Pricing*, Academic Pr., San Diego, 2008.
[128] E. Bacry, A. Kozhemyak, J. Muzy, *J. Economic Dynamics and Control* **32** (2008) 156.
[129] E. Bacry, A. Kozhemyak, J. Muzy, in *Encyclopedia of quantitative finance* Wiley, New York, 2010.
[130] L. Calvet, A. Fisher, B. Mandelbrot, *The multifractal model of asset returns*, Cowles Foundation Discussion Papers: 1164, 1997.
[131] G. Chavarria, C. Baudet, S. Ciliberto, *Phys. Rev. Lett.* **74** (1995) 1986.
[132] A. Kolmogorov, Pages 151–161 of *Turbulence, Classic Papers on Statistical Theory*, S. Friedlander, L. Topper, eds., Interscience, New York, 1941.
[133] A. Arneodo *et al.*, *Europhys. Lett.* **34** (1996) 411.
[134] A. Arneodo, S. Manneville, J. Muzy, *Eur. Phys. J. B* **1** (1998) 129.
[135] B. Castaing, Y. Gagne, E. Hopfinger, *Physica D* **46** (1990) 177.
[136] B. Castaing, Y. Gagne, M. Marchand, *Physica D* **68** (1993) 387.
[137] A. Chhabra *et al.*, *Phys. Rev. A* **40** (1989) 1327.
[138] S. Pietropinto, *et al.*, *Physica C* **386** (2003) 512.
[139] B. Lashermes, S. Roux, P. Abry, S. Jaffard, *Euro. Phys. J. B* **61** (2008) 201.
[140] C. Meneveau, *J. Fluid Mech.* **232** (1991) 469.
[141] L. Bachelier, *Annales de l'École Normale Supérieure* **3** (1900) 21.
[142] C. Granger, *J. of Econometrics* **14** (1980) 220.

[143] C. Granger, R. Joyeux, *Journal of Time Series Analysis* **1** (1980) 15.
[144] I. Dittman, C. Granger, *J. of Econometrics* **110** (2002) 113.
[145] G. Teyssière, P. Abry, Pages 173–238 of *Long Memory in Econometrics*, G. Teyssière, A. Kirman, eds., Springer-Verlag, Berlin, 2007.
[146] B. Mandelbrot, *Mandelbrot1997*, Springer-Verlag, New York, 1997.
[147] B. Mandelbrot, *J. of Business* **36** (1963) 394.
[148] B. Mandelbrot, *Rev. of Econometrics and Stat.* **53** (1971) 225.
[149] A. Fisher, L. Calvet, B. Mandelbrot, *Multifractality of Deutsche Mark/US Dollar Exchange Rate*, Cowles Foundation Discussion Paper: 1165, 1997.
[150] E. Bacry, L. Duvernet, J. Muzy, *J. Appl. Prob.* **49** (2012) 482.
[151] L. Calvet, A. Fisher, *Rev. Econ. and Stat.* **84** (2002) 381.
[152] J. Barral, X. Jin, B. Mandelbrot, *Ann. Appl. Prob.* **20** (2010) 1205.
[153] J. Barral, B. Mandelbrot, *Probab. Theory Rel.* **124** (2002) 409.
[154] S. Seuret, *Adv. Math.* **220** (2009) 936.
[155] S. Jaffard, B. Mandelbrot, *Math. Intell.* **19** (1997) 21.
[156] P. Abry, *et al.*, *32nd Annual Int. Conf. IEEE Eng. in Medicine and Biology, Buenos Aires* (2010) 106.
[157] M. Doret, H. Helgason, P. Abry, P. Gonçalvès, C. Gharib, P. Gaucherand, *Am. J. Perinatol.* **28** (2011) 259.
[158] S. Akselrod, D. Gordon, F. Ubel, D. Shannon, A. Berger, R. Cohen, *Science* **213** (1981) 220.
[159] MAWI Traffic Archive, http://mawi.wide.ad.jp/mawi/
[160] P. Borgnat, G. Dewaele, K. Fukuda, P. Abry, K. Cho, *INFOCOM2009*, IEEE 711.
[161] J. Berger, B. Mandelbrot, *IBM J. Res. Dev.* **7** (1963) 224.
[162] W. Leland, M. Taqqu, W. Willinger, D. Wilson, *ACM/IEEE Transactions on Networking*, **2** (1994) 1.
[163] W. Leland, M. Taqqu, W. Willinger, D. Wilson, Pages 183–193 of *SIGCOMM '93: Conf. Proc. on Communications Architectures, Protocols and Appl.* ACM Press, New York, 1993.
[164] K. Park, W. Willinger, Pages 1–33 in *Self-Similar Network Traffic and Performance Evaluation*, K. Park, W. Willinger, eds., Wiley, New York, 2000.
[165] V. Paxson, S. Floyd, *IEEE/ACM Transactions on Networking* **3** (1995) 226.
[166] P. Abry, R. Baraniuk, P. Flandrin, R. Riedi and D. Veitch, *IEEE Signal Processing Magazine* **3** (2002) 28.
[167] A. Feldmann, A. Gilbert, W. Willinger, *SIGCOM98*, (1998) 42.
[168] H. Hohn, D. Veitch, P. Abry, *Computer Network J.* **48** (2005) 293.
[169] R. Riedi, M. Crouse, V. Ribeiro, R. Baraniuk, *IEEE Trans. Info. Theory* **45** (1999) 992.
[170] K. Cho, K. Mitsuya, A. Kato, Cho, K., K. Mitsuya, and A. Kato. *Traffic data repository maintained by the MAWI Working Group of the WIDE Project.* URL http://mawi.wide.ad.jp/mawi, (2000).

[171] G. Dewaele, K. Fukuda, P. Borgnat, P. Abry, K. Cho, *ACM SIGCOMM 2007 Workshop on Large-Scale Attack Defense* (2007) 145.
[172] M. Taqqu, W. Willinger, R. Sherman, *SIGCOMM Comput. Commun. Rev.* **27** (1997) 5.
[173] P. Abry, P. Borgnat, F. Ricciato, A. Scherrer, D. Veitch, *Telecom. Syst., Spec. Issue on Traffic Modeling, Its Computations and Applications* **43** (2009) 147.
[174] N. Hohn, D. Veitch, P. Abry, *IEEE Trans. Signal Proc.* **51** (2003) 2229.
[175] P. Loiseau, P. Gonçalves, G. Dewaele, P. Borgnat, P. Abry, P. Primet, *IEEE/ACM Trans. on Networking* **18** (2010) 1261.
[176] B. Mandelbrot, *J. Fluid Mech.* **72** (1975) 401.
[177] B. Mandelbrot, *Comptes Rendus Acad. Sci. (Paris)* **280** (1975) 1075.
[178] H. Biermé, F. Richard, *J. Math. Imaging Vision* **36** (2010) 227.
[179] D. Schertzer, S. Lovejoy, *J. Geophys. Res.* **92** (1987) 9693.
[180] A. Arneodo, N. Decoster, P. Kestener, S. Roux, Pages 1–98 of *Advances in Imaging and Electron Physics*, P. Hawkes, B. Kazan, T. Mulvey, eds., Academic Press, New York 2003.
[181] H. Wendt, P. Abry, S. Jaffard, H. Ji, Z. Shen, Pages 3829–3832 of *Proc. IEEE Int. Conf. Image Proc. (ICIP)*, IEEE, 2009.
[182] P. Abry, S. Jaffard, H. Wendt, *Signal Processing* **93** (2013) 554.
[183] S. Orey, S. Taylor, *Proc. London Math. Soc.* **28** (1974) 174.
[184] E. Perkins, *Z. Wahrsch. Verw. Gebiete* **64** (1983) 369.
[185] A. Fraysse, S. Jaffard, *Revista Matematica Iberoamericana* **22** (2006) 663.
[186] S. Jaffard, *J. Math. Pures et Appl.* **79** (2000) 525.

Chapter 4

Three-dimensional fractal homeomorphisms

Michael F. Barnsley and Brendan Harding

Abstract. We provide a simple introduction to fractal transformations and fractal homeomorphisms. We introduce tri-affine iterated function systems on $\mathbb{R}^3$, and illustrate associated fractal homeomorphisms applied to three dimensional graphical data sets. We also comment on the algorithms used.

4.1 Introduction

Fractal transformations relate the attractors of pairs of hyperbolic iterated function systems. They are defined using code spaces and masks. They are interesting and useful because they have the following properties: (i) they may alter the dimensions of sets and measures, even when the attractors themselves have integer dimension; (ii) they may transform geometrically simple pictures and objects into complicated ones; (iii) they can convert deterministic fractal objects into seemingly random fractal objects; and

(iv) they can be quite simple to handle and compute. Previously published related work [1–5] comprise mathematical underpinnings and generalizations, and applications to digital imaging [6] and to artwork, via synthesis of fractal transformation and digital photography [7, 8].

We believe that fractal transformation provides a reasonable continuation of part of Benoit B. Mandelbrot's vision of the fractal geometry of nature, by extending his idea of the iterative formation of self-referential objects to the usage of inherent transformations between such objects, to define new geometrical models for nature. An example of a fractal transformation of a classical Euclidean object, a set of concentric annuli, is shown in Fig. 4.1. An example of a self-similar two dimensional set before and after a fractal transformation is shown in Fig. 4.2. Fractal transformation is a fascinating, beautiful subject for theoretical and experimental investigation: mathematically one may seek topological and measure-theoretic invariants, while experimentally one may seek to characterize fractal transformations that preserve given classes of objects, for example, ones that map pictures of clouds to pictures of clouds or snowflakes to snowflakes.

So in this brief paper, written in honor of Benoit, we explain how simple fractal transformations work, we introduce tri-affine functions, and use them to construct three-dimensional fractal homeomorphisms. Since

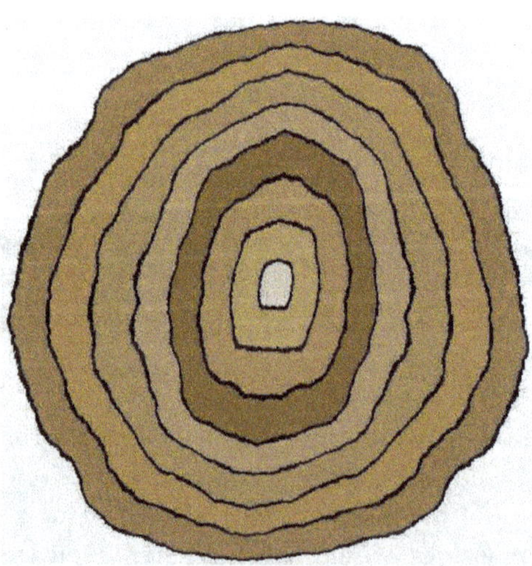

Figure 4.1 The result of applying a simple fractal homeomorphism to a set of colored concentric annuli. The result is far from being a classical Euclidean object.

Figure 4.2 The image on the left represents the attractor of an IFS of three similitudes. On the right is shown the result of applying a simple fractal homeomorphism. Both represent deterministic fractals, but the one on the right appears to be more random.

the mathematical formalization and algorithms are direct generalizations of previous work, mainly two-dimensional, here we provide a basic introduction to three-dimensional systems which we illustrate. We include some comments on computational algorithms.

We think that Benoit would have approved of this approach. This paper was written in his memory. He inspired us with his vision: we seek to describe the world around us in the language of fractal geometry.

4.2 Development of a simple example of fractal homeomorphism

4.2.1 Iterated function systems and their attractors

Throughout this paper a (hyperbolic) *iterated function system* (IFS) is a finite set of strictly contractive injective transformations acting on a complete metric space. A typical IFS is denoted

$$F = (X, f_1, f_2, \ldots, f_N) \tag{4.1}$$

where X is a complete metric space, N is a positive integer, and $f_n : X \to X$ for $n = 1, 2, \ldots, N$ are injective contractions. An example of an IFS is

$$(\mathbb{R}, f_1(x) = x/2, f_2(x) = x/2 + 1/2). \tag{4.2}$$

It is well known [9] that the IFS F possesses a unique *attractor*, a compact nonempty set $A \subset X$, with the property

$$A_F = \bigcup_{n=1}^{N} f_n(A).$$

For the example IFS in (4.2), the attractor is $A = [0,1] \subset \mathbb{R}$. One goal in this Sec. 4.2 is to explain how to construct a fractal transformation from $[0,1]$ to itself. We note in advance that although the venue of this transformation is a classical Euclidean line segment, how it behaves is fractal and rough. To construct this transformation, and fractal transformations in general, we need the concepts of *code space, coding map, shift-invariant section of an IFS, address space,* and *mask.*

4.2.2 Code spaces and coding maps

Let $I = \{1, 2, \ldots, N\}$ and let I^∞ be the set of all infinite sequences of symbols in I. For $\sigma \in I^\infty$ we write $\sigma = \sigma_0\sigma_1\sigma_2\ldots$ where $\sigma_k \in \{1, 2, \ldots, N\}$ for all $k = 0, 1, 2, \ldots$. We endow I^∞ with the metric $d : I^\infty \times I^\infty \to [0, \infty)$ defined by $d(\sigma, \omega) = 0$ if $\sigma = \omega$ and, otherwise, $d(\sigma, \omega) = 1/2^k$ where k is the least index such that $\sigma_k \neq \omega_k$. Equipped with this metric, I^∞ is a compact metric space that we refer to as *the code space for the IFS F*.

It is well known [9] that the IFS F possesses a *coding map*

$$\pi_F : I^\infty \to A$$

with these properties: (i) π_F is continuous; (ii) π_F is onto; (iii) the following diagram commutes, for $n = 1, 2, \ldots, N$,

$$\begin{array}{ccc} I^\infty & \overset{s_n}{\rightarrow} & I^\infty \\ \pi_F \downarrow & & \downarrow \pi_F \\ A_F & \underset{f_n}{\rightarrow} & A_F \end{array} \tag{4.3}$$

where $s_n : I^\infty \to I^\infty$ is defined by $s_n(\sigma) = \omega$ where $\omega_0 = n$ and $\omega_{k+1} = \sigma_k$ for $k = 0, 1, \ldots$. See [10] for discussion of relationships between the topology of an attractor and the equivalence class structure induced by a coding map.

For example, for the IFS in (4.2), the coding map is $\pi : \{1, 2\}^\infty \to [0, 1]$ with

$$\pi(\sigma) = \sum_{n=0}^{\infty} \frac{(\sigma_k - 1)}{2^{k+1}}.$$

It is readily verified that π has the properties (i), (ii) and (iii).

In general, the coding map for the IFS F in (4.1) is given by

$$\pi_F(\sigma) = \lim_{k\to\infty} f_{\sigma|_k}(x)$$

where $x \in X$ is fixed, $\sigma|_k = \sigma_0\sigma_1\cdots\sigma_k$, and $f_{\sigma_0\sigma_1\cdots\sigma_k} = f_{\sigma_0} \circ f_{\sigma_1} \circ \cdots \circ f_{\sigma_k}$ where $\circ$ means composition of functions.

4.2.3 Sections and address spaces

Concerning the IFS F, we say that $\sigma \in I^\infty$ is *an address* of $x \in A_F$ when $\pi_F(\sigma) = x$.

We say that $\tau_F : A_F \to I^\infty$ is a *section* for F when

$$\pi_F \circ \tau_F(x) = x \quad \text{for all } x \in A_F,$$

that is, when $\pi_F \circ \tau_F$ is the identity map on A. For example, a section of the IFS in (4.2) is $\tau : [0,1] \to \{1,2\}^\infty$ such that

$$\tau(1) = \overline{2}, \quad \text{and for } x \in [0,1), \tau(x) = \sigma_0\sigma_1\cdots \tag{4.4}$$

where $\sigma_k = x_k{+}1$ for all k and $x = 0.x_0x_1\cdots$ is the unique binary expansion of $x \in (0,1)$ whose tail does not consist of an infinite sequence of twos. We write $\overline{1} = 111111\ldots$, $\overline{12} = 121212\ldots$, and so on. Note that the range of τ is strictly contained in the code space.

The range of the section $\tau_F : A_F \to I^\infty$ for the IFS F is called an *address space* for A_F. We use the notation

$$\Omega_F = \tau_F(A_F).$$

for an address space. For example, the range of the section in (4.4) is the address space

$$\Omega := \tau(A) = \{\sigma \in \{1,2\}^\infty |\ \sigma_k\sigma_{k+1}\cdots \neq \overline{2} \text{ for all } k = 1,2,\ldots\} \cup \{\overline{2}\} \tag{4.5}$$

for the IFS in (4.2). Figure 4.3 illustrates, on the left, the attractor of the IFS in (4.2) together with some addresses belonging to the address space Ω.

4.2.4 Shift-invariant sections

Let $S : I^\infty \to I^\infty$ be the shift operator defined by $S(\sigma) = \omega$ where $\omega_n = \sigma_{n+1}$ for all $n = 0,1,2,\ldots$. We say that the section $\tau_F : A_F \to \Omega_F \subset I^\infty$ is *shift-invariant* when $S(\Omega_F) \subset \Omega_F$. For example, the section $\tau : [0,1] \to \Omega \subset \{1,2\}^\infty$, defined in (4.4), obeys $S(\Omega) = \Omega$ and so it is shift-invariant.

4.2.5 Pictures on attractors

Define a *picture* to be a function of the form

$$P : D \subset \mathbb{R}^n \to \mathfrak{C}$$

where $\mathfrak{C}$ is a color space. The *support* of the picture is the domain D of the function. We are concerned with situations where D is a subset of an attractor of an IFS. For example, Fig. 4.3 illustrates, on the right, a one-dimensional picture supported on the attractor of the IFS in (4.2). The support of this picture is $[0, 1]$.

Below we consider two and three-dimensional pictures and transformations of pictures.

4.2.6 Fractal transformations

We introduce a second IFS

$$G = (X, g_1, g_2, \ldots, g_N). \tag{4.6}$$

For clarity we have supposed that the two IFSs F and G comprise the same number of functions and that the underlying space, X, is the same for both. Let the attractor of G be A_G and let the coding map be $\pi_G : I^\infty \to A_G$.

As in [11] we restrict attention to shift-invariant sections.

Definition 4.2.1 Let F and G be IFSs as above. Let A_F and A_G be the attractors of F and G, respectively. A transformation $T = T_{FG} : A_F \to A_G$ is called a **fractal transformation** if it can be written in the form

$$T_{FG} = \pi_G \circ \tau_F$$

where $\tau_F : A_F \to I^\infty$ is a shift-invariant section for F and $\pi_G : I^\infty \to A_G$ is the coding map for G. A fractal transformation $T : A_F \to A_G$ is called a **fractal homeomorphism** when it is a homeomorphism between the compact metric spaces A_F and A_G.

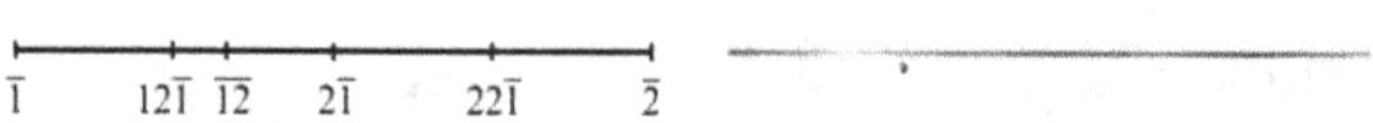

Figure 4.3 (Left) The attractor of the IFS in (4.2) with the addresses of some points marked on it. These addresses belong to the address space in (4.5). (Right) This illustrates a one-dimensional picture, supported on [0,1], the attractor of the IFS in (4.2).

4.2.7 Example of a one-dimensional fractal transformation

As an example, let F be the IFS in (4.2) and let τ be the section for F described in (4.4). Let G be the IFS

$$(\mathbb{R}, g_1(x) = x/3, g_2(x) = 2x/3 + 1/3).$$

The attractor of G is $[0,1]$. The coding map for G is $\pi : \{1,2\}^\infty \to [0,1]$ defined by

$$\pi(\sigma) = \frac{(\sigma_0 - 1)}{3} + \sum_{k=1}^{\infty} \frac{\sigma_0\sigma_1 \cdots \sigma_{k-1}(\sigma_k - 1)}{3^{k+1}}.$$

On the bottom-left in Fig. 4.4 is shown the attractor of A_G with the addresses of some points marked. The corresponding fractal transformation in this example is $T : [0,1] \to [0,1]$ defined by

$$T(x) = \pi \circ \tau. \tag{4.7}$$

We have illustrated the action of T in Fig. 4.4 (Left) showing how points with the same addresses correspond.

4.2.8 When is a fractal transformation a homeomorphism?

The example discussed in Fig. 4.4 is actually a fractal homeomorphism, as a consequence of the following general theorem. We write $\overline{S}$ to denote the closure of a set S.

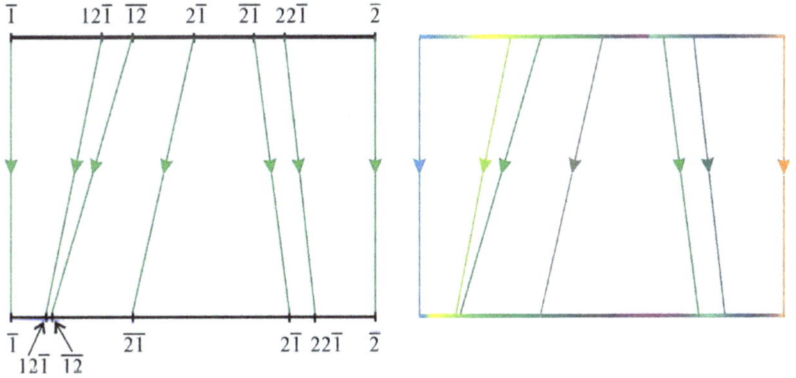

Figure 4.4 (Left) This illustrates the correspondence between the attractors of two one-dimensional IFSs provided by the fractal transformation in equation (4.7). (Right) This illustrates the action of T on a one-dimensional picture.

Theorem 4.2.1 [6] *Let F and G be iterated function systems, as above. Let $\tau_F : A_F \to I^\infty$ be a section for F and let $\pi_G : I^\infty \to A_G$ be the coding map for G. The fractal transformation $T_{FG} = \pi_G \circ \tau_F : A_F \to A_G$ has the following properties.*

(i) *If, whenever $\sigma, \omega \in \overline{\Omega_F}$, $\pi_F(\sigma) = \pi_F(\omega) \Rightarrow \pi_G(\sigma) = \pi_G(\omega)$, then $T_{FG} : A_F \to A_G$ is continuous.*
(ii) *If $\Omega_G := \Omega_F$ is an address space for G, and if, whenever $\sigma, \omega \in \overline{\Omega_F}$, $\pi_F(\sigma) = \pi_F(\omega) \Leftrightarrow \pi_G(\sigma) = \pi_G(\omega)$, then T_{FG} is a homeomorphism and $T_{FG}^{-1} = T_{GF}$.*

4.2.9 How fractal transformations act on pictures

For precision, we describe fractal transformations of (mathematical) pictures as follows. Let F and G be iterated function systems as above, with attractors A_F and A_G contained in $\mathbb{R}^n$. Define a mapping from the space of pictures on A_G into the space of pictures on A_F by

$$P_G \mapsto P_F := P_G \circ \pi_G \circ \tau_F = P_G \circ T_{FG}. \tag{4.8}$$

In other words, the picture P_F is the result of applying the fractal transformation T_{FG} to P_G, and we may write, by slight abuse of notation,

$$P_F = T_{FG}(P_G).$$

In Fig. 4.4(Right) we have illustrated the action of T on the one-dimensional picture in Fig. 4.3. In Secs. 4.3 and 4.4 we apply fractal transformations to two and three-dimensional pictures.

4.2.10 Masks and masked dynamical systems

Let F be an IFS as above. A *mask* for F is a partition of A_F of the form $\{M_1, M_2, \ldots, M_N\}$ such that $M_n \subset f_n(A_F)$ for $n = 1, 2, \ldots, N$. For example, for the IFS (4.2) the only possible masks are $\{M_1 = [0, 1/2], M_2 = (1/2, 1]\}$ and $\{[0, 1/2), [1/2, 1]\}$. Given a mask $\mathcal{M} := \{M_1, M_2, \ldots, M_N\}$ for F we define the corresponding *masked dynamical system* $W_{\mathcal{M}}: A_F \to A_F$ by

$$W_{\mathcal{M}}(x) = \begin{cases} f_1^{-1}(x), & x \in M_1, \\ f_2^{-1}(x), & x \in M_2, \\ \vdots & \\ f_N^{-1}(x), & x \in M_N. \end{cases} \tag{4.9}$$

Since the functions of F are one-to-one, $A_F = \cup f_n(A_F) = \cup M_n$ and $M_n \cap M_m = \emptyset$ for $n \neq m$, it follows that $W_{\mathcal{M}}$ is well defined.

For example, the masked dynamical system $W : [0,1] \to [0,1]$ for the IFS (4.2), with mask $\{M_1 = [0,1/2], M_2 = (1/2,1]\}$, is defined by

$$W(x) = \begin{cases} 2x & x \in [0,1/2], \\ 2x-1, & x \in (1/2,1]. \end{cases} \qquad (4.10)$$

4.2.11 How to construct a shift-invariant section by using a mask

It is shown in [6] that if we are given a mask $\mathcal{M} := \{M_1, M_2, \ldots, M_N\}$ for F then we can define a shift-invariant section $\tau_F : A_F \to I^\infty$ for F by

$$(\tau_F(x))_k = i \quad \text{if } W^k_{\mathcal{M}}(x) \in M_i,$$

where $W^k_{\mathcal{M}}$ means $W_{\mathcal{M}}$ composed with itself k times. In this case we say that τ_F is the shift-invariant section for F derived from the mask $\mathcal{M}$ and we also refer to τ_F as a masked section. It turns out that masked sections and shift-invariant sections are one and the same concept, because of the following theorem.

Theorem 4.2.2 [6, 11] *Let F be an IFS as above. Then τ_F is a shift-invariant section for F if and only if τ_F is a masked section for F.*

For example, the mask $\{M_1 = [0,1/2), M_2 = [1/2,1]\}$ for the IFS (4.2) corresponds to the shift-invariant section defined in (4.4).

4.2.12 Comment: topological conjugacies are interesting objects

We make the following observation. In the theory of dynamical systems it is often the case that one seeks to understand a complicated system by establishing that it is topologically conjugate to a simpler one. Our interest is in the nature of the transformations that achieve such simplifications: the homeomophisms that yields conjugacies can themselves be interesting objects of study. This leads us into the next section where we illustrate two-dimensional fractal homeomorphisms and show by means of an example how they can enrich our view of topologically conjugate dynamical systems.

4.3 Two-dimensional examples, starting from our one-dimensional example.

4.3.1 A family of two-dimensional IFSs

Let $\square = [0,1] \times [0,1] \subset \mathbb{R}^2$ denote the unit square. Throughout this Sec. 4.3 we will take the two IFSs F and G to be

$$F = \left\{ \square : \begin{array}{l} f_1(x,y) = (x/2, y/2), f_2(x,y) = ((x+1)/2, y/2), \\ f_3(x,y) = ((x+1)/2, (y+1)/2), f_4(x,y) = (x/2, (y+1)/2) \end{array} \right\} \tag{4.11}$$

$$G = \left\{ \square : \begin{array}{l} g_1(x,y) = (ax, by), g_2(x,y) = (a + (1-a)x, by), \\ g_3(x,y) = (a + (1-a)x, b + (1-b)y), \\ g_4(x,y) = (ax, b + (1-b)y) \end{array} \right\}$$

where $0 < a, b < 1$. See Fig. 4.5.

Note that F and G are the same when $a = b = 1/2$. The ranges of the functions of each IFS overlap only on the boundaries of their rectangular ranges. To obtain a shift-invariant section τ_F for F we define a mask $\mathcal{M}^F = \{M_1, M_2, M_3, M_4\}$ by

$$\begin{aligned} &M_1 = f_1(\square), \quad M_2 = f_2(\square) \backslash M_1, \\ &M_3 = f_3(\square) \backslash (M_1 \cup M_2), \quad M_4 = f_4(\square) \backslash (M_1 \cup M_2 \cup M_3). \end{aligned} \tag{4.12}$$

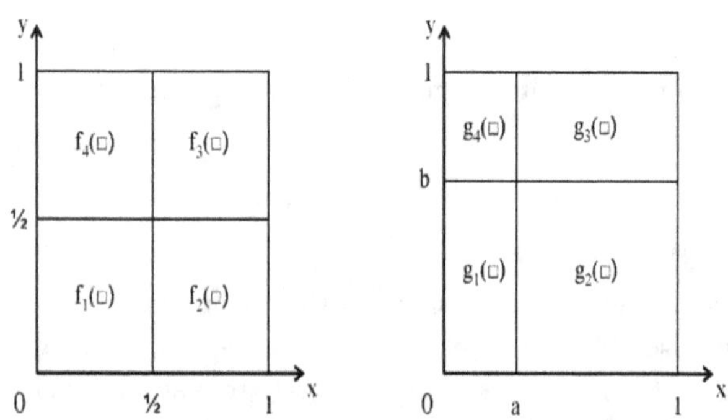

Figure 4.5 These diagrams illustrate the actions of the IFSs, F and G, in equation (4.11) on the unit square. The "address structure" of both IFSs is the same.

Analogously we define a mask $\mathcal{M}^G$ for G. The masks $\mathcal{M}^F$ and $\mathcal{M}^G$ yield shift-invariant sections, τ_F for F and τ_G for G. We call these choices the *top* masks for F and G; they yield what we call the top address spaces for A_F and A_G, see [6].

With the aid of Theorem 4.2.1 it follows that the fractal transformation $T_{FG} = \pi_G \circ \tau_F$ is a homeomorphism, see for example [3]. The inverse of T_{FG} is

$$T_{FG}^{-1} = T_{GF} = \pi_F \circ \tau_G.$$

Note that two masks for F may differ only on the common boundaries of the squares in the left-hand image in Fig. 4.5. In this case, any choice of mask leads to the same fractal homeomorphism. All of the examples in this paper involve IFSs such that the "overlapping regions" $f_i(A_F) \cap f_j(A_F)$ for $i \neq j$ have empty interior. But there exists a great diversity of non-trivial examples for which at least one of these intersections contains a nonempty open set, see [6, 12].

The family of examples $T_{FG} : \square \to \square$ depends on the parameters $0 < a, b < 1$. Each member of the family maps any directed horizontal line segment into a directed horizontal line segment and similarly preserves vertical line segments. As an example, let $a = b = 1/3$. The corresponding homeomorphism $T_{FG} : \square \to \square$ obeys

$$T_{FG} = T \times T, \tag{4.13}$$

where $T\colon [0, 1] \to [0, 1]$ is (4.7), our main example in 4.2.7. This is interesting because $T \times T$, in addition to preserving horizontal and vertical line segments, also maps the diagonal line segment, that joins the points $(0, 0)$ and $(1, 1)$, onto itself. Hence $T \times T$ preserves the well-known "cobweb" diagrams associated with dynamical systems on $[0, 1]$. This is because such diagrams comprise sequences of directed horizontal and vertical line segments, the graph of the dynamical system, and the diagonal. The dynamical system changes, but the space of all cobweb diagrams for one system is mapped homeomorphically onto the space of all cobweb diagrams for the other system. This notion is illustrated in Fig. 4.6.

In Fig. 4.7 we illustrate bi-affine fractal homeomorphisms applied to a digital photograph. In this case the IFSs comprise bi-affine functions [11]. They are two-dimensional versions of the tri-affine functions that we introduce in Sec. 4.4.

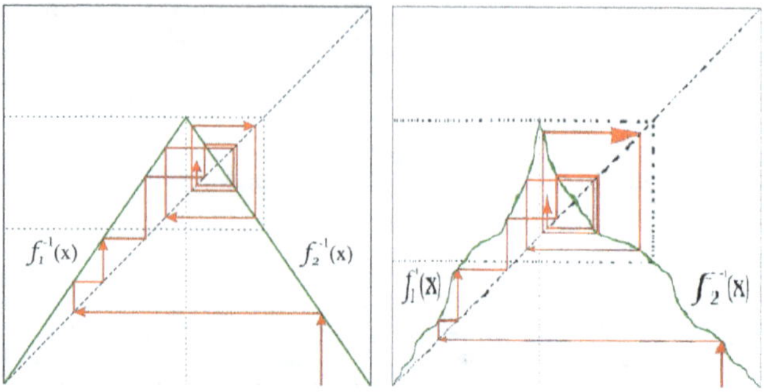

Figure 4.6 Left: a "cobweb" diagram for an orbit of a piecewise linear dynamical system on the unit interval. Right: the result of applying the special fractal homeomorphism $T \times T : \square \to \square$ in equation (4.13). The transformed cobweb is correct, but the dynamical system is now a non-differentiable one! Note the fascinating manner in which the various objects are transformed.

Figure 4.7 (Top) A digital photograph of flowers, sky and mountain. (Middle) A gentle bi-affine fractal homeomorphism has been applied, accentuating the amount of purple, but still quite recognizable. (Bottom) A more extreme bi-affine fractal homeomorphism has been applied: the resulting digital picture is greener, but we may be entranced by the artistic fractal patterns.

4.4 Three-dimensional fractal homeomorphisms

4.4.1 Tri-affine functions

Let $\blacksquare = [0,1] \times [0,1] \times [0,1] \subset \mathbb{R}^3$ denote the unit cube. We will work with the following family of transformations. They have wonderful properties, generalizing work in [11].

Definition 4.4.1 A **tri-affine** function $f : \mathbb{R}^3 \to \mathbb{R}^3$ is a mapping such that

$$\begin{aligned} f(x,y,z) = p_0 &+ (p_1 - p_0)x + (p_3 - p_0)y + (p_4 - p_0)z \\ &+ (p_0 + p_2 - p_1 - p_3)xy + (p_0 + p_7 - p_3 - p_4)yz \\ &+ (p_0 + p_5 - p_1 - p_4)xz \\ &+ (p_1 + p_3 + p_4 + p_6 - p_0 - p_2 - p_5 - p_7)xyz \end{aligned} \tag{4.14}$$

for some set of vectors $p_i \in \mathbb{R}^3$ for $i = 0,1,2,\ldots,7$, for all $(x,y,z) \in \mathbb{R}^3$.

A tri-affine function has these properties: (i) lines parallel to the coordinate axes are mapped to lines; (ii) planes parallel to the coordinates axes are mapped into ruled surfaces; (iii) proportions are preserved along the images of lines parallel to the coordinate axes; (iv) for the function defined by (4.14) we have

$$\begin{aligned} &f(0,0,0) = p_0, \quad f(1,0,0) = p_1, \quad f(0,1,0) = p_3, \quad f(0,0,1) = p_4, \\ &f(1,1,0) = p_2, \quad f(1,0,1) = p_5, \quad f(0,1,1) = p_7, \quad f(1,1,1) = p_6; \end{aligned}$$

(v) each edge of $\blacksquare$ is mapped affinely onto the line segment that joins the images of the endpoints of the edge (for example the line segment that connects $(0,0,0)$ to $(1,0,0)$ is mapped onto the line segment that connects p_0 to p_1;) (vi) each face of the unit cube is mapped to a two-dimensional (but not in general flat) object whose position in space depends only on the images of the four corners of the face; (vii) each point $\mathbf{x}$ on a face of the unit cube is mapped to a point $f(\mathbf{x}) \in \mathbb{R}^3$ whose location is completely determined by the location of the point $\mathbf{x}$ relative to the face and the images of the four corners of the face (for example, $f(x,y,0) = p_0 + (p_1 - p_0)x + (p_3 - p_0)y + (p_0 + p_2 - p_1 - p_3)xy$ which depends only on the locations of the images (p_0, p_1, p_2, p_3) of the four vertices of the face.)

In particular, concerning (vii), if f_i and f_j are two tri-affine functions such that $f_i(c_u) = f_j(c_u) = p_u$ for $u = 0,1,2,3$ where the c_u are the corners of a face $\square$ of $\blacksquare$, then $f_i(\mathbf{x}) = f_j(\mathbf{x})$ for all $\mathbf{x} \in \square$ (in fact, for

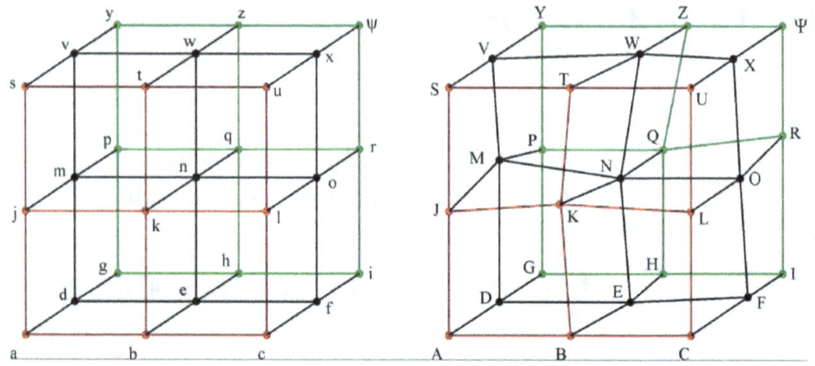

Figure 4.8 Two labellings of points in the unit cube, used to define two tri-affine IFSs. The smaller cubes on the left represent a three-dimensional "tiling" of the cube by copies of itself. The corresponding objects on the right have ruled surfaces but fit together perfectly. See text.

all $\mathbf{x}$ in the plane in $\mathbb{R}^3$ that contains $\square$.) This remarkable fact enables us to construct three-dimensional fractal homeomorphisms using IFSs of tri-affine functions.

We define a pair of IFSs of tri-affine functions as follows. Let

$$F = (\mathbb{R}^3, f_1, f_2, \ldots, f_8), \quad G = (\mathbb{R}^3, g_1, g_2, \ldots, g_8)$$

where, using the notation in Fig. 4.8, the tri-affine functions are uniquely defined by

$$f_1(acusgi\psi y) = abkjdenm, \quad f_2(acusgi\psi y) = cbklfeno,$$
$$f_3(acusgi\psi y) = utklxwno, \quad f_4(acusgi\psi y) = stkjvwnm,$$
$$f_5(acusgi\psi y) = ghqpdenm, \quad f_6(acusgi\psi y) = ihqrfeno,$$
$$f_7(acusgi\psi y) = \psi zqrxwno, \quad f_8(acusgi\psi y) = yzqpvwnm,$$

and

$$g_1(ACUSGI\Psi Y) = ABKJDENM, \quad g_2(ACUSGI\Psi Y) = CBKLFENO,$$
$$g_3(ACUSGI\Psi Y) = UTKLXWNO, \quad g_4(ACUSGI\Psi Y) = STKJVWNM,$$
$$g_5(ACUSGI\Psi Y) = GHQPDENM, \quad g_6(ACUSGI\Psi Y) = IHQRFENO,$$
$$g_7(ACUSGI\Psi Y) = \Psi ZQRXWNO, \quad g_8(ACUSGI\Psi Y) = YZQPVWNM.$$

where, for example, $f_1(acusgi\psi y) = abkjdenm$ means $f_1(a) = a$, $f_1(c) = b$, $f_1(u) = k$, $f_1(s) = j$, $f_1(g) = d$, $f_1(i) = e$, $f_1(\psi) = n$, $f_1(y) = m$. It can be shown, using similar methods to those in [11], that any mask for F yields the same fractal homeomorphism $T_{FG} : \blacksquare \to \blacksquare$. The reason is that the corresponding masked dynamical system $W_F : \blacksquare \to \blacksquare$ is continuous: since two masks can differ only on the internal faces of the small cubes, such as the internal faces of *abkjdenm* in Fig. 4.8, the continuous function W_F must remain unaltered. Indeed, G too is associated with a unique continuous masked dynamical system that is topologically conjugate to W_F. The homeomorphism that provides the conjugacy is precisely T_{FG}, and we have

$$T_{FG} \circ W_F = W_G \circ T_{FG}.$$

In Fig. 4.9 we show the result of applying such a tri-affine fractal homeomophism $T_{FG} : \blacksquare \to \blacksquare$ to a three dimensional picture, supported inside $\blacksquare$, of a collection of concentric colored spherical shells, something like a gob-stopper. We have cut the before and after images in half to reveal the action of the transformation on the internal structure of the spherical picture.

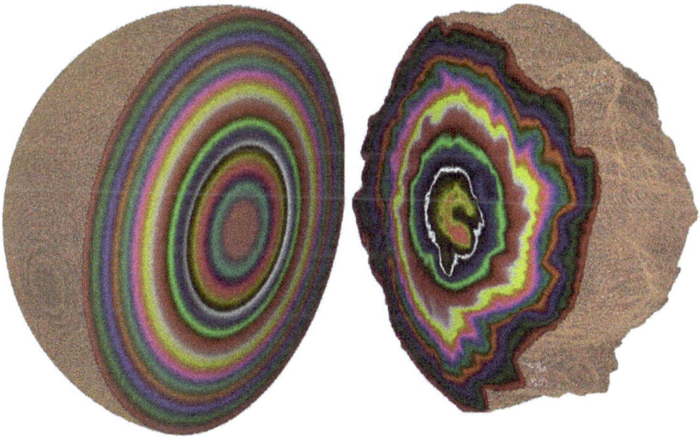

Figure 4.9 This illustrates the action of a tri-affine fractal homeomorphism on a three-dimensional picture that resembles a gob-stopper, one of those candies that have multiple different colored layers that are revealed by sucking on the candy. The before and after images have been cut in half to reveal the effects of the transformation on the internal structure of the spherical picture. It is wonderful to think of the warps and twists hidden inside the image on the right. Digital technology used in connection with tomography can be used to explore such three dimensional pictures.

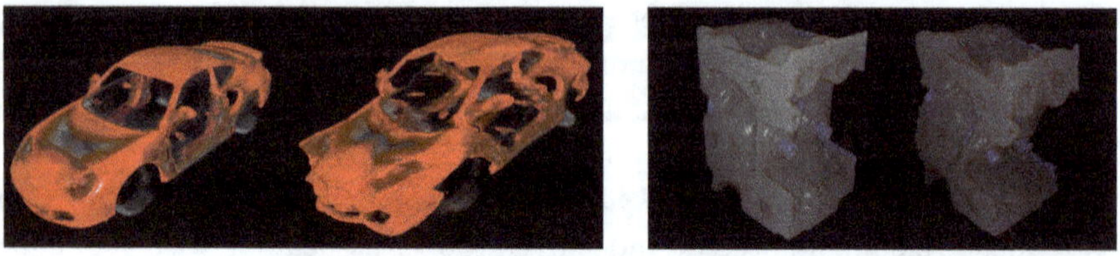

Figure 4.10 Left: A Porche, before and after a tri-affine fractal homeomorphism has been applied. The engine block, the internals of the wheels, everything is warped in a non-differentiable manner. Note the fractal texture of the transformed paintwork. Right: Three-dimensional picture of an engine block, before and after a tri-affine fractal homeomorphism has been applied. The engine has been rendered to be faintly translucent and some of the internal structure can be discerned.

In Fig. 4.10 we illustrate tri-affine fractal homeomorphisms applied to a model car and an engine block. It is important to understand that the transformations act throughout the objects: the cylinders and gaskets of the engine block are also transformed, warped and changed; you can just discern some of the blue internal structure of the engine, both before and after. Note too the fractal textured paintwork of the transformed car.

4.5 Algorithms

4.5.1 General discussion

We discuss algorithms for computing digital approximations to fractal transformations of pictures. To simplify the presentation we mainly restrict attention to the pair of IFSs F and G introduced in Eqs. (4.11); we suppose that the goal is to compute the action of a masked fractal transformation

$$T_{FG} : A_F \to A_G,\ T_{FG}(x) = \pi_G \circ \tau_F(x);$$

in particular, we want to compute a transformed picture $P_F = T_{FG}(P_G)$, given a picture P_G supported on $\square$. Our discussion applies equally well to three-dimensional pictures.

In digital space we need only compute the action of T_{FG} on a finite number of points, corresponding to pixels in the before and after images. The choice of representation and precision for these points depends upon properties of the fractal transformation and the type of algorithm. Most algorithms have both a contractive (stable) part, corresponding to the coding map π_G, and an expanding (chaotic) part, corresponding to the action of the masked dynamical system for F,

$W_F : \square \to \square$, needed to compute τ_F. Each point $x \in \square$, representing a pixel, is associated with an infinitely long address string $\tau_F(x)$. Since we cannot compute infinite addresses we must understand the errors produced by truncation. Shorter addresses require less memory and lead to faster computation. We need to find a balance between truncation and accuracy.

The length of the address which results in visually accurate transformation depends on three factors: the size of the dataset to be transformed, the rate of convergence of $\pi_G(\sigma) = \lim_{k\to\infty} g_{\sigma|_k}(x)$, and how difficult it is to describe and iterate the masked dynamical system W_F, which involves the inverse functions f_n^{-1}. For a visually accurate transformation, truncated addresses need to be long enough, say of length k, such that the diameter of $g_{\sigma|_k}(\square)$ is sufficiently small for all σ, say less than half the width, ϵ, of a pixel. If λ_G is the largest of the contractivity factors for the functions of the IFS G, we require k to be chosen so that

$$\frac{\lambda_G^{k+1}}{1-\lambda_G} < \frac{\epsilon}{2}.$$

A second type of error accumulates under iterative application of the masked dynamical system $W_F : \square \to \square$ to generate addresses. An 8 byte double has 15 digits so we start with a rounding error of approximately $\delta = 10^{-15}$. Since the masked dynamical system is expansive this error increases by a factor bounded above by d_F^{-1} at each iteration, where d_F is such that $d_F \|x-y\| \leq \|f_n(x) - f_n(y)\|$ for all $x, y \in \square$, for all $f_n \in F$. If we assume that at each iteration there is an expansion by this factor, plus an additional rounding error, then the error accumulates according to $\delta \mapsto d * \delta + 10^{-15}$. For example if $d_F = 4$ then after 20 iterations the error accumulates to almost $1.47 * 10^{-3}$.

Rather than determine the best truncation level k in advance, the determination of appropriate code length may be done dynamically. For example, if most of the maps have a much smaller contraction factor than the largest, λ_G, then the address length will not need be as long for most of the pixels, while still yielding an accurate result.

4.5.2 Chaos game and color-stealing algorithm

One of the fastest and most easily implemented algorithms for computing an incomplete image of a fractal transformation of a digital picture is via a modification of the chaos game algorithm, [13].

The chaos game algorithm for calculating the attractor of an IFS involves computing and plotting a sufficiently long random orbit $\{x_m\}_{m=0}^{L}$ on the attractor of the IFS, obtained at each iterative step by randomly selecting, according to some fixed set of probability rules, a function of the IFS and applying it to the previously computed point, starting from any point x_0 on the attractor. Given enough iterations L and appropriate probability rules the random orbit will eventually hit every point on the discretized attractor, even in cases where there may be no metric with respect to which the IFSs are contractive [14]. A version of this algorithm was used by Benoit when he computed some of the first pictures of Julia sets, see [15].

In order to apply the chaos game algorithm to compute a fractal transformation, the basic idea is apply the chaos game to the IFS

$$F \times G := (A_F \times A_G, f_1 \times g_1, f_2 \times g_2, f_3 \times g_3, f_4 \times g_4)$$

whose attractor has this property: its projection onto A_F is A_F and onto A_G is A_G: its attractor is something like the union of the graphs of all fractal transformations from A_F to A_G. From this point of view, the goal is to use the description of the mask for F to enable one to extract from the random orbit $\{x_m \times y_m\}$ a subsequence $\{x_{m_n} \times y_{m_n}\}$ that lies exactly on the graph of $T_{FG} : A_F \to A_G$. In particular, if there is only one fractal transformation from A_F to A_G, independently of the choice of mask, then the attractor of $F \times G$ is precisely the graph of the transformation. This is the situation in all of the examples presented in this paper.

In practice, two chaos games are "played" simultaneously, one generating a random orbit $\{x_m\}$ on A_F and one generating a random orbit $\{y_m\}$ on A_G; however, at each iteration the same random index is chosen, so that if f_n is applied to x_m on A_F then g_n is applied to y_m on A_G and so on. To compute the picture $T_{FG}(P_G)$, upon each iteration, the point x_m is assigned the color of the point y_m in P_G, *after the mask is taken into account.* This is the basic notion of "color-stealing," where the top mask is chosen, which simplifies the computation because one only has to check, at each step, the address of the latest value of y_m for a given value of x_m and see if it is the "top" one so far encountered; if it is then the point x_m is assigned the color of y_m.

Outline of chaos game color-stealing in C:

```
#include⟨random⟩ // this defines the pseudo random number generator
  rand()
int i; // our loop variable for counting iterations
int itmax = enter number here; // the maximum number of iterations
double x1, y1; // color stealing coordinates
double x2, y2; // mapping coordinates
for (i = 0;i < itmax;i++) do
  ran = rand()%N;
  for (j = 1;j ≤ N;j++) do
    if(ran== j − 1) then
      // apply f_j to (x1,y1), apply g_j to (x2,y2) (setting (x1,y1) and
        (x2,y2)
      to be the new results respectively)
      break;// once we have applied a function we move on
    end if
  end for
  // now take color from the point (x1,y1) and apply to the point (x2,y2)
end for
// finished at the end of the last for loop
```

In practice the use of the pseudo random number generator here is not ideal and should be replaced with something more reliable. For some values of N, rand()%N has a period which is insufficient for the chaos game to fill an entire image. In the example above each random number is chosen with equal probability but in most circumstances it is more efficient if the probability of choosing a particular pair of functions is proportional to the area given by the range of the function acting on the target image (e.g. functions from the IFS G in the example above).

The chaos game is fast initially. Its complexity depends on the number of functions in each IFS and the complexity of function evaluation at each iteration. The chaos game can therefore fill a good portion of the target pixels quite quickly. Roughly, if it takes L iterations to fill up half the pixels, then it takes $2L$ iterations to fill up three quarters of the pixels, and so on. In practice, the chaos game algorithm is often combined with other algorithms, first running the chaos game for a certain number of iterations, and then switching to another approach. The advantages of the chaos game are that it requires minimal memory, it is initially fast, and it is accurate, because errors are repeatedly reduced by application of contraction mappings.

4.5.3 Perpixel algorithms

The perpixel method is applied as follows. For each pixel we find an address corresponding to that pixel by reverse iterating the corresponding point x by successively applying the appropriate inverse function f_i^{-1}. This is equivalent to running the masked dynamical system W_F for the given point. This is done for a number of iterations determined by the desired address truncation length.

To find the top address [6] of each point the following method may be used.

First apply g_1^{-1} to the pixel co-ordinates. If the result lies in $\square$ then we keep the result and assign 1 as the first code in the sequence. If not then we try g_2^{-1} and so on. At least one of the N inverse functions must yield a result in $\square$ since the attractor of the IFS is $\square$. We then take the resulting point from this procedure and repeat the process to find out the next code in the sequence and continue to repeat until an appropriate address length, k, has been reached (where the determination of this length has been previously discussed). Once the address of this point has been determined ($\sigma|_k = \tau_G(x)$) it can then be applied to the functions in F for some initial point on A_F, that is we apply coding map, π_F, to the finite address sequence. The resulting point $y = \pi_F(\sigma|_k)$ corresponds to a pixel on the picture from which we take the colors and apply to our starting pixel on the target image. After performing this calculation for every pixel we will clearly have a complete image of the transformation.

This method is slower than the chaos game to begin with since there are many more calculations required for each pixel. Where the chaos game needs to apply only one function from each IFS in order to plot a point, this algorithm must apply a function from each IFS a number of times, corresponding to the desired address length k. However as an image completes and the chaos game tends to repeatedly go over points it has already filled, the perpixel algorithm will continue to steadily fill pixels and as a result it can be quicker to complete an image. It also tends to yield more consistent runtimes. The speed of the algorithm can also be improved by dynamically determining the appropriate address length. The perpixel method also has the advantage that it works with masks other than the top mask.

Outline of the calculation for a single pixel being transformed using masked IFSs:

```
double x1, y1; // color stealing coordinates
double x2, y2; // mapping coordinates
x2 = x coordinate of target pixel;
y2 = y coordinate of target pixel;
int i;
int codelength = the desired length of addresses;
char code[codelength]; // to record addresses
// this next for loop runs the masked dynamical system
for (i = 0;i <codelength;i + +) do
  for (j = 1; j ≤ N; j + +) do
    if ((x2,y2) is in M_j (the jth element of the mask)) then
      // apply g_j^{-1} to (x2,y2) (setting (x2,y2) to be the new result)
      code[i]= j − 1;
      break; // we exit and move onto the next i
    end if
  end for
end for
x1 = 0;
y1 = 0;
// this next loop applies the coding map
for (i =codelength−1;i > −1;i − −) do
  for (j = 1; j ≤ N; j + +) do
    if(code[i]== j − 1) then
      // apply f_j to (x1,y1) (setting (x1,y1) to be the new result);
      break; // we exit and move onto the next i
    end if
  end for
end for
// find the pixel on the image corresponding to the new point (x1,y1);
// take the color data and apply it to the target pixel coordinates;
// we repeat this for every target pixel;
```

4.5.4 Pixel chaining

Pixel chaining is similar to the perpixel method, except the pixels are handled in chains rather than one at a time. It is described in the context of fractal image compression algorithms in [16]. It can be implemented as a modification of the perpixel method described above. It involves following a (deterministic) orbit of the masked dynamical system on a digital picture until it becomes periodic, forming a chain on A_F. We continue to iterate for some k additional steps. This gives us one long address to which we can repeatedly apply the shift operator, yielding the addresses to length k of a chain of points on $A_F = \square$, to each of which we apply π_G, to assign the color to each point on the chain. After this chain of points on $A_F = \square$ has been assigned colors, we pick another point in A_F that has not been colored, and start a new chain. This new chain may intersect previous chains (revealed because it intersects a point that already has color assigned) at which stage we iterate a further k times and then terminate the chain. Again we apply π_G, to assign color to each point on the new chain.

This pixel chaining algorithm fills in most of the image relatively quickly to start with but slows down towards the end as chains become shorter. At its slowest it fills pixels no slower than the perpixel method. Hence it has the advantage of being faster than the perpixel method. However it is also a little more difficult to implement and requires more memory to run (the extra memory is the result of an extra buffer to store long addresses corresponding to the chains.)

4.5.5 Combinations of algorithms

One of the fastest ways of completing the fractal transformation of an image consists of combining the chaos game and the perpixel algorithms. We start with the chaos game and as the image begins to fill out the chaos game will start to continually hit pixels that have already been calculated. At this point the perpixel algorithm is implemented to fill in remaining pixels. The complicated part of combining the two algorithms is determining the optimal moment to switch between the two. Optimally we want to switch to the perpixel algorithm as soon as the chaos game starts filling pixels at slower rate than the perpixel method would. The easiest way to implement the switch is to simply run the chaos game for a set number of iterations. Experimentally a good number of iterations in many circumstances seems to be N times the total number of pixels in the target image. Alternatively this could be done dynamically. By monitoring how many times the chaos game hits pixels that have been filled before hitting new ones we can program the shift in algorithm to occur when this number is such that the perpixel algorithm is faster. Whilst giving us something closer to the optimum shift the continuous monitoring may slow the chaos game part slightly.

The pixel chaining algorithm can also be combined with the chaos game but generally it is slightly quicker when the perpixel method is used in this scenario. This is because if the chaos game fills some initial pixels it prevents the pixel chaining algorithm from building the long chains that make it effective. The extra overhead involved in combining the two also acts to slow the chaos game part down a little. It also becomes much more difficult to determine the optimal switching point. A combination of the chaos game and perpixel method also requires less memory to run than combining the chaos game with the pixel chaining algorithm.

4.5.6 Deterministic and approximation algorithm

The approximation algorithm is used when we have already performed a fractal transformation of an image and want to change the parameters, such as a and b, in the underlying pair of IFSs, F and G, to yield the perturbed fractal transformation. The unperturbed transformation may be the identity.

The approximation algorithm is based on the deterministic algorithm for computing the attractor of an IFS. The deterministic algorithm for computing the attractor of F in Eq. 4.1 is as follows. Let $\mathbb{H}$ denote the compact metric space that comprises the nonempty compact subsets of X together with the Hausdorff metric. Define, by slight abuse of notation $F : \mathbb{H} \to \mathbb{H}$ by

$$F(S) = \bigcup_{n=1}^{N} f_n(S) \quad \text{for all } S \in \mathbb{H}.$$

Then we have that $F : \mathbb{H} \to \mathbb{H}$ is a contraction mapping and hence by the Banach fixed-point theorem $\{F^k(S)\}_{k=0}^{\infty}$ converges to A_F in the Hausdorff metric, regardless of the choice of nonempty compact set $S \subset X$. The translation of this iterative procedure into the digital realm yields the deterministic algorithm for computing the attractor of F. In the present setting we suppose that the fractal homeomorphism is unique, regardless of the choice of mask, as is the case for F and G defined in sections 4.3 and 4.4, so that the attractor of the iterated function system $F \times G$ (which can be readily proved to be hyperbolic in the naturally implied metric) is the graph of the fractal homeomorphism T_{FG}, namely,

$$A_{F \times G} := \{x \times y \in X \times X : y = T_{FG}(x)\}.$$

For any compact nonempty set $S \subset X \times X$ we have that

$$\lim_{k\to\infty} (F \times G)^k (S) = A_{F\times G}.$$

Suppose now that we already know $A_{F\times G}$, i.e. we know the transformation, and we want to calculate the tranformation for the perturbed system. Then we simply take the starting set S in the above expression to be $A_{F\times G}$ (or a digital approximation to it) and apply the operator $\widetilde{F} \times \widetilde{G}$, yielding a first approximation to $A_{\widetilde{F}\times\widetilde{G}}$ the graph of the perturbed homeomorphism. The translation of this notion to the digital realm is what we call the approximation algorithm. By repeating this you get successive approximations of the new transformation. After applying a number of approximations corresponding to the desired code length k, the resulting transformed picture should become visually accurate for the perturbed new transformation; however if the change in parameters is only small, then fewer iterations may be required for an accurate result.

The major advantage of the approximation algorithm is that the user can very quickly see updated approximations. A disadvantage is that by the time an accurate image of the new transformation has been generated by the approximations the same image generally could have already been generated by one of the other faster algorithms. The use of extra memory buffers also means this algorithm uses more memory than the others.

4.5.7 Parallel algorithms

The above algorithms may be parallelised. The chaos game can be parallelised by running it on multiple threads, making sure that the random number generator has a different seed on each thread and that the random number sequences do not overlap on any of the threads. This will fill in lots of points quickly to begin with but each thread will eventually have a lot of overlap. It is also feasible to force a different sequence on each thread, but this requires a lot of communication between threads. Alternatively, DeBruijn sequences may be used, the sequence being split into smaller sections of equal length and distributed across the threads.

The perpixel method has the most potential for parallelisation, by domain decomposition of the image over the threads with each thread applying the perpixel algorithm to the assigned subdomain. There may be minor problems with distributing workload but these may be fixed by distributing the pixels dynamically. That is, after a thread completes the calculation for a data element, it starts on a new one. The resulting computation time is approximately inversely proportional to the number of threads.

The pixel chaining algorithm is not particularly suitable for running on multiple threads. The dataset may be divided up much like the perpixel algorithm but the data element being calculated by one thread could end up in the chain of another pixel on another thread making it a redundant calculation. With some additional overhead it would be possible have threads share information so that when one thread chain intersects a chain on another thread then one of the two threads can take over linking the two chains whilst the other thread starts again at a different point. Given the additional overhead and the complexity it generally seems that this is not the best approach to take.

The approximation algorithm can also be easily run on multiple threads by again dividing up the pixels between the threads. The number of threads used for this algorithm then affects the run-time in a similar manner to the perpixel method. This requires that each thread have sufficient memory available.

Algorithms can still be combined when running parallel threads. The chaos game and the perpixel method can be combined in the same way as before, but here determining the optimal moment to switch algorithms becomes a little more complicated. If done correctly this yields an extremely fast method.

Acknowledgments

We thank Louisa Barnsley for producing the illustrations and much other help.

Bibliography

[1] M. F. Barnsley, *Fractals in Engineering: New Trends in Theory and Applications* (Springer-Verlag, London, 2005), pp. 3–20.
[2] M. F. Barnsley, Superfractals (Cambridge University Press, Cambridge, 2006).
[3] M. F. Barnsley, *Amer. Math. Monthly* **116** (2009) 291.
[4] M. F. Barnsley, *Progress in Probability* **61** (2009) 227.
[5] K. B. Igudesman, *Russian Mathematics (Iz. VUZ)* **53** (2009) 67.
[6] M. F. Barnsley, B. Harding, K. Igudesman, *SIAM J. Imaging Sciences* **4** (2011) 1001.

[7] M. F. Barnsley, *Notices Am. Math. Soc.* **57** (2010) 10.
[8] S. Nikiel, *Iterated Function Systems For Real Time Image Synthesis* (Springer-Verlag, London, 2007).
[9] J. E. Hutchinson, *Indiana Univ. Math. J.* **30** (1981) 713.
[10] C. Bandt, K. Keller, *Math. Nachr.* **154** (1991) 27.
[11] M. F. Barnsley, A. Vince, Fractal homeomorphism for bi-affine iterated function systems, *Int. J. Applied Nonlinear Science* **1**(1) (2013) 3.
[12] M. F. Barnsley, B. Harding, A. Vince, The entropy of a special overlapping dynamical system, *Ergod. Th. & Dynam. Sys.* **0** (2012) 1, doi:10.1017/etds.2012.140
[13] M. F. Barnsley, *Fractals Everywhere — New Edition* (Dover Publications, Mineola, NY, 2012).
[14] M. F. Barnsley, A. Vince, *Ergod. Th. & Dynam Sys.* **31** (2011) 1073.
[15] B. B. Mandelbrot, *The Fractal Geometry of Nature* (W. H. Freeman, San Francisco, 1983).
[16] N. Lu, *Fractal Imaging* (Academic Press, New York, 2003).

Chapter 5

Mandelbrot's cascades: a legendary destiny

Julien Barral and Jacques Peyrière

To embrace Mandelbrot's whole work would necessitate knowledge and skills in many scientific fields. This is why this article deals only with the topics we are the most acquainted with: multiplicative processes and multifractal analysis. Nevertheless, to isolate them is somewhat artificial because in Mandelbrot's eyes all his research topics were closely related and the parts of the same design.

Mandelbrot has inspired not only us but also a large number of mathematicians during almost fifty years. We are now presenting our thoughts on Mandelbrot's contributions to these fields and the stream of works to which they led.

5.1 Multiplicative processes and intermittency

5.1.1 The genesis

Mandelbrot's main motivation to introduce multiplicative processes comes from his careful study of Kolmogorov's work on turbulence [1] in which he had introduced the "lognormal hypothesis". As a matter of fact, this hypothesis raised quite a few interrogations. During the early 60's, several attempts, not quite convincing, were made to amend it.

Mandelbrot published his own study [2] entitled "Possible refinement of the lognormal hypothesis concerning the distribution of energy dissipation in intermittent turbulence" in 1972. But, despite the forethought with which he chose the title, this work was met by reluctance to the point that it took almost three years before it could be accepted for publication. We see two reasons to this: on the one hand a dogma was questioned, on the second hand this work brought so much novelty that it was not immediately understood. Yet this article was the germ of a whole new theory. Mandelbrot modelled the dissipation of energy in a turbulent fluid by what he called "limit lognormal random processes" the construction of which he sketched as follows:

> Under these conditions, one can relate ε to a sequence of random functions (r.f.'s) $F'(x, \lambda, L)$ such that the $\log F'(x, \lambda, L)$ are Gaussian, with the variance $\mu \log(L/\lambda)$, the expectation $-(\mu/2)\log(L/\lambda)$ and a spectral density equal to $\mu/2k$ for $1/L < k < 1/\lambda$, and to 0 elsewhere. Consequently, the covariance $C(s, \lambda)$ of $\log F'(x, \lambda, L)$ will be assumed to satisfy $C(s) = \lim_{\lambda\to\infty} C(s, \lambda) = -\mu \log(2\pi e^{\gamma} s/L)$. (Here, γ is the Euler constant, whose value is about 0.577.)

This was a bold intuition: $\log \varepsilon$ should be a gaussian process of covariance $\log t$ (which, as is well known, cannot exist). The true mathematical construction remained to be done. It is only ten years later that Kahane [9] built his theory of multiplicative chaos: these objects do exist, not as random scalar functions, but as random operators on measures.

Mandelbrot was aware of the fact that his model was difficult to understand and wrote another version, in which he gave a definition better tractable to mathematician concerns, in two notes [3], which also better met the intuition of turbulence community.

Given an integer $c \geq 2$, one starts from the cube $Q = [0,1]^d$. This cube is divided into $b = c^d$ congruent sub-cubes $Q_1, Q_2, \ldots, Q_b$. Each one of those is in turn divided into b congruent sub-cubes; for instance, the cube Q_j is divided into cubes $Q_{j,1}, Q_{j,2}, \ldots, Q_{j,b}$, and so on. One is given a sequence $\{W_{j_1,j_2,\ldots,j_n}\}_{n\geq 1, 1\leq j_k \leq b}$ of independent random variables, all equidistributed with a nonnegative random variable W of expectation 1. Then, one considers the random measure μ_n whose density with respect to the Lebesgue measure is the step function assuming the value $W_{j_1}W_{j_1,j_2}\cdots W_{j_1,j_2,\ldots,j_n}$ on the cube $Q_{j_1,j_2,\ldots,j_n}$.

This sequence of measures is a nonnegative martingale which weak-$*$ converges with probability 1 towards a measure μ.

Here is the heuristic interpretation of this construction. When $d = 3$, this hierarchy of nested cubes models the hierarchy of vortices in a turbulent fluid and the quantity $\mu_n(Qj_1, \ldots, j_n)$ is an approximation to the dissipation of energy within this cube. The random multiplication which leads from one step to the next accounts for the energy transfer between neighboring vortices. Mandelbrot calls this model *the canonic cascade.* Other cascades had already been considered before, namely by Novikov and Stewart [4] and Yaglom [5], but, in contrast to the canonical case, energy tranfers occured only towards the smaller scales.

Let Y_n stand for the total mass of μ_n:

$$Y_n = b^{-n} \sum_{j_1,j_2,\ldots,j_n} W_{j_1}W_{j_1,j_2}\cdots W_{j_1,j_2,\ldots,j_n}, \tag{5.1}$$

which can also be written as

$$Y_n = b^{-1}\sum_{j=1}^{b} W_j Y_{n-1}(j). \tag{5.2}$$

By taking the limit, one gets the almost sure following equality

$$Y = b^{-1}\sum_{j=1}^{b} W(j)Y(j) \tag{5.3}$$

where all the variables in the right hand side are independent, the $W(j)$ being equidistributed with W and the $Y(j)$ with Y.

When the limit Y is nonzero this construction solves the Eq. (5.3) in which the unknown is the distribution of Y. This equation is a generalization of the one which defines the semi-stable laws.

These canonical cascades can degenerate: it may happen that the sequence μ_n converges towards 0. If W has a finite moment of order 2, Eq. (5.2) yields

$$\mathbb{E}Y_n^2 = b^{-1}\mathbb{E}W^2\mathbb{E}Y_{n-1}^2 + b - 1.$$

So, if $\mathbb{E}W^2 < b$, then the martingale Y_n is bounded in L^2, and therefore the expectation of its limit Y is 1. Mandelbrot conjectured that the condition $\mathbb{E}W^p < b^{p-1}$ implies that this martingale converges in L^p and that the condition $\mathbb{E}W \log W < \log b$ is equivalent to its uniform integrability. Kahane [6, 7] solved this conjecture by exploiting Eqs. (5.2) and (5.3) in a very subtle way.

These two notes by Mandelbrot contain many other remarks and conjectures. One of them is that, with probability 1, the measure μ is carried by a Borel set of Hausdorff dimension $D = d\,(1-\mathbb{E}W \log_b W)$, and that, any Borel set of dimension less than D has a zero μ-measure with probability1. This has been proved to hold by Peyrière [7, 8] under the extra hypothesis $\mathbb{E}Y \log^+ Y < \infty$, later removed by Kahane.

These notes immediately aroused the interest of mathematicians; in contrast the turbulence community needed more time to grasp and value them.

The study of these random measures and of their generalizations as well as of functional Eq. (5.3) has generated an important subsequent literature.

From the beginning, Mandelbrot thought of considering weights of non constant signs, or even complex. It is only recently that this study has been done (see Sec. 5.3). These measures are also at the origin of multifractal analysis.

5.1.2 The T-martingales

Kahane introduced [9, 10] the following formalism. Let $(\Omega, \mathcal{B}, \mathbb{P})$ be a probability space and T a locally compact space the Borel σ-field of which is denoted by $\mathcal{B}(T)$. The space $T \times \Omega$ is endowed with the σ-field $\mathcal{B}(T) \otimes \mathcal{B}$. Consider $(\mathcal{B}_n)_{n\geq 1}$ a non-decreasing sequence of sub-σ-fields of $\mathcal{B}$ and $(Q_n)_{n\geq 1}$ a sequence of nonnegative measurable functions on $\big(T\times\Omega, \mathcal{B}(T)\otimes\mathcal{B}\big)$ such that, for all $t \in T$, $\big(Q_n(t,\cdot), \mathcal{B}_n\big)_{n\geq 1}$ be a nonnegative martingale of expectation 1. Such a sequence is called a *T-martingale*. Given λ a bounded Borel measure on T, for $n \geq 1$, let μ_n be the random measure, also denoted by $Q_n \cdot \lambda$, absolutely continuous with respect to λ with density Q_n. The sequence $(\mu_n)_{n\geq 1}$ converges with probability 1, according to the weak$-*$ topology, towards a measure μ, denoted by $Q\cdot\lambda$ or more simply by $Q\lambda$. This is a straightforward consequence of the nonnegative martingales theorem and of the Riesz representation theorem.

The random measure so obtained from λ, if not zero with probability 1, will be in general singular with respect to λ. We shall see that, in most interesting cases, the processes $\left(Q_n(t)\right)_{t\in T}$ are stationary. Then with Mandelbrot we expect that, due to self-similarity, the geometric properties of μ inherits both those of λ as well as the large deviation properties of Q_n.

5.1.3 From cascades to multifractals

These cascades as a model of turbulence enable some quantization of the fact that turbulence (energy dissipation) takes place on a thin set. Also, when W takes value a with probability $1/a$, iterated multiplications result in the building of a random Cantor set: this is the essence of the Frisch-Parisi β-model [11] which publicized the Mandelbrot martingales among turbulence scientists.

Of course, to describe a turbulent fluid by one only number, namely the dimension of the support of energy dissipation, could appeared somewhat simplistic. Then Frisch and Parisi [12] introduced the notion of multifractal. They consider the velocity field rather than energy dissipation. They view turbulence, so to say, as carried by a superposition of sets of different dimensions; on each of these sets the local Hölder exponent of the velocity has a fixed value.

Shortly after, Halsey *et al.* [13] generalized this idea to other dynamical systems; for instance, they consider measures on strange attractors. In both cases, the link with the thermodynamical formalism and the stationary phase is clear: the Hausdorff dimension of level sets of the local Hölder exponent is given by the Legendre transform of a partition function τ, known under several different names (structure function, L^q spectrum, ...).

Previously, Hentschel and Procaccia [14] had already considered the Rényi dimensions, i.e., the quantities $\tau(q)/(q-1)$, where τ is the structure function, but they did not mention any link with the dimension of level sets.

If the word multifractal and the associated concept appear for the first time in [12], long before, Mandelbrot had described such multifractal behaviors, e.g. in his article [15] on errors in transmission lines. Also there is a paragraph in his 1974 notes [3] where large deviation statistics are mentioned. The link with [12, 13] is the Cramér theorem.

This was the starting point of numerous works. To our knowledge, the first one which was mathematically clear is [16], followed by [17].

A viewpoint closer to geometric measure theory is due to Olsen [18], but we shall not develop this aspect. The interested reader can consult [19, 20].

Also, we shall not describe applications, which are important, of multifractals to other fields, in particular in finance.

It is also worth mentioning works which are not unrelated to multifractals, although they are not exactly multifractal analysis. They deal with the use of thermodynamical formalism to compute certain Hausdorff dimensions, namely in the case of conformal repellers [21–23] or of limit sets of kleinien groups [24, 25].

5.2 Nonnegative multiplicative cascades

5.2.1 b-adic cascades

5.2.1.1 Construction, non-degeneracy and moments

Now let us return to Mandelbrot's construction in a slightly more general setting.

Definitions Let $b \geq 2$ be an integer and $\mathscr{A}$ stand for the alphabet $\{0, \ldots, b-1\}$. Set $\mathscr{A}^* = \bigcup_{n \geq 0} \mathscr{A}^n$ (by convention, $\mathscr{A}^0$ has the empty word for only element). If $w \in \mathscr{A}^n$ the number n is the length of w and is denoted by $|w|$.

$\mathscr{A}^*$ is a monoid when endowed with the catenation, which will be denoted by a dot or simply by juxtaposition if this does not generate ambiguity.

Let T stand for the product $\mathscr{A}^{\mathbb{N}}$. It elements are considered as infinite words. If $t \in T$ and $n \in \mathbb{N}$, one denotes by $t_{|_n}$ the prefix of length n of t, i.e., the word in $\mathscr{A}^n$ formed with the n first letters of t.

In the same way, if $w \in \mathscr{A}^m$ and $n \leq m$, $w_{|_n}$ stands for the prefix of length n of w.

Each $w \in \mathscr{A}^*$ defines the cylinder $[w] \subset T$, whose elements are the infinite words of which w is a prefix.

The space T is endowed with the usual ultrametric: the balls are the cylinders, and the diameter of $[w]$ is $b^{-|w|}$.

Now we use the T-martingale language to explain the Mandelbrot construction [3, 26]. Let $W = (W_0, \ldots, W_{b-1})$ be a random vector with nonnegative components of expectation 1. Consider $\{W(w)\}_{w \in \mathscr{A}^*}$ a family of copies of W, all independent. The components of $W(w)$ will be so denoted: $(W_{w\cdot 0}, \ldots, W_{w\cdot(b-1)})$.

For $t \in [0,1)$ and $n \geq 1$, define

$$P_n(t) = W_{t_{|n}} \quad \text{and} \quad Q_n(t) = \prod_{k=1}^{n} P_k(t).$$

Then Q_n is a T-martingale.

We are also given a probability vector $\lambda = (\lambda_0, \ldots, \lambda_{b-1})$ on $\mathscr{A}$. Consider $\boldsymbol{\lambda}$ the Bernoulli product on T defined by λ: $\boldsymbol{\lambda}([w]) = \lambda_{w_1} \cdots \lambda_{w_n}$ for all $n \geq 1$ and $w \in \mathscr{A}^n$. With Mandelbrot we consider the random mesure $\mu = Q\boldsymbol{\lambda}$, and the function

$$\varphi(q) = -\log_b \mathbb{E}\left(\sum_{j=0}^{b-1}(\lambda_j W_j)^p\right) \quad (q \in \mathbb{R}) \tag{5.4}$$

which plays a central rôle in this study as well as the following equation which expresses the self-similarity of this construction. If Y is the total mass of μ and $Y(j)$ the total mass of the copy of μ obtained with the family $\left(\widetilde{W}(w) = W(j \cdot w)\right)_{w\in\mathscr{A}^*}$, with probability 1, one has

$$Y = \sum_{j=0}^{b-1} \lambda_j W_j Y(j), \tag{5.5}$$

where variables $Y(j)$, $0 \leq j \leq b-1$ are independent and independent of W.

The canonical cascades correspond to the simpler case when $\boldsymbol{\lambda}$ is the Lebesgue measure and the components of W are independent and identically distributed. The following theorem, conjectured by Mandelbrot, was proven by Kahane [7] for the canonical cascades by exploiting 5.5. Later Durrett and Liggett [27], by using more probabilist an approach, based on random walk and renewal theorem, extended it to the general case.

When $b = c^d$, with $d \geq 1$, base c expansions define a mapping from T onto $[0,1]^d$ which transforms cylinders in c-adic cubes. The Mandelbrot martingales then define random measures on $[0,1]^d$. The convergence of these martingales, after projection on $[0,1]$ is shown on Fig. 5.1.

In case when $\mathbb{P}(\lambda_j W_j \in \{0,1\}) = 1$ for all j, we get a critical Galton-Watson process and, if the measure μ is not zero it is a Dirac δ mass.

From now on we exclude this later case.

Theorem 5.2.1 *The following are equivalent.*

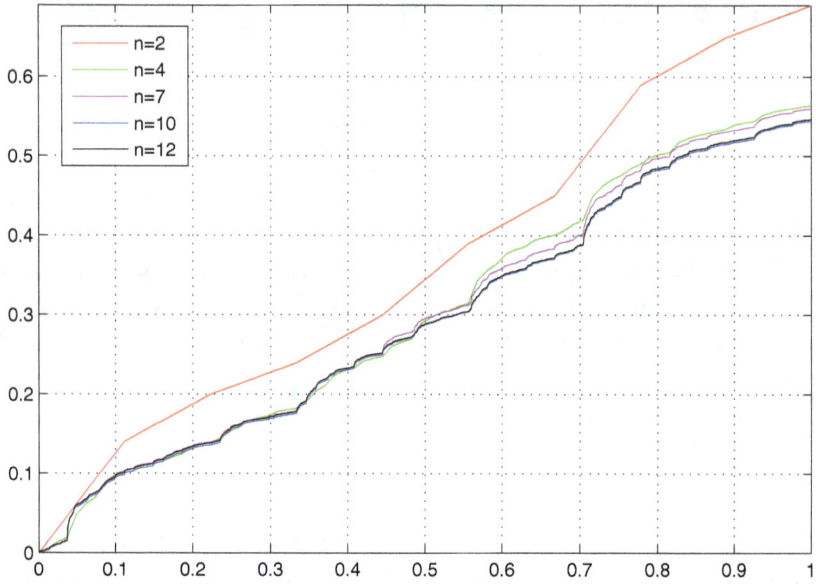

Figure 5.1 Integral $\int_0^t \mathrm{d}\mu_n(u)$ of a triadic Mandelbrot cascade, for $n \in \{2, 4, 8, 12\}$.

1. $\mathbb{P}(\mu \neq 0) > 0$;
2. *The martingale* $(\|\mu_n\|, \mathcal{B}_n)_{n\geq 1}$ *is uniformly integrable*;
3. $\varphi'(1_-) > 0$.

Furthermore, Mandelbrot conjectured that the number $\varphi'(1_-)$ bears a precise geometric information on μ: this is the Hausdorff dimensions of μ. Peyrière [7, 8] answered this question by proving the following result (with the extra hypothesis $E(Y \log^+(Y)) < \infty$ which Kahane [28] get rid of by using a percolation technique) by introducing the probability $\mathcal{Q}$, $\mathcal{Q}(A\times B) = \mathbb{E}(\mathbf{1}_B \mu(A))$, on $\mathcal{B}(T) \otimes \mathcal{B}$.

Theorem 5.2.2 *Let* $D = \varphi'(1_-)$ *and suppose* $D > 0$. *Then, with probability* 1, *if* $\mu \neq 0$, *one has*

$$\lim_{n\to\infty} \frac{\log(\mu([t_{|_n}]))}{\log |[t_{|_n}]|} = D \quad \mu\text{-a.e.},$$

where $|[t_{|_n}]|$ *stands for the diameter* b^{-n} *of* $[t_{|_n}]$. *In other words,* μ *is carried by a set of Hausdorff dimension* D *and all Borel set of Hausdorff dimension* *less than D is of μ measure zero (one then says than the measure μ is exact dimensional of dimension D).*

One fundamental issue of this model is that it allows the total mass Y of μ to have a distribution with a fat tail. Therefore it is well fitted to model intermittent phenomena which present extreme events. Mandelbrot conjectured the following result (with a less precise formulation of the second assertion).

Theorem 5.2.3 *Suppose $\varphi'(1_-) > 0$.*

1. *Let $q > 1$. One has $\mathbb{E}(Y^q) < \infty$ if and only if $\varphi(q) > 0$.*
2. *If it exists $q > 1$ such that $\varphi(q) = 0$, and if $\mathbb{E}\sum(\lambda_j W_j)^q \log^+(\lambda_j W_j) < \infty$, then*

$$0 < \liminf_{x\to\infty} x^q\, \mathbb{P}(Y > x) \leq \limsup_{x\to\infty} x^q\, \mathbb{P}(Y > x) < \infty.$$

   Moreover, the equality holds if and only if the subgroup of $\mathbb{R}$ generated by the supports of the laws of variables $\log(\lambda_j W_j)$ conditioned by $\lambda_j W_j > 0$ is dense.

The first assertion, proven by Mandelbrot for moments of integral order, is due to Kahane for canonical cascades, and by Durrett and Liggett in the general case; the proof relies on a rather direct manipulation of 5.5. The second assertion is due to Guivarc'h [29] for canonical cascades, and to Liu [30] in general. The proof exploits the rewriting of 5.5 as an equation viewed from the probability $\mathcal{Q}$, and uses the renewal theorem.

There are many variants and generalizations of this construction. For instance one can replace the homogeneous tree $\mathscr{A}^*$ by a random Galton-Watson tree [31], or by a colored tree [32] which is the same as making the T-martingale act on a Markov measure [33]. See also [34].

5.2.1.2 Large deviations and multifractal analysis

Mandelbrot soon become aware, by relying on canonical cascades, that the large deviations properties of the products defining the densities $Q_n(t)$ had to be reflected in the Hölder behavior of μ on b-adic cylinders [3].

For $\alpha \in \mathbb{R}_+$, $n \geq 1$ and $\epsilon > 0$ he considers the logarithmic frequency of occurrences of cylinder $[w]$ of generation n for which $\frac{\log \mu([w])}{|[w]|}$ is equal to α

up to ϵ:

$$f(\alpha, n, \epsilon) = \frac{\log \operatorname{card}\{w \in \mathscr{A}^n : b^{-n(\alpha+\epsilon)} \leq \mu([w]) \leq b^{-n(\alpha-\epsilon)}\}}{\log \operatorname{card}(\mathscr{A}^n)}.$$

Let us assume we are in the canonical case, i.e., components of W are i.i.d.; we also assume, for sake of simplicity, that these weights are positive. Also suppose that $\varphi(q) > -\infty$ for all $q \in \mathbb{R}$, and of course that $\varphi'(1) > 0$.

The μ-mass of any cylinder $[w]$ writes

$$\mu([w]) = Y(w) b^{-n} \prod_{k=1}^{|w|} W_{w_{|k}}, \tag{5.6}$$

where $Y(w)$ is the copy of Y obtained from the family $\{\widetilde{W}(v)\}_{v \in \mathscr{A}^*} = \{W(w \cdot v)\}_{v \in \mathscr{A}^*}$. The asymptotic behavior of frequencies $f(\alpha, n, \epsilon)$ should be related to large deviations properties of the random walk $W_{t_{|1}} \cdots W_{t_{|n}}$ whose law does not depend on $t \in [0, 1)$, and for which Cramér theorem (cf. [35]) says that, for all $\alpha \in \varphi'(\mathbb{R})$, one has

$$\lim_{\epsilon \to 0} \lim_{n \to \infty} \frac{1}{n} \log_b \mathbb{P}\big(b^{-n(\alpha+\epsilon)} \leq b^{-n} W_{t_{|1}} \cdots W_{t_{|n}} \leq b^{-n(\alpha-\epsilon)}\big) = \varphi^*(\alpha) - 1,$$

where $\varphi^*(\alpha) = \inf_{q \in \mathbb{R}} \alpha q - \varphi(q)$ is the Legendre transform of φ (this transform takes its values in $[-\infty, 1]$ and reaches its maximum value 1 at point $\alpha = \varphi'(0)$).

It follows, since $\operatorname{card} \mathscr{A}^n = b^n$, that, when $\varphi^*(\alpha) < 0$ and for $\epsilon > 0$ small enough, one has

$$\sum_{n \geq 1} \mathbb{P}\big(\exists\, w \in \mathscr{A}^n : b^{-n(\alpha+\epsilon)} \leq b^{-n} W_{w_{|1}} \cdots W_{w_{|n}} \leq b^{-n(\alpha-\epsilon)}\big) < \infty.$$

Then, by using the Borel-Cantelli lemma and taking for granted that the contribution of $Y(w)$ to $\mu([w])$ is null (this can be proven by controlling the moments of negative as well as positive orders), that $f(\alpha, n, \epsilon) = -\infty$ for n large enough. This means that from some generation one cannot observe the exponent α. But, if $0 \leq \varphi^*(\alpha) \leq 1$ we expect to observe asymptotically about $b^{n\varphi^*(\alpha)}$ intervals for which the local dimension of μ is approximately α.

At that time Mandelbrot did not go so far as to adopt the geometric viewpoint consisting in studying the level sets

$$E_\mu(\alpha) = \left\{t : \lim_{n\to\infty} \frac{\log(\mu([t_n]))}{\log|[t_n]|} = \alpha\right\} \quad (\alpha \geq 0)$$

other than $E_\mu(D)$; these level sets, introduced in [12] and [13], describe in a geometric way the heterogeneity of the distribution of μ. We shall see that the measure μ complies to the multifractal formalism: the Hausdorff dimension of $E_\mu(\alpha)$ is the Legendre transform of the function

$$\tau_\mu(q) = \liminf_{n\to\infty} \frac{-1}{n} \log_b \sum_{w\in\mathscr{A}^n} \mu([w])^q.$$

Sticking to the large deviation viewpoint, Mandelbrot later defined the notions of latent exponent and negative dimension. Indeed, even if we do not observe the exponent α asymptotically because $\varphi^*(\alpha) < 0$, nevertheless it still has some existence in the sense that it has a non-zero probability to be observed at a fixed generation. Furthermore, if one considers many (around $b^{-n\varphi^*(\alpha)}$ times) independent realizations of μ, this exponent will end to be observed at generation n for large enough n. (This is written in a formal way in [36]).

This led Mandelbrot to interpret $1 - \varphi^*(\alpha)$ as a codimension according to the following heuristic: if a fractal set E contained in $[0,1]^d$ has D as box dimension, it can be covered by around b^{nD} cylinders of generation n, so to be sure to get one cylinder of generation n which intersects E, one must draw at random at least $b^{n(d-D)}$ of them ($d - D$ is the codimension of E). Then, in the case when the measure μ is constructed in $[0,1]$, since to observe a dyadic interval on which the exponent of μ is about α, around $b^{-n\varphi^*(\alpha)}$ realizations of μ are needed, so to say $b^{n(1-\varphi^*(\alpha))}$ intervals, the dimension of the virtual set on which the Hölder exponent is α is the negative number $\varphi^*(\alpha)$.

Also Mandelbrot explained the notion of latent exponents in a more geometrical way. If the cascade construction is done on $[0,1]^d$ with $d \geq 1-\varphi^*(\alpha)$, then for n large enough an exponent around α will be observed on roughly speaking $b^{n(d-1+\varphi^*(\alpha))}$ cubes, but at the same time if one considers a random one dimensional section the exponent α will not be observed.

These large deviation properties have a geometrical counterpart given by the multifractal analysis of μ as well as of its variants [37–43]. In the following theorem the cascade is no longer assumed to be canonical, but it is still assumed that the components of W do not vanish and that $\varphi > -\infty$ on $\mathbb{R}$ (this imposes $\lambda_j > 0$ for all j).

Theorem 5.2.4 *With probability* 1, *for all* $\alpha \in \mathbb{R}_+$, $E_\mu(\alpha)$ *is non-empty if and only if* $\varphi^*(\alpha) \geq 0$, *and, when it is so, one has*

$$\dim E_\mu(\alpha) = \lim_{\epsilon\to 0}\lim_{n\to\infty} f(\alpha, n, \epsilon) = \varphi^*(\alpha) = \tau_\mu^*(\alpha),$$

where dim *stands for the Hausdorff dimension.*

We are going only to outline the proof. First, if, as for instance in [43], we set

$$q_+ = \sup\{q > 0 : \varphi^*(\varphi'(q)) \geq 0\} \quad \text{and} \quad q_- = \inf\{q < 0 : \varphi^*(\varphi'(q)) \geq 0\},$$

with probability 1, one has (see Fig. 5.2 for an illustration)

$$\tau_\mu(q) = \lim_{n\to\infty} -\frac{1}{n}\log_b \sum_{w\in\mathscr{A}^n} \mu([w])^q = \begin{cases} \varphi(q) & \text{if } q \in [q_-, q_+], \\ q\,\dfrac{\varphi(q_+)}{q_+} & \text{if } q > q_+, \\ q\,\dfrac{\varphi(q_-)}{q_-} & \text{if } q < q_-. \end{cases} \tag{5.7}$$

(it is not difficult to see that $\varphi'(1) > 0$ imposes $q_+ > 1$). When μ is not the uniform measure on $\partial\mathscr{A}$, the function τ_μ is linear near $+\infty$ or $-\infty$ if and only if there exist latent exponents. If one rewrites $\sum_{w\in\mathscr{A}^n}\mu([w])^q$ as $\sum_{w\in\mathscr{A}^n} b^{-nq\log(\mu([w])/\log(|[w]|)}$, then the equality 5.7 and the Gartner-Ellis theorem (cf. [35]) yield

$$\lim_{\epsilon\to 0}\lim_{n\to\infty} f(\alpha, n, \epsilon) = \varphi^*(\alpha),$$

with probability 1, simultaneously for all $\alpha \in \mathbb{R}_+ \setminus \{\varphi'(q_-), \varphi'(q_+)\}$.

To study the sets $E_\mu(\alpha)$, for all $q \in]q_-, q_+[$, one considers the vectors $W_q(w) = \left(\frac{W^q_{w\cdot 0}}{\mathbb{E}(W_0^q)}, \dots, \frac{W^q_{w\cdot b-1}}{\mathbb{E}(W_{b-1}^q)}\right)$, $w \in \mathscr{A}^*$, and the Bernoulli measure $\boldsymbol{\lambda}_q$ associated with the probability vector

$$\boldsymbol{\lambda}_q = b^{\varphi(q)}\left(\lambda_0^q\mathbb{E}(W_0^q), \dots, \lambda_{b-1}^q\mathbb{E}(W_{b-1}^q)\right).$$

Any one of these families gives birth to a martingale $Q_{q,n}\cdot\lambda_q$ which converges with probability 1 towards a limit $\mu_q = Q_q\cdot\lambda_q$. Moreover, as shown by a direct computation, the condition $\varphi'(1) > 0$ writes $\varphi^*(\varphi'(q)) > 0$. Therefore the limit is non-degenerate. But, to obtain a result

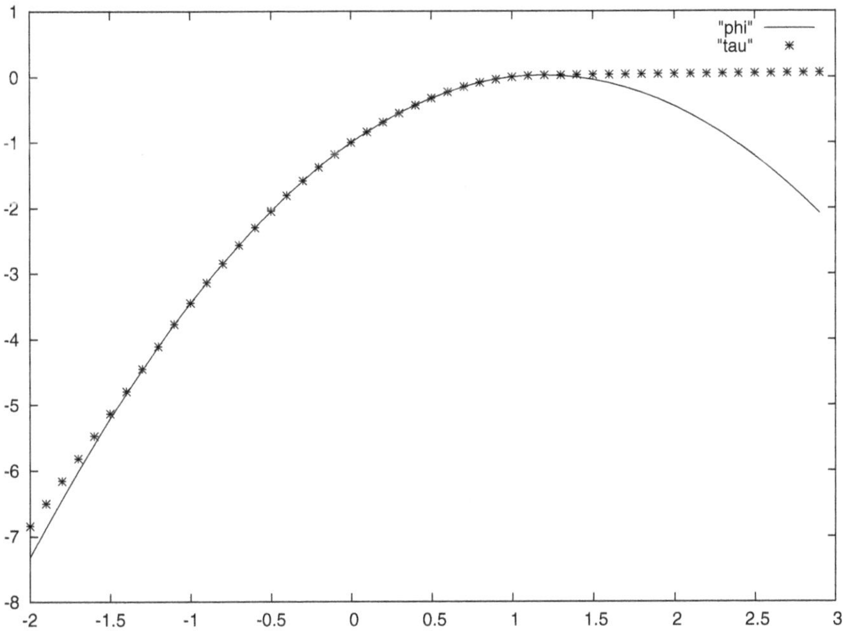

Figure 5.2 Example of couple (φ, τ_μ) in the case of a lognormal cascade. Here $b = 2$, $(W_0, W_1) = \left(\frac{e^{\xi_0}}{\mathbb{E}(e^{\xi_0})}, \frac{e^{\xi_1}}{\mathbb{E}(e^{\xi_1})}\right)$, where ξ_0 and ξ_1 are two independent standard normal laws, and $\lambda_0 = \lambda_1 = 1/2$. One has $\varphi(q) = (q-1)\left(1-\frac{q}{2\log(2)}\right)$, $D = \varphi'(1) = 1 - \frac{1}{2\log(2)}$, $q_+ = \sqrt{2\log(2)}$, $q_- = -q_+$, $\tau_\mu(q) = \left(1 + 1/\sqrt{2\log(2)}\right)^2 q$ if $q < q_-$, $\tau_\mu(q) = \varphi(q)$ if $q \in [q_-, q_+]$, and $\tau_\mu(q) = \left(1 - 1/\sqrt{2\log(2)}\right)^2 q$ if $q > q_+$.

uniform with respect to α, it is necessary to have, with probability 1, all these measures simultaneously for all $q \in]q_-, q_+[$. This has been achieved in [42] by exploiting an idea coming from the large deviations for branching random walks [44].

Then, one considers, for all $q \in \mathbb{R}$, the logarithmic generating functions

$$L_{q,n}(s) = -\frac{1}{n} \log_b \sum_{w \in \mathscr{A}^n} \mu([w])^s \mu_q([w])$$

and

$$\widetilde{L}_{q,n}(s) = -\frac{1}{n} \log_b \sum_{w \in \mathscr{A}^n} \mu_q([w])^s \mu_q([w]),$$

and show that, with probability 1, for all $q \in]q_-, q_+[$, and for s small enough, one has

$$\lim_{n\to\infty} L_{q,n}(s) = \varphi(q+s) - \varphi(q) \quad \text{and} \quad \lim_{n\to\infty} \widetilde{L}_{q,n}(s) = \varphi(qs) - s\varphi(q).$$

Set $L_q(s) = \varphi(q+s) - \varphi(q)$ and $\widetilde{L}_q(s) = \varphi(qs) - s\varphi(q)$; then $L_q'(0) = \varphi'(q)$ and $\widetilde{L}_q'(0) = \varphi^*(\varphi'(q))$. Then, it follows from the fact that $L_{q,n}$ and $\widetilde{L}_{q,n}$ converge and from the Gartner-Ellis theorem that, with probability 1, for all $\epsilon > 0$,

$$\limsup_{n\to\infty} \frac{1}{n} \log \mu_q \left(\left\{ t \in T : \frac{\log(\mu([t_{|n}]))}{\log |[t_{|n}]|} \notin [L_q'(0) - \epsilon, L_q'(0) + \epsilon] \right\} \right) < 0$$

and

$$\limsup_{n\to\infty} \frac{1}{n} \log \mu_q \left(\left\{ t \in T : \frac{\log(\mu_q([t_{|n}]))}{\log |[t_{|n}]|} \notin [\widetilde{L}_q'(0) - \epsilon, \widetilde{L}_q'(0) + \epsilon] \right\} \right) < 0.$$

Then, due to the Borel-Cantelli lemma the measure μ_q is concentrated on $E_\mu(L_q'(0))$ and $E_{\mu_q}(\widetilde{L}_q'(0))$. So, it is carried by $E_\mu(\varphi'(q))$ and is exact dimensional of dimension $\varphi^*(\varphi'(q))$. This yields the expected lower bound for $\dim E_\mu(\varphi'(q))$ uniformly with respect to $q \in]q_-, q_+[$. To obtain the upper bound $\dim E_\mu(\alpha) \le \lim_{\epsilon\to 0} \liminf_{n\to\infty} f(\alpha, n, \epsilon)$ is easy. The study of points $\varphi'(q_-)$ et $\varphi'(q_+)$, more delicate, is done in [42] and [45].

5.2.1.3 Other works on b-adic cascades

Equation (5.5), valid with probability 1 for multiplicative cascades, induces an equation whose the unknown is a probability distribution. This equation gave rise to many works (in particular [27, 29]). The solutions to this equation generalize the semi-stable laws. This leads to include in a same class the generalized semi-stable processes, the stable Lévy subordinators and the indefinite integrals of Mandelbrot measures, $F(t) = \int_0^t \mathrm{d}\mu(u)$, by considering the more general object $L_\beta \circ F$, where L_β is a stable Lévy subordinators of index $\beta \in]0, 1]$ (L_1 is the identity). The multifractal analysis of these processes [46] raises real difficulties and requests the introduction of the notion heterogeneous ubiquity, which yields a non trivial generalization of the Jarnik theorem concerning the metric number theory [47].

Also, some limit theorems are obtained in this context of cascades. For the canonical cascades there are a law of large number, a central limit theorem, a law of iterated logarithm and a large deviation principle [48, 49] for measures μ_n when the branchind index b goes to infinity. In [50], the cascade construction is iterated and give birth to a functional CLT whose limit objet is an exotic gaussian process.

Self-similar fragmentation which is a close cousin to the cascades on trees [51] is also worth mentioning.

5.2.2 Random coverings and cascades

5.2.2.1 Random coverings and compound Poisson cascades

Random coverings

In 1956, A. Dvoretzky [52] asks the following question: given a sequence $(\omega_n)_{n\geq 1}$ of random variables with values in the torus $\mathbb{T}$, uniform, and independent, and $(\ell_n)_{n\geq 1}$ a non increasing sequence of numbers in $]0,1[$, under which condition on $(\ell_n)_{n\geq 1}$ does one has $\bigcup_{n\geq 1}(\omega_n, \omega_n+\ell_n) = \mathbb{T}$ with probability 1?

First the case $\ell_n = \alpha/n$ has been considered: if $\alpha < 1$, the circle is a.s. not covered (Billard, cf. [53]); if $\alpha > 1$ a.s. the circle is covered (Kahane, cf. [53]); the case $\alpha = 1$ has been solved independently by Mandelbrot [54] and Orey [55]. The complete solution is due to Shepp [56].

Compound Poisson cascades

Let ν be a Radon measure on $(0, +\infty)$ and λ the Lebesgue measure on $\mathbb{R}$. Let $\mathcal{P}$ be a Poisson process of intensity $\Lambda = \lambda \otimes \nu$ and W be a positive random variable. With points M of $\mathcal{P}$ we associate independent random variables W_M equidistributed with W, also independent of $\mathcal{P}$. Let $b \geq 2$ be an integer. For $n \geq 1$ and $t \in [0,1]$, consider the truncated cone

$$\mathcal{C}_n(t) = \left\{(t', r) : b^{-n} < r \leq b^{1-n},\ t - r/2 \leq t' < t + r/2\right\}.$$

Then we construct a $[0,1]$-martingale: for $t \in [0,1]$ and $n \geq 1$, set

$$P_n(t) = e^{-\Lambda\left(\mathcal{C}_n(t)\right)(\mathbb{E}(W)-1)} \prod_{M\in\mathcal{P}\cap\mathcal{C}_n(t)} W_M,$$

and

$$Q_n(t) = \prod_{k=1}^{n} P_k(t).$$

When $\nu(\mathrm{d}r) = \delta \mathrm{d}r/r^2$, with $\delta > 0$, the measure $Q \cdot \lambda_{|[0,1]}$ appears to be independent of the base b. Mandelbrot [57] proposed it as a model for turbulence. This has the advantage on the usual cascades to be free of any grid (such a grid has no physical meaning; to achieve this goal one could also define a multiplicative chaos in continuous time [57]). The weights $P_k(t)$ are i.i.d. and the results concerning the non-degeneracy, the bounds of moments, and the multifractal analysis match those of the b-adic canonical cascades. The corresponding φ function still is given by 5.4, but has to be computed with the random weight $P_1(t)$. This gives

$$\varphi(q) = q - 1 + \delta(q(\mathbb{E}(W) - 1) - (\mathbb{E}(W^q) - 1)).$$

The martingales $Q_n \cdot \lambda$ are called compound Poisson cascades. Besides their own interest, these martingales play a major rôle when finely analysing random coverings [51, 58]. For instance, in the Dvoretzky problem with $\ell_n = \alpha/n$, let $N_n(t)$ stand for the number of arcs among the n-th first which cover t. One has $\mathbb{E}(N_n(t)) = \ell_1 + \cdots + \ell_n \sim \alpha \log(n)$ for each $t \in \mathbb{T}$; then the Fubini theorem gives $\lim_{n\to\infty} \frac{N_n(t)}{\alpha \log(n)} = 1$ with probability 1, almost surely with respect to the Lebesgue measure. By using the connection between the Dvoretzky problem and the Poisson coverings due to Mandelbrot and considering a suitable family of Poisson cascades, one can prove that with probability 1, for all $\beta \geq 0$, the Hausdorff dimension of the set of points t of $\mathbb{T}$ such that $\lim_{n\to\infty} \frac{N_n(t)}{\alpha \log(n)} = \beta$ is given by $1 - \alpha$ $(1 - \beta + \beta \log \beta)$.

5.2.2.2 Log-infinitely divisible cascades

The compound Poisson cascades can be embedded within the larger framework of log-infinitely divisible cascades introduced in [59] to model turbulence and financial data.

We use the notation of the previous section. Let ψ be a Lévy characteristic exponent defined on $\mathbb{R}$, i.e.,

$$\psi : \xi \in \mathbb{R} \mapsto \mathrm{i}a\xi - \sigma^2\xi^2/2 + \int_{\mathbb{R}} \left(1 - e^{\mathrm{i}\xi x} + \mathrm{i}\xi x \mathbf{1}_{|x|\leq 1}\right)\right) \pi(\mathrm{d}x), \tag{5.8}$$

where $a \in \mathbb{R}$, $\sigma \in \mathbb{R}$, and π is a Radon measure on $\mathbb{R} \setminus \{0\}$ such that $\int (1 \wedge |x|^2)\pi(\mathrm{d}x) < \infty$.

Let ρ be an infinitely divisible signed random measure on $\mathbb{R} \times \mathbb{R}_+^*$, of control measure Λ and of Lévy exponent ψ (see [60] for its construction).

So, for all Borel set $B \in \mathbb{R} \times \mathbb{R}_+^*$ an all $\xi \in \mathbb{R}$, one has

$$\mathbb{E}(e^{\mathrm{i}\xi\rho(B)}) = \exp\left(\psi(\xi)\Lambda(B)\right)$$

and, for all finite family $\{B_i\}$ of mutually disjoint Borel sets in $\mathbb{R}\times\mathbb{R}_+^*$ such that $\Lambda(B_i) < \infty$, the random variables $\rho(B_i)$ are independent.

Let I be the interval consisting of those $\xi \in \mathbb{R}$ such that $\int_{|x|\geq 1} e^{\xi x}\pi(\mathrm{d}x) < \infty$. The function ψ can be naturally extended on $-\mathrm{i}I$. If one assumes $1 \in I$, then, without loss of generality, one may suppose $\psi(-\mathrm{i}) = 1$, and define, for $n \geq 1$ and $t \in [0,1]$, the random weight (of expectation 1)

$$P_n(t) = \exp\left(\rho(\mathcal{C}_n(t))\right),$$

and the $[0,1]$-martingale $Q_n(t) = \prod_{k=1}^n P_n(t)$.

When $\nu(\mathrm{d}r) = \mathrm{d}r/r^2$, one has the same results as for the compound Poisson cascades. The proofs use the same ingredients (see [61] which deals with a class of abstract $[0,1]$-martingales containing those described in this article as well as other such objects statistically self-similar). The corresponding φ function is

$$\varphi(q) = q - 1 - \psi(-\mathrm{i}q).$$

Another important innovation brought by [59] is a modification of the weight $P_1(t)$ in order to obtain a remarkable property of scale invariance for the masses of sub-intervals of $[0,1]$.

The attention on the gaussian log-infinitely divisible cascades ($a = 0$ and $\pi = 0$), as well as the log-normal canonical cascades, has recently been drawn in connection with quantum gravity and the proof of the "KPZ formula", where $\log(Q_n(t))$ is the free gaussian field [62–64]. They also come into play to construct processes close to curves SLE_κ by the conformal welding method [65].

5.3 Signed multiplicative cascades

Certain intermittent phenomena, as the change of speed within a turbulent fluid or fluctuations of stock prices chronological series, are naturally modeled by functions. So, models of random multifractal functions are needed. One can obtain some by composing a function B of constant irregularity (said to be monofractal), such as a fractional brownian motion or a Weierstrass function, and a non-decreasing function F (for instance the primitive of a multiplicative cascade).

This model has been introduced and promoted by Mandelbrot [66] (then some variants have been also introduced, which consist in integrating a $[0,1]$-martingale against a fractional brownian motion). His motivation was to try and write a complex function under the form $B \circ F$ where B is monofractal (i.e., there exists $H \in (0,1)$ such that, for all t, $\liminf\limits_{r\to 0^+} \frac{\log \mathrm{Osc}_B([t-r,r+r])}{\log(r)} = H$) and F is a change of time which bears the whole multifractal behavior.

This kind of representation works for certain self-similar or self-affine functions, whose multifractal analysis then becomes easier ([66–68]). It is also valid in a generic sense in some classes of functions [69].

Another construction consists in using wavelets expansions with suitably chosen coefficients [70–72].

But, from the very beginning of his work on multiplicative processes, Mandelbrot was aware of the fact that multiplicative cascades with weights allowed to take negative or even complex values [3] would lead to interesting random functions. It is only recently that this idea has been developed. Let us now give a short glimpse at the last studies impulsed by Mandelbrot's work in this direction. In particular, it appears that almost-sure limits of signed multiplicative cascades can themselves decompose into a monofractal motion with multifractal time, and that the study of certain degenerate cases has yielded a description of brownian motion in multifractal time as the limit process of a functional TCL. We are going to give an account of this for the case of b-adic cascades.

5.3.1 Construction and convergence theorems

Now $W=(W_0,\dots,W_{b-1})$ is a random vector with complex and integrable components such that $\mathbb{E}(\sum_{j=0}^{b-1} W_j) = 1$. Let $\{W(w)=(W_{w\cdot 0},\dots, W_{w\cdot b-1})\}_{w\in\mathscr{A}^*}$ be independent copies of W. For $n \geq 1$, consider the following function on $[0,1]$:

$$F_{W,n}(t) = \int_0^t b^n W_{u_{|1}} W_{u_{|2}} \cdots W_{u_{|n}} \, du,$$

where $u_{|n}$ is the word whose letters are the n first digits of the b-adic expansion of u.

In case the weights are non-negative, $F_{W,n}$ is the primitive of the measure μ_n associated with vectors $(W_{w\cdot 0}/\mathbb{E}(W_0),\dots,W_{w\cdot b-1}/\mathbb{E}(W_{b-1}))$

and with the probability vector $\lambda = (\mathbb{E}(W_0), \ldots, \mathbb{E}(W_{b-1}))$ according to Sec. 5.2.1.1.

Now, we set

$$\varphi_W(q) = -\log_b \mathbb{E}\left(\sum_{j=0}^{b-1} |W_j|^q\right).$$

5.3.1.1 Strong convergence and decomposition

There are conditions sufficient to imply the convergence of $F_n = F_{W,n}$, which in view of the non-negative case seem optimal (Theorems 5.2.1 and 5.2.3). The following results, illustrated by Figs. 5.3 and 5.4, are proved in [73, 74]. It is natural to study separately the *conservative* case (for which $W_0 + \cdots + W_{b-1}$ is 1 a.s.) and the non conservative case.

Theorem 5.3.1 *Suppose* $\mathbb{P}\big(\sum_{i=0}^{b-1} W(i) \neq 1\big) > 0$ *and assume that there exists* $q > 1$ *such that* $\varphi_W(q) > 0$. *Also suppose* $q \in]1, 2]$ *or* $\varphi_W(2) > 0$.

1. $(F_n)_{n\geq 1}$ *uniformly converges, with probability* 1, *and in* L^q *norm, when* n *goes to* ∞, *towards a Hölder function* $F = F_W$.
2. $F = \sum_{i=0}^{b-1} \mathbf{1}_{[i/b,(i+1)/b]}(F(i/b) + W(i)\, F_i \circ S_i^{-1})$, *where* $S_i(t) = (t+i)/b$, W, $F_0, \ldots, F_{b-1}$ *are independent,* F_i *is equidistributed with* F, *and the equality holds with probability* 1.

In the non-conservative case, we obtain a non-trivial generalisation of the classical construction of auto-affine functions [75] which corresponds to deterministic weights. In particular, the second assertion, somewhat unexpectedly, yields functions, statistically self-affine, nowhere locally Hölder, which exhibit new multifractal behaviors (cf. Sec. 5.3.2).

Theorem 5.3.2 *Suppose* $\mathbb{P}\big(\sum_{i=0}^{b-1} W(i) = 1\big) = 1$.

1. *If there exists* $q > 1$ *such that* $\varphi_W(q) > 0$, *the same conclusion as in the preceding theorem holds.*
2. (Critical case) *Suppose* $\lim_{q\to\infty} \varphi_W(q) = 0$, *i.e.,* $\mathbb{P}(\forall\, 0 \leq i \leq b-1, |W(i)| \leq 1) = 1$ *and* $\sum_{i=0}^{b-1} \mathbb{P}(|W(i)| = 1) = 1$.

   Also assume $\mathbb{P}(\mathrm{card}\{i : |W(i)| = 1\} = 1) < 1$ *and that there exists* $\gamma \in (0, 1)$ *such that, with probability* 1, *for all* i *with* $0 \leq i < b$, *one has*

$$|W_i| \leq \gamma$$

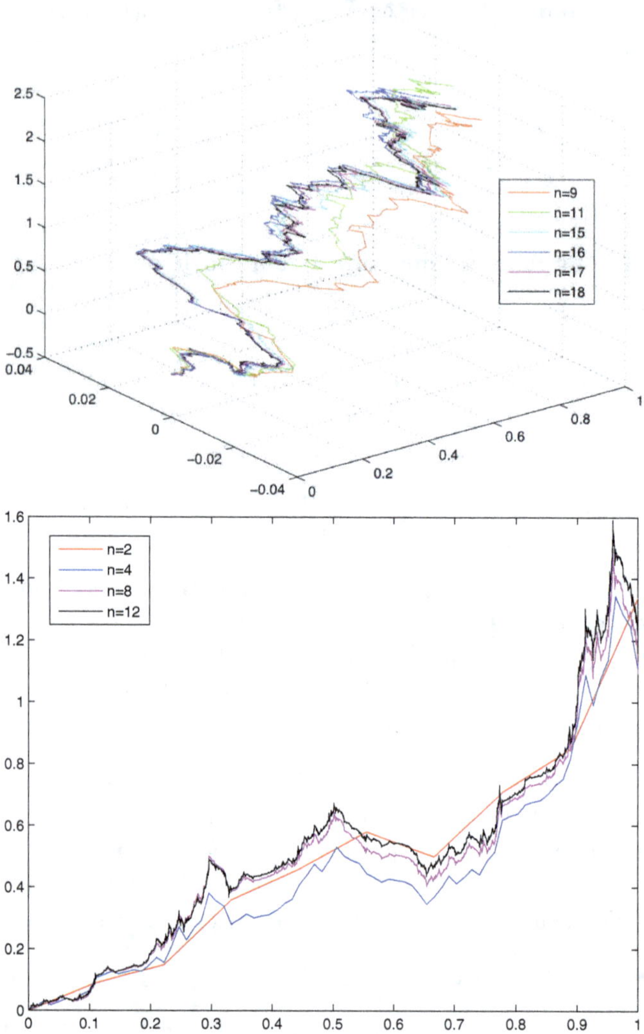

Figure 5.3 Top: uniform convergence of a complex dyadic conservative cascade: F_n for $n \in \{9, 11, 15, 16, 17, 18\}$. Bottom: uniform convergence of a non-canonical, non-conservative real cascade: $F_{W,n}$ for $n \in \{2, 4, 8, 12\}$ $(b = 3)$.

or

$$|W_i| = 1 \ et \ \left(\sum_{k=0}^{i-1} W(i), \sum_{k=0}^{i} W(i) \right) \in \{(0,1), (1,0)\}.$$

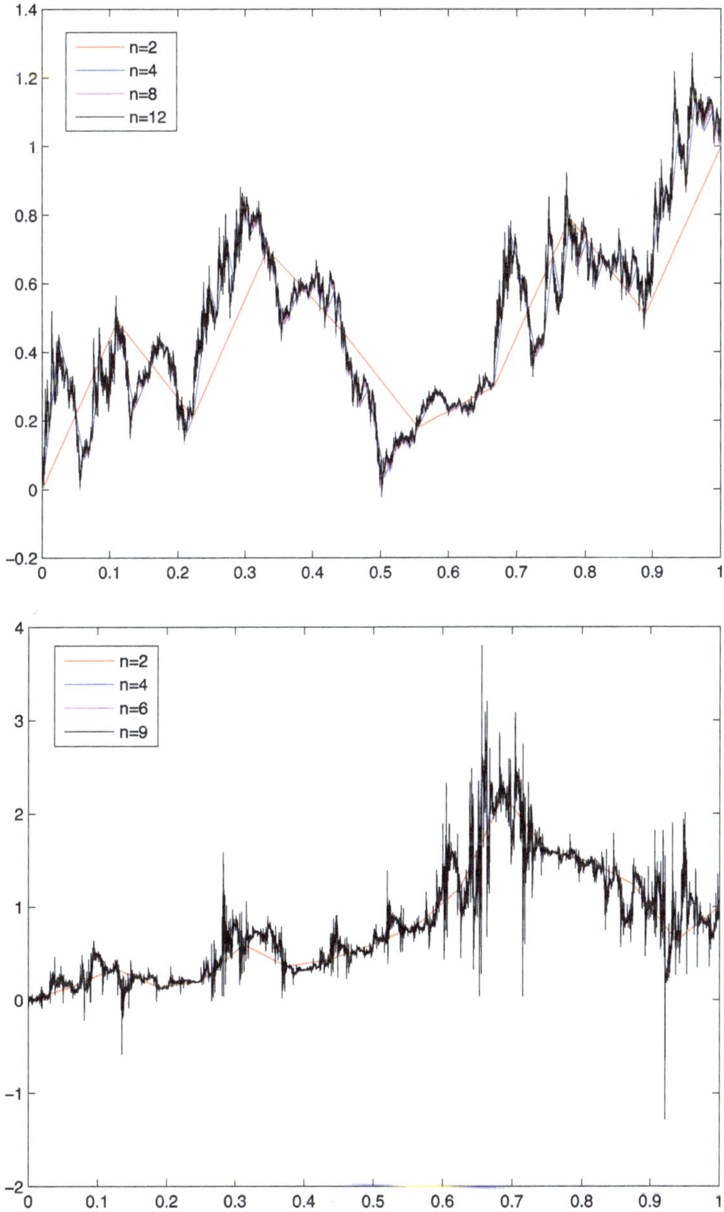

Figure 5.4 Top: uniform convergence of a real valued conservative cascade, in the non-critical case: $F_{W,n}$ pour $n \in \{2,4,8,12\}$ $(b=3)$. Bottom: uniform convergence of a real valued conservative cascade, in the critical case: $F_{W,n}$ for $n \in \{2,4,6,9\}$ $(b=4)$.

Then, with probability 1, $(F_n)_{n\geq 1}$ *uniformly converges towards a limit* $F = F_W$ *which is nowhere locally Hölder and which fulfills the second part of Theorem* 5.3.1.

The statistical self-affinity expressed by Theorem 5.3.1 could be considered as an extension of Eq. (5.5). Coming back to a point already mentioned in Sec. 5.2.1.3, we remark that the equality in law which results is the same as the one that is satisfied by a Lévy process composed with the integral of an independent Mandelbrot measure.

A natural question is: Can these limit functions be decomposed as a monofractal process in multifractal time? Since the answer is affirmative in the deterministic case [66], this is expected to be true under suitable hypotheses. Nevertheless it is thought that this does not hold in general (see the discussion of the critical conservative case in [74]). We need a new definition.

Definition 5.3.1 If $\beta > 0$ and $\varphi_W(\beta) > -\infty$, then for $w \in \mathscr{A}^*$ set

$$W^{(\beta)}(w) = b^{\varphi_W(\beta)}(|W_{w0}|^\beta, \dots, |W_{w\cdot b-1}|^\beta).$$

We have $\varphi_{W^{(\beta)}}(q) = \varphi_W(\beta q) - q\varphi_W(\beta)$ for all $q > 0$. In particular $\varphi_{W^{(\beta)}}(1) = 0$, and if there exists $q \in]1, 2[$ with $\varphi_{W^{(\beta)}}(q) > 0$, we denote by $F_{W^{(\beta)}}$ the non-decreasing function, a.s. uniform limit of

$$F_{W^{(\beta)},n}(t) = \int_0^t W^{(\beta)}_{u_{|1}} W^{(\beta)}_{u_{|2}} \cdots W^{(\beta)}_{u_{|n}} b^n du$$

in Theorems 5.3.1 and 5.3.2. The function $F_{W^{(\beta)}}$ is a.s. increasing if the $|W_j|$ are a.s. positive.

Under suitable hypotheses, one has the following result [74] illustrated by Fig. 5.5.

Theorem 5.3.3 *Suppose that we have* $\mathbb{P}(W \in \mathbb{C}^b \setminus \mathbb{R}^b_+) > 0$ *and* $\varphi_W > -\infty$ *on* $\mathbb{R}$ *and that the hypotheses of Theorem* 5.3.1 *or Theorem* 5.3.2(1) *are fulfilled.*

Let $\beta = \min\{q : \varphi_W(q) = 0\}$. *Suppose* $\varphi_W(q) > 0$ *for all* $q > \beta$. *Then* $\beta > 1$, $F_{W^{(\beta)}}$ *is non degenerate, and, with probability* 1, *the function* $B_{1/\beta} = F_W \circ F^{-1}_{W^{(\beta)}}$ *is monofractal with Hölder exponent* $1/\beta$.

When the weights W_j are real and such that a.s. $|W_j| = b^{-1/\beta}$, the change of time is the identity and the limit function F is a monofractal

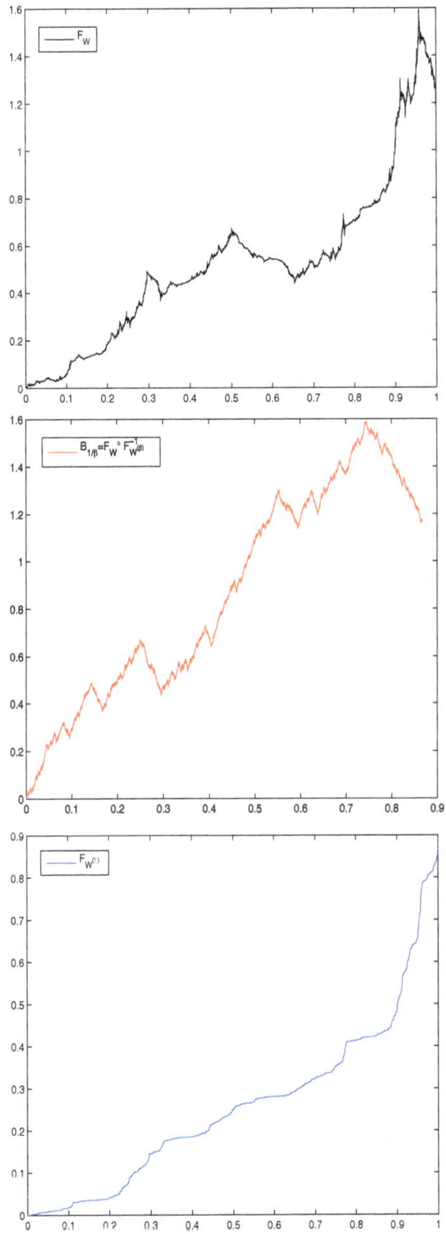

Figure 5.5 The limit function F_W on the right of Fig. 5.2 (top) writes as the monofractal function $B_{1/\beta} = F_W \circ F^{-1}_{W^{(\beta)}}$ of exponent $H = 1/\beta \approx 0.7168$ (middle) in multifractal time $F_{W^{(\beta)}}$ (bottom).

object obtained by a multiplicative cascade. In the non conservative case, one gets a process which shares lots of properties with the fractional brownian motion of the same index, but its construction is more straigthforward. This remarkable fact was one of Mandelbrot's first motivations to consider the signed cascades. Nevertheless there is a constraint on the exponent $1/\beta$: due to Theorem 5.3.1, it should lie within the interval $]1/2, 1]$ (see [76] for the study of the monofractal case).

5.3.1.2 Weak convergence towards a brownian motion in multifractal time

When one cannot use Theorem 5.3.1, it is natural to seek for a suitable normalisation of the process $F_{W,n}$ in order it be convergent in distribution. When the weights are non-negative, such a normalisation is known in the critical situation where $\varphi'(1) = 0$: $F_{W,n}/F_{W,n}(1)$ converges in probability but not almost surely, conditionally to the fact that $F_{W,n}(1)$ remains positive as $n \to \infty$ [77]; the limit process is the indefinite integral of the measure (normalized by its total mass) naturally associated with 5.3 when $\varphi'(1) = 0$ (see [78, 79]) and studied in [42]. For the signed cascades, there is also a natural normalization, which sometimes makes the process to converge weakly towards a brownian motion in multifractal time. We assume the following hypotheses.

1. The cascade is non conservative and the components of W are real.
2. $\varphi_W > -\infty$ on $\mathbb{R}_+$.
3. $\varphi_W(2) \leq 0$ and φ_W is non-decreasing.

It follows from these hypotheses that $|W_j| \leq 1$ a.s. for all j, and also that the weights are not positive (otherwise one would have $\varphi_W(1) = 0 \geq \varphi_W(2)$, which would contradict the increase of φ_W since this function is concave); in particular the hypotheses of Theorem 5.3.1 are not fulfilled, and a direct computation shows that the martingale $F_n(1)$ is not bounded in L^2. More precisely

$$\mathbb{E}(|F_n(1)|^2) \sim \begin{cases} \sigma^2 b^{-n\varphi_W(2)} \quad \text{with } \sigma^2 = \dfrac{\mathbb{E}\left(\left|\sum_{i=0}^{b-1} W_i\right|^2\right)-1}{\mathbb{E}\left(\sum_{i=0}^{b-1} |W_i|^2\right)-1} & \text{if } \varphi_W(2) < 0 \\ \sigma^2 n \quad \text{with } \sigma^2 = \sum_{i\neq j} \mathbb{E}(W_i W_j) & \text{if } \varphi_W(2) = 0 \end{cases}.$$

Moreover, (3) implies $\varphi_{W^{(2)}}(q) > 0$ for all $q > 1$. So, the non decreasing function $F_{W^{(2)}}$ (cf. Definition 5.3.1) is not degenerate.

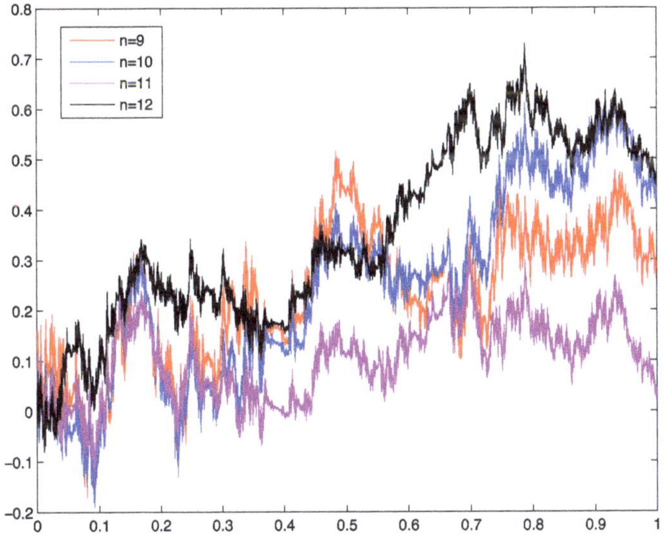

Figure 5.6 Weak convergence of $F_n/\sqrt{\mathbb{E}(F_n(1)^2)})$ towards a brownian motion in multifractal time (non conservative case).

Theorem 5.3.4 *With the same hypotheses, the random continuous function $F_n/\sqrt{\mathbb{E}(F_n(1)^2)}$ converges in distribution towards $B \circ F_{W^{(2)}}$, where B is a standard Brownian motion independent of $F_{W^{(2)}}$.*

This theorem is illustrated by Fig. 5.6.

5.3.2 Multifractal analysis of roughness in the graph of F

The multifractal analysis of the limit functions F in Theorems 5.3.1 and 5.3.2 is a refinement of the one of μ. Now, one considers the Hölder exponent associated with the oscillations of order 1 of F,

$$h_F(t) = \liminf_{r \to 0^+} \frac{\log \mathrm{Osc}_F([t-r, r+r])}{\log r}$$

and the corresponding level sets

$$E_F(h) = \{t \in [0,1] : h_F(t) = h\} \quad (h \geq 0).$$

The exponent $h_F(\cdot)$ should be considered as a measure of the aspect more or less rough of the graph of F at each point: the smaller $h_F(t)$, the larger roughness at t is.

Let

$$\tau_F(q) = \liminf_{r\to 0} \frac{\log \sup\left\{\sum_i \mathrm{Osc}_F(B_i)^q\right\}}{\log r},$$

where the supremum is taken over all families of pairwise disjoint closed intervals B_i, of diameter $2r$, and centered in $\mathrm{supp}(F')$, where F' is the derivative of F in the distribution sense. This is a grid free version of the free energy function τ_μ which is fitted to studying the sets $E_F(h)$. The following statement is, on purpose, sloppy on the hypotheses to avoid tedious discussions. The precise theorem can be found in [45], where a finer multifractal analysis of F is described (also see [51, 72, 80–82] to have a complete overview of the multifractal formalism for functions).

Theorem 5.3.5 *Under suitable hypotheses, with probability* 1, *for all* $h \geq 0$, $E_F(h)$ *is non-empty if and only if* $\tau_F^*(h) \geq 0$, *and when it is so* $\dim E_F(h) = \tau_F^*(h)$.

This result is illustrated on Fig. 5.7.

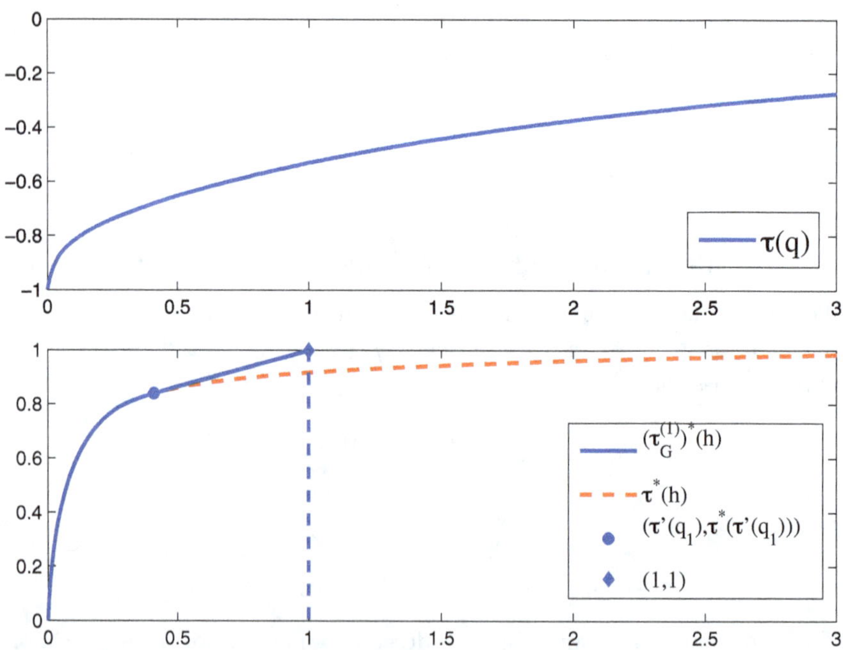

Figure 5.7 Function τ_F (top) and its Legendre transform (bottom) for a limit of a critical and conservative cascade.

It follows from this theorem that, for the limits of conservative critical signed cascades (Theorem 5.3.2), the multifractal spectrum of F, i.e., the map $h \mapsto \dim E_F(h)$, vanishes at 0 and has an infinite right derivative at 0. It seems that up to now no natural [70–72] or generic [81, 83] construction exhibits this behavior. Moreover one can also get a non-decreasing spectrum whose range is the whole interval $\varphi'_W(0+) = -\infty$: see Fig. 5.6.

Nevertheless, the sets $E_F(h)$ are only indirectly linked to the apparent roughness of the graph because they lie in the support of F. This leads to consider the *roughness spectrum* which gives the Hausdorff dimension of the sets obtained by lifting on the graph of F the sets $E_F(h)$. In other terms, this spectrum is the Hausdorff dimensions of the sets

$$G_F(h) = \{(t, F(t)) : t \in E_F(h)\} \quad (h \geq 0).$$

The following result [84] is the first of this kind for multifractal stochastic processes.

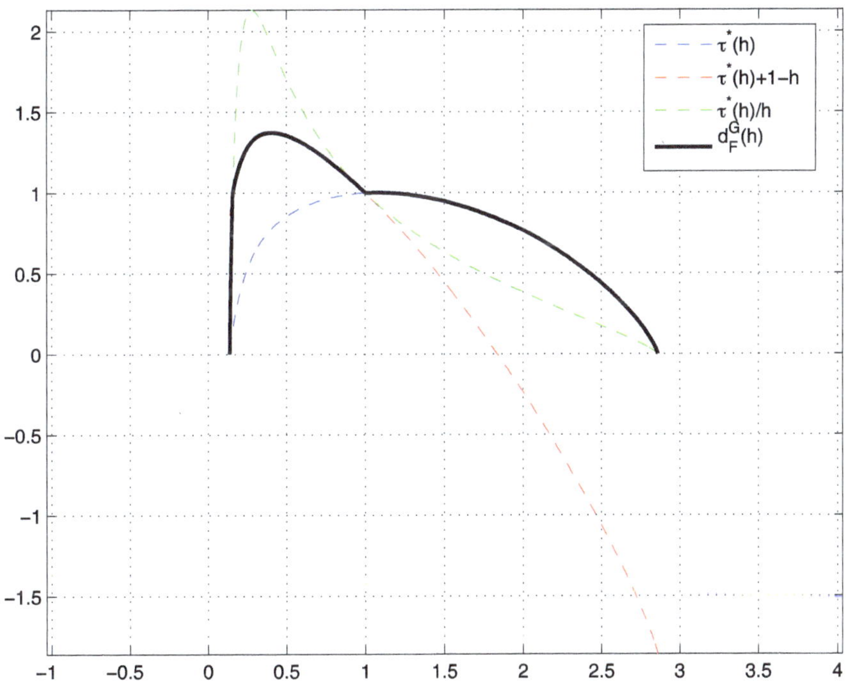

Figure 5.8 Example of roughness spectrum $h \mapsto \dim G_F(h)$.

Theorem 5.3.6 *In the the real valued non-conservative case and under suitable hypotheses, with probability* 1, *one has*

$$\dim \operatorname{Graph}(F) = 1 - \tau(1),$$

and, for all $h \geq 0$ *such that* $E_F(h) \neq \emptyset$, *one has*

$$\dim E_F(h) = \tau_F^*(h)$$

and

$$\dim G_F(h) = \left(\frac{\dim E_F(h)}{h} \wedge (\dim E_F(h) + 1 - h) \right) \vee \dim E_F(h).$$

This result is illustrated on Fig. 5.8.

Acknowledgment

The authors are grateful to Françoise Axel for valuable suggestions.

Bibliography

[1] A. Kolmogorov, *Mécanique de la turbulence, Colloq. intern. CNRS Marseille* 1961, Editions CNRS 1962, 447.
[2] B. Mandelbrot, Pages 333–351 of *Statistical Models and Turbulence*, M. Rosenblatt, C. Atta, eds. (Springer-Verlag, New York, 1972).
[3] B. Mandelbrot, *C. R. Acad. Sci. Paris* **278** (1974) 289 and 355.
[4] E. Novikov, R. Stewart, *Isvestia Akademii Nauk SSR Earia Geofizicheskaia* **3** (1964) 408.
[5] A. Yaglom, *Doklady Akademii Nauk SSSR* **16** (1966) 49.
[6] J. Kahane, *C. R. Acad. Sci. Paris* **278** (1974) 567.
[7] J.-P. Kahane, J. Peyrière, *Adv. Math* **22** (1976) 131.
[8] J. Peyrière, *C. R. Acad. Sci. Paris* **278** (1974) 567.
[9] J.-P. Kahane, *Ann. Sci. Math. Québec* **9** (1985) 105.
[10] J.-P. Kahane, *Chi. Ann. Math* **8B1** (1987) 1.
[11] U. Frisch, P. Sulem, M. Nelkin, *J. Fluid Mech.* **87** (1978) 719.
[12] U. Frisch, G. Parisi, pgs 84–88 of *Turbulence and Predictability in Geophysical Fluid Dynamics and Climate Dynamics* (M. Ghil, ed., North Holland, 1985).
[13] T. Halsey, M. Jensen, L. Kadanoff, I. Procaccia, B. Shraiman, *Phys. Rev. A* **33** (1986) 1141.
[14] H. Hentschel, I. Procaccia, *Physica D* **8** (1983) 435.
[15] B. Mandelbrot, *IEEE Trans. Commun. Tech.* **13** (1965) 71.
[16] P. Collet, J. Lebowicz, A. Porzio, *J. Stat. Phys.* **47** 1987, 609.
[17] G. Brown, G. Michon, J. Peyrière, *J. Statist. Phys.* **66** (1992) 775.
[18] L. Olsen, *Adv. Math.* **116** (1995) 92.

[19] F. Ben Nasr, I. Bhouri, Y. Heurteaux, *Adv. Math.* **165** (2002) 264.
[20] J. Peyrière, *Proc. Symp. Pure Math.* **72** (2004) 217.
[21] D. Ruelle, *Ergod. Th. & Dynam. Sys.* **2** (1982) 99.
[22] T. Bedford, *J. London Math. Soc.* **33** (1986) 89.
[23] T. Bohr, D. Rand, *Physica D* **25** (1987) 387.
[24] R. Bowen, *Publ. Math. IHES* **50** (1979) 11.
[25] D. Sullivan, *Acta Mathematica* **147** (1981) 259.
[26] B. Mandelbrot, *J. Fluid. Mech.* **62** (1974) 331.
[27] R. Durrett, R. Liggett, *Z. Wahrsch. verw. Gebiete* **64** (1983) 275.
[28] J.-P. Kahane, *Ann. Inst. Henri Poincaré Probab. Stat.* **23** (1987) 289.
[29] Y. Guivarc'h, *Ann. Inst. H. Poincaré Prob. Stat.* **26** (1990) 261.
[30] Q. Liu, *Stoch. Proc. Appl.* **95** (2001) 83.
[31] J. Peyrière, *Duke Math. J.* **44** (1977) 591.
[32] F. Ben Nasr, *C. R. Acad. Sc. Paris Sér. I Math.* **304** (1987) 255.
[33] A. Fan, *J. Math. Pures Appl.* **81** (2002) 967.
[34] Q. Liu, *Adv. Appl. Prob.* **29** (1997) 353.
[35] R. Ellis, *Entropy, Large Deviations, and Statistical Mechanics* (Springer-Verlag, New York, 1985).
[36] E. Bacry, A. Gloter, M. Hoffmann, J. Muzy, *Ann. Appl. Prob.* **20** (2010) 1729.
[37] P. Collet, F. Koukiou, *Commun. Math. Phys.* **147** (1992) 329.
[38] R. Holley, E. Waymire, *Ann. Appl. Probab.* **2** (1992) 819.
[39] K. Falconer, *J. Theor. Prob.* **7** (1994) 681.
[40] G. Molchan, *Commun. Math. Phys.* **179** (1996) 681.
[41] M. Arbeiter, N. Patszchke, *Math. Nachr.* **181** (1996) 5.
[42] J. Barral, *J. Theoret. Probab.* **13** (2000) 1027.
[43] M. Ossiander, E. Waymire, *Ann. Stat.* **28** (2000) 1.
[44] J. Biggins, *Ann. Prob.* **20** (1992) 137.
[45] J. Barral, X. Jin, *Commun. Math. Phys.* **219** (2010) 129.
[46] J. Barral, S. Seuret, *Adv. Math.* **214** (2007) 437.
[47] J. Barral, S. Seuret, *Bull. Braz. Math. Soc.* **38** (2007) 476.
[48] Q. Liu, A. Rouault, *Ann. Appl. Probab.* **10** (2000) 218.
[49] Q. Liu, E. Rio, A. Rouault, *J. Theoret. Prob.* **16** (2003) 971.
[50] J. Barral, J. Peyrière, Z.-Y.Wen, *Probab. Th. Relat. Fields* **144** (2009) 615.
[51] J. Barral, J. Berestycki, J. Bertoin, A.-H. Fan, B. Haas, S. Jaffard, G. Miermont, J. Peyrière, *Quelques Interactions entre Analyse, Probabilités et Fractals* (Société Mathématique de France, 2010).
[52] A. Dvoretzky, *Pro. Nat. Acad. Sci. USA* **42** (1956) 199
[53] J.-P. Kahane, *Some random series of functions* (Cambridge Univ. Pr., Cambridge, 1985).
[54] B. Mandelbrot, *Z. Wahrsch. v. Geb.* **22** (1972) 145.
[55] S. Orey, "Random arcs on the circle," University of Minnesota, 1971.
[56] L. Shepp, *Israel J. Math.* **11** (1972) 328.
[57] J. Barral, B. Mandelbrot, *Probab. Th. Relat. Fields* **124** (2002) 409.
[58] J. Barral, A. Fan, *Bull. Sci. Math.* **129** (2005) 275.

[59] E. Bacry, J. Muzy, *Commun. Math. Phys.* **236** (2003) 449.
[60] B. Rajput, J. Rosinski, *Prob. Th. Relat. Fields* **82** (1989) 451.
[61] J. Barral, B. Mandelbrot, pgs. 3–90 In M. Lapidus, M. Frankenhuijsen, eds. *Fractal geometry and applications: a jubilee of Benoît Mandelbrot* (Amer. Math. Soc, Providence, 2004).
[62] I. Benjamini, O. Schramm, *Comm. Math. Phys.* **289** (2009) 653.
[63] R. Rhodes, V. Vargas, *ESAIM Prob. Stat.*, doi:10.1051/ps/2011110 (2010).
[64] B. Duplantier, S. Sheffield, *Phys. Rev. Lett.* **102** (2009) 150603.
[65] K. Astala, P. Jones, A. Kupiainen, E. Saksman, *Acta Math.* **207** (2011) 203.
[66] B. Mandelbrot, *Fractals and Scaling in Finance: Discontinuity, Concentration, Risk* Springer, New York, 1999.
[67] S. Jaffard, B. Mandelbrot, *Math. Intelligencer* **19** (1997) 21.
[68] S. Seuret, *Adv. Math.* **220** (2009) 936.
[69] Z. Buczolich, S. Seuret, *J. Fourier Analysis Appl.* **17** (2011) 457.
[70] A. Arneodo, E. Bacry, J. Muzy, *J. Math. Phys.* **39** (1998) 4142.
[71] J. Aubry, S. Jaffard, *Commun. Math. Phys.* **227** (2002) 483.
[72] J. Barral, S. Seuret, *J. Fourier Analysis Appl.* **11** (2005) 589.
[73] J. Barral, X. Jin, B. Mandelbrot, *Ann. Appl. Probab.* **20** (2010) 1205.
[74] J. Barral, X. Jin, B. Mandelbrot, *Ann. Appl. Probab.* **20** (2010) 1219.
[75] N. Kôno, *Japan J. Appl. Math.* **3** (1986) 259.
[76] J. Barral, B. Mandelbrot, *Ann. Inst. H. Poincaré, Probab. Stat.* **45** (2009) 1116.
[77] E. Aidekon, Z. Shi, Martingale ratio convergence in the branching random walk, arXiv:1102.0217.
[78] J. Biggins, A. Kyprianou, *Ann. Probab.* **25** (1997) 337.
[79] Q. Liu, *Stoch. Proc. Appl.* **86** (2000) 263.
[80] S. Jaffard, *J. Math. Phys.* **39** (1998) 4129.
[81] S. Jaffard, *J. Math. Pures Appl.* **79** (2000) 525.
[82] S. Jaffard, Pages 91–151 in M. Lapidus, M. Frankenhuijsen, eds. *Fractal geometry and applications: a jubilee of Benoît Mandelbrot* (Amer. Math Sco., Providence, 2004).
[83] S. Jaffard, Y. Meyer, *J. Funct. Analysis.* **175** (2000) 415.
[84] X. Jin, *Adv. Math.* **226** (2011) 4987.

Chapter 6

Benoit Mandelbrot and art

Javier Barrallo

The loss of Benoit Mandelbrot has left us without one of the great scientific visionaries, not only of the twentieth century but of the entire history of science. His revolutionary interpretation of the world situates him at the height of genius like Archimedes, Kepler, Galileo, Newton or Einstein, geniuses who, armed only with a privileged imagination could discover what no one had seen before.

Benoit's major work has been called extraordinarily beautiful, iconic and evocative: Fractal. A word that has gone beyond the scientific jargon and today is part of everyday language in many idioms. It is contained even in the elitist dictionary of the Royal Spanish Academy of Language, reluctant to take non-Spanish terms.

Benoit has not only left his written work, but has also excelled as a teacher, communicator, master and inspiration for other scientists. But above all, has managed to broadcast his work to the public. He probably is the only major twentieth-century mathematician who has managed to contact the average citizen and transmit his discoveries outside the usual channels of scientific disclosure: journals, seminars or conferences. Today half the world knows the term fractal, and moreover is able to recognize its forms in nature, and correctly identify them on posters, disks and shirts.

Surprisingly, this populist side seems to irritate some mathematicians. I remember waiting over 40 minutes while Benoit patiently signed autographs and had his picture taken with fans after a conference in Madrid. He was more like a rock star or a Hollywood actor. In fact it is a unique case of popular affection to a scientist. And the key to that admiration must be sought not only in his enormous charisma, but the ease with which he illustrated and published his theories and conjectures, avoiding unnecessary mathematical jargon.

In an increasingly technological society, which urges interdisciplinarity and knowledge transfer, Benoit clung to a speaking and writing style accessible to the public. His work was a perfect example of what should be the role of applied mathematics. Actually, more than a mathematician, Benoit was a scientific generalist, whose worldview of science allowed him to weave a beautiful web that linked mathematics to physics, biology, economics, engineering, physiology,... matters in which he was a true scholar.

Looking at the awards won by Benoit throughout his long career highlights the enormous range of scientific fields and areas in which he was honored. In his last visit to Bilbao, a fascinated and slightly surprised Benoit told me that an Italian university had granted him a Doctorate in Medicine and Surgery. Even he was amazed at how over time fractal geometry had become a common link between so many different disciplines.

But Benoit not only had extensive scientific training, he also dazzled with his humanistic and artistic knowledge. I remember that while we walked in a beautiful autumn morning in San Sebastian, Benoit looked at a sculpture located on the rocks of the sea. It was *The Comb of the Wind* by Eduardo Chillida. Immediately Benoit recognized the author of the sculpture, and also told me that a few miles away, at Loyola, San Ignacio, founder of the Jesuits, was born and in the nearby port of Guetaria was born in the fifteenth century Juan Sebastian Elcano, the first person to complete the circumnavigation. He had an unheard of culture for a foreign citizen.

From a young age, Benoit was an atypical mathematician. He moved away from the Bourbaki School, omnipresent in post-war France, and to which his uncle Szolem Mandelbrot belonged. He sought solutions to the problems of the time using geometry instead of plunging into the dominant formalism. Thanks to its peculiar way of understanding mathematics, many researchers understand his work in an intuitive and visual way, as opposed to the abstract mathematics prevalent in the late twentieth century. Thanks to Benoit, mathematics recovered its lost appeal among many young people.

Benoit's great discovery, fractal geometry, was the missing puzzle piece to complete the map of twentieth century mathematics. Until then, geometry did not seem very useful for modeling a tree, a cloud, or our own bodies. But from the findings of Benoit, the map of the geometry which began to be drawn in Greece two thousand years ago is now completed. Thanks to Fractal Geometry now everything seems to make sense and looks incredibly logical. And geometry, one of the most underrated branches of mathematics in the second half of the twentieth century, has started a great revival of popularity.

Fractal geometry was also generous with its discoverer. A curious combination of perseverance, intuition and luck led Benoit to the discovery of the simple formula, $z \rightarrow z^2 + c$, that produces the set which bears his name. In fact, for many people, this is the most brilliant contribution of his career. But let me disagree. There is no doubt of the great importance of the Mandelbrot set, not only for its incredible beauty, complexity and ability to serve as a testing bench to iterate polynomial functions with a complex variable. But do not forget that the same formula was discussed by Gaston Julia, and that Benoit had significant help from his programmers in his IBM years.

Certainly the intuition and the works of Benoit advanced by ten or twenty years the study of such functions. But the analysis of quadratic functions in complex variable was in the travel log of Mathematics from the 1920's. And probably, had not been discovered by Benoit, another scientist soon would have discovered the Mandelbrot set.

The brilliant mind of Mandelbrot knew not only to discover the Mandelbrot set and convert it in the icon for his new geometry, but also how to relate it to his fractal geometry. To establish that the process that generates the Mandelbrot set shares some aspects with processes that govern the growth of coral in the Pacific, the formation of a cloud over Paris or fluctuations in the Tokyo Stock Exchange, needed a clairvoyance only possessed by the great visionaries of science.

Let us congratulate Benoit for discovering this wonderful gem of Mathematics, undoubtedly the most beautiful ever revealed, and that today is part of one the most exciting adventures of science: the Fractal Geometry. Without Benoit's insight about the visual appeal of the Mandelbrot set and his efforts to present these images to the general public, probably the

Mandelbrot set would have disappeared after being published in a computer graphics journal, without securing a place in science. Indeed, this happened with the Koch curve, the Sierpinski Triangle or many other wonderful jewels lost until Benoit brought the into the scientific mainstream.

Benoit achieved with his toy, fractals, how to interpret the geometry of nature, much as Johannes Kepler used his toy, the ellipse, to explain the solar system. Once Benoit explained to me with his usual sense of humor, "Not only should the toy be built, but also we must know how to play with it."

Strikingly, mathematics has not always been fair to Benoit. If we examine the long list of awards he received during his life, we can see that only a small part come from mathematical societies or organizations. This certainly is strange for one of the great mathematicians of the twentieth century. But this happened because Benoit, always faithful to his principles, has been considered an outsider in the world of mathematics and his greatest recognition has come from other disciplines whose practicioners found that access to fractal geometry was simple, multidisciplinary and easily implemented in other areas of knowledge. Arguably, Benoit's direct way of spreading fractal geometry has helped him to reach the great public, but also it earned him earn the distrust of many colleagues.

My first contact with Benoit was the invitation I sent to him to chair the International Fractal Art Contest that bears his name. Since he accepted the honorary presidency, he contributed with a commendable intensity. He opened exhibitions, wrote prefaces and took the trouble to sign large piles of catalogs for the participants. Despite his dedication to the contest, he did not take part in the voting, nor did he show preferences for any work. He just looked at the images and found a source of praise and encouragement for each participant, probably knowing the influence his personal preferences could have had over the jury.

When we visited the first exhibition contest, Benoit was surprised to see a long line of people at the Exhibition Hall Conde Duque, one of the most important in Madrid. The audience had crowded the hall and for several days it was necessary to wait to enter the exhibition hall. "I am an inveterate optimist, but never expected to see a crowd standing in a long line to admire mathematics in any of its forms," Benoit said that night.

Before his death, Benoit chaired three editions of the Fractal Art Contest. In each, 25 images were selected. In the third contest, the 25 winners belonged to 17 different nationalities, and the pictures were exhibited at Bilbao (Spain), Buenos Aires (Argentina) and Hyderabad (India), which shows the high level of international expansion of the competition. The organizers and the jury, wisely directed by Benoit, oriented our efforts to discover new ways of expression in fractal art. Thus, the typical filaments and spirals became reduced towards an aesthetic closer to contemporary art rather than usual fractal structures. Not everyone welcomed this effort. Looking at the last exhibition contest Benoit said, "Many will prefer the old images, but compared with these, they look like antiques." He always conveyed the importance of advancing. "Society is reluctant to change, but sometimes you have to help them make that leap."

The International Fractal Art Contest also helped Benoit participate for the first time in the ICM (International Congress of Mathematicians). He entered through the back door, only as honorary director of the Fractal Art Exhibition to be exhibited in the space of the congress. But when he confirmed that he would attend, he raised an unexpected excitement, well above any other guest speaker. The organization soon sensed that could have an audience of thousands of people and located him in a privileged space of the program despite the reticence of some mathematicians.

Thus, Benoit was invited for the closing lecture of ICM 2006 with several thousand people attending the main auditorium. In his speech he congratulated Wendelin Werner as recepient of one of the Fields Medal and very especially because of being able to demonstrate one of Benoit's conjectures. In fact, he said with some mild sarcasm, "This is is the third time a Fields Medal is awarded for proving one of my conjectures." Thus, the old lion stated the importance of his work at one of the institutions that had ignored most of his contribution to mathematics.

For some people, this story may portait a smug man, but Benoit was not smug. As an example I remember that that night Benoit turned down a prestigious dinner with some of the best mathematicians in the world to join my group of young colleagues who had planned a beer and tapas tasting in the bustling Plaza of Santa Ana "Could we join you?" he asked. That night we drank and laughed, but mostly listen to Benoit tell fantastic stories and anecdotes from his life, science, history, art.... It was an unforgettable moment that reveals a much more down to earth and charming person than one might expect.

I also remember the satisfaction of Aliette, who told me how important it was for Benoit to enjoy in life such recognition. And it is true, because looking to the past very few mathematicians enjoyed in his life the reputation like Benoit's. I remember people greeting him as we walked through Bilbao or approaching our table to congratulate him as we ate in Salamanca. This usually happens with media artists, but rarely with scientists. Benoit left his life work almost finished, except for his memoirs that he could not finish despite that being his goal during his last months of life. It is perhaps his only unfinished major work, but Aliette and Benoit's assistant Merry Morse have finished Benoit's memoirs.

My last conversations with Benoit dealt with the Mandelbrot Set in 3D, the so-called *Mandelbulbs*. These sets have a series of algebraic irregularities due to questionable decisions while developing the formula in the third dimension, which from a strictly mathematical point of view is not viable. However, the visual aspect of the set is really spectacular. Benoit never wanted to enter the Mandelbulb question. Seen now with adequate perspective, Benoit understood that this task already belonged to another generation. And although he truly admired the gorgeous animations of the Mandelbulb, and other graphics experiments like the Mandelboxes, he never entered the debate on them. His cycle was already closed and a new era was being opened in which he was conscious he should not participate.

When today we speak of Fractals, everything is incredibly logical and sometimes we wonder how we could not have seen a theory that now looks so obvious, and that even children discover with ease. Benoit's prescience will be remembered forever in the annals of science and I'm sure that over time he will rise to the height of the greatest visionaries of science and be recognized as the last indomitable Renaissance spirit.

To confirm this, Benoit also provided another formidable vision. He strengthened the relationship between Art and Mathematics by means of fractal geometry. Benoit Mandelbrot told me once that Eugène Delacroix used to teach his students that to paint a tree it was necessary to draw inside another smaller tree, and inside another and another.... Later he endorsed his theory by means of the beautiful carving of the Great Wave of Kanawaga, by the Japanese painter Katsushika Hokusai which shows with extraordinary sensitivity a wave with a fractal structure. It was fascinating to listen to Benoit. He could talk about Hokusai's style and immediately illustrate the Japanese character by relating an anecdote that happened while he was dinning with the Empress of Japan.

And to prove once more his interdisciplinary ability, I remember Benoit Mandelbrot gave me a signed copy of his book *The (Mis) Behaviour of Markets: A Fractal View of Risk, Ruin and Reward.* When I read it, it seemed to be extremely apocalyptic. I told my colleagues that I really did not understand why he had dedicated so much effort to the field of economics in which, in my opinion, he was not as well-known as in other topics. I even compared Benoit's work in finance with the last years of Carl Friedrich Gauss, fully devoted to the measurement of the earth's magnetic field, a task that squandered his brilliant mind. But the collapse of the banking system in 2008 showed me again the brilliant intuition of Benoit, and how he warned in his book of the imminent financial disaster under the current economic model. Once again he saw farther than the rest of the world.

Benoit Mandelbrot never was a conventional mathematician, but certainly the most brilliant mind I've had the chance to meet.

Some winning entries from the Benoit Mandelbrot Fractal Art Contest in 2007, 2009, and 2011 are shown in Figs. 6.1–6.5. These samples were selected to give a sense of the range of the styles represented. The images are:

Figure 6.1: "Modern Times" by Ulli Behrendt (2007), "Finery" by Chris Oldfield (2007), and "Sunset Mood" by Stefan Vitanov (2007)

Figure 6.2: "Shuttered Windows" by Susan Cambless (2007), "Heaven" by Eileen Friend (2007), and "310509" by Ramon Pasternak (2009)

Figure 6.1 Work of Susan Cambless, Chris Oldfield, and Stefan Vitanov.

Figure 6.2 Work of Ulli Behrendt, Eileen Friend, and Ramon Pasternak.

Figure 6.3 Work of Esin Turkakin, Yvonne Mous, and Alexey Ermushev.

Figure 6.4 Work of Victor Carbajo, Jessica Orr, and Therese Aasrud.

Figure 6.5 Work of Torsten Stier, Ramon Pasternak, and Hal Tenny.

Figure 6.3: "Sandcastles" by Esin Turkakin (2009), "Round the Block" by Yvonne Mous (2009), and "Autumn" by Alexey Ermushev (2009)

Figure 6.4: "Cells Alive" by Victor Carbajo (2009), "Vincent" by Jessica Orr (2011), and "Worn Out" by Therese Aasrud (2011)

Figure 6.5: "the harvest" by Torsten Stier (2011), "Ramifications" by Ramon Pasternak (2011), and "Gordian Twist" by Hal Tenny (2011)

Color images of these and the other winning entries, and additional information about the contests, can be found at

http://www.fractalartcontests.com/2007/
http://www.fractalartcontests.com/2009/
http://www.fractalartcontests.com/2011/

Chapter 7

The nature of fractal music

Harlan J. Brothers

Abstract. Benoit Mandelbrot always had a strong feeling that music could be viewed from a fractal perspective. However, without our eyes to guide us, how do we gain this perspective? Here we discuss precisely what it means to say that a piece of music is fractal.

7.1 Introduction

In a sense, Benoit Mandelbrot was the Midas of mathematics. Every field he touched blossomed, richly sprouting tendrils that went on to link other often disparate fields.

He was a born geometer, realizing as a young adult that he could "see" the solutions to mathematical problems as an interaction of geometric constructions [1]. It is no wonder that Benoit endlessly extolled the virtue and power of the eye in discerning mathematical truth. This innate skill, and his precise yet open mind, guided him relentlessly through an eclectic array of natural, cultural, and scientific phenomena where issues of scaling, texture, and roughness arise. Everywhere he looked, in nature and in a wide range of historical records and mathematical research, he found the signature of self-similarity that is born out of power-laws. He always had a strong feeling that music also shared this quality in a measurable sense [2]. Here too, Benoit's intuition served him well.

It should be mentioned that over the last three decades, there have been many impressive efforts at transliterating fractal images or converting fractal algorithms into audio, sometimes with surprisingly musical results. This is often referred to as "fractal music" in the sense that it is derived from something directly related to fractals. However, there is an important distinction to be made: this type of derivation does not in any way ensure a composition with measurable fractal characteristics [3]. Here we will discuss music that is actually fractal; it possesses measured properties reflecting an underlying power-law, consistent with Benoit's intuition.

7.2 Music and time

Around the time that Benoit published *Les Objets Fractals: Forme, Hasard et Dimension* [4], the precursor to his seminal work *The Fractal Geometry of Nature* [5], two researchers, Richard Voss and John Clarke at the University of California, published a fascinating finding. While examining the power spectrum of $1/f$ noise in electronics, Voss serendipitously discovered strong indications of $1/f$ scaling in the distribution of both the power and the set of zero-crossings for audio signals of several different genres of music [6].

$1/f$ noise belongs to a category of phenomena Benoit referred to as *scaling noise*. It was a subject to which he devoted considerable attention [5, pp. 249–250, 254, 374–375] [7–9]. The idea that music generally seemed to possess this specific type of inherent power-law relation led Voss, other researchers, and composers including Charles Wuorinen and Charles Dodge, to explore the algorithmic generation of compositions using $1/f$ scaling as a guideline [10–13].

It turns out that music composed in this fashion sounds relatively pleasing to most listeners [10, 14]. For the fractional Brownian exponent β and music with a spectral density distribution $1/f^{\beta}$, melodies generated using $\beta = 0$ (associated with white noise) sound too random, while melodies derived using $\beta = 2$ (associated with Brownian motion) sound too monotonous (see Fig. 7.1). Thus, it appears that the "proper" amount of correlation in a melody should lie somewhere in between ($1/f$ scaling has a value of $\beta = 1$). A cognitive study by Michael Beauvois [15] indicated that a value of $\beta = 1.5$ produces "optimally preferred" melodies. Regardless

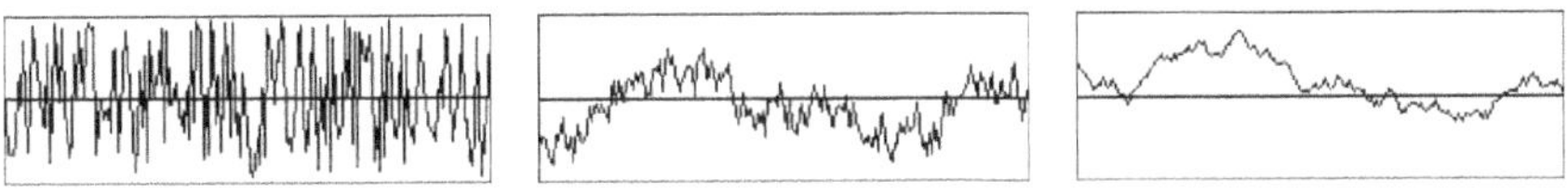

Figure 7.1 Amplitude vs. time plots of white noise (left), $1/f$ noise (center), and brown noise (right).

of a precise value for β, the significance of Voss' and Clark's work remains unchanged; music generally possesses an inherent self-similarity as it unfolds with respect to time. Martin Gardner, while discussing music's essential mixture of surprise and expectation, summed it up by writing "...what Voss has done is to suggest a mathematical measure of the mixture" [14].

More recent findings indicate that $1/f$ scaling in music is not limited to variation in audio power and to pitch correlation. Summer Rankin *et al.* [16] investigated temporal fluctuations in piano performance and the prediction of these fluctuations by listeners. Their findings indicated that not only did the sample performances show fractal structure with respect to tempo fluctuation, but also that listeners appeared capable of predicting the variations, consistent with $1/f$ correlation. Using musical scores, Daniel Levitin *et al.* [17] examined rhythmical timings in an extensive selection of Western classical music. They found that, across both genre and composer, the spectra of merged note onset markers obeyed a $1/f^{\beta}$ power law, with β falling in the range of approximately 0.5 to 1.

Finally, musical pulse itself can be viewed from a fractal perspective. Across a wide spectrum of music, the simultaneous propagation of distinct pulses of different frequency often forms a scale-invariant structure [18].

Thus, in the time domain, with respect to audio power fluctuation, melodic pitch correlation, tempo/pulse variation, rhythmical timing, and pulse itself, fractal structure appears to be a widespread attribute of music.

7.3 Music and space

In the visual domain, the eye can quickly guide us in identifying fractal structure. Looking at the Mandelbrot set, we are instantly taken with the infinite detail of its boundary rather than the uniform interior (see Fig. 7.2, left). On the other hand, a glimpse of the Sierpinski triangle draws us not to its outer boundary, but directly to its gasket-like interior region (see Fig. 7.2, right). Sometimes we discern multiple possibilities for measurement, as in the case of a tree: we can investigate the structure of its branches, root

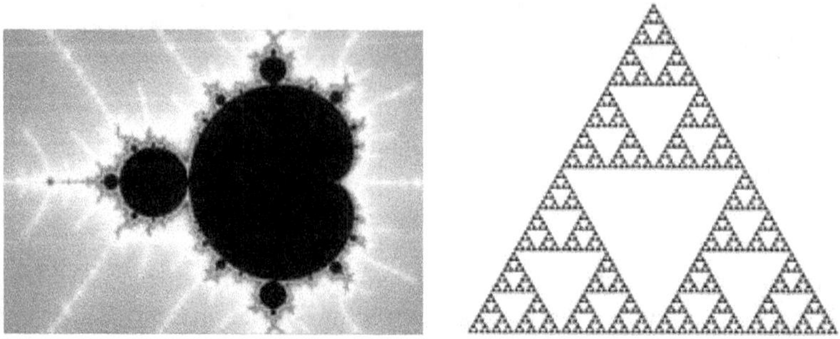

Figure 7.2 Classic examples of fractals from the visual domain. Left: the Mandelbrot set. Right: the Sierpinski triangle.

system, canopy, or perhaps its mass distribution; the view from a distance might suggest assessing its relation to the forest as a whole.

In a similar sense, music can be analyzed in terms of the pitch, location, duration, velocity and overall structural attributes of its constituent notes.[1] In effect, we examine the properties of the composition as reflected in the 2-dimensional representation of a musical score or the transcription (or MIDI recording) of a performance.[2]

As with graphics, to find fractal structure in musical space we must first decide what characteristic(s) we wish to examine. However, given music's ephemeral nature, it is more difficult for the ear to guide us in identifying potentially fractal structure. Musical compositions can indeed manifest a wide variety of the power-law relationships that form the foundation of fractal geometry. Echoing the self-similarity of cloud-like structures, these relationships often arise in a statistical sense that requires careful measurement.

Some examples of measurable self-similarity include:

1. *Duration scaling*: The distribution of individual note durations is statistically self-similar [19, 20]. A detailed treatment appears in the following section.

---

[1]Here, the term "velocity" refers to note *attack*; that is, the speed with which a note is engaged. Higher velocity primarily translates to greater volume.

[2]MIDI is the acronym for **M**usical **I**nstrument **D**igital **I**nterface. It is a communications protocol that encodes characteristics such as timing, pitch, velocity, and instrument selection. This data is interpreted by a synthesizer or computer and played live, analogous to the manner in which a scroll-driven player piano performs.

2. *Pitch scaling*: The distribution of pitches is statistically self-similar [19, 21].
3. *Melodic interval scaling*: The distribution of melodic intervals (changes in pitch) is statistically self-similar [19, 22, 23].
4. *Melodic moment scaling*: The distribution of the changes in melodic intervals is statistically self-similar [19, 23].
5. *Harmonic interval scaling*: The distribution of harmonic intervals (those formed by simultaneously occurring pitches) is statistically self-similar [19].
6. *Structural scaling*: An aspect of the compositional structure, such as song form or dynamics, displays a nested or recursive form [24–27].
7. *Motivic scaling*: A melodic or rhythmic motif is repeated simultaneously at different time scales [28–30].

While motivic scaling could also be classified as fractal with respect to time, it is listed here because it deals with the construction of distinct musical phrases whose patterns with respect to pitch and duration are directly identifiable and easy to discern.

It is interesting to note that in the case of motivic scaling, musicians have been *consciously* composing this form of fractal music for over six centuries. It was familiar to many of the great Flemish composers such as Johannes Ockeghem and Josquin des Prez [30] who developed the art of the *mensuration* or *prolation* canon.[3] This type of canon is characterized by a melody or rhythmic motif that is repeated in different voices simultaneously, but at different tempos [31].

Modern composers, including Martin Bresnick, Conlon Nancarrow [28], and Arvo Pärt [32], have developed sophisticated compositional styles that often incorporate forms of motivic scaling. Many algorithmic artists also utilize this concept in their compositions. To be clear, not all mensuration canons are fractal. See [33] for a discussion of the fundamental requirements that must be met in order to classify an object as such.

Whether working in the time or space domain, there are many approaches to analyzing music from a fractal perspective, both with regard to the source material and the mathematical tools employed. A few

[3] *Le Ray Au Soleyl*, composed by Johannes Ciconia (ca. 1335–1411) is one of the earliest known examples of a three part mensuration canon.

examples include:

1. Maxence Bigerelle and Alain Iost [34] employed two distinct methods for calculating the fractal dimension of digitized audio data as a means for classifying musical genre.
2. Using elemental musical characteristics, Bill Manaris *et al.* [19] used MIDI files to explore the classification of both genre and aesthetic appeal by analyzing rank-frequency distributions for a wide range of musical metrics.
3. Zhi-Yuan Su and Tzuyin Wu [35] converted the set of pitches and the set of durations for several melodies into respective sets of points distributed along a straight line. They then used these linear distributions to calculate multifractal spectra and surmised that the shape and width of the resulting $f(\alpha)$ plots might be used to distinguish different styles of music.
4. Jean Pierre Boon [36] examined the time series of digitized pitch variations in multi-voice compositions from a dynamical perspective. He obtained box-counting dimensions from the resulting phase space of the combined voices. Boon went on to compare these dimensions to a characterization of the complexity of the pieces obtained by applying a variation of the information theoretic concept of Shannon entropy.

Some researchers assert that the fractal analysis of audio files is preferable, offering a certain richness in that it preserves information regarding instrument timbre and harmonic interplay. On the other hand, it has been argued that timbre can cloud the picture when looking for melodic scaling in audio files [37]. There is also the broader consideration that, though a composer may initially arrange a piece with a specific instrumentation in mind, the composition is nevertheless recognized and enjoyed regardless of whether it is played by an orchestra, by a solo piano, or even on a kazoo. It is reasonable then to explore the fractal analysis of compositional characteristics independent of audio data; aspects that, in some sense, reside at the core of our listening pleasure.

While the body of research is growing rapidly, not everything written on the subject is accurate. Fractal geometry caught on quickly because it is intuitively simple to grasp; it is also easy to misunderstand. In the rush to search for fractal characteristics, even established researchers have been known to make errors by overlooking essential principles.

Regardless of the field of study, fractal geometry is, first and foremost, geometry, a subject whose name implies measurement. For music to be considered "fractal," it must possess some measurable property that qualifies it as such.

7.4 New results: Duration scaling

One afternoon, while driving and listening to Maurice Ravel's *Le Tombeau de Couperin*, I was struck by the natural way in which notes of longer duration occurred less frequently, while conversely, those of shorter duration occurred more often. Duration scaling appeared to be an obvious place to explore the corpus of musical composition for power-laws.

Intuitively, this makes sense. The simply constructed example in Fig. 7.3 provides an illustration.[4] Comparing the four simultaneous, monophonic parts, we see that the lowest staff contains the longest durations. Because they are quarter notes, within each common time measure there is room for fewer of them (a maximum of 4).[5] The top staff contains the shortest notes and conversely, because they are sixteenth notes, within each measure there is room for more of them (a maximum of 16).

Table 7.1 shows the size and count distribution for note durations in the four measures of music. Figure 7.4 is a plot of log(count) vs. log(size), confirming a power law relation with an exponent of 1.

Because each note is followed by a rest of the same duration, this example exhibits a second fractal symmetry in that the power-law also happens to hold for the set of rests. In addition, it is the simplest case of motivic scaling with respect to rhythm: the pattern [note, rest], where

Figure 7.3 Example illustrating duration scaling.

[4]Audio files of examples appearing in this paper can be found online at [39].
[5]The term "common time" refers to a time signature of 4/4 wherein each measure contains the durational equivalent of 4 quarter notes.

Table 7.1 Summary of data for Fig. 7.3 (without repeat). Staff 2 contains "eighth note triplets"; each note lasts for 1/3 the duration of a single quarter note.

Staff	Duration Size (quarter notes)	Count
1	1/4	32
2	1/3	24
3	1/2	16
4	1	8

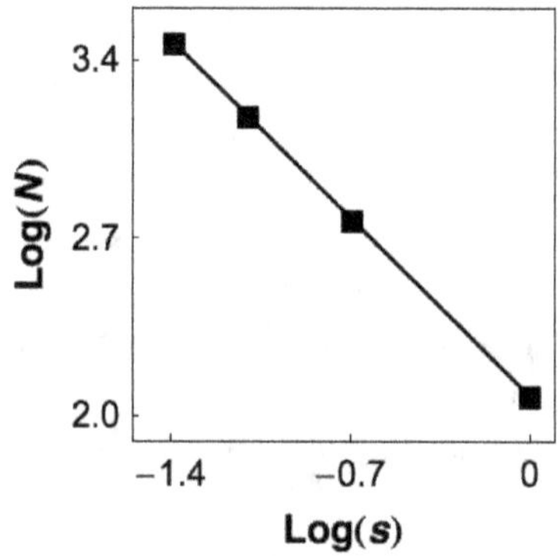

Figure 7.4 Log-log plot of the duration and count data in Table 7.1.

the durations of each are equal, is repeated at expanding time scales from top staff to bottom. Thus, Fig. 7.3 provides us with three distinct fractal perspectives.

There are two important considerations in this example. The first is that it is heterogeneous. For our purposes, heterogeneity means that everywhere we look we find roughly the same distribution of sizes (see [38] for a simple example and counterexample). In the world of music, a perfectly even mixture of different-sized elements is generally boring. As with natural objects that exhibit rough though recognizable self-similarity, music is much the same. We therefore work in the context of a relaxed interpretation of heterogeneity. For the present purposes, if visual inspection of the data set for a composition shows no obvious, large-scale clumping of same-sized elements, we will consider the composition "sufficiently heterogeneous." In the current example, the written score serves as an adequate visual representation. In general, scores and MIDI "piano roll" views can be useful but are cumbersome in size. In the following examples we will use a pin-type plot of duration against time line (note onset). The value of placing the "pinhead" at the top of the line representing duration is that when multiple notes of different duration occur simultaneously, they can all be seen. In a 2-dimensional plot, to guarantee that all notes are seen, even simultaneous notes of the same duration, it is necessary to plot the duration data in the order in which it appears; that is, duration against location in a sequence. These two types of plots can be very similar (see Fig. 7.5).

The second consideration is the small note count. It means two things: (1) after one repetition, the example becomes inarguably boring, and (2) it is too small a sample space on which to base any meaningful statistical analysis. It is important to emphasize that any assertion of statistical self-similarity carries the assumption that the sample size under consideration is sufficiently large to offer a meaningful result. In comparison to the world of scientific data analysis where data sets can be many orders of magnitude greater, the number of notes that constitute most classical piano music is relatively small, in the neighborhood of 1,000 to 20,000 notes. While the construction of this particular example ensures a heterogeneous distribution, most music is not composed in so rigid a manner.

Particularly in the case of smaller compositions, we will assert simply that there is a strong relation between the durational "texture" of a song and its underlying distribution of durations (i.e., a fast-moving sprinkle-like collection of short durations sounds substantially different in texture from a languid passage consisting of long durations). From this standpoint, it is not unreasonable to expect that, in cases where duration scaling exists, a composition that continued in roughly the same style with respect to durational texture would also continue to exhibit a self-similar distribution of durations across several scales of measurement.

7.5 Method for analysis

To perform our analysis, we will use MIDI piano data chosen from different musical periods. We will group data into equal-size bins; the width of each covers an equal range of values. Assigning a single size to a group of neighboring values can help to reveal trends that might not otherwise be apparent. To make a rough analogy, like the popular *Magic Eye* autostereograms [40] of the 1990's, when we loosen our focus by binning like-sized elements, otherwise hidden patterns can emerge. For details on the technique of "coarse graining" see [41]. For application in music, see [23].

The representative size of a bin is typically assigned using either the lowest, highest, or mean values with the choice applied consistently across all bins. In the case of pitch-related data, the choice does not affect the measured result; pitch is fundamentally logarithmic in nature (see [23] for a detailed explanation).

However, unlike pitch, duration is linear in nature. Because we will be using log-log plots of this data, the assigned bin size *will* affect our measurement. It therefore makes the most sense to assign bin sizes based on the mean value of the bin width, w, measured in note durations (e.g., "$w = 1/4$" signifies a bin width of one quarter note).

If N_i represents the number of elements in bin B_i and s_i represents the assigned size of its elements, then plotting $\text{Log}(N_i)$ against $\text{Log}(s_i)$ for all B_i reveals the presence of a power-law if the plotted points fall along a straight line. As we discuss in the following paragraph, the absolute value of the slope of this line can be interpreted as the dimension of the object under examination. In practice, we find a line-of-best-fit by performing a linear regression on the binned data and look at its R^2 value to assess how well the line fits the data. For reference, an R^2 value in the neighborhood of 0.9 (1 being the highest) is, by most standards, considered an excellent fit.

Because we are analyzing a collection of different-size elements that unfold in time and often occur simultaneously, we can neither treat the collection as a simple time-series, nor obtain a box-counting dimension. We will therefore refer to its measure of dimension as the *frequency dimension*, d_f, of the set of elements, where "frequency" refers to class frequency.

As stated in the previous section, the appearance of heterogeneity is a fundamental requirement when seeking fractal structure in a statistical sense. Specifically, one must first define and establish the existence of heterogeneity before attempting to derive a frequency dimension. In the following section, the measurements for fractal dimension will be given to 2 d.p.a. and the R^2 value will be given to 4 d.p.a.

A data set can sometimes be effectively binned in more than one manner. Because the number of plotted points corresponds to the number of levels of scaling, it is desirable to coarse grain data using the smallest bins that offer a meaningful result.

We therefore have two major considerations in our analysis: to obtain convincing evidence of a power-law, we want many bins and, at the same time, a reasonably high R^2 measurement. Simply choosing the binning schemes with the highest R^2 values often results in schemes with only

3 or 4 bins (3 being the absolute minimum requirement for evidence of linearity). However, a smaller number of bins often translates to a higher dimension; a relative scarcity of the longest durations can carry undue influence, increasing the slope of the regression line. In the following section, the first example will illustrate this effect.

It is important to emphasize that binning data cannot magically make a power-law appear where none exists; no configuration of binning schemes will provide convincing evidence without the appropriate underlying distribution.

In addition to facilitating trend analysis, binning duration values makes sense, particularly when analyzing performance data. Whether by means of direct key pressure or use of the sustain pedal, the actual length of time a pianist holds a note of specified duration can vary considerably with respect to the written score or even to the way it is played earlier in the same performance.

The example in Fig. 7.3 was analyzed on the basis of the score only. To analyze music recorded using MIDI, we need to first extract duration information from the recorded file. MIDI timing resolution is based on the number of "Pulses Per Quarter Note" (PPQ), a number that is measured in "ticks" and generally ranges from 96 to 1024. It determines the framework for the relative measure of note durations along with their location in time. The examples in the following section have all been standardized to 480 PPQ.

Unlike pitch, which we hear in an absolute sense as a specific frequency, we experience duration relative to the perceived "pulse" of the music. Not to be confused with PPQ, here we use term *pulse* to describe the underlying and non-notated (i.e., unwritten in the score) flow of rhythm that can be perceived at different levels of periodicity in the same passage of music. To be clear, consider listening consecutively to two renditions of a song and tapping a foot on every quarter note. If the first performance is played at a tempo of 120 BPM (Beats Per Minute) and the second at 150 BPM, we notice that we are tapping faster to the second version but we are not directly aware of the fact that a quarter note has changed in absolute duration from 500 milliseconds to 400 milliseconds. In both versions we experience a quarter note as "one tap of the foot."

To offer a musical context for our measurement, we will therefore show examples using bin widths relative to the duration of the primary pulse of the piece as indicated by the composer's metronome marking; that is, values of w that correspond to a musically encountered fraction or multiple of the pulse (e.g., 1/3 pulse, 1/2 pulse, 1 pulse, etc.).[6]

In cases where the performance is transcribed or recorded using a metronome, assigning bins relative to the pulse duration is easy to accomplish. However, the task is complicated in the case of pieces where the tempo is unconstrained. Tempo can fluctuate substantially, and often by intent, particularly in the classical genre [16]. This, in effect, changes the absolute duration of corresponding durations *in the same performance.* Even when there are no intended tempo changes, without a "click track" to keep the performance in sync with the recorder, there is no steady relation between the performer's notion of a quarter note and the recorder's preprogrammed value for a quarter note. In such cases, through visual inspection and appropriate rescaling of the MIDI recording, we can attempt to match sections of the MIDI data to the score and thereby arrive at reasonable estimate for the duration of a quarter note consistent with the composition.

Finally, it should be mentioned that, with regard to performed data, we will be dealing with durations as measured by the length of time the pianist physically holds down a given key; any data for the sustain pedal (generally used at the discretion of the performer) is not applied. The results are revealing, if only regarding the primary aspect of the manner in which the music is actually performed.

7.6 Detailed examples

We will examine three works, all substantially different in musical style: jazz, classical, and impressionist.

7.6.1 Ellington and Tatum

We begin with jazz, analyzing Art Tatum's 1942 piano arrangement of Duke Ellington's *Don't Get Around Much Anymore.* The data is reasonably heterogenous as seen in Fig. 7.5.

The data for this piece was sequenced directly from a published arrangement (Robbins Music Corporation, New York, 1946). Thus, there is no tempo variation and no opportunity for interpretation or error on the

[6]The metronome marking sets the tempo and is often assigned using the value of a quarter note for reference.

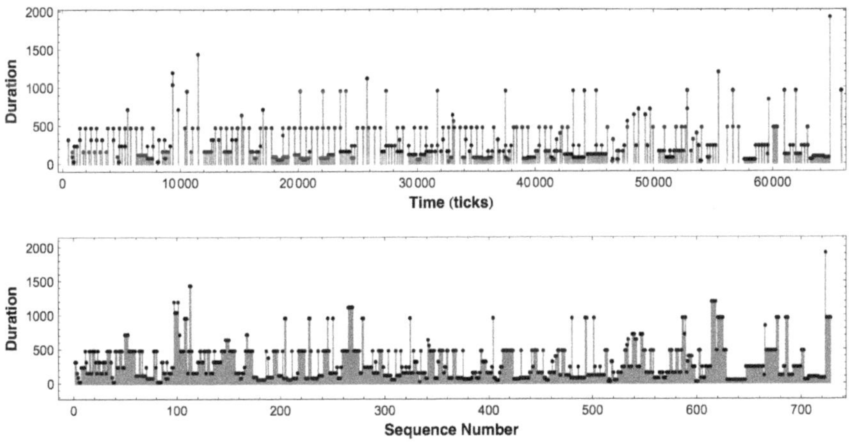

Figure 7.5 Time line plot (top) and sequence plot (bottom) of the duration data for a transcription of Tatum's arrangement of *Don't Get Around Much Anymore.* Durations are measured in ticks.

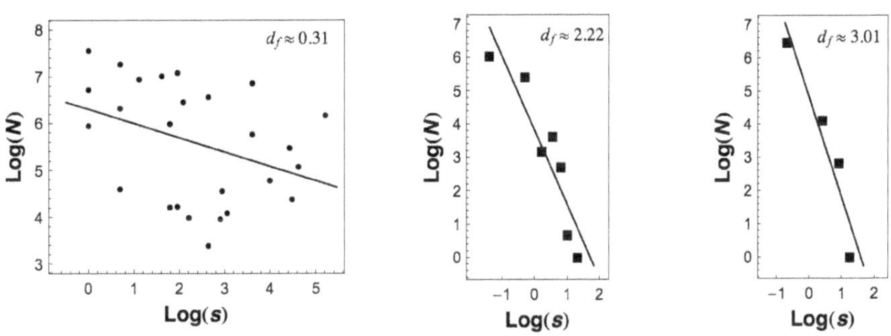

Figure 7.6 Log-log plots of raw data (left), binned data $w = 1/8$ (center), and binned data $w = 1/4$ (right) for Tatum's arrangement of *Don't Get Around Much Anymore.*

part of a performer.[7] As a result, out of a total of 728 notes there are only 26 distinct duration values (see [39] for the complete data sets of the pieces analyzed in this paper).

Figure 7.6 shows a substantial disparity between the log-log plots of the raw data and the two sets of binned data in Table 7.2. Given the extremely

[7] The lack of a live performer does not necessarily rule out the possibility of error in the process of sequencing.

Table 7.2 Binned duration data for Tatum's arrangement of *Don't Get Around Much Anymore.* On the left $w = 1/8$ and on the right $w = 1/4$.

Bin	**Size** (quarter notes)	**Count**
1	1/4	423
2	3/4	225
3	5/4	24
4	7/4	38
5	9/4	15
6	11/4	2
7	13/4	0
8	15/4	1

Bin	**Size** (quarter notes)	**Count**
1	1/2	648
2	3/2	62
3	5/2	17
4	7/2	1

Table 7.3 Note count, dimension, and R^2 data for Tatum's arrangement of *Don't Get Around Much Anymore.*

Note Count	**Distinct Durations**	d_f/R^2 Raw Data	d_f/R^2 $w=1/8$	d_f/R^2 $w=1/4$ (4 bins)	d_f/R^2 $w=1/4$ (3 bins)
728	26	0.31/0.1364	2.22/0.8389	3.01/0.9015	2.21/0.9990

low R^2 value for the raw data (as seen in Table 7.3), this stark contrast provides a good example of how binning data can reveal hidden order.

This piece also illustrates the value of understanding the music that is being analyzed and that a meaningful measurement cannot necessarily be judged on the basis of either the number of bins or the R^2 value alone. It is important to remember that what we seek in approximating the frequency dimension is a measure that roughly describes the distribution across any given portion of the composition. The dimension obtained in the right hand plot of Fig. 7.6 is substantially higher than the center plot due to the influence of a single large element in bin 4 (bottom of right plot). Specifically, if we were to add this one longest duration, occurring at the very end of the piece, to the third bin with the other largest durations, the result would be a dimension $d_f = 2.21$ with an R^2 value of 0.9990. Though this produces a mere three bins, the associated dimension agrees remarkably well with the middle plot of Fig. 7.6 while its R^2 measure is exceptional.

Considering the improvisatory nature of jazz, it is important to point out that the interpretation of most songs is wholly left to the performer. In such cases, the effects of musical and personal performance styles largely define the distribution of both pitches and their respective durations.

7.6.2 Mozart

We now step back to the late 18th century and the first movement of Mozart's Sonata in D Major (K.284). Also known as the "Dürnitz" Sonata, it was written in 1775 for Baron Thaddeus von Dürnitz. Figure 7.7 shows two views of its time line plot. In the first, we notice three extreme duration values (each lasting approximately six beats) that have the effect of crowding most of the data into the lower half of the plot. The second view limits the vertical range to approximately three beats so that we can better appreciate the bulk of the distribution which appears reasonably heterogeneous.

In Fig. 7.8 we see log-log plots for the raw data and two sets of binned data. Because this is a recorded performance, there is more variety in the raw data; 465 distinct duration values out of a total 3795 notes. Table 7.4 shows the distribution of the binned data. Being that the log of zero is undefined, corresponding bins are not plotted.

While the raw plot is far more suggestive of a trend than its counterpart in the previous example, the approximate R^2 value is just 0.6229, compared to $R^2 = 0.9778$ for the middle plot and $R^2 = 0.9847$ for the right-hand plot.

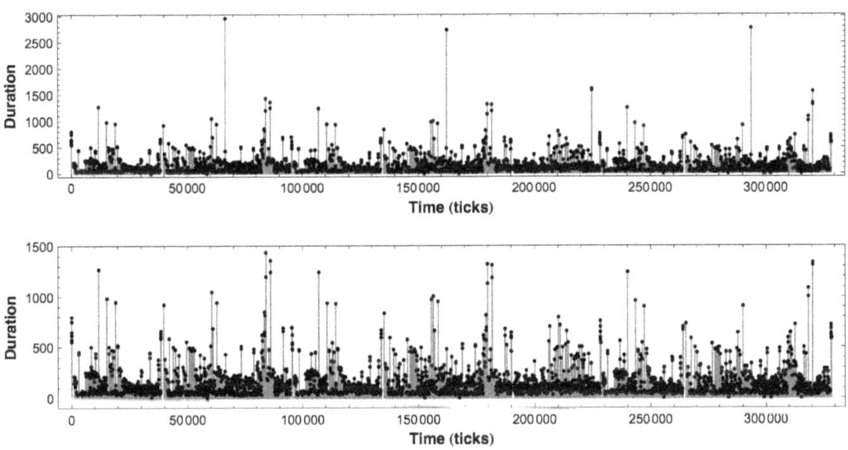

Figure 7.7 Time line plots of duration data for the first movement of Mozart's *Sonata in D Major*. The lower plot is a vertical zoom into the lower half of the top plot.

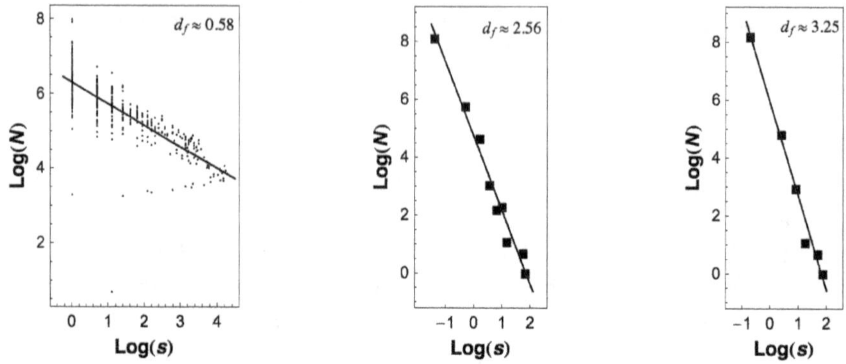

Figure 7.8 Log-log plots of duration data for the first movement of Mozart's *Sonata in D Major*. From left to right: raw data, binned data ($w = 1/8$), binned data ($w = 1/4$).

Table 7.4 Binned duration data for the first movement of Mozart's *Sonata in D Major*. On the left $w = 1/8$ and on the right $w = 1/4$.

Bin	**Size** (quarter notes)	**Count**
1	1/4	3322
2	3/4	323
3	5/4	104
4	7/4	21
5	9/4	9
6	11/4	10
7	13/4	3
8	15/4	0
9	17/4	0
10	19/4	0
11	21/4	0
12	23/4	2
13	25/4	1

Bin	**Size** (quarter notes)	**Count**
1	1/2	3645
2	3/2	125
3	5/2	19
4	7/2	3
5	9/2	0
6	11/2	2
7	13/2	1

As in the previous example, the binned data appears to be a more accurate reflection of the underlying power-law. In Table 7.5 we see that the R^2 values for both binning schemes are close. Because a higher number of bins is desirable and lowering the number of bins tends to inflate the measured dimension, it is reasonable to accept the scheme with more bins and lower dimension as being more accurate.

Table 7.5 Note count, dimension, and R^2 data for the first movement of Mozart's *Sonata in D Major.*

Note Count	Distinct Durations	d_f/R^2 Raw Data	d_f/R^2 $w=1/8$	d_f/R^2 $w=1/4$
3795	465	0.58/0.6229	2.56/0.9778	3.25/0.9847

7.6.3 Ravel

We return finally to the late 19th- and early 20th-century composer, Maurice Ravel. His *Le Tombeau de Couperin* was composed first for piano between 1914 and 1917 and commemorates friends who had died in World War I. We examine three different versions of the third movement, *Forlane.*

Figure 7.9 shows time line plots for each version. Similar to the bottom plot of Fig. 7.7, these plots show only the lower 2/3 of the full range, clipping

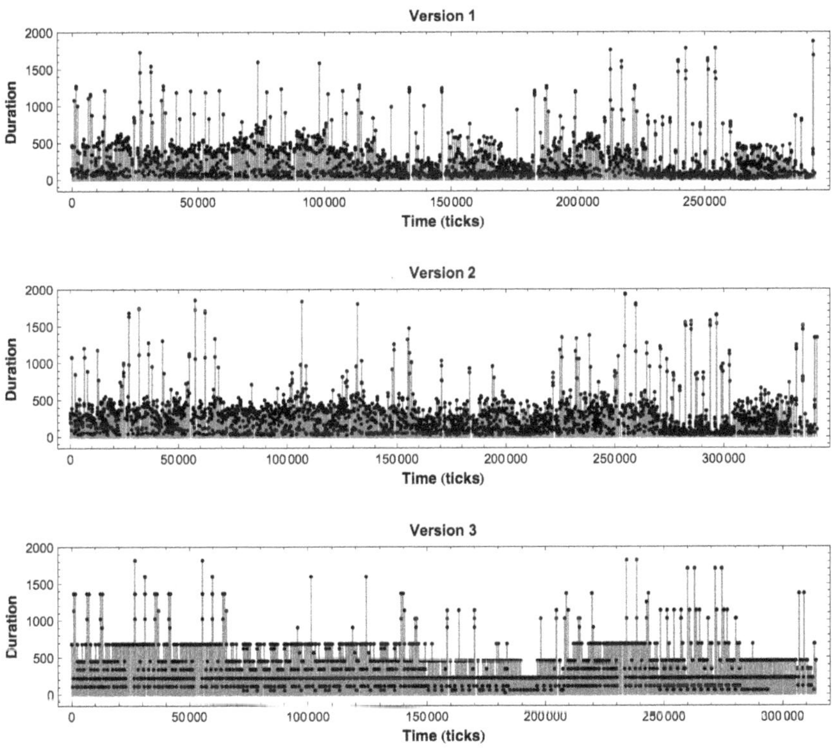

Figure 7.9 Time line plots of duration data for three versions of Ravel's *Forlane.* The vertical scale is clipped at 2000 ticks, hiding the few extreme duration values.

the few longest durations and thereby allowing us to see greater detail in each of the reasonably heterogenous distributions.

Unlike the two preceding examples which are written in 4/4 time (4 quarter notes per measure), this piece is in 6/8 time (6 eighth notes per measure). Its metronome marking is given in terms of a dotted quarter note, equivalent to 3 eighth notes. We will examine each rendition from two perspectives: using the shorter, eighth note pulse and also the primary pulse (a dotted quarter note).

Table 7.6 shows the binned data as represented in the middle column of Fig. 7.10 while Table 7.7 shows the binned data as represented in the last column. Here again, bins with a count of zero are not plotted.

Of the three versions, the third was manually sequenced directly from a score (Durand & Cie., Paris, 1918). As with the jazz example, this explains the lack of variety regarding distinct note durations as seen in Table 7.8.

In each version, we see convincing evidence of an underlying power-law, even in the case of the sequenced score whose raw data appears least persuasive. Despite having three differently realized data sets, for three different arrangements, at three different tempos, using two different recording

Table 7.6 Binned duration data for three versions of Ravel's *Forlane*. The bin width is one eighth note, equal to one third of the primary pulse. The mean value of the size of the first bin is therefore one sixth.

Bin	**Size** (dotted quarters)	**Count** Version 1	**Count** Version 2	**Count** Version 3
1	1/6	1738	1669	1526
2	3/6	694	1050	836
3	5/6	239	190	562
4	7/6	65	61	9
5	9/6	24	38	88
6	11/6	25	17	73
7	13/6	17	17	6
8	15/6	5	11	16
9	17/6	2	2	4
10	19/6	0	0	0
11	21/6	0	0	0
12	23/6	0	0	4
13	25/6	4	2	0

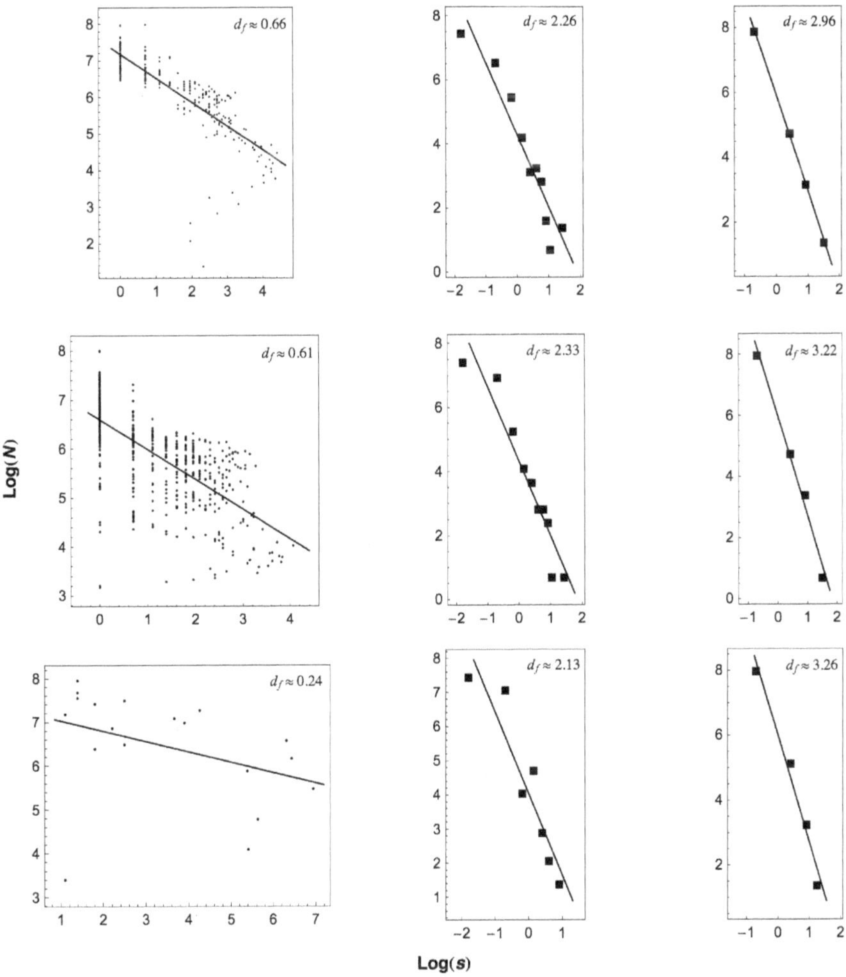

Figure 7.10 Log-log plots of duration data for three versions of Ravel's *Forlane*, showing raw and binned data. Top row: Version 1, Middle row: Version 2, Bottom row: Version 3.

techniques (unsynchronized and sequenced), the measured dimensions for the plots in the center column of Fig. 7.10 all agree closely. The first two versions, whose raw data give some hint of a power-law, agree with high R^2 values. It is also interesting to note that, at least for these three renditions, the more suggestive the raw data is of a power-law, the less of a spread

Table 7.7 Binned duration data for three versions of Ravel's *Forlane.* The bin width is a dotted quarter, equivalent to three eighth notes.

Bin	Size (dotted quarters)	Count Version 1	Count Version 2	Count Version 3
1	1/2	2671	2909	2924
2	3/2	114	116	170
3	5/2	24	30	26
4	7/2	0	0	4
5	9/2	4	2	0

Table 7.8 Note count, dimension, and R^2 data for three versions of Ravel's *Forlane.*

	Note Count	Distinct Durations	d_f/R^2 Raw Data	d_f/R^2 $w=1/8$	d_f/R^2 $w=3/8$
Version 1	2813	246	0.66/0.6337	2.26/0.9087	2.96/0.9996
Version 2	3057	534	0.61/0.4770	2.33/0.9120	3.22/0.9860
Version 3	3124	19	0.24/0.1485	2.13/0.7602	3.26/0.9737

there is between the measured dimensions for each of the two binning schemes.

As with the previous examples, this piece illustrates the general tendency for less robust binning schemes to inflate the measure of dimension. Although the R^2 values are higher in the last column of Table 7.8, as in the previous jazz and classical examples, the more robust binning scheme (appearing in the adjacent column) provides a more convincing measurement.

What is encouraging from a fractal standpoint is that the arrangements are indeed different. The fact that we can, in the case of the first version, remove a section without substantially changing the overall distribution implies a certain level of heterogeneity, reinforcing the evidence for fractal structure.

7.7 Discussion

While only a few select examples are offered here, the multiple styles of music surveyed in the course of this research indicate that power-laws with respect to note duration occur naturally across many genres in Western music. Whether they also occur frequently in music across cultures is a compelling question.

As with natural phenomena which can exhibit fractal scaling only over a limited range, with duration scaling there is often an excellent power-law fit for the mid-range bin sizes that only deteriorates in the realm of the very shortest or longest durations. Even in the best examples though, it is important to keep in mind that the dimensions obtained here are rough estimates of the relationship that describes the overall distribution of elements from smallest to largest. The technique of binning appears to be a useful method for obtaining such estimations and makes sense from a musical standpoint. This is especially so in the case of performance data where inaccuracies arise and result in ranges of values, each range intended by the performer to correspond to a duration of a specific length.

In performances where a metronome is not used, to achieve greater accuracy in the analysis, it would be necessary to derive a "tempo map" that continuously tracks changes in tempo. This technique allows one to account for the variation in durations that are, relative to the composition, intended by the composer to be the same size. Separately, the integration of piano pedal information into duration data would add an additional perspective.

An interesting extension of this research would be to analyze a large body of music using only written scores. It would then be possible to compare this type of analysis to the analysis of respective performances and examine the effect of interpretation on composition. Given that interpretation is, to a great extent, a matter of timing, this might provide a perspective on the relation between the "idealized" vision of the composer and the realization of that vision through performance.

With regard to jazz in particular, in cases where the score is not explicitly written and the arrangement is therefore left to the discretion of the performer(s), duration scaling may prove useful in classifying musical or personal styles.

Whether fractal structure with respect to duration is a property primarily of performance, or also of composition, is an open question. The fact that three versions of Ravel's *Forlane* all offer similar results suggests that composition plays its part.

In the interest of analyzing larger data sets, with carefully captured or sequenced orchestral scores, one could investigate large musical structures for evidence of durational power-laws. In the context of entire compositions, it may be that measuring power-laws is more generally useful than measuring fractal structure, whose requirement of heterogeneity might be too restrictive for broad application. On the other hand, here we have only examined a small sample of piano music data. The results for a greater selection of work and for larger ensembles would contribute to a broader perspective.

7.8 Conclusion

As with many new frontiers, interest in the fractal aspects of music has grown steadily. The lines branch through elementary education and Academia, amateurs and professionals. Indeed, the fact that students often inquire about fractals in relation to music delighted Benoit.

In all likelihood, a mature approach to the fractal analysis of music will need to be nuanced, as it is when we assess the fractal characteristics of many organic structures. Looking at a tree, from leaves to branches, bark to root system, each part has its own fractal characteristic. There is no single measure for the tree in its entirety. Well-developed and intricate music will demand a similar approach to understanding.

In the end, the existence of multiple forms of fractal structure in music underscores the extent of Benoit's extraordinary insight. Through music, we affirm his vision of the universe and our place in it as physiologically fractal beings whose creative processes simultaneously reflect and give rise to the world around us.

Acknowledgments

I owe the deepest debt of gratitude to Benoit Mandelbrot who first suggested I pursue a mathematically rigorous exploration of the ways in which music can exhibit fractal structure. It was *his* intuition that served *me* well, instantly providing a vehicle that would unite my passion for both music and mathematics. Over the years, Benoit continued to share his overwhelming expertise, his humor, and his wisdom in practical matters. His generosity of spirit, sincere interest, and fundamental good nature provided me with a constant source of inspiration.

My sincere thanks to Anatoly Larkin for sharing his expertise in MIDI analysis and piano performance, and for pointing me to the Minnesota International Piano-e-Competition data. Thanks also to Summer Rankin for sharing the work of Michael Beauvois.

Appendix

List of Works and Performers

Duke Ellington: *Don't Get Around Much Anymore*
Arranged by Art Tatum, sequenced by Harlan J. Brothers

Wolfgang Amadeus Mozart: *Sonata in D Major (K.284)*
Performed by T. J. Keanu Tario
http://www.yamaha.com/ecompetition/contestantbios2008/tariotj.asp

Maurice Ravel: *Forlane* from *Le Tombeau de Couperin* (Version 1)
Performed by Gary Lloyd

Maurice Ravel: *Forlane* from *Le Tombeau de Couperin* (Version 2)
Performed by Anonymous
Performer will be posted at [39], should the information become available.

Maurice Ravel: *Forlane* from *Le Tombeau de Couperin* (Version 3)
Sequenced by Pink Mozart and Harlan J. Brothers
https://soundcloud.com/pinkmozart

Bibliography

[1] N. Lesmoir-Gordon, *et al.*, *Clouds Are Not Spheres*, VHS/DVD, Gordon Films, 2000 (VHS) & 2011 (DVD).
[2] B. Mandelbrot, H. Brothers, Private conversation, August 2003.
[3] H. Brothers, "Fractal Music: Common Errors." http://www.brotherstechnology.com/fractal-music/errors.html
[4] B. Mandelbrot, *Les Objets Fractals: Forme, Hasard et Dimension*, Flammarion, Paris, 1975.
[5] B. Mandelbrot, *The Fractal Geometry of Nature*, W. H. Freeman, New York, 1982.
[6] R. Voss, J. Clarke, *Nature* **258** (1975) 317.
[7] B. Mandelbrot, *Water Resources Research* **7** (1971) 543.
[8] B. Mandelbrot, *Multifractals and 1/f Noise: Wild Self-affinity in Physics*, Springer, New York, 1998.
[9] B. Mandelbrot, "A maverick's apprenticeship," in *The Wolf Prize for Physics*. ed. David Thouless, World Scientific, Singapore, 2004.
[10] R. Voss, J. Clarke, *J. Acoust. Soc. Am.* **63** (1978) 258.
[11] T. Bolognesi, *Computer Music J.* **7** (1983) 25.
[12] C. Dodge, *Computer Music J.* **12** (1988) 10.

[13] M. Bulmer, *Proc. of the Mathematics 2000 Festival, Melbourne*, January 2000.
[14] M. Gardner, *Sci. Am.* **238** (April, 1978) 16.
[15] M. Beauvois, *Music Perception* **24** (2007) 247.
[16] S. Rankin, E. Large, P. Fink, *Music Perception* **26** (2009) 401.
[17] D. Levitin, P. Chordiab, V. Menonc, *Proc. Nat. Acad. Sci. USA* **109** (2012) 3716.
[18] H. Brothers, "Fractal music: Pulse trains." http://www.brotherstechnology.com/yale/FractalMusic/Pulse/Pulse.html
[19] B. Manaris, J. Romero, P. Machado, D. Krehbiel, T. Hirzel, W. Pharr, R. Davis, *Comput. Music J.* **29** (2005) 55.
[20] H. Brothers, "Fractal music: Duration scaling." http://www.brotherstechnology.com/yale/FractalMusic/FracMusicBground/DurScaling/DurScaling.html
[21] H. Brothers, "Fractal music: Pitch scaling." http://www.brotherstechnology.com/yale/FractalMusic/FracMusicBground/PitchScaling/PitchScaling.html
[22] K. Hsu, A. Hsu, *Proc. Natl. Acad. Sci. USA* **87** (1990) 938.
[23] H. Brothers, *Fractals* **17** (2009) 537.
[24] H. Brothers, *Fractals*, **15** (2007) 89.
[25] C. Wuorinen, "Music and Fractals," this volume.
[26] D. Kormann, "About the golden ratio." http://bowerbird-studios.com/aicaramba/page2.html#gold
[27] H. Brothers, "Fractal music: Structural scaling." http://www.brotherstechnology.com/yale/FractalMusic/StrScaling/StrScaling.html
[28] J. Scrivener, *Proc. of Bridges 2000*, (2000) 185.
[29] H. Brothers, "Examples: Motivic scaling." http://www.brotherstechnology.com/math/fractal-music.html#motivic
[30] H. Brothers, "Fractal music: Josquin des Prez." http://www.brotherstechnology.com/fractal-music/josquin.html
[31] D. Grout, C. Palisca, *A History of Western Music*, 7th ed., Norton, 2005; p. 197–198.
[32] Wikipedia, "Cantus in Memoriam Benjamin Britten." http://en.wikipedia.org/wiki/Cantus_in_Memoriam_Benjamin_Britten
[33] H. Brothers, "Fractal Music: Prerequisites for fractal classification." http://www.brotherstechnology.com/yale/FractalMusic/FracMusicBground/Frac.html
[34] M. Bigerelle, A. Iost, *Chaos, Solitons, Fractals* **11** (2000) 2179.
[35] Z.-Y. Su, T. Wu, *Physica D* **221** (2006) 188.
[36] J. Boon, *Adv. in Complex Systems* **13** (2010) 155.
[37] N. Nettheim, *J. New Music Research* **21** (1992) 135.
[38] H. Brothers, "Fractal Music: Duration scaling: Are the melodies fractal?" http://www.brotherstechnology.com/yale/FractalMusic/FracMusicBground/DurScaling/DurScalingFrac.html

[39] H. J. Brothers, "The Nature of Fractal Music: Supplementary Material." http://www.brotherstechnology.com/papers/nature-of-fractal-music.html
[40] Wikipedia, "Autostereogram." http://en.wikipedia.org/wiki/Autostereo gram
[41] M. Frame, "Fractal Geometry: Driven IFS." http://classes.yale.edu/Fractals/IntroToFrac/DrivenIFS/DataIFS/CoarseGrain/CoarseGrain.html

Chapter 8

Fractal antenna and fractal resonator primer

Nathan Cohen

Abstract. Self-similarity and fractals have opened new and important avenues for antenna and electronic solutions over the last 25 years. This primer provides an introduction to the benefits provided by fractal geometry in antennas, resonators, and related structures. Such benefits include, among many, wider bandwidths, smaller sizes, part-less electronic components, and better performance. Fractals also provide a new generation of optimized design tools, first used successfully in antennas but applicable in a general fashion.

8.1 Introduction

Antennas are commonly devised as linear wires, planar slabs, cylindrical tubes and other quasi-arbitrary Euclidean shapes invoked by familiarity, and corrected for performance by add-on components, such as coils, capacitors, dielectrics, and matching networks. While antennas are, literally, geometric shapes that radiate, antenna engineering has failed to appreciate and exploit the full technology advantage that may be gleaned by putting geometry first.

That perspective changed [1], and continues to change, with the recognition of the benefits of fractals and self-similarity in antennas and electronics. Here, we provide a brief introductory primer (although not an exhaustive review) highlighting the issues, and advantages, in what amounts to a technological paradigm shift in antennas and related electronics, using the geometry of self-similarity and fractals.

8.2 General antenna attributes

Antennas convey current distributions that excite moving electrons via a repetition of alternating current waves, manifest as RF. Antennas are thus defined by current distributions, guided and imposed by their structure ([2] see additional general antenna information therein).

Borders of the antenna structure invoke boundary conditions where the current must fall to zero, and the length/area of the radiating structure, relative to a fraction of a wavelength, determines the peak magnitude of the current and the respective '**radiation resistance**' by parsing out the RF sinusoid. The higher the radiation resistance the more efficient the RF wave generation and the wider the **bandwidth** (usually defined as the passband or adjacent frequencies near resonance).

Although antennas are debatably an arcane intellectual pursuit, they fundamentally come down to the magnitudes of current maxima; how many there are; and where they are spaced relative to one another in 3-dimensional space. They are analogously Young's multi-slit experiment [3] in application, with interference dominating (although not uniquely; diffraction is a secondary consideration in most cases) the interaction of waves in the far field, producing directionality, beams, and so on. There is a somewhat gray area between a bona fide 'antenna' or '**antenna element**' as opposed to an '**array**' but in general array components each have separate excitation sections, as opposed to an antenna (element) which is a singly-excited structure.

When antennas comprise several wavelengths in aperture and/or have discrete elements that are spaced across such a larger aperture, then an array of alternating currents is formed. It is this array of current maxima which produce the 'gain' and 'directivity' of the antenna, and narrow its power pattern or 'beam.' In general, such gain antennas are of order a wavelength or multi-wavelength in size.

When the antenna is substantially **sub-wavelength** in size then there is usually less opportunity to carry much more than a fraction of the amplitude of a single current wave, thus producing a deterioration in radiation resistance, and the antenna becomes a poor radiator with ever decreasing size. Such antennas are dubbed with the technical term '**electrically small**' [2]. Typically electrically small antennas are approximately 1/10 of a wave or smaller, in maximum extent.

The principle objective of antenna design is to play off bandwidth or multiple passbands, size, and gain/directive characteristics. Although other issues arise, especially with use of components to 'fix' the antennas characteristics, it is these three that dominate the perspective and thus the opportunity for fractals to be useful, usually uniquely so, as a tool. What makes a useful antenna paradoxically, lies with who, in part, is asking. For the perspective of this review, **a better antenna design** is one that yields an antenna that is smaller, cheaper in materials, has a better performance matrix of multiples in bandwidth/size/gain and has fewer, if any external components.

An unusual emphasis of antenna engineering focuses on making **very electrically small** antennas (roughly smaller than $1/20 - 1/25$th of a wavelength). These do not perform well: poor performance on very small electrical sizes is a pluralistic, equal opportunity for any antenna approach, Euclidean or not. Curiously, very electrically small antennas periodically get reexamined with (often irrational) exuberance in some studies. In reality, the 'very electrically small antenna' is a 'holy grail' pursuit with no basis for belief of realization. Oddly, fractals have been accused of making such holy grail claims, although the facts present an appropriate disclaimer in the infancy of the fractal antenna field [4], and the general rule of thumb is that fractals shrink antennas a factor of $2 - 4$ times without major tradeoffs in performance. To wit: very electrically small antennas beget feeble performance, irrespective of geometry. Although there is undue interest in very electrically small antennas, actual applications neither assume nor invoke such regimes, so their discussion is of some intellectual interest, but no practical interest, and has no impact on application.

With the understanding that it is the currents and their form and placement that define antenna, we may now proceed to understand advantages gleaned by fractal geometry.

8.3 Early fractal antenna history and a definition

The origin of fractal antennas lies with arrays. DuHamel and Isbell [5] invented the log periodic antenna, whose structure is defined by opening angles and has the attribute of being frequency independent or invariant over a wide bandwidth, a most sought after performance attribute. Use of an opening angle is a geometric constraint, a rare 'before-fractal' invocation of geometry in antennas. In fact, the log periodic is clearly self-similar about a point. Figure 8.1 shows a log periodic antenna: self-similar about a point. Most antenna engineers still draw a distinction between 'log periodic' antennas as being separate from 'fractal antennas' for reasons that cannot be rationally explained. Hence an early opportunity to fully explore, and understand the value, of fractal geometry in antennas was missed for generations [1].

Another early fractal approach was also tied into arrays. By 1986, the notion of fractals had gained some traction and spacing arrangement of antenna elements in an array showed some advantage in reducing sidelobes, again through the use of geometry, compared to 'random' approaches [6] We shall see later that fractal arrays can be exploited to produce virtually no sidelobes, while attaining frequency invariance.

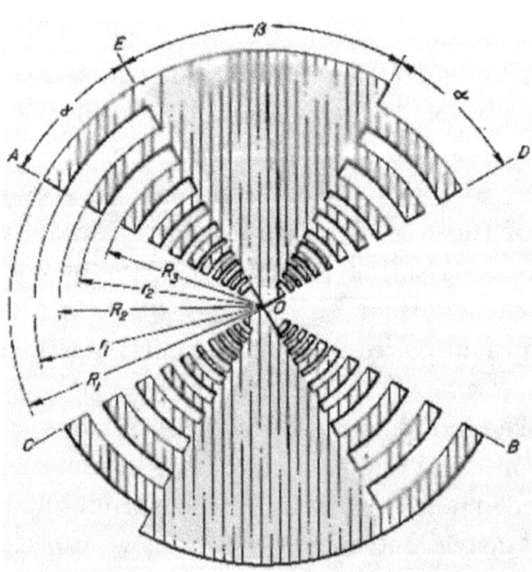

Figure 8.1 A log periodic antenna.

The first true creation and definition of fractal antennas occurred in 1987/88 [1], after inspiration in 1987 from a conference talk by Benoit Mandelbrot and subsequent discussions. Publication, after numerous rejections, occurred in 1995 in a ham radio and electronics journal — essentially the first publication on fractal antennas. Even at this early stage [7], a practical definition of a 'fractal antenna' was one with **at least portions with two levels of scaling and offset, of a generator**. This avoided the mathematical contrivance of infinite iterations in a finite world, and various issues with different fractal dimension formula and various definitions invoking fractal dimension. By defining fractal antennas as a class a priori, their overall attributes could be studied as a geometric approach — rather than getting blindsided by a specific and narrow example, such as log periodic antennas. This definition allows us to thus describe these attributes.

8.4 Fractals in action. Loading and multiband

Bandwidths are determined by the ability of an antenna to be tuned (at or close to resonance) at the desired frequency or frequencies. It is surprising how difficult this can be, particularly if, as is now commonly the case, more than one frequency 'band' is desired. This arises because the antenna may be sized appropriately for the lowest frequency, but other higher frequencies may not have the additional desired resonances. For example, a simple dipole of 1/2 wave has a multitude of resonances — but they are harmonic: they are resonant at modulo 1/2 waves. In real world applications, few desired bands are harmonic. Real needs require antennas with passbands that are not determined merely by the tuned characteristics of the fundamental (lowest resonance) frequency. Conventional approaches to this problem require adding parts (capacitors or inductors) to achieve one or two additional resonances, but such an approach adds demonstrable insertion loss (attenuation), cost, and degrades reliability.

A more appropriate approach is to **shape the antenna** so that the geometry itself yields the desired multiple bands. In addition, the size of the antenna is dominated by the requirement of achieve transmission with its fundamental resonance, so any method that also achieves a smaller size is extremely desirable.

Fractals have an unusual advantage in these regard, called 'fractal loading.' Here we will demonstrate the loading caused by space filling fractals as illustrative example, as shown in the accompanying diagrams [8].

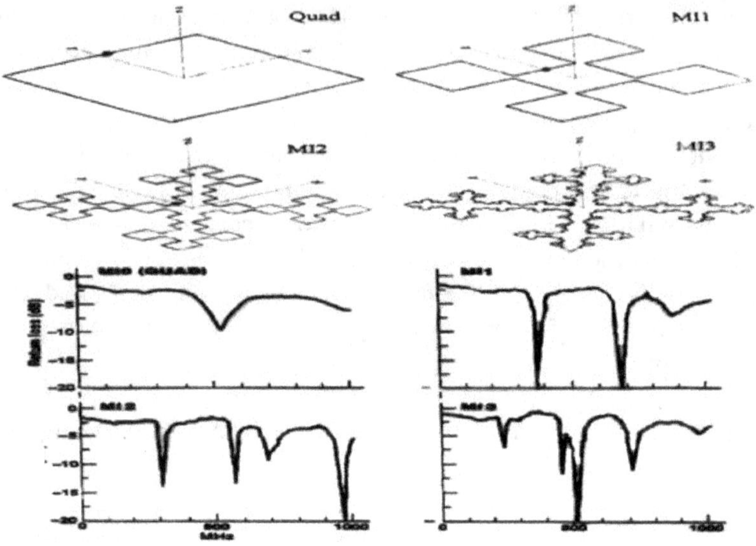

Figure 8.2 Four iterations of a Minkowski antenna.

Figure 8.2 shows four antennas with various iteration of a Minkowski (Koch rectangular) motif imposed upon a loop antenna, all with the same 6 inch footprint. Fractalists recognize them as 'islands' (Minkowski Islands (MI)). The higher the iteration, the more a shifting of resonances **down** in frequency and **new ones produced** non-harmonically, as shown in the respective spectra.

Note that of the lowest frequency is desired to remain the same, then the effect produces a smaller antenna for the lowest resonance. Thus the antenna is shrunk **purely by its fractal shape**. This is an attribute that is not limited to space filling fractal curves, but **all fractals as antennas**, some better than others. In this particular case, the result is not itself surprising, because an obvious attribute of fractal islands is their ability to produce long perimeters in small areas, thus rendering long antennas in electrically small areas [8].

What is surprising is that, in reasonable limits, the electrical performance is maintained. Figure 8.3 shows the MI2 island current distribution amplitude, and power pattern (beam). It has a dipole figure-8 pattern, but is only 1/4 the size of the halfwave dipole for essentially the same gain — a copper dipole is about 1.9 dBi in 'freespace.' To wit: it is a dipole equivalent gain antenna, in 1/4 the footprint [1]. The losses attributable to the

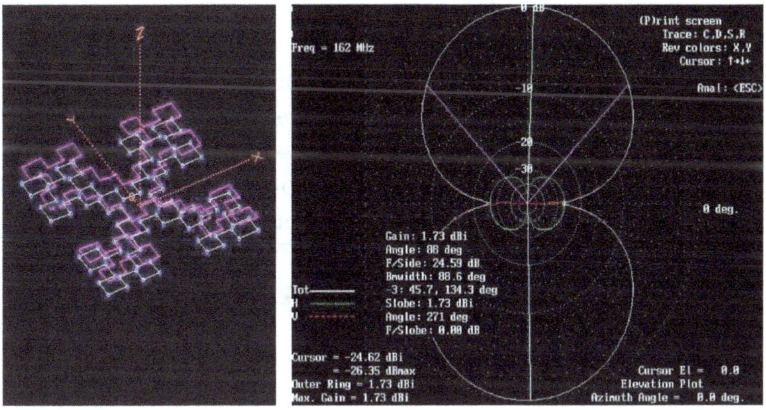

Figure 8.3 The current distribution amplitude of a Minkowski antenna.

conductive perimeter are approximately 0.2 dB, or 4% loss of power to 'ohmic' loss. Note that the proximity of two current maxima in one antenna yields a modest amount of constructive interference, even on this small size, an attribute of antennas that are otherwise dubbed 'electrically long' [9]. The electrical length also allows for the two current maxima, which increases the radiation resistance and essentially counterbalancing the ohmic loss caused by a longer perimeter length. This indicates that fractals can easily be used to produce small, high performance, low loss antennas in small form factors that are not very electrically small.

Of course, practical antenna problems almost never invoke single frequencies in modern times, and the second attribute of fractals is as a physical method to realize multiple frequencies which can be chosen *a priori*, unlike a dipole which has harmonic resonances.

In other words, the fractal can have multiple resonances which are formed by the geometric shaping and not by the addition or replacement of components. This is a breakthrough ability to control the size of the antenna and the bands, without added cost and compromised reliability invoked by using parts, such as capacitors or inductors.

In practical RF and wireless applications, multiple band performance is desired, and such bands seldom have any arithmetic, especially harmonic, relation. With a choice of fractals it **is possible to find the fractal shape or multi-fractal shapes that tailor the bands** made available by the antenna. Thus fractal shaping not only shrinks the size of the antenna, but provides, in an additional unique combination, multiple bands as desired, no matter where such bands lie in frequency. If one has a need for a smaller size and performance at specific frequency bands, one can find a fractal or multi-fractal that has the smaller size and bands of interest. This is a very powerful and, for many multiple bands, unique property of fractal antennas.

Of course, further iterations do shrink the antenna, but the advantage is asymptotic [1]. Typically 4 – 5 iterations prove the depth of the benefit for shrinking. Also, it is possible to use large numbers of iterations to make a very electrically small fractal antenna with a performance that boasts the same non-performance as a Euclidean. Thus the reminder that geometry can be abused, whether it is Euclidean or fractal. From standpoint of shrinkage, the 'sweet spot' for fractals is roughly 2 – 4 times smaller than the conventional design it replaces [4].

As the impedance is studied above the fundamental passband of a fractal antenna, another unusual attribute occurs. Figure 8.4 shows several fractal dipoles and their extended spectrum's standing wave ratio (SWR) — a conventional measure of impedance matching, in this case to 50 ohms. The standard straight dipole on the upper left shows a 'picket fence' set of bands from harmonics. The fractal ones all show and extremely stable and wide passbands with a transition from wideband to multiband. Fractals are also unique in this attribute of having a wide passband at higher frequencies. We shall also see that such wide performance also extends to gain and power pattern characteristics in frequency invariance for a special subset of fractal shapes.

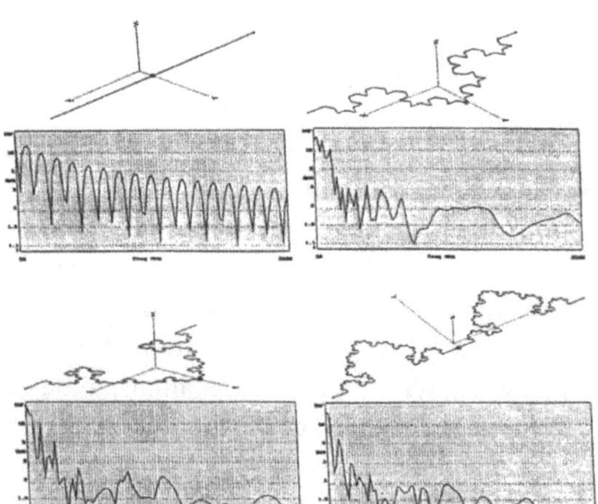

Figure 8.4 Several fractal dipoles and their standing wave ratios.

8.5 FRAGO: Optimization and solution degeneracy

Optimized antennas need geometry to best realize their performance. This was demonstrated by Landstorfer [10], who arrived at optimized antennas that, with no preconception as to the appearance of the results, ended up being tortuous fractals, although not recognized by the author as such. Examples are shown in Fig. 8.5. These are complex structures; they are tortuous paths. It takes some thought to understand that **modern antennas that are optimized are complex structures**. It took another decade to turn the question around and ask how fractals could most easily produce optimized antennas [1].

The advantage posed by fractal geometry in antenna optimization is vast. This arises from the ability of a complex fractal shape to be described by just a handful of variables, such as: motif; iteration; base; and so on. A complex shape without a fractal description must regard the structure as a piecewise function with length; position; and angular rotation for each segment. Thus optimization in a non-fractal scheme must look at a huge number of variables that have to be considered and varied to find convergence.

In contrast, and is known explicitly for 'fractal forgery scenes' of mountains and coastlines, or the fractal collages in fractal compression [11] a few fractal variables make for a complex surface or structure. A complex fractal antenna can be arrived at by a few variables in 'fractal coding.'

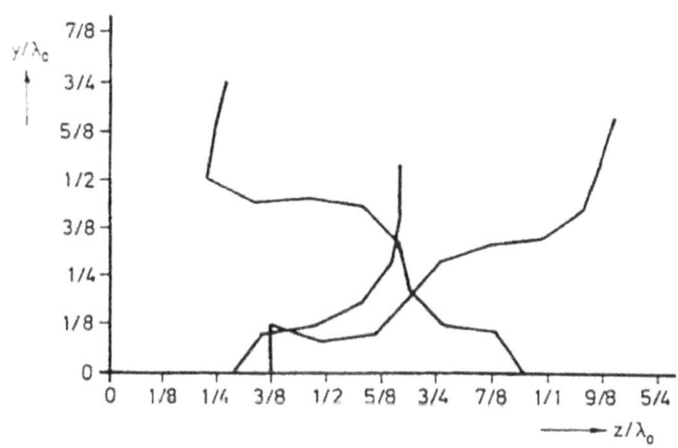

Figure 8.5 Tortuous antennas.

Fractal coding can be summarized by describing a complex structure as one or a combination of fractals. In simplest form, the variables describing the structure are thus:

1. One or more fractal motifs (say, two variables)
2. Iteration number for each fractal (say, two variables)
3. Fractal to fractal offset for concatenation (one variable)
4. Magnification factor (say, two variables)

Thus a handful of variables fully characterize the structure. Compare to a structure of complexity that may require, in contrast, 500+ variables to describe its shape.

A straightforward way to invoke the optimization is with genetic algorithms, using a fractal **genome**, hence Fractal Genetic Optimization (FRAGO) [12]. An objective function, or cost function, is defined a priori so that the best candidates may be compared. The cost function combines the performance values (such as gain, bandwidth, size) in a simple mathematical way that allows a numerical rating of candidate antennas. Using a comparison of cost functions, a 'natural selection' is done which culls out the worst candidates (those with the lowest cost functions). Hybrid collection of new antenna candidates are then made by mixing the genes (variables) of the survivor candidates, and allowing for random variation to mimic mutation. Culling and mixing of variations of variables assures new candidates, and the process is repeated through many generations. Eventually the difference in cost function values exceeds a defined 'high pass,' and the process is successfully terminated. A flow chart is shown in Fig. 8.6.

Genetic algorithms are known as powerful optimizers that essentially act as a guided trial and error approach. FRAGO has proven powerful and rapidly converging too, easily implemented with parallel computing: there is no need to figure out the cost function of candidates sequentially, and this can be determined in parallel.

The most surprising outcome of optimization using fractal coding is that it works at all: in essence, the definition and difference between variable values gives very different samplings of the phase space with which the objective function is assessed. What has been found is that the **solutions are degenerate** — several very different looking fractal antennas often achieve the 'best' in the objective function. Differences in which is better are often so minor as to make them indistinguishable. In other words, practical optimization invites **local maxima** without a requirement for an absolute

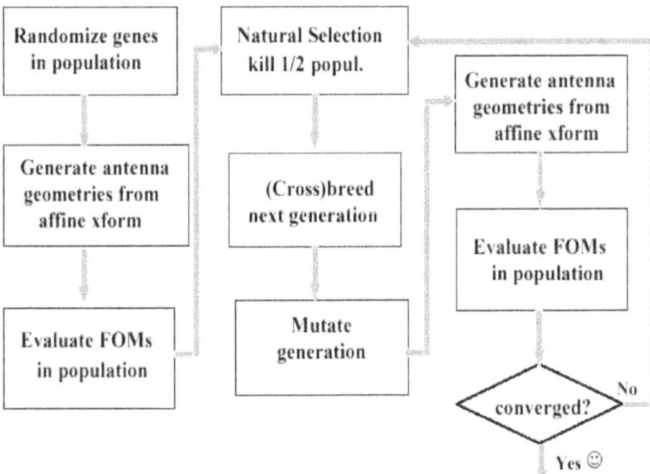

Figure 8.6 A flowchart for fractal genetic optimization.

maximum. This relieves the computational problem dramatically and makes it tractable. This demonstrates that, at least in the case for antennas, fractal optimization schemes are tremendously powerful in a practical sense, for finding a unique 'best' is almost never advantageous over a 'many' that are almost as good. Even finding just a single one of the many that meets the 'goodness' cutoff of the cost function solves the practical need.

It has not escaped attention that this approach also promises advantages to many practical problems, many remote from electromagnetics; FRAGO is a powerful and rapid optimization tool that applies across a very broad swath of such problems, and is not limited to antennas. Surprisingly, as a rapid optimization tool, this advantage has not been exploited in other realms beyond antennas.

8.6 Frequency invariance and fractals

Invariance properties of Maxwell's Equations are hard-earned nuggets of insight into nature, at some deep level. As mentioned, the notion of frequency invariance was explored in some depth over 60 years ago with log periodic arrays, culminating in Rumsey's Principle [2]: antennas with the same radiation properties result from the geometric stipulation of being defined only by angles.

Such antennas (for example, Fig. 8.1) are clearly self-similar (about a point). The fact that this was not recognized until [1] only obscured some deeper aspect of the importance of fractal geometry in electromagnetics. However, by 1999, the matter had been fully resolved via an analytic solution for frequency invariance of Maxwell's equations [13], briefly summarized here.

$$\nabla' \times \vec{E} = -Z_0 \vec{H}$$

$$\nabla' \times \vec{H} = Z_0^{-1} \vec{E}$$

$$Z_0 = \sqrt{\mu/\epsilon}$$

$$\vec{x} \to s\vec{x} \quad \text{and} \quad \omega \to \omega/s$$

$$0 < |s| < \infty$$

Here, $\nabla' \times$ coordinates scaled in units of wavelength; μ and ϵ are permeability and permittivity and s is a scaling factor. Maxwell's equations are invariant under these transformations, which are solved by self-similar, origin symmetric constructs only.

That analysis [13], now dubbed the HCR (Hohfeld-Cohen-Rumsey) Principle, showed conclusively that **self-similarity along with origin symmetry about a point** are the unique and closed form constraints for frequency invariance. However, 'defined by angles' is not the requirement for frequency invariant antennas, although clearly some HCR antennas may be trivially described by the simpler and inclusive description. Since few self-similar structures have these two constraints, frequency invariant fractal antennas are a subset of the overall family of fractal antennas. To wit: not all fractal antennas are frequency invariant, but **every frequency invariant antenna is fractal (self similar)**. This includes log periodic structures, but is not exclusive to them. Figure 8.7 is a handy Venn diagram that shows these requirements.

Note the two examples shown in Fig. 8.7: the left (log periodic) is 'defined by angles;' the Sierpinski bowtie (SB) on the right is not. The SB has triangles defined by scaling, angular rotation, and offset. The SB is frequency independent at discrete frequencies while the log periodic is quasi continuous in its invariance.

A special aspect must be appreciated as an attribute of some log periodic fractal arrays: they are also **self-complementary**. This led to some fundamental missteps in understanding frequency invariance by

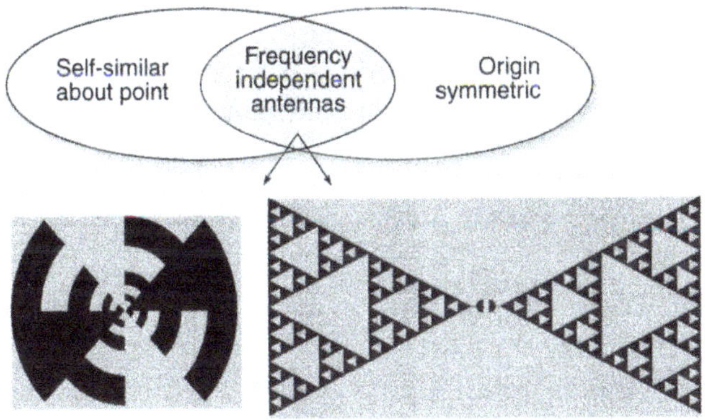

Figure 8.7 Some antenna geometries.

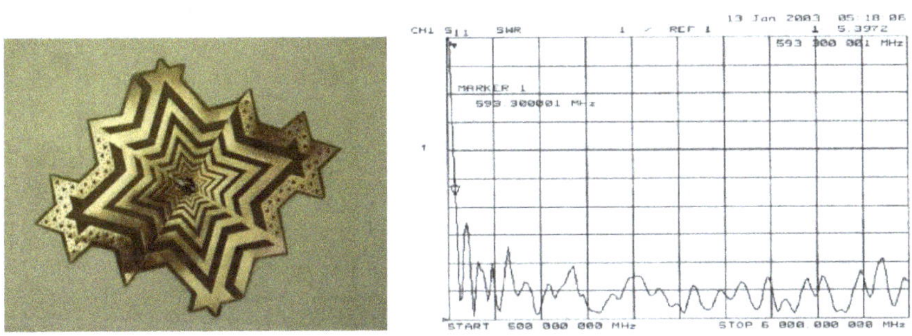

Figure 8.8 A fractal dipole.

Mushiake [14], and others. In fact, none of those efforts recognized self similarity as the underlying theme for frequency invariance; DuHamel and Ore [15] understood that a theory based analysis was needed to understand the geometric requirements for frequency independence, and that self complementarity did not provide the answer. Hohlfeld and Cohen [13] resolved the issue, showing self-similarity and origin symmetry yield frequency invariance, and if only the impedance aspect is desired, then self-complementarity may be used. Self-similarity, origin symmetry, and self-complementarity combined produce antennas that are strictly frequency independent and also impedance-similar at frequencies that lie outside of the other two constraints.

Using the HCR Principle, a simple wideband, invariant fractal dipole is shown in Fig. 8.8 along with its SWR in unit steps, over a 12 : 1 range, a portion of the overall performance invariance. Its gain is dipole equivalent across at least a 20 : 1 range of frequencies.

Unfortunately antenna engineering has been slow to accept the fractal nature of frequency invariance despite the solution being analytic and close-form. An excellent example of the fractal utility is discussed in the fractal array section.

8.7 Fractal resonators

Resonators are another name for tuned RLC circuits (filters) and by definition all antennas are radiating resonators. Much of RF electronics come down to making tuned circuits that convey power, and leaving the radiating part to the antenna. Analog filter theory is a hoary subject based upon reasonably simple solutions to differential equations, that predicts 'poles' and 'zeroes' [16]. It is a subtle but important point that such solutions do not suggest specific geometric constructs, and as such there has been an historical preference to reduce the embodiment of said solutions with discrete capacitors, inductors, and resistors. This has regrettably rendered filters to tried and true component-forms for many generations. Analogously to antennas, filters failed to exploit geometric approaches with any great degree.

The creation and invention of the first bona fide fractal filters/resonators was by Cohen [1, 7]. Subsequent benefits from lower loss; smaller size; greater bandwidth; and part-less application are now highly documented [17, 18]. These benefits mirror those of fractal antennas.

The main benefit of using fractals as filters is that it frees up the design space to allow, just as with antennas, deterministic, complex shapes that can be used to reduce size, increase isolation and rejection, increase and control passbands, and steepen passband skirts. An example of a fractal filter is given below [18] in Fig. 8.9, which clearly shows better rejection and smaller size with iteration. The MI fractal of Fig. 8.2, that is apparent in Fig. 8.9, is not accidental or coincidental, but an unattributed example from [1]–.

In very recent times, fractal resonators have rejuvenated analog RF electronics especially at microwave, and coming years should see near ubiquity of fractal designs in filters, from cost and size savings: from a cost standpoint, an etch or trace on a circuit board is always preferable than discrete components. The tuned circuit becomes controllable in performance and part-less with fractals.

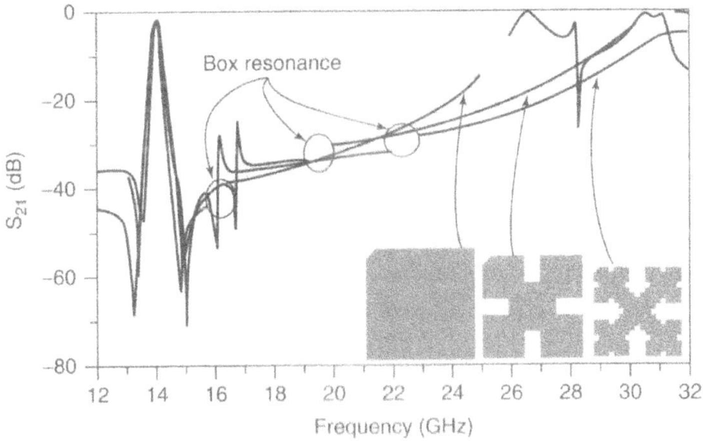

Figure 8.9 A fractal filter.

8.8 Fractal plasmonics and metamaterials

In addition to understanding benefits of individual fractal shapes as resonators, an interesting behavior arises when resonators are placed in a **very close array**, of roughly 1/10 of a wave separation or less. Marconi and Franklin [19] were the first to consider the placement of close spaced resonators and used them to make a reflector. This was the first bona fide example of 'metamaterials.' Fractals obviously allow closer spacing, tighter placement/coupling and wider bandwidth in close packing, thus opening opportunities for a plethora of fractal metamaterial applications. An example of such a 'fractal metamaterial' array is shown in Fig. 8.10 as a sandwich, for a wideband flat antenna.

Figure 8.10 A flat wideband antenna made of fractal metamaterial.

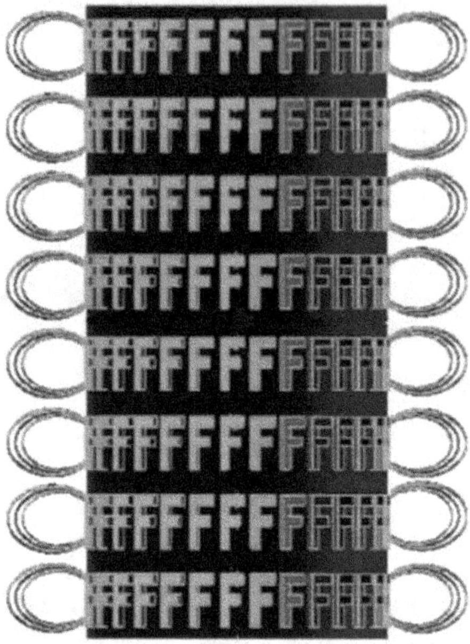

Figure 8.11 A cylinder of fractal resonators.

Although it should be obvious from the construction of transformers and capacitors [16], evanescent waves propagate well in the very near field, and in such an aforementioned closed-spaced array, the evanescent waves are exploited. Recently this has been dubbed 'plasmonic' coupling [20], where surface waves exist from close spacing. Each resonator couples via a plasmonic wave to the next, and each resonator essentially duplicates the current distribution of its neighbor. While the current on a continuous sheet experiences the usual sinusoidal variations, with phase inversion, across the aperture, a plasmonic surface, with a superposition of nearly identical currents, closely spaced, has the pleasing effect of superposing to a slightly rippled plane wave in the far field. Figure 8.11 shows a schematic cylinder with same, each F corresponding to a fractal resonator and showing the current distribution for the fractal resonators on the sides.

The 'fractal plasmonic surfaces' (FPS) have several non-intuitive features. Unlike a regular aperture, the resonance characteristics are NOT determined by the physical aperture in wavelengths of the sheet — they are determined by the **individual** resonators, cloned as an array. Next, the current distribution is cloned across the sheet, so there are little sinusoidal

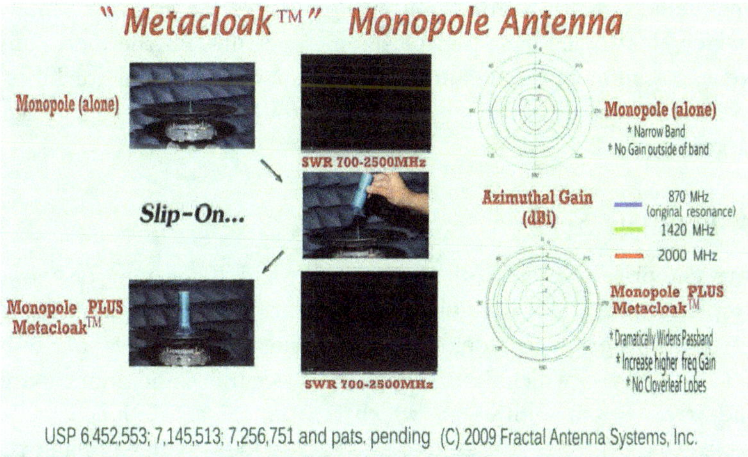

Figure 8.12 Metacloak monopole antenna.

variations in the far field. Thus the power pattern has minimal sidelobes and is concentrated as one main lobe or beam. Finally, like a very large aperture, FPSs share the ability to have their effective aperture determined by their geometric aperture. So the resonance is determined by the single, clones resonators, but the gain and power pattern beam size is determined by the area of the cloned array aperture. This is unlike a conventional current sheet, where the resonance is determined by the aperture size, and the number of waves conveyed by the aperture sets the power pattern. Put simply, FPSs emulate the characteristics of geometric optics even though most of the actual apertures are sub wavelength in size. This is a remarkable and unexpected result.

An example of exploiting this lies with a wideband fractal cloak, discussed in Chapter 9. It will be appreciated that spacing of fractal metamaterial layers, and the placement of a boundary condition, very much changes the characteristics of the very near field, so that superficial appearances of similar structures can yield very different results.

Another useful example entails a 'fractal collar' [21] a parasitic antenna made from layers of fractal plasmonic surfaces, in which the layers are excited by a conventional antenna, such as a monopole, upon which it is placed by not physically connected. Figure 8.12 shows the antenna add-on, making no direct electrical connections. The fractal collar dramatically extends the operational bandwidth of the antenna, typically increasing the passband by 3 to 5 times, and also increases the gain with increasing frequency, due to the increased plasmonic aperture. Furthermore, unlike a conventional monopole, the antenna does not break into major lobes with increasing frequency, that change the elevation power pattern and decrease peak gain.

8.9 Fractal arrays

Arrays comprise discrete collections of currents. They may be viewed as incomplete apertures with gaps and truncation in the spatial frequency spectrum. This produces sidelobes — off angle gain responses that resemble 'ghost' artifacts — which limit the imaging, accuracy, and gain capabilities of the array — capabilities for which is what it was designed for in the first place. Add to this the issue of making the array wideband and the full

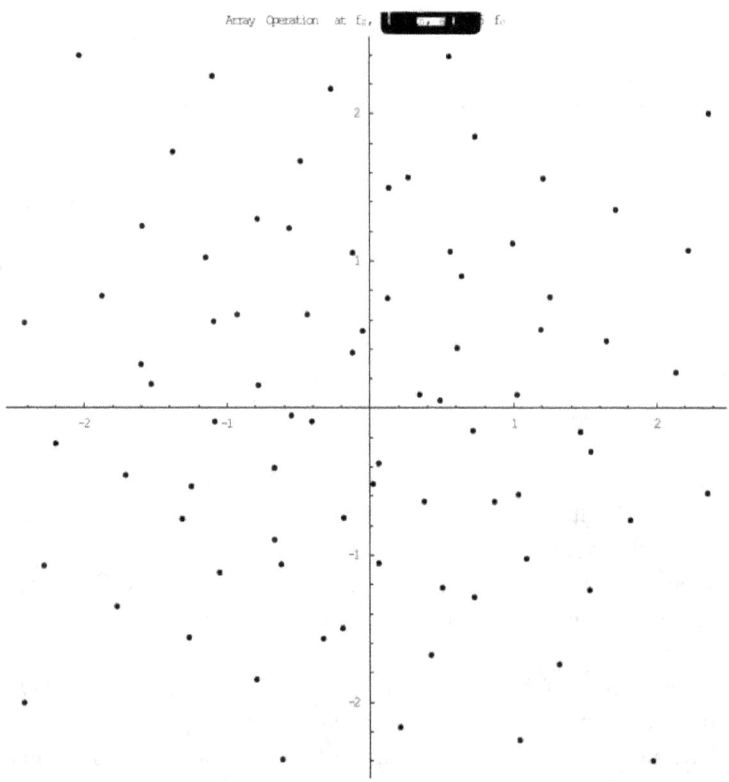

Figure 8.13 A sparse antenna array.

play of frustration ensues, along with a century of trying to mitigate the problems. Such is the objective and history of arrays [22].

As for fractals, there were certainly early attempts to show self similarity in arrays configurations, over a large range of wavelength baselines. For example, scaled sampling of the spatial frequency spectrum, as is natural from multiple Cantor iterations, produces a scaled set of sidelobes [23]. Using fractals to do something new in discrete arrays followed from Kim and Jaggard [6], who showed that a fractal arrangement could do as well as a random one in mitigating sidelobes, when sidelobe mitigation was desired. However, arrays usually seek to maximize wideband use in addition to sidelobe reduction. That solution was achieved after the discovery of HCR [24].

Figure 8.13 shows an array of positions of a plurality of discrete antenna elements as a sparse array across many wavelengths in a plane, at three

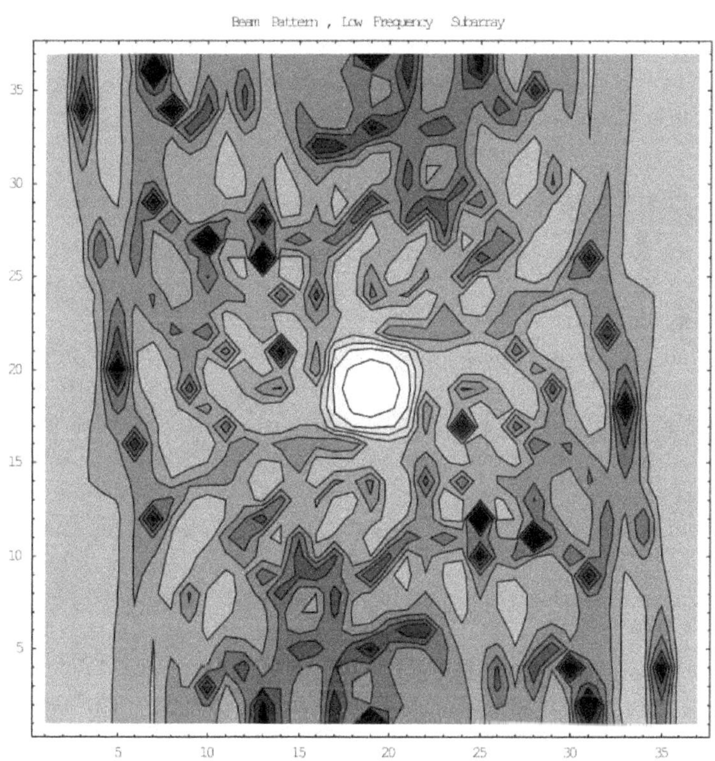

Figure 8.14 Antenna power pattern.

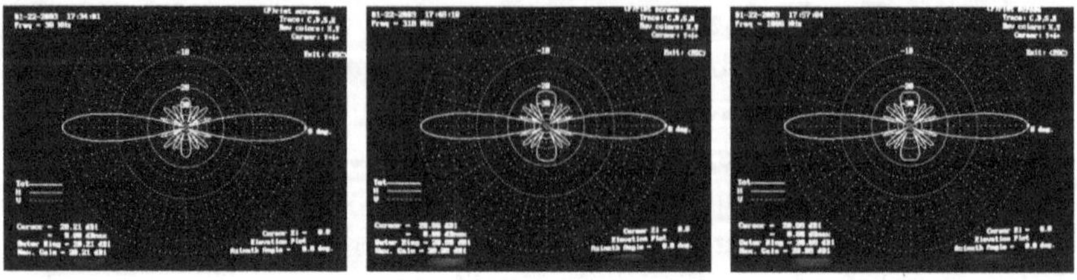

Figure 8.15 Power pattern elevation at different frequencies.

different frequencies. It hardly appears, by eye, to be fractal, but was constructed purely from self-similarity and HCR constraints. It is not a scatter diagram.

Without a backplane the array is bidirectional. Figure 8.14 shows the power pattern — note near absence of lobes, and Fig. 8.15 shows the power pattern in elevation a several different frequencies. The power pattern remains the same over, with the three representative beams being the same over a factor of 1000 in frequency. Thus is principle, arrays with a near absence of lobes and invariance of gain and beams can be easily constructed, despite being sparse, using the fractal nature of HCR.

8.10 Summary

As shown here, fractals have proven to not only be a tool for making better antennas and resonators, but in many cases they are an essential and enabling technology in antenna and electronic design. Despite limited acceptance to date, rapidly growing use from more and comprehensive telecommunications and wireless needs will ultimately make fractal antennas and electronics required, rather than disruptive [25]. This acceptance grows daily with a dramatic increase in published research efforts, now executed globally.

Antennas as a technological art are truly only a few generations old: the tried and true pendulum in paradigm shifts favors the fractally prepared. Thomas Kuhn's adage [26] applies — paradigm acceptance lies with the demise of the hold outs, and the overwhelming evidence for the obviousness of the benefits of the shift. With the case of fractal antennas and electronics, the firm foundation has pre-dated widespread use by roughly 25 years, so the shift in the paradigm is ongoing and perhaps like the structure of fractals themselves, a world with interesting answers and further fascinating questions.

Bibliography

[1] N. Cohen, "Fractal antennas part 1," *Communications Quarterly* **5** (1995) 7–22.

[2] J. Kraus, *Antennas*, McGraw-Hill, New York, 1988.

[3] D. Halliday and R. Resnick, *Fundamentals of Physics*, Wiley, New York, 1971.

[4] N. Cohen, "Exploring a fractal dipole," *Applicd Computational Electromagnetics Society Newsletter* **13** (1998) 23–27.

[5] R. du Hamel and D. Isbell, "Broadband logarithmically periodic antenna structures," *IRE National Convention Record, Part 1* (1957) 119–128.

[6] Y. Kim and D. Jaggard, "The fractal random array," *Proc. IEEE* **74** (1986) 1278–1280.

[7] N. Cohen, *Fractal antennas and fractal resonators*, U.S. Patent 6,452,553 (2002; filed 1995).

[8] N. Cohen, "Fractal antennas part 2," *Communications Quarterly* **6** (1996) 53–66.

[9] C. Puente, J. Romeu, R. Pous, J. Ramis, A. Hijazo, "Small but long Koch fractal monopole," *IEEE Electronics Letters* **34** (1998) 9–10.

[10] F. Landstorfer and R. Sacher, *Optimisation of Wire Antennas*, Wiley, London, 1985.

[11] M. Barnsley and L. Hurd, *Fractal Image Compression*, Peters, Boston, 1992.

[12] N. Cohen, "Genetic antenna optimization with fractal chromosomes," *Proc. Applied Comp. Electromag. Soc.* (1997) 306–308.

[13] R. Hohlfeld and N. Cohen, "Self-similarity and the geometric requirements for frequency independence in antennae," *Fractals* **7** (1999) 79–84.

[14] Y. Mushiake, *Self-Complementary Antennas*, Springer, New York, 1996.

[15] R. DuHamel and F. Ore, "Logarithmically periodic antenna designs," *IRE National Convention Record* **1** (1958) 139–152.

[16] P. Horowitz and W. Hill, *The Art of Electronics*, Cambridge Univ. Pr., Cambridge, 1980.

[17] J. Hong and M. Lancaster, *Microwave Filters for RF/Microwave Applications*, Wiley, New York, 2001.

[18] P. Jarry and J. Beneat, *Designs and Realization of Miniaturized Fractal RF and Microwave Filters*, Wiley, New York, 2009.

[19] G. Marconi and C. Franklin, *Reflector for use in wireless telegraphy and telephony*, U.S. Patent 1,301,473 (1919).

[20] S. Maier, *Plasmonics*, Springer, London, 2007.

[21] N. Cohen and M. Chin, *Methods and Apparatus for Enhanced Radiation Characteristics From Antennas and Related Components*, U.S. Patent pending (20110063189; filed 2009).

[22] B. Steinberg, *Microwave Imaging with Large Antenna Arrays*, Wiley, New York, 1986.

[23] C. Puente and R. Pous, "Fractal design of multiband and low sidelobe arrays," *IEEE Trans. Antenn and Prop.* **44** (1996) 730–739.

[24] R. Hohlfeld and N. Cohen, "Genetic Optimization of Sparse, Frequency Invariant Arrays Using the HCR Principle," *IEEE Int. Conf. Phase Arrays Systems and Technology*, IEEE Press, Waltham, 2003, pp. 588–593.

[25] C. Christensen, *The Innovator's Dilemma*, Harvard Business School Press, Cambridge, 1997.

[26] T. Kuhn, *The Structure of Scientific Revolutions*, Univ. of Chicago Pr., Chicago, 1962.

Chapter 9

Fractal-based wideband invisibility cloak

Nathan Cohen, Obinna Okoro, Dan Earle, Phil Salkind, Barry Unger, Sean Yen, Daniel McHugh, Stefan Polterzycki and A. J. Shelman-Cohen

Abstract. A wideband invisibility cloak (IC) at microwave frequencies is described. Using fractal resonators in closely spaced (sub wavelength) arrays as a minimal number of cylindrical layers (rings), the IC demonstrates that it is physically possible to attain a 'see through' cloaking device with: (a) wideband coverage; (b) simple and attainable fabrication; (c) high fidelity emulation of the free path; (d) minimal side scattering; (d) a near absence of shadowing in the scattering. Although not a practical device, this fractal-enabled technology demonstrator opens up new opportunities for diverted-image (DI) technology and use of fractals in wideband optical, infrared, and microwave applications.

9.1 Introduction

Close spaced (<1/10 wave) resonator array structures (CSR's) have been identified to have unusual benefits, with the first evidence for such provided by Gugliemo Marconi and Charles Franklin [1]. Recent evidence for remarkable attributes have manifest under the rubrics of 'plasmonics' [2] and 'metamaterials' [3], but all share the commonality of close-spaced resonators with evanescent waves, with unusual properties often unattainable by other means.

A key example of these attributes is an invisibility cloak (IC). Such a device must have the ability to divert electromagnetic waves around an 'object' so that measurement from the opposing side closely replicates a free-path, without such an obstruction. Although the physics is 'see around,' to the observer, the invisibility mimics 'see through.' Thus the object is hidden. In 2006, construction of the first invisibility cloak [4] was done at microwaves, and showed results that could be interpreted as meeting these invisibility conditions. However, this IC had considerable restrictions, which included: (1) narrow frequency coverage; (2) measureable shadowing in front and in back of the object; (3) reduced intensity of the path compared to the free-path (that is, some inefficiency). This first IC also invoked transformational optics as the unique method by which the phenomenon was produced.

Invocation of transformational optics led to a theoretical study [5] that asserted an IC with wideband capabilities was not possible. Such wideband behavior is necessary, for example, if visual light invisibility is ever to be achieved, with its bandwidth approaching roughly 3:1, as opposed to a narrow band system with, for example, a ratio of bandwidth of 0.1:1 Thus transformational optics motivated the formation of an IC with limited capabilities that were shown to be inherently limited by the physics so invoked, and in doing so rendered narrow band exploration for all future efforts, and the assumption that ICs must be narrow band. We believe the invisibility cloak approach of CSRs as being one of a wave guide approach, enabled by plasmonic coupling via evanescent waves rather than transformational optics. Indeed, this is what guided us to pursue the construct described here, which elucidates construction of a wideband IC using CSRs of a simple fractal design. This wideband IC must be considered a new approach that renders previous wideband objections and shadowing/scattering limitations as moot, and demonstrates the first example of a true invisibility cloak (see also [6]).

9.2 A fractal approach

Fractals, at least in simple forms, have been shown to have some powerful applications within the realm of antenna and resonator fields [7]. As is the convention of the term 'fractals' in antennas, the self similarity is applied at least partially across the structure and is finite in iterations. Key amongst these advantages is the ability to produce multiband or wideband behavior in a shrunken form factor, simply from a geometric structure, and without additional components or materials, for example.

In the case of invisibility cloak construction, fractals used in CSRs produce continuous 'bed of nails' current cells concurrent with each individual fractal resonator. This effectively acts as a mechanism for radiative transmission and transport across an aperture with evanescent surface waves coupling to the free space waves. Such radiative transport is a guide for controlling the placement of waves and is a mode typical of waveguides. This fact was adduced to be key to construction of a wideband invisibility cloak.

The advantage that fractals allow in an IC are: smaller size of the resonator cells; multiple frequency resonance for CSR layer or sheet (and thus wideband when several layers); more efficient plasmonic coupling via closer packing; and more geometrically constrained plasmonic coupling. None of these are achieved, for example, with the conventional use of split ring resonators (SRRs) as with the first IC [4]. The approach difference is clearly ascertained by comparing CSRs used in this new wideband IC using fractals, and the previous example using SRRs. Figure 9.1 shows a typical collection of fractal (Minkowski [7]) cells (resonators), in comparison to split ring resonator cells, such as used in [4]. Note that a Minkowski cell is a loop structure with rectangular Koch iteration along at least one sides. Trace width on the Minkowski is approximately 1.3 mm.

Figure 9.1 Sections of CSRs used in cloaking. The shows fractal patterns of a second iteration Minkowski, the right is what was previously used as split ring resonators (SRRs).

The construction basis for this IC is a series of concentric rings or shells made from these fractal cells/CSRs, with an additional inner boundary layer defining a region, the 'object,' to be made invisible. All invisibility cloaks, including this one, are essentially two dimensional demonstration devices, the cloak being a series of rings rather than three-dimensional shells. In the first IC [4] the number of rings is considerable, and can be a dozen of more.

The inner boundary layer imposes a boundary condition that prevents transmission through the object being cloaked. Outside waves impinge on the rings and are guided around the boundary layer. Anything inside the boundary layer is thus cloaked. Essentially the waves reconverge to form a free space far field waveform at the antipodal point to their entry.

The practicality of producing such an unusual waveguide is dictated by the desire to produce reasonable fidelity of the 'direct path' between, say two antennas, with limited attenuation or scattering compared to the passband characteristics of the direct path. In other words, the view through an invisibility cloak must closely match that seen when there is nothing there.

9.3 IC construction and testing

Our objective was to construct an IC with characteristics that had good to excellent fidelity of reproduction of the direct path gain; had the fewest cloak rings to achieve that, and, most keenly, was wideband.

The general criteria for fidelity was set by the experimental measurement of a cloaked response that fell within a 3 dB peak to peak reproduction of the passband response of the direct path.

Wideband, although somewhat arbitrary in definition, may be set by analogy to the visible light spectrum, where the eye is a detector across a roughly 2.5:1 passband. The IC here was, for reasons of practicality of scale and cost, designed at microwave frequencies, from 500 to 1300 MHz, (wavelengths 60 to 25 cm) to be consistent with the wideband analogy of a visual system. To wit: demonstrating a wideband IC at microwaves indicates the physical ability to do so at visual light, when technology allows for the appropriate scaling of wavelengths.

The wideband IC mechanical platform consisted of a polycarbonate RF transparent comb in radial circles, that supported the CSR rings of the cloak as well as the boundary layer. Additional mechanical support was provided by styrene foam and a polyethylene drum. Measurements were made in a low-reflection environment with anechoic absorbing material at potential multipath reflection points.

A direct path was defined as a roughly 1.3 meter displacement between two log periodic antennas, each identical to each other with forward gains of 6 dbi $\pm$ 0.5 dB across 500 – 1300 MHz passband. Back rejection on the LPs was F/B in excess of 25 dB. The two antennas were placed so that the roughly 80° beams of each log periodic antenna were filled by the angular extent of the cloak. An Agilent E5071C network analyzer provided S12 gain measurements between the two antennas, with normalization done for the direct path with all mechanical supports present. The signal to noise was in excess of 50 dB and the frequency resolution was roughly 50 MHz, with Gaussian smoothing applied. Measurement uncertainties were negligible. Variations measured were essentially true variations caused by the IC or object themselves.

Considerable guided trial and error experimentation, emulating a genetic algorithm, led to a wideband, working IC consisting of only 6 rings and a boundary layer, with specifications given in Table 9.1.

The rings were fabricated from conformal circuit board of 10 mil thickness. The substrate material was RF35 with a DK of 3.5. Note that the boundary layer is also, and not coincidentally, a fractal made from a Sierpinski motif, as shown in Fig. 9.2, and is covered in RF transparent tape to show visually its distinction from the other rings. The wideband IC and object with log periodic antennas is shown schematically in Fig. 9.3 and in actual view in Fig. 9.4.

Measurements comprise the normalized 500 – 1300 MHz gain from the direct path, which was then compared to the passband gain with the object present, and then with the cloak-surrounded object. In addition, copper rings were placed as a substitute for the cloak rings to act as a control for comparison, thereby evincing the effects of cloaking are not spurious nor caused by some other effect, such as multipath reflection.

Table 9.1 Specification of the Cloak Rings.

Ring/Object	**Radius (cm)**	**Cell/Type**	**Footprint/Separation (cm)**
Object	14.2	Sierpinski ribbon	7.8 (width)
1	17.0	Minkowski	1.9 × 2.4/0.2
2	17.9	Minkowski	2.1 × 2.8/0.3
3	19.5	Minkowski	1.9 × 2.4/0.2
4	20.7	Minkowski	2.1 × 2.8/0.3
5	22.7	Minkowski	2.1 × 2.8/0.3
6	24.8	Minkowski	2.1 × 2.8/0.3

Figure 9.2 Sierpinski ribbon of 3 iterations used to make a boundary layer.

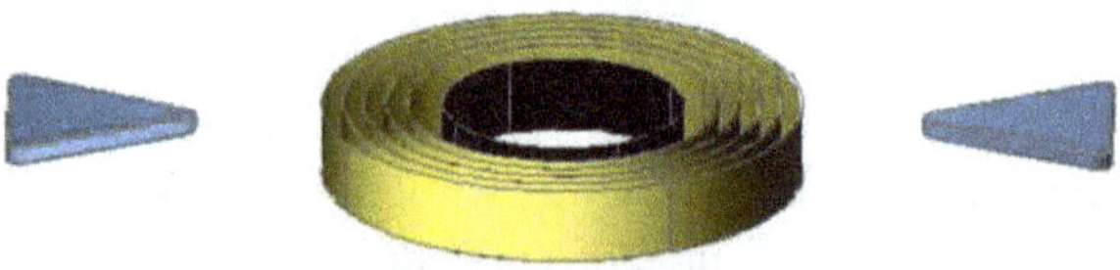

Figure 9.3 Schematic of object with IC rings and placement between two log periodic antennas. Removal of the IC and the object gives the 'direct' path.

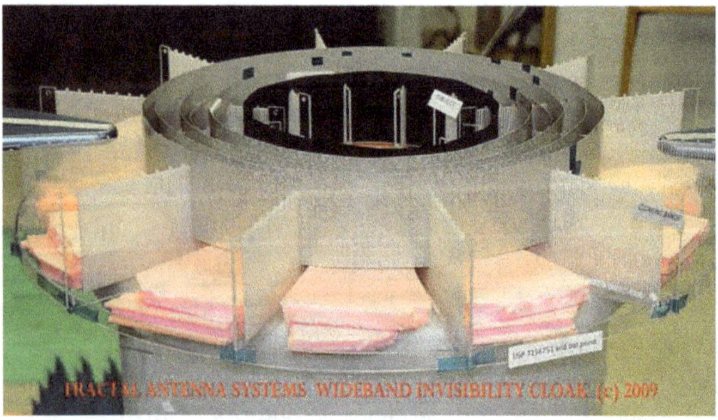

Figure 9.4 Wideband microwave invisibility cloak.

9.4 Results: Antipodal and scattering gains

Figure 9.5 shows the passband gain for the direct path (so normalized to zero gain in dB) and with the object present from the opposite (antipodal)

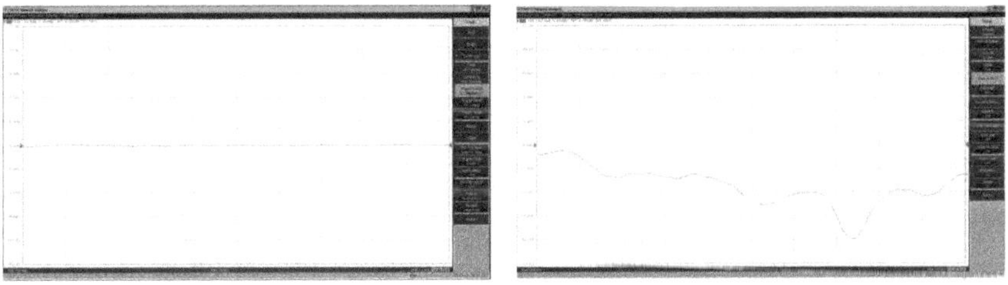

Figure 9.5 Left: normalized direct path gain passband from 500 – 1300 MHz, in 3 dB increments. Right: occultation by object (boundary condition) causes gain reduction. Passband from 500 – 1300 MHz, in 3 dB increments.

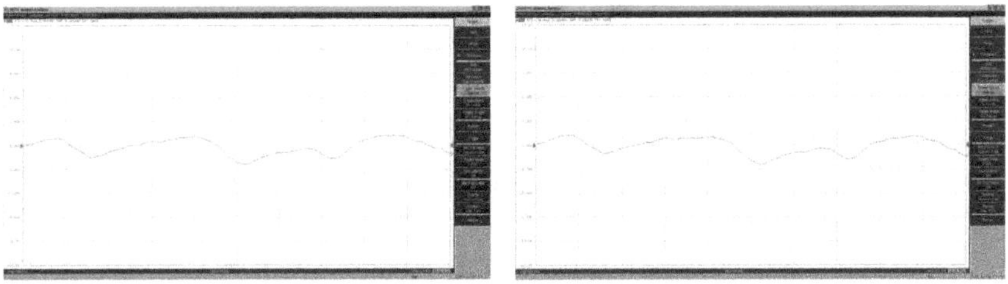

Figure 9.6 Left: IC rings around object restore direct path response with better than a 3 dB peak to peak variation, and thus high fidelity. Same passband. Right: with a half dozen metallic objects, including a cell phone, inside the object, the response is virtually identical to the left image. They are thus 'cloaked.'

side. Note that as expected in Mie regime scattering transition, as seen here, the occultation of the direct path by the presence of the object yields 4 to 12 dB reduction over the passband. In other words, the depth of the occultation seen gives only about 6% to 40% transmission of the direct path, depending on frequency. Also, because the log periodic beamwidths are three dimensional and not two dimensional, some leakage beyond the physical extent of the rings must be expected. Such partial gain occultation is typical of what one sees when any conductive object of roughly a few wavelengths in size is present in an RF path.

When the cloak's rings of CSAs were placed as layers over the object, a dramatic restoration of the passband gain is experienced, as shown on the left of Fig. 9.6. The variation in gain is approximately 3 dB peak to peak or less across the passband and as such is a high fidelity rendering of the direct path passband. However, the boundary layer now acts as a cylinder in which anything inside is cloaked over the passband. Indeed, no change in this antipodal response is seen when a variety of conductive devices and things are placed within the object, including resonant wires, things in excess of a wavelength, and a cell phone, which operates within the passband so measured, shown on the right of Fig. 9.6.

In the sense that invisibility cloaking is a diverted imaging scheme to 'see through to the other side,' the left side of Fig. 9.6 demonstrates this effect handily. In addition, the DI cloaking is clearly of a substantial bandwidth, thus rendering the objections and consideration against a wideband cloak possibility as incorrect.

Although an IC has interest as a DI device, the question remains on how an invisibility cloak may betray its presence via scattering at angles that differ from the antipodes. For example in [4], narrow band cloak clearly evinced shadows and back scattering reflections at other angles. Thus the scattering becomes a key phenomenon to measure in the IC context and is important as a true indication of DI invisibility.

Measuring scattering without timing information, that is, an incoherent source as is the case here, means that true forward back scatter cannot be measured, because the receive and transmit antennas are at the same location and have sufficiently high crosstalk that no such backscatter reflection is measureable. However, the case of measuring backscatter at the same angle of incidence is only applicable in a radar or lidar application, and is a departure from the true application of an invisibility cloak, which is to be a 'see through' device, caused by DI. With this caveat, we measured the scattering of the invisibility cloak, not just at the antipodes, but with a nearly a full 360° of azimuth.

To see the value of the scattering measurements, a control is required. In this case, it was a series of rings of copper in place of the cloak, with the object boundary layer present. These copper rings essentially act partially as a curved mirror for the RF. This shows to what degree the cloak acts like a cloak rather than a mirror-like structure. And, of course, the IC objective is to achieve scattering that emulates that of the direct path – no object, no cloak, no copper rings. If the control achieved the same results then clearly this IC does not exist as a true DI system.

Figure 9.7 shows the scattering gain at a full range of scattering azimuth angles, save for the front, with the direct path. Not surprisingly, there is nothing to scatter, so the azimuthal gain response is thus that of two antennas at various angles of skewness. Also seen, as shown in Fig. 9.7, are the copper rings in place of the cloak, the azimuthal gain response clearly indicates a deep occultation at the antipodes, as expected (there is

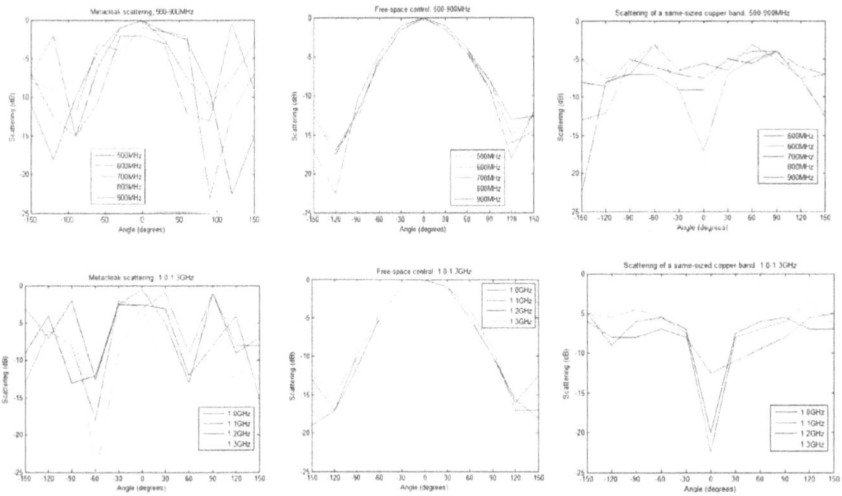

Figure 9.7 Azimuthal scattering for cloak, direct, and copper ring- control. Zero degrees is antipodal. Low and High shown for convenience of display. Note attenuation and deep shadowing with the control copper rings. Scattering reproduction for the low bands of the IC have excellent emulation of direct over most azimuthal angles.

a mirror in the way), with a consistent attenuation at other angles. Finally, in Fig. 9.7, we show the scattering with the cloak present.

Note that we have broken the data sets into two groups, high and low, so that the effects can be more easily discerned. The azimuthal scattering with the IC evinces excellent emulation of the direct path scattering. In particular, there are no obvious shadows towards antipodal angles, and no deep occultation at the antipodes, when compared to the control. The scattering loses robustness at higher frequencies but does maintain the antipodal fidelity, as already shown in the left of Fig. 9.6.

9.5 Discussion and conclusions

The measurements clearly show that a wideband invisibility cloak has been realized, using a fractal approach. The antipodal gain response across a wide passband, with a 3 dB peak to peak variation from true, is an excellent reproduction of the direct path, indicating the diversion of waves around the object. In addition the scattering azimuthal response closely matches the direct path over most of the passband range. There is minimal shadowing, for example.

DI research now has a basis for further exploration and development. In particular, the objective of this effort was to demonstrate wideband gain fidelity as a 'see through' invisibility cloak. In most cases that in itself will emulate real situations where the background is desired to shine through the object. However, DI transmission around an object does not necessarily mean it is undetectable. For example, a glass filled with water, although not DI, is an example where the gain of the direct path is reproduced with high fidelity, but minor changes in angular distortion easily render the glass as present. In other words, the fundamental issue of whether DI is viable has been demonstrated, but the subtlety of optical-like precision must now be entertained and, in more than one way, resolved.

Finally, there is no reason why the use of a fractal approach in DI cannot be scaled to work at infrared and visual light wavelength. The limitation is a technological one, where the ability to fabricate micron and submicron structures determines the ability to have such wideband invisibility cloaks, with 'see through' capability under DI, available.

The advantage of fractals was clearly present here: the fact that a wideband high fidelity IC had not been achieved before the use of fractals, and has not been demonstrated with other means since, indicates the added value presented by fractal shapes in this example. As with other applications in electromagnetics, the small size, wide and multiband capabilities of fractal shapes are a continued asset in the exploration of electromagnetic applications.

Bibliography

[1] G. Marconi, C. Franklin, *Reflector for Use in Wireless telegraphy and Telephony*, U.S. Patent 1301473 (1919).

[2] S. Maier, *Plasmonics*, Springer, New York, 2007.

[3] B. Munk, *Metamaterials: Critique and Alternatives*, Wiley, New York, 2009.

[4] D. Schurig, J. Mock, B. Justice, S. Cummer, J. Pendry, A. Starr, D. Smith, "Metamaterial Electromagnetic Cloak at Microwave Frequencies," *Science* **314** (2009) 977–980.

[5] H. Chen, Z. Liang, P. Yao, X. Jiang, H. Ma, C. Chan, "Extending the Bandwidth of Electromagentic Cloaks," *Phys. Rev. B* **76** (2007) 241104-1–241104-4.

[6] N. Cohen, *Wideband Electromagnetic Cloaking Systems*, U.S. Patent 8253639 (2012).

[7] N. Cohen, "Fractal Antennas Part 1," *Communications Quarterly* **5** (1995) 7–22.

Chapter 10

From fractional Brownian motion to multifractional and multistable motion

Kenneth Falconer

Abstract. Fractional Brownian motion, introduced by Benoit Mandelbrot and John Van Ness in 1968, has had a major impact on stochastic processes and their applications. We survey a few of the many developments that have stemmed from their ideas. In particular we discuss the local structure of fractional and multifractional Brownian, stable and multistable processes, emphasising the 'diagonal' construction of such processes. In all this, the ubiquity and centrality of fractional Brownian motion is striking.

10.1 Fractional Brownian motion

In a seminal paper published in 1968, Benoit Mandelbrot and John Van Ness [1] introduced 'fractional Brownian motion', a generalization of Brownian motion which continues to play an central rôle both in the theory of stochastic processes and in applications to areas such as economics, surface modelling, hydrology and internet traffic. Indeed, over 650 papers listed on *MathSciNet* have the words 'fractional Brownian motion' in the title, not to mention its frequent occurrence in less-mathematically oriented literature. In this article we survey a range of processes whose origins can be traced back ultimately to Mandelbrot and Van Ness's paper, including multifractional Brownian and multistable motions. We follow a direction taken by the author in collaboration with Jacques Lévy Véhel, Ronan Le Guével and Lining Liu [2–6], to whom grateful thanks are due. Some of the results discussed here involve quite complicated technical conditions, so we outline rather than give full statements of these as well as illustrating them by examples.

Mandelbrot and Van Ness defined index-h *fractional Brownian motion* (fBm) on $\mathbb{R}$ for a parameter $0 < h < 1$ as a stochastic integral with respect to Wiener measure W:

$$B_h(t) = c(h)^{-1} \int_{-\infty}^{\infty} \left((t-x)_+^{h-1/2} - (-x)_+^{h-1/2} \right) W(dx), \tag{10.1}$$

where $(a)_+ = \max\{0, a\}$ and $c(h) = \Gamma(h + \frac{1}{2})$ is a normalising constant that ensures that the variance of $B_h(1)$ is 1. When $h = \frac{1}{2}$ we regard the integrand as reducing to the indicator function of the interval $[0, t)$ so $B_{1/2} = B$ is standard Brownian motion. (Note that Paul Lévy [7] had previously mentioned a related process

$$B_h^0(t) = \int_0^t (t-x)^{h-1/2} W(dx),$$

but this does not have stationary increments and gives excessive weight to the origin.)

The paper [1], together with many subsequent papers authored or coauthored by Mandelbrot that may be found in his *Selecta* Volume H [8], develops properties and applications of fBm. The process B_h is Gaussian, i.e. its finite dimensional distributions are multivariate Gaussian vectors, it has stationary increments, but crucially, if $h \neq \frac{1}{2}$, the increments are not independent and the process is non-Markovian; this last fact makes fBm far harder to work with than standard Brownian motion. The means, variances and covariances of fBm may be deduced from (10.1):

$$\mathbb{E}(B_h(t)) = 0, \qquad \mathbb{E}(B_h(t)^2) = |t|^{2h} \qquad (t \in \mathbb{R}),$$
$$\mathbb{E}\big((B_h(t_2) - B_h(t_1))^2\big) = |t_2 - t_1|^{2h} \qquad (t_1, t_2 \in \mathbb{R}),$$

or equivalently

$$\mathbb{E}\big((B_h(t_1)B_h(t_2))\big) = \frac{1}{2}\left[t_1^{2h} + t_2^{2h} - |t_2 - t_1|^{2h}\right]. \tag{10.2}$$

It also follows that, for $t_1 < t_2 < t_3 < t_4$,

$$\mathbb{E}\big((B_h(t_2) - B_h(t_1))(B_h(t_4) - B_h(t_3))\big)$$

is positive or negative according to whether $h > \frac{1}{2}$ or $h < \frac{1}{2}$. In particular, if $h > \frac{1}{2}$ fBm exhibits *long range dependence*, that is

$$\sum_{n=1}^{\infty} \mathbb{E}\big((B_h(1) - B_h(0))(B_h(n+1) - B_h(n))\big) = \infty,$$

implying strong predictability of future behaviour of the process, see [9]. It would be enormously significant if, say, such a process provided a good model for stock market prices, see the discussions in Mandelbrot's *Selecta* Volume E [10]; however, Rogers [11] showed that an index other than $\frac{1}{2}$ was inconsistent with the standard 'non-arbitrage' assumption of finance.

Not surprisingly, as well as the importance of fBm as a stochastic process and for its applications, the fractal nature of the sample paths was central to Mandelbrot's thinking. In [1] it is shown that the sample paths of B_h are almost surely continuous, but nowhere differentiable. More quantitatively, almost surely the graph of fBm has Hausdorff and packing dimensions $2-h$, and at almost t has Hölder exponent h, that is

$$\liminf_{r \to 0} \frac{\log |B_h(t) - B_h(t+r)|}{\log r} = h,$$

see Fig. 10.1; for dimensional aspects, see, for example, [12–16].

It is clear from the definition of fBm that B_h has *stationary increments*, that is

$$B_h(t) - B_h(u) \stackrel{d}{=} B_h(t-u) - B_h(0) \qquad (t, u \in \mathbf{R}) \tag{10.3}$$

where $\stackrel{d}{=}$ denotes equality in distribution. A further crucial property, which follows from making a substitution in (10.1), is that the process scales about 0 according to a power law of exponent h, that is fBm is *self-similar*. This self-similarity is highlighted in [8] and in many other works, and we will focus on this here. [Note that whilst Mandelbrot originally referred

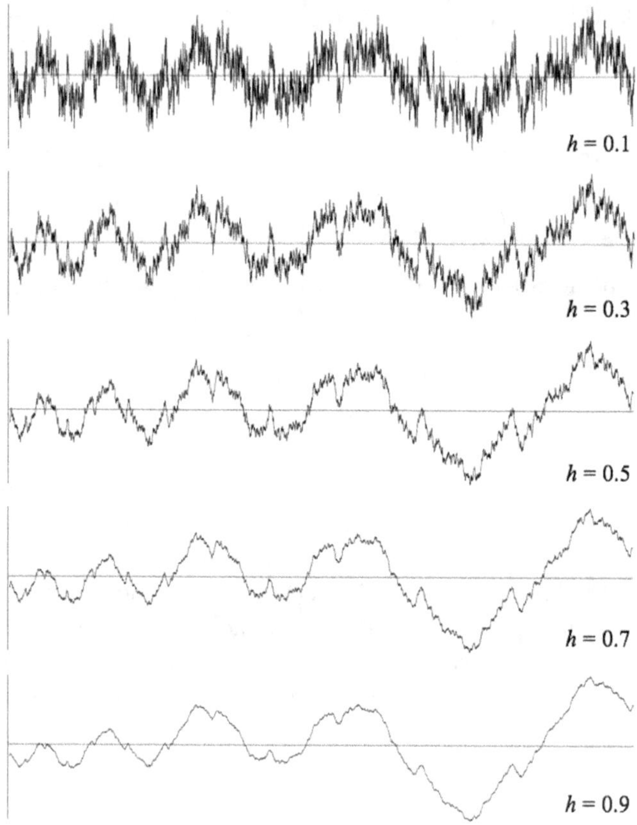

Figure 10.1 Sample paths of fractional Brownian motion for various h.

to 'self-similar' processes, in his Selecta [8] he adopts the arguably more appropriate term,'self-affine'.] Thus

$$B_h(rt) \stackrel{d}{=} |r|^h B_h(t) \qquad (t, r \in \mathbb{R}). \tag{10.4}$$

Together with (10.3), this implies that the process is self-similar around any $t_0 \in \mathbb{R}$, that is

$$B_h(rt) - B_h(rt_0) \stackrel{d}{=} |r|^h (B_h(t) - B_h(t_0)) \qquad (t, r \in \mathbb{R}). \tag{10.5}$$

An immediate consequence of (10.5) is that fBm is 'localisable', that is the limit (in the sense of finite dimensional distributions) of enlarged and appropriately scaled copies of fBm about any point u is just fBm itself, that is

$$\frac{B_h(u+rt)-B_h(u)}{|r|^h} \xrightarrow{\text{fdd}} B_h(t) \qquad (t\in\mathbb{R}) \tag{10.6}$$

as $r \to 0$, where fdd denotes convergence in finite dimensional distribution, see, for example, [1, 17, 18] for further discussion of this as well as the next section.

10.2 Localisability and multifractional Brownian motion

A number of authors, e.g. [19–22], have considered the local form of random processes, with some variation in definitions and terminology. Given a process $Y(t)(t\in\mathbb{R})$ we call a random process W a *scaling limit of* Y *at* $u\in\mathbb{R}$ if there exist sequences $r_n \searrow 0$ and $c_n \searrow 0$ such that

$$\frac{Y(u+r_nt)-Y(u)}{c_n} \xrightarrow{\text{fdd}} W \tag{10.7}$$

as $n\to\infty$, where convergence is in finite dimensional distributions. If Y and W have versions in $C(\mathbb{R})$ (the space of continuous functions) or in $D(\mathbb{R})$ (the space of càdlàg functions, i.e. functions that are continuous on the right and have limits on the left) and convergence in (10.7) is in distribution (with respect to the metric of uniform convergence on bounded intervals, respectively the Skorohod metric), we call W a *strong scaling limit of* Y *at* u.

We call the family of possible scaling limits the *tangent space* of Y at u:

$$\mathrm{Tan}(Y,u) = \{W : W \text{ is a (strong) scaling limit of } Y \text{ at } z\}.$$

Clearly, if $W\in\mathrm{Tan}(Y,u)$ then $cW\in\mathrm{Tan}(Y,u)$ for all $c>0$. In the nicest case there is some $W\neq 0$ such that

$$\mathrm{Tan}(Y,u) = \{cW\}_{c\geq 0},$$

and we say that Y is *(strongly) localisable* at u with *local form* W, and write $Y'_u = W$. In particular, if for some $h>0$,

$$\frac{Y(u+rt)-Y(u)}{r^h} \to Y'_u(t)$$

as $r\searrow 0$ we call Y *(strongly) h-localisable* at u with *(strong) local form* Y'_u. Of course, strongly localisable processes are localisable. (Note that localisability is referred to by other names, for example, local-asymptotic self-similarity [19, 20].)

Clearly, every process Y that is h-self-similar with stationary increments (sssi) is h-localisable, since

$$\frac{Y(u+rt)-Y(u)}{r^h} \stackrel{d}{=} \frac{Y(rt)-Y(0)}{r^h} \stackrel{d}{=} \frac{Y(rt)}{r^h} \stackrel{d}{=} Y(t) \xrightarrow{\text{fdd}} Y(t),$$

where d denotes equality in distribution. More formally:

Proposition 10.2.1 *Let $Y(t)\,(t \in \mathbb{R})$ be a process that is h-self-similar with stationary increments. Then Y is h-localisable at all $u \in \mathbb{R}$ with $Y'_u = Y$. If in addition Y is in $C(\mathbb{R})$ or $D(\mathbb{R})$ then Y is strongly h-localisable at all $u \in \mathbb{R}$.*

In particular, index-h fBm is h-localisable with $B'_{h,u} = B_h$ for all u.

Both for theoretical reasons and for applications it may be desirable to work with processes with local form Y'_u that varies with u; for example in financial modelling volatility may vary considerably with time, perhaps suggesting that the local form at u should be index-$h(u)$ fBm for a non-constant $h(u)$. This raises two natural questions. Firstly, are there restrictions on the families of local forms $\{Y'_u : u \in \mathbb{R}\}$ that are possible for Y, and secondly how might one construct construct a process with a prescribed local form for each $u \in \mathbb{R}$?

With regard to the first question, we have noted that an sssi process everywhere has an sssi process (namely itself) as its local form. By considering tangent spaces it turns out that this is true in much greater generality, that strong local forms of *any* (reasonable) process are in general sssi. We first note the invariance properties of tangent spaces, summarised in the next theorem.

Theorem 10.2.1 *Let $Tan(Y, u)$ be the strong tangent space of a process Y at $u \in \mathbb{R}$.*

1. *If $W(t) \in \mathrm{Tan}(Y,u)$ then*

$$cW(t) \in \mathrm{Tan}(Y,u) \quad \textit{for all } c > 0 \qquad \textit{(vertical scaling)},$$

2. *If $W(t) \in \mathrm{Tan}(Y,u)$ then*

$$W(rt) \in \mathrm{Tan}(Y,u) \quad \textit{for all } r > 0 \qquad \textit{(horizontal scaling)},$$

3. *For almost all $u \in \mathbb{R}$, if $W(t) \in Tan(Y,u)$ then*

$$W(z+t) - W(z) \in \mathrm{Tan}(Y,u) \quad \textit{for all } z \in \mathbb{R} \qquad \textit{(translation invariance)}.$$

Proof Parts (1) and (2) are immediate from the definitions. Part (3) is more subtle, see [2, 23]. □

In the localisable situation, where all the processes in $\mathrm{Tan}(Y,u)$ are constant multiples of one another, it follows easily from Theorem 10.2.1 that the local forms are sssi almost everywhere.

Corollary 10.2.1 *Let Y be a random process in $C(\mathbb{R})$ or $D(\mathbb{R})$. For almost all $u \in \mathbb{R}$ at which Y is strongly localisable, the local form Y'_u is h-self-similar for some $h > 0$ and has stationary increments.*

The corollary above places considerable restrictions on the possible families of local forms of localisable processes. The following proposition about h-sssi Gaussian processes follows easily from the sssi conditions and the fact that a continuous Gaussian process is determined by the covariances $\mathbb{E}\big(X(t_1)X(t_2)\big)$, see [18, 23].

Proposition 10.2.2 *Let X be a Gaussian process in $C(\mathbb{R})$ that is h-sssi, where $h > 0$. Then either $0 < h < 1$ and $X \stackrel{d}{=} cB_h(t)$ where c is a constant and $B_h(t)$ is index-h fBm, or $h = 1$ and $X \stackrel{d}{=} tW$ where W is a (not-necessarily zero mean) Gaussian random variable.*

The following theorem is now immediate from Corollary 10.2.1 and Proposition 10.2.2.

Theorem 10.2.2 Gaussian processes. *Let Y be a Gaussian process in $C(\mathbb{R})$. For almost all $u \in \mathbb{R}$ at which Y is strongly localisable, the local form Y'_u is either index-h fractional Brownian motion for some $0 < h < 1$, or $Y'_u(t) \equiv tW$ where W is a Gaussian random variable (not necessarily with mean 0).*

Theorem 10.2.2 demonstrates the remarkable ubiquity of Mandelbrot and Van Ness's fractional Brownian motion: essentially the only possible local forms of *any* continuous Gaussian process are index-h fBm for some h, or a trivial exceptional alternative.

We now address the question of finding a process with prescribed local forms at each $u \in \mathbb{R}$. In view of Theorem 10.2.2 we can only expect to achieve this if the local forms are fBm (or exceptionally of the form tW). Thus it is natural to construct *multifractional Brownian motion* (mBm) where the local Y'_u is index-$h(u)$ fBm, where h now varies with u; this was introduced in different ways in [20] and [21].

One approach to constructing localisable funtions with prescribed local form is to 'join together' localisable processes $X(t,v)(t,v \in \mathbb{R})$ over a range of v. Specifically, we might hope that, under suitable conditions, the 'diagonal process' $Y(t) = X(t,t)(t \in \mathbb{R})$ will locally 'look like' $X(t,u)(t \in \mathbb{R})$ when t is close to u. More precisely, we want $Y(\cdot)$ and $X(\cdot,u)$ to have the same local forms at u, that is $Y'_u(\cdot) = X'_u(\cdot,u)$ where $X'_u(\cdot,u)$ is the local form of $X(\cdot,u)$ at u. The following proposition see [3] gives sufficient conditions for this to occur.

Theorem 10.2.3 *Let $X(t,v)\,(t,v \in \mathbb{R})$ be a random field (i.e a stochastic process on $\mathbb{R}^2$) and let Y be the diagonal process $Y(t) \equiv X(t,t)(t \in \mathbb{R})$. Suppose that for some $0 < h < \eta$ the process $X(\cdot,u)$ is h-localisable at u with local form $X'_u(\cdot,u)$. If*

$$\mathbb{P}(|X(v,v) - X(v,u)| \geq |v-u|^{\eta}) \to 0$$

as $v \to u$ then $Y(t) = X(t,t)\,(t \in \mathbb{R})$ is h-localisable at u with $Y'_u(\cdot) = X'_u(\cdot,u)$.

In particular, this conclusion holds if for some $p > 0$ and $\eta > h$

$$\mathbb{E}(|X(v,v) - X(v,u)|^p) = O(|v-u|^{\eta p}) \tag{10.8}$$

as $v \to u$.

There are similar, but stochastically tighter, conditions relating to strong localisability. By putting mBm into this context, its localisability and local forms are an immediate consequence of Theorem 10.2.3.

Example 10.2.1 Multifractional Brownian motion. Let $h : \mathbb{R} \to (0,1)$ satisfy an η-Hölder condition where $h(u) < \eta \leq 1$. The random field defined by the stochastic integral

$$X(t,v) = \int_{-\infty}^{\infty} \left((t-x)_+^{h(v)-1/2} - (-x)_+^{h(v)-1/2}\right) W(dx) \quad (t,v \in \mathbb{R}) \tag{10.9}$$

satisfies (10.8). Since for each fixed u we have that $X'_u(\cdot,u)$ is (to within a constant) index $h(u)$-fractional Brownian motion $B_{h(u)}$, the process

$$Y(t) \equiv X(t,t) = \int_{-\infty}^{\infty} \left((t-x)_+^{h(t)-1/2} - (-x)_+^{h(t)-1/2}\right) W(dx) \quad (t \in \mathbb{R})$$

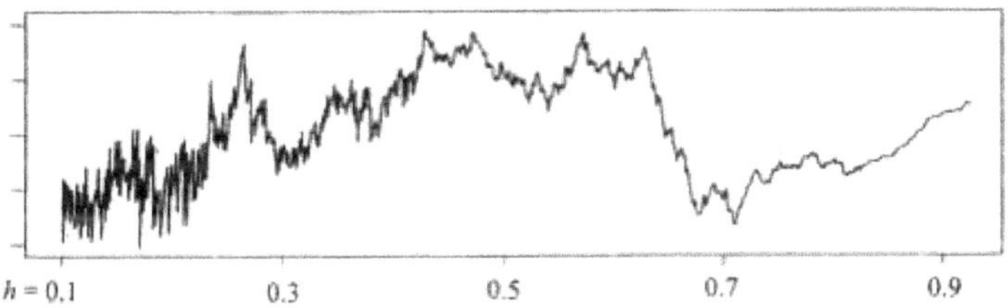

Figure 10.2 Sample path of multifractional Brownian motion with $h(t)$ increasing from 0.1 to 0.9

is $h(u)$-localisable (indeed is strongly localisable) at each u, with $Y'_u = c(h(u))B_{h(u)}$, where $c(h)$ is a normalisation constant, see Fig. 10.2.

Note that this is one of a number of possible constructions of mBm, see [17, 20, 21, 24, 26, 27] for other approaches — indeed, not all constructions are equivalent [28].

We mention briefly the problem of simulating sample paths of fBm and mBm — since these processes do not have independent increments this is not straightforward. An early paper by Mandelbrot and Wallis [29] discusses the (equivalent) problem of simulating fractional Gaussian noise, and many algorithms have since been proposed. The two main methods are (i) obtaining a discrete approximation using a multivariate Gaussian random vector — this involves inverting the covariance matrix given by (10.2), (ii) approximating the stochastic integral (10.1) by sums using Gaussian random increments — since the increments are not independent, this involves re-evaluating the integral for each t. There are many other methods, including using wavelet sums [30] and spectral simulation [31]. A further challenging problem is to assume that an fBm, or perhaps even an mBm, underlies a (large) data set, and to seek estimates of h or of the function $h(t)$ from the data. For direct approachs see [32–35] and for a Bayesian treatment see [37].

10.3 Multifractional stable processes

As well as his work on fractional Brownian motion, Benoit Mandelbrot's insights into α-stable (originally termed L-stable) processes have had great influence, notably in the area of finance. Whilst such processes date back to Paul Lévy [38] it was Mandelbrot, who studied under Lévy at the École Polytechnique in the mid-1940s, who realised that such processes without finite second moments might be appropriate for financial modelling [10, 39].

As far as localisability goes, there are natural parallels with fBm and mBm considered in the previous section; again there are considerable restrictions on the local forms achievable, and we can construct multifractional stable processes (where the Hölder exponent varies), multistable processes (where the stability index varies) and even multifractional multistable processes.

Recall that a random vector $\mathbf{Y}$ is *α-stable* if $\mathbf{Y}_1 + \mathbf{Y}_2$ has the same distribution as $2^{1/\alpha}\mathbf{Y} + \mathbf{D}$ for some fixed vector $\mathbf{D}$, where $\mathbf{Y}_1$ and $\mathbf{Y}_2$ are independent copies of $\mathbf{Y}$; if $\mathbf{D} = \mathbf{0}$ then $\mathbf{Y}$ is *strictly-stable.* An α-stable vector has finite p^{th} moment if and only if $0 < p < \alpha$, see [18].

For given $0 < \alpha < 2$, an *α-stable process* is a process Y for which the finite dimensional distributions $(Y(t_1), \ldots, Y(t_k))$ are α-stable vectors for all $t_1, \ldots, t_k$. Such processes may be constructed analogously to fBm but with stochastic integrals taken with respect to α-stable measure rather than Wiener measure, see [18] for definitions and properties of α-stable measures and integrals. For example, *linear stable fractional motion* may be defined as

$$L_{\alpha,h}(t) = \int_{-\infty}^{\infty} \left((t-x)_+^{h-1/\alpha} - (-x)_+^{h-1/\alpha} \right) M(dx), \tag{10.10}$$

with stability index $\alpha(0 < \alpha < 2)$, Hölder exponent $h(0 < h < 1)$ and M a symmetric α-stable random measure with constant skewness β and where, as throughout, we take M to have Lebesgue measure as control measure, see [18]. The process (10.10) is h-sssi and is h-localisable at all $u \in \mathbb{R}$ with $(L_{\alpha,h})'_u = L_{\alpha,h}$, so like fBm has itself as its local form. Provided that $h > 1/\alpha$ the process (10.10) has a version in $C(\mathbb{R})$, and is strongly localisable. However, if $h < 1/\alpha$ then almost surely Y is unbounded on every interval and so is not a process of $D(\mathbb{R})$, though it is nevertheless localisable.

We may interpret Theorem 10.2.1 and Corollary 10.2.1 in the context of α-stable processes.

Corollary 10.3.1 α-stable processes. *Let Y be an α-stable process in $C(\mathbb{R})$ or $D(\mathbb{R})$ where $0 < \alpha < 2$. For almost all $u \in \mathbb{R}$ at which Y is strongly localisable, the local form Y'_u is α-stable and h-sssi for some $0 < h \leq \max\{1, 1/\alpha\}$. This imposes considerable restrictions on the possibilities for the scale, skewness and shift parameters of $Y'_u(t)$ at each $t \in \mathbb{R}$. In particular, the scale parameter of the increments satisfies*

$$\sigma_{Y'_u(s)-Y'_u(t)} = |s-t|^h \sigma_{Y'_u(1)} \qquad (s, t \in \mathbb{R}),$$

see [2] *and* [18, *Sec.* 7.3] *for more details.*

As with mBm, we may use Theorem 10.2.3 to obtain a multifractional version of linear stable motion (10.10) (see [22, 40] for related constructions).

Example 10.3.1 Linear multifractional stable motion. Let $0 < \alpha < 2$ and $h : \mathbb{R} \to (0, 1)$. Define a random field by the stochastic integral

$$X(t, v) = \int_{-\infty}^{\infty} \left((t - x)_+^{h(v)-1/\alpha} - (-x)_+^{h(v)-1/\alpha}\right) M(dx) \quad (t, v \in \mathbb{R}).$$

where M is an α-stable random measure $(0 < \alpha < 2)$ with constant skewness intensity β and control measure Lebesgue measure, with $h(t) \in (0, 1)$ for all $t \in \mathbb{R}$. For each fixed v the process $X(\cdot, v)$ is just a linear stable fractional motion (10.10) so is $h(v)$-localisable, with $X'_u(\cdot, v) = L_{\alpha,h(v)}(\cdot)$ for all $u \in \mathbb{R}$. Set

$$Y(t) \equiv X(t, t) = \int_{-\infty}^{\infty} \left((t - x)_+^{h(t)-1/\alpha} - (-x)_+^{h(t)-1/\alpha}\right) M(dx) \quad (t \in \mathbb{R}).$$

(1) If $0 < \alpha < 2$ and h satisfies a η-Hölder condition at u, where $h(u) < \eta \leq 1$, it follows using Theorem 10.2.3 that Y is $h(u)$-localisable at u with local form $Y'_u = L_{\alpha,h(u)}$.

(2) If $1 < \alpha < 2$ and h is differentiable with $1/\alpha < h(u) < 1$ and $h'(v)$ satisfies an η-Hölder condition where $1/\alpha < \eta \leq 1$, it may be shown that Y is strongly $h(u)$-localisable at u.

To end this section we mention processes with independent increments. Recall that an *α-stable Lévy motion* $(0 < \alpha < 2)$ is a process in $D(\mathbb{R})$ with stationary independent increments which have a strictly α-stable distribution. It may be represented as

$$L_\alpha(t) = M([0, t]) \tag{10.11}$$

where M is an α-stable random measure on $\mathbb{R}$ with constant skewness intensity, see [18]. Then L_α is $1/\alpha$-sssi, and so is strongly $1/\alpha$-localisable.

The local forms of processes with independent increments must themselves have independent increments and Corollary 10.2.1 implies that they are sssi almost everywhere, so must be stable Lévy processes.

Corollary 10.3.2 Independent increments. *Let Y be a process in $C(\mathbb{R})$ or $D(\mathbb{R})$ with independent increments. For almost all $u \in \mathbb{R}$ at which Y is strongly localisable, the local form Y'_u is an h-self-similar $1/h$-stable Lévy process for some $h \geq \frac{1}{2}$.*

10.4 Multistable processes

The processes discussed in Sec. 10.3 are defined in terms of integration with respect to an α-stable measure for some fixed α. However, we may wish to construct processes with local form an $\alpha(u)$-stable process where $\alpha(u)$ varies with $u \in \mathbb{R}$. Of course, we remain constrained by Theorem 10.2.1 and Corollary 10.2.1.

Several formulations of α-stable processes for fixed α are presented in the book [18], in terms of stochastic integrals defined by characteristic functions, in terms of sums over Poisson processes, and in terms of series of random terms. These formulations may be adapted to allow varying α.

A representation by sums over Poisson processes allows $X(\cdot, v)$ to be given using the same underlying Poisson process for different v. Let Π be a Poisson point process on $\mathbb{R}^2$ with respect to plane Lebesgue measure; thus the number of points (X, Y) in Π in a plane region of measure λ has Poisson distribution with mean λ, with the number of points in disjoint regions independent. For $0 < \alpha < 2$ and $f \in \mathcal{F}_\alpha$, we may express the stochastic integral with respect to α-stable measure M as a Poisson process sum

$$\int f(x)M(dx) = c(\alpha) \sum_{(\mathsf{X},\mathsf{Y})\in\Pi} f(\mathsf{X})\mathsf{Y}^{\langle -1/\alpha \rangle} \tag{10.12}$$

where $c(\alpha)$ is a normalising constant, with the sum almost surely convergent, see [18]. (We write $a^{\langle b \rangle} = \text{sign}(a)|a|^b$.) Note that α in (10.12) occurs only as an exponent of Y since the underlying Poisson process does not depend on α.

Let $0 < a \le b \le 2$ and write

$$\mathcal{F}_{a,b} = \left\{ f : \int_{-\infty}^{\infty} |f(x)|^a dx, \int_{-\infty}^{\infty} |f(x)|^b dx < \infty \right\}.$$

If $f(t, v, \cdot) \in \mathcal{F}_{a,b}$ for all $t, v \in \mathbb{R}$ and $\alpha(t) \in [a, b]$ for all $t \in \mathbb{R}$, we may define a random field

$$X(t, v) = \sum_{(\mathsf{X},\mathsf{Y})\in\Pi} f(t, v, \mathsf{X})\mathsf{Y}^{\langle -1/\alpha(v) \rangle} \tag{10.13}$$

which is almost surely convergent for all (t, v). As in the multifractional settings of Secs. 10.2 and 10.3, we seek conditions that ensure that the

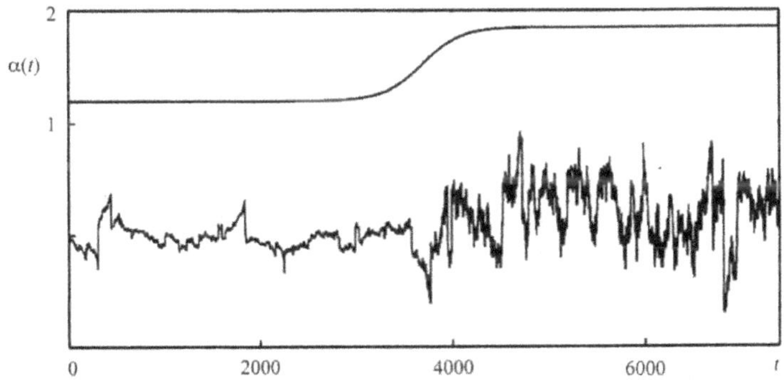

Figure 10.3 Sample path of a multistable motion with $\alpha(t)$ indicated on the upper graph (from [4]).

diagonal process

$$Y(t) \equiv X(t,t) = \sum_{(\mathsf{X},\mathsf{Y})\in\Pi} f(t,t,\mathsf{X})\mathsf{Y}^{\langle -1/\alpha(t)\rangle} \tag{10.14}$$

has local form at u that is an $\alpha(u)$-stable process; if $\alpha(u)$ is non-constant we term such a process a *multistable* process, see Fig. 10.3.

Under various reasonable but quite technical conditions on $f(t,v,x)$, it may be shown that, with these representations, if $X(\cdot,u)$ is h-localisable at u with local form $X'_u(\cdot,u)$ then $Y(t) \equiv X(t,t)$ $(t \in \mathbb{R})$ is h-localisable at u with $Y'_u(\cdot) = X'_u(\cdot,u)$. Indeed, h need not be constant either: if $0 < h(t) < 1 (t \in \mathbb{R})$ and $X(\cdot,u)$ is $h(u)$-localisable at u then $Y(t)$ is $h(u)$-localisable at u. Detailed sufficient conditions on f for this to be valid are specified in [3] where this model is considered in detail; here we just give two important examples which satisfy these conditions.

Example 10.4.1 Multistable Lévy motion. Let $\alpha : \mathbb{R} \to (1,2)$ and let $h : \mathbb{R} \to (0,1)$ be continuously differentiable. The process

$$Y(t) = c(\alpha(t)) \sum_{(\mathsf{X},\mathsf{Y})\in\Pi} 1_{[0,t]}(\mathsf{X})\mathsf{Y}^{\langle -1/\alpha(t)\rangle} \quad (t \in \mathbb{R})$$

is $1/\alpha(u)$-localisable at u, with $Y'_u = L_{\alpha(u)}$, where L_α is α-stable Lévy motion.

To see this, we consider the random field

$$X(t, v) = \sum_{(\mathsf{X},\mathsf{Y})\in\Pi} \mathbf{1}_{[0,t]}(\mathsf{X})\mathsf{Y}^{\langle -1/\alpha(v)\rangle} \quad (t, v \in \mathbb{R})$$

and observe that $X(\cdot, u)$ is $1/\alpha(u)$-localisable at u with local form $X'_u(\cdot, u) = L_{\alpha(u)}$, noting that L_α has the representation

$$L_\alpha(t) = M([0, t]) = c(\alpha) \sum_{(\mathsf{X},\mathsf{Y})\in\Pi} \mathbf{1}_{[0,t]}(\mathsf{X})\mathsf{Y}^{\langle -1/\alpha\rangle} \quad (t \in \mathbb{R}),$$

where M is α-stable measure.

In the following example, both the stability index α and the Hölder exponent h may vary with T.

Example 10.4.2 Linear multistable multifractional motion. Let $\alpha : \mathbb{R} \to (0, 2)$ and $h : \mathbb{R} \to (0, 1)$ be C^2. The process

$$Y(t) = c(\alpha(t)) \sum_{(\mathsf{X},\mathsf{Y})\in\Pi} \mathsf{Y}^{\langle -1/\alpha(t)\rangle} \left(|t - \mathsf{X}|^{h(t)-1/\alpha(t)} - |\mathsf{X}|^{h(t)-1/\alpha(t)}\right) \quad (t \in \mathbb{R})$$

is $h(u)$-localisable at $u \in \mathbb{R}$, with $Y'_u = L_{\alpha(u),h(u)}$, where $L_{\alpha,h}$ is linear α-stable motion (10.10). Moreover, if u is such that $h(u) > 1/\alpha(u)$ then Y is strongly $h(u)$-localisable in $C(\mathbb{R})$.

In this case we work with the random field

$$X(t, v) = \sum_{(\mathsf{X},\mathsf{Y})\in\Pi} \mathsf{Y}^{\langle -1/\alpha(v)\rangle} \left(|t - \mathsf{X}|^{h(v)-1/\alpha(v)} - |\mathsf{X}|^{h(v)-1/\alpha(v)}\right) \quad (t, v \in \mathbb{R}),$$

noting that $X(\cdot, u)$ is $h(u)$-localisable at u with local form $X'_u(\cdot, u) = L_{\alpha(u),h(u)}$.

An alternative way of constructing multistable processes, developed in [6], uses a random series representation. Let $\alpha \in (0, 2)$ and m be a probability measure on a space E. Let $(\Gamma_i)_{i\geq 1}$ be a sequence of arrival times of a Poisson process, let $(V_i)_{i\geq 1}$ be a sequence of random variables on E with distribution given by m and let $(\gamma_i)_{i\geq 1}$ be random variables such that $P(\gamma_i = 1) = P(\gamma_i = -1) = \frac{1}{2}$, with all the random variables $(\Gamma_i)_{i\geq 1}$,

$(V_i)_{i\geq 1}$ and $(\gamma_i)_{i\geq 1}$ independent. It is shown in [18] that

$$Y(t) = \sum_{i=1}^{\infty} \gamma_i \Gamma_i^{-1/\alpha} f(t, V_i) \quad (t \in \mathbb{R})$$

defines an α-stable integral of $f \in \mathcal{F}_\alpha$ where $\mathcal{F}_\alpha(m)$ denotes those functions $f : E \to \mathbb{R}$ such that $|f|^\alpha$ is integrable with respect to m.

This formulation may be used to define a random field and hence, using the diagonal construction again, a random process. Let $\alpha : \mathbb{R} \to [a, b] \subset (0, 2)$ be continuous and let $f(t, u, .) \in \mathcal{F}_{a,b}(m)$ for all $(t, u) \in \mathbb{R}^2$ (where $\mathcal{F}_{a,b}(m)$ is defined with respect to m). Consider the random field

$$X(t, v) = \sum_{i=1}^{\infty} \gamma_i \Gamma_i^{-1/\alpha(v)} f(t, v, V_i) \qquad (t, v \in \mathbb{R}),$$

where V_i, γ_i and Γ_i are as above. Assume that $X(\cdot, u)$ is h-localisable at u with local form $X'_u(\cdot, u)$. As with the Poisson sum construction we may find conditions on f that ensure that the diagonal process

$$Y(t) \equiv X(t, t) = \sum_{i=1}^{\infty} \gamma_i \Gamma_i^{-1/\alpha(t)} f(t, t, V_i) \quad (t \in \mathbb{R})$$

is h-localisable at u with $Y'_u(\cdot) = X'_u(\cdot, u)$.

An apparent disadvantage of this construction is that m needs to be a probability measure on E to specify the distribution of the V_i, whereas it may be desirable to work with Lebesgue measure on $\mathbb{R}$. Nevertheless, as with the constant α case, see [18], if m is a σ-finite measure, one can transform m to a probability measure, at the expense of multiplying the summands by functions of V_i and u, and this can then be treated as above.

Under some technical conditions on f we get that Y is h-localisable at u with $Y'_u(\cdot) = X'_u(\cdot, u)$, both in the cases where m is a probability measure and a σ-finite measure, see [6]. Again, we just give one of many examples.

Example 10.4.3 Multistable Lévy motion. Let $\alpha : \mathbb{R} \to [a, b] \subset (1, 2)$. Let $(\Gamma_i)_{i\geq 1}$ and $(\gamma_i)_{i\geq 1}$ be as above, and let $(V_i)_{i\geq 1}$ be a sequence of iid random variables on $\mathbb{R}$ with distribution $m(dx) = \sum_{j=1}^{\infty} 2^{-j} 1_{[j-1,j)}(x)dx$.

Then the process

$$Y(t) = c(\alpha(t))^{1/\alpha(t)} \sum_{i=1}^{\infty}\sum_{j=1}^{\infty} \gamma_i \Gamma_i^{-1/\alpha(t)} 2^{j/\alpha(t)} \mathbf{1}_{[0,t]\Pi[j-1,j)}(V_j) \quad (t \in \mathbb{R}^+)$$

is $h(u)$-localisable at $u \in \mathbb{R}^+$, with $Y'_u = L_{\alpha(u)}$, where L_α is α-stable Lévy motion.

This construction utilises the random field

$$X(t,v) = c(\alpha(v))^{1/\alpha(v)} \sum_{i=1}^{\infty}\sum_{j=1}^{\infty} \gamma_i \Gamma_i^{-1/\alpha(v)} 2^{j/\alpha(v)} \mathbf{1}_{[0,t]\Pi[j-1,j)}(V_j) \quad (t \in \mathbb{R}^+)$$

noting that $X(\cdot,u)$ is symmetrical α-stable motion, and so is $1/\alpha(u)$-localisable at u with local form $X'_u(\cdot,u) = L_{\alpha(u)}$.

Finally we mention an alternative approach to multistable processes, by defining the integral of functions in $\mathcal{F}_{a,b}$ with respect to a symmetric $\alpha(x)$-*multistable measure* M in terms of the characteristic function of finite dimensional distributions. Thus, given $0 < a \le b < 2$ and $\alpha(t) \in [a,b]$, set

$$\begin{aligned}\mathbb{E}&\left(\exp i\left(\sum_{j=1}^{d}\theta_j \int f_j(x)dM(x)\right)\right)\\ &= \exp\left\{-\int\left|\sum_{j=1}^{d}\theta_j f_j(x)\right|^{\alpha(x)} dx\right\} \quad (\theta_1,\dots,\theta_d) \in \mathbb{R}^d\end{aligned}$$

for $f_j \in \mathcal{F}_{a,b}$. It may be shown, using the Kolmogorov consistency conditions and the Lévy continuity theorem, that these characteristic functions define a consistent probability distribution on the functions of $\mathcal{F}_{a,b}$. Moreover, the integral thus defined is linear, and integrals of functions with disjoint supports are independent. Thus we can consider processes of the form

$$Y(t) = \int f(t,x)M(dx),$$

where $f(t,\cdot) \in \mathcal{F}_{a,b}$, and where the integral is with respect to $\alpha(x)$-multistable measure M. By suitable choice of f the process Y may display multistable features, see [5] for details.

The question of simulation has also been considered in the stable and multistable context, see [4, 41] for some recent approaches.

Bibliography

[1] B. B. Mandelbrot, J. Van Ness, *SIAM Rev.* **10** (1968) 422.
[2] K. J. Falconer, *J. London Math. Soc.* **67** (2003) 657.
[3] K. J. Falconer, J. Lévy Véhel, *J. Theoret. Probab.* **22** (2009) 375.
[4] K. J. Falconer, R. Le Guével, J. Lévy Véhel, *Stochastic Models* **25** (2009) 648.
[5] K. J. Falconer, L. Liu, Stochastic Models **28** (2012) 503.
[6] R. Le Guével, J. Lévy Véhel, Bernoulli **18** (2012) 1099.
[7] P. Lévy, *University of California Publications in Statistics* **1** (1953) 331.
[8] B. B. Mandelbrot, *Gaussian Self-affinity and Fractals* (Springer-Verlag, New York, 2002).
[9] P. Doukhan, G. Oppenheim, M. S. Taqqu (eds.) *Theory and Applications of Long Range Dependence* (Birkhäuser, Basel, 2000).
[10] B. B. Mandelbrot, *Fractals and Scaling in Finance* (Springer-Verlag, New York, 1997).
[11] L. C. G. Rogers, *Math. Finance* **7** (1997) 95.
[12] K. J. Falconer, *Fractal Geometry — Mathematical Foundations and Applications*, 3rd edn. (Wiley, Chichester, 2014).
[13] R. J. Adler, *Random Fields* (Wiley, New York, 1981).
[14] J.-P. Kahane, *Some Random Series of Functions* (Cambridge Univ. Pr, Cambridge, 1985).
[15] R. W. Reidi, in [9], pp. 625–716.
[16] Y. Xiao, *Math. Proc. Cambridge Philos. Soc.* **122** (1997) 565.
[17] P. Embrechts, M. Maejima, *Selfsimilar Processes* (Princeton Univ. Pr., Princeton, 2002).
[18] G. Samorodnitsky, M. S. Taqqu, *Stable Non-Gaussian Random Processes* (Chapman and Hall, London, 1994).
[19] A. Benassi, S. Cohen, J. Istas, *Bernoulli* **8** (2002) 97.
[20] A. Benassi, S. Jaffard, D. Roux, *Rev. Mat. Iberoamericana* **13** (1997) 19.
[21] R. F. Peltier, J. Lévy Véhel, "Multifractional Brownian motion: definition and preliminary results," *Rapport de recherche de l'INRIA*, No. 2645 (1995). Available at: http://www-rocq1.inria.fr/fractales/index.php?page=publications
[22] S. Stoev, M. S. Taqqu, *Adv. Appl. Probab.* **36** (2004) 1085.
[23] K. J. Falconer, *J. Theoret. Probab.* **15** (2002) 731.
[24] A. Ayache, J. Lévy Véhel, in [25], pp. 17–32.
[25] M. Dekking, J. Lévy Véhel, E. Lutton, C. Tricot, eds., *Fractals — Theory and Applications in Engineering* (Springer, Berlin, 1999).
[26] A. Ayache, *J. Fourier Anal. Appl.* **8** (2002) 581.
[27] A. Ayache, J. Lévy Véhel, *Stat. Inference Stoch. Process.* **3** (2000) 7.
[28] S. Stoev, M. S. Taqqu, *Stochastic Process. Appl.* **116** (2006) 200.

[29] B. B. Mandelbrot, J. R. Wallis, *Water Resources Research* **5** (1969) 228.
[30] V. Pipiras, *Appl. Comput. Harmon. Anal.* **19** (2005) 49.
[31] A. B. Dieker, M. Mandjes, *Probab. Engrg. Inform. Sci.* **17** (2003) 417.
[32] A. Ayache, J. Lévy Véhel, *Stochastic Processes Appl.* **111** (2004) 119.
[33] A. Benassi, S. Cohen, J. Istas, *Statist. Probab. Lett.* **39** (1998) 337.
[34] S. Cohen, in [25], pp. 3–16.
[35] J. Lévy Véhel, R. Riedi, in [36], pp. 185–202.
[36] J. Lévy Véhel, E. Lutton, C. Tricot, eds., *Fractals in Engineering* (Springer-Verlag, Berlin, 1997).
[37] K. J. Falconer, C. Fernández, *Biometrika* **94** (2007) 313.
[38] P. Lévy, *Processus stochastiques et mouvement Brownien* (Gauthier Villars, Paris, 1948).
[39] B. B. Mandelbrot, *International Economic Review* **1** (1960) 79.
[40] S. Stoev, M. S. Taqqu, *Fractals* **13** (2005) 157.
[41] S. Cohen, C. Lacaux, M. Ledoux, *Stoch. Process. App.* **118** (2008) 1489.

Chapter 11

Watching the markets misbehave

Richard L. Hudson

My first encounter with Benoit Mandelbrot was, in many ways, typical: He came to tell me I didn't know what I was talking about. And that was his message, for much of his life, to much of the financial world. It is a pity he wasn't heeded more quickly.

It was 1997. He showed up at the Brussels office of the Wall Street Journal in Europe, where I was at the time managing editor. I had vaguely heard of him — something about the Mandelbrot Set; but a contact at his employer, IBM, insisted he had a story we should hear. At the appointed hour, a Hollywood caricature of a scientist walked into the office: fly-away white hair, distracted air, heavy accent. But his message got through: The standard theory of finance is wrong, and he can prove it. Because of it, he said, people can underestimate how risky financial markets really are — and lose money unnecessarily.

Well, his argument was timely; and we published something about it. But his passion was the impression that lingered in my mind. Shortly after, when I was looking for an extra speaker at a conference on risk, I thought of him. This was a typical Fortune 500-style event: CEOs, successful entrepreneurs, global investors. Not the place for a lecture on math. But as Mandelbrot spoke, I could see the initial bemusement turn to fascination: Could he be right? Could those know-it-alls in finance be wrong?

Those were the questions that dogged Mandelbrot during his frequent forays into finance. He was often in the position of an outsider looking in, critiquing the financial theorists — not from a respected university chair, but from a workaday industrial lab. Partly as a result, his impact on finance has been a long, slow burn; resistance was high. After all, great effort has gone into building up the theory of Modern Finance over three generations; based partly on its teachings, trillions of dollars, yen and pounds have changed hands. Clearly, Mandelbrot's critics said, the standard models are 'good enough' for most purposes. So why all the fuss?

The answer came in the financial meltdown of 2008. The very risk-modelling tools Mandelbrot had criticized were being used to help price the mortgage certificates that precipitated the crash, to help set the minimum bank reserves that proved so inadequate. Clearly, something was wrong — and it wasnt greed alone. The lucky timing of another, eloquent ally helped, too: Nassim Nicholas Taleb's best-seller, *The Black Swan* [1], had appeared just a year before the crash — and helped popularize one of Mandelbrot's key ideas, that extreme events — outliers — are far more common in financial markets than the standard models assume. (Taleb dedicated his book to Mandelbrot, calling him "a Greek among Romans.")

So at least one of Mandelbrot's ideas has gotten through to the world of finance. But there was much more in Mandelbrot's work than that. Since his first finance paper in 1962, he doggedly pursued an independent and iconoclastic course of study — dipping in and out of the field, skirmishing with academic critics, drawing on other disciplines for new ideas. Indeed, finance was one of the primary inspirations for his development of fractal geometry, as its rich data sets were a gold mine for him to hunt the scaling patterns on which his math was built.

Taken together, his insights point to a new view of finance — a roadmap to where financial researchers should look next for useful ideas, tools and strategies. This was the message that we tried to convey in a 2004 book on the topic, *The (Mis)Behavior of Markets* [2]. By then, he had so won me over to his view that I agreed to work with him to express it in terms accessible to any curious, educated investor. The education started with me: On our first working session at his home in Scarsdale, NY, he pointed me to a two-foot stack of documents — past papers of his, published and unpublished — and ordered me to read. In those papers, and in our weeks of conversation that followed, were the seeds of a new way of looking at financial markets — one altogether more disturbing than the professionals on Wall Street wished to hear.

11.1 The background

Before going further, a quick reminder of modern financial theory is in order. The story begins in 1900, when Louis Bachelier, a young French mathematician, devised some formulae to price options on government bonds. To do that, he made a few simplifying assumptions that have underpinned orthodox financial theory ever since. Chief among them: that the market can be viewed as a 'fair game' — like tossing a coin, the simplest kind of random process. With a fair coin, a head is as likely as a tail to come up. And the coin does not remember what came up at the last toss; each toss is independent of the last. The same with prices, Bachelier reasoned. To all appearances, the next price change could be as likely to rise as to fall, and each change happens independently of the last. Further, he posited, these price changes fit a common probability pattern, a normal distribution, with most changes very small and very few of them large. The many small changes cluster in the center of the familiar bell curve; the few large changes sit out on the edges, or tails, of the curve.

Taken together, this was a formidable intellectual achievement: It was the first logically consistent theory of finance, and it came with a familiar mathematical toolkit. Of course, Bachelier's point wasn't that price changes really are determined by a coin toss; it's people who set prices, based on their imperfect understanding or motivations. But, just as in a related field that he studied, thermodynamics, one didn't need to know the detailed dynamics of exactly how and why one molecule excites another in order to design an efficient radiator; so in markets one didn't need to know exactly why or how each investor acts as he does in order to estimate the value of an option. "The dynamics of the Exchange will never be an exact science. But it is possible to study mathematically the state of the market at a given instant," [3] Bachelier wrote.

On that thought is built the entire edifice of financial theory. Modern Portfolio Theory, developed in the 1950s, took the two main parameters of the bell curve, mean and variance, and translated them into measures of a stock's return and risk. With it, an investment manager could use price records to calculate just two numbers on each of a range of possible investments, plug them into an optimisation formula, and come up with the best possible stock portfolio for a client. That was the start of financial engineering, and even today is a standard tool of personal investment advisors. In the 1960s, the Capital Asset Pricing Model simplified things further, viewing the entire stock market as the optimum portfolio — and the performance of any individual stock as characterized by how it varies from the market. That was the start of index funds, and a range of new tools for corporate finance officers trying to decide whether to build a new plant or buy another company. In the 1970s, two economists, Fischer Black and Myron Scholes, updated Bachelier with a powerful new formula to calculate the value of any kind of option in real time. That was the start of global derivatives trading. Then in the 1990s, the Value at Risk method was developed to estimate how much a bank or fund could reasonably expect to lose if markets turned against it; it continues to be one of the most widely used risk-management tools.

By now, the toolkit of modern financial theory is stuffed full of handy formulae, many of them easy enough for an individual investor to play with on a home computer. They are taught to thousands of MBAs across the globe; indeed, a facility with these tools is needed for most jobs in finance these days. And a strong record of further research in this field is a prerequisite for academic appointments in financial economics. But most of this work still rests on the mathematical foundation laid by Bachelier. Alas, Mandelbrot came to conclude, the foundation is sand.

11.2 Cotton and scaling

Mandelbrot's interest in finance began by chance, in 1961. At the time, he was an applied mathematician working in IBM's main lab in Yorktown Heights, NY. The horizons of the nascent computer industry seemed limitless, and IBM had a lab full of brilliant researchers seeking cool new things to do with computers. Mandelbrot had been looking at economics — in particular, analysing charts of how incomes are distributed in a society from rich to poor. His work, building on that of Italian economist Vilfredo Pareto, had drawn some interest from other economists and thus, an invitation to Harvard to give a talk.

When he entered the classroom, he was taken aback. There, on the blackboard, was a diagram that looked very like one of his own at IBM. How was that possible? The professor, Hendrik S. Houtthaker, was equally mystified. Before Mandelbrot arrived, Houtthaker had drawn the chart on the board — but it had nothing to do with income distributions; it was about cotton prices. How could the two charts look alike? "Now here was a puzzle that grabbed me," Mandelbrot later related [4].

He left Harvard with Houtthaker's cardboard boxes full of punch cards recording cotton prices, then gathered more from the New York Public Library and the US Department of Agriculture — more than a century of trading records. With the help of a programmer, he ran the numbers through IBM's computer center. What he found: The price changes in that market did not fit a bell curve, at all. The proportion of small price changes to big ones — crashes or booms — was not normal; in fact, there were many more big changes than expected. An entirely different mathematical property appeared to be in action: a power law — just as he had seen in his income data. The cotton and income charts looked similar because they shared some key mathematical properties. But there was more. His data were from diverse sources. When he charted them, he found the daily price records seemed to vary the same way as the monthly data. This was scaling: there was some inner symmetry, over time, about the way prices changed.

Scaling, of course, became the central feature of what he would, a decade later, call fractal geometry — and the episode illustrates the creative importance of finance for him. But his work had broad implications in finance itself. First, standard theory said there should be no scaling or power function at all; there should be a nice, neat bell curve. But he could measure the power (1.7). An exponent of that magnitude had many disturbing implications — not least, that a market that could be characterized that way would have many violent price swings. In fact, he later calculated, a participant in such a market would face odds of ruin at about 10%; by contrast, the standard theory would suggest it was a negligible risk, on the order of 10^{-20}. Given the turbulent history of cotton trading, especially in the 19th century, which model reflected reality best?

The academic community did not thank him for his efforts. An MIT economist, Paul H. Cootner, included Mandelbrot's paper in a collection of more conventional financial research, but appended a commentary criticizing the work as too simple, and cotton as too peculiar a commodity. "Surely," he sarcastically wrote, "before consigning centuries of work to the ash pile, we should like to have some assurance that all our work is truly useless." Mandelbrot obliged. He went on to study other markets — commodities, currencies and stocks — and found a similar story. But the common view was that, while Mandelbrot may be right in some cases, it didn't much matter: Every model has simplifying assumptions, and the standard financial models generally worked well enough most of the time to justify their continued teaching and use.

11.3 Floods and dependence

A bit of luck led Mandelbrot to his next discovery. After a talk he gave on cotton, a listener told him it reminded him of some research he had read on Nile flooding. At the turn of the last century, a British hydrologist, H.E. Hurst, was trying to calculate how tall a dam would have to be if it were successfully to control the Nile's periodic floods. The standard engineering manuals were wrong when applied to the Nile; they underestimated the necessary dam height. The reason, Hurst found, was subtle. If one looks at the annual flood levels over many years, they do tend to follow a nice, normal distribution — so one might think that standard bell-curve math could be used in dam design. But in fact, the high-water marks from one year to the next are not independent of one another. A high flood one year leaves more water in the basins for the next year, and that has a cumulative effect. There is a type of dependency from one year to the next — and, in theory, if you had enough data you could trace it all the way back in time to the Pharaohs. The Nile has a memory, of sorts.

Mandelbrot saw an analogy to price changes. Contrary to the standard models, professional traders know that the precise sequence of events in a market matters. A vertiginous drop in a stock one day is often followed by a few more days of turbulence, up or down; the market has some kind of memory, as well. Mandelbrot travelled to England and met the by-then elderly Hurst. He developed a statistical test, called R/S, to measure the degree of this long-term dependence. Applying it to market data, the results of the test were intriguing. He found good evidence of long-term dependency in some financial records — typically, those with strong linkage to the underlying economy, such as bank loan rates. With other records — typically, those in heavily arbitraged markets — the test came up without much evidence of dependency.

In his published report, he didn't speculate on why; but in our later conversations while working on the book, he tried. First, he expected the degree of long-term 'memory' in a financial price series to differ by instrument class — and that is exactly what some researchers reported in the 1990s as Mandelbrot's ideas spread and began to be tested more widely. But second, he felt, there is often a kind of memory in a stock. "How could the price of IBM stock today depend on the price of IBM stock 20 years ago? The answer is elementary. A decision by an IBM executive 20 years ago created two different companies, called Intel and Microsoft. Nobody would say that the (IBM) price we are getting today and the variability does not depend on Microsoft and Intel .... Think of economic relations as reflections in a mirror. The price of IBM is reflected."

11.4 The synthesis

Mandelbrot took his two main objections to the standard model — extreme events and dependence — and turned them into a colorful metaphor: the Noah and Joseph effects. Noah's Flood was an extreme, seemingly impossible event — yet did happen. Just so, a market crash like 1929 or 1987 is unthinkable by the standard models of finance, yet they happened, too. And Joseph interpreted Pharaoh's dream to forecast a series of seven fat years and seven lean years — just as in price records Mandelbrot saw long-term dependency. In mathematical terms, Mandelbrot was saying that the guideposts that all prior financial theory had used were in fact illusory. It was impossible to calculate a reliable value for price variance (often used by financial modellers as a proxy for volatility); the nature of the data was such that, whatever estimate one made for a given set of price changes, a rare and catastrophic event could come along that would completely change the picture and force the modeller to start over again. In short, the variance was infinite. And likewise, Mandelbrot said, not only were price changes not independent of one another as the modellers assumed, but their *dependence* was so long as to be infinite.

He developed some models around these two concepts, using a variable he called α to measure the scaling power, and H to measure the dependency. But they were — and still are — controversial among orthodox financial theorists. As Mandelbrot put it: "Infinite memory and infinite variance generate many paradoxes." [5]

11.5 Multifractals and finance

Mandelbrot developed these ideas further as he worked on the broader problems of fractal geometry — from anatomy to cosmology. By the 1970s, most important for finance was the interplay with his work on turbulence, which he studied in records of wind gusts and many other phenomena. As in a wind tunnel, so with currency markets he could see certain wild but measurable hallmarks of a turbulent system. He could also recognize patterns from his old cotton charts. Once again, a common mathematical toolkit — in this case, the multifractal geometry he was developing in the 1970s and 1980s — could show a deep symmetry between two seemingly unrelated phenomena.

In the 1990s, he developed his Multifractal Model of Asset Returns further. The idea behind it is almost poetic. In markets, the flow of time is like water in a turbulent river; the current appears to speed up or slow down, to cluster or diffuse, due to a whole range of factors hidden from sight. In the same way, market trading appears to speed up or slow down from one minute or day to the next due to any number of imperfectly guessed causes — a rogue computer program, or a disappointing jobs report. Nevertheless, a constantly varying 'market time' can be measured; this is a mathematical transformation of linear clock time. That in turn affects prices. With two Yale students, Laurent Calvet and Adlai Fischer, Mandelbrot began testing his model on tick-by-tick market data gathered from around the world: in dollar-deutschmark rates, in dollar-yen rates, in the prices of individual stocks such as General Motors, Lockheed and Motorola. Some fit the model well, some not so well. He felt it a promising start, but saw much more research yet to be done.

11.6 The warning

The reaction among finance theorists was then, and remains now, mixed. His math was unfamiliar to them — and his message wasn't calculated to please. Still, his work, from cotton to multifractals, is frequently cited in the academic literature. His data-intensive approach, and his frequent resort to physics for analogies and techniques, helped spur a movement in the field that came to be called econophysics (though it was a label Mandelbrot didn't welcome.) Generally, however, he had more time for market professionals than theorists. Good traders, he often said, rely more on their own experience than theory; and as a researcher, he preferred to watch what they do rather than what they say. "I know some people there (on Wall Street) who are clearly very, very skilled, shrewd risk-takers," Mandelbrot told me. "What I'm against is the business school finance departments' attitude that this (Modern Finance) is the last word."

So what was Mandelbrot's impact on finance? Clearly, at least one big idea got through: Markets are riskier than the orthodox financial religion assumes. When writing our book, we calculated that the conventional models would have predicted that, in the long run of Dow Jones Industrial Average records from 1916 into this century, there should have been 58 days when the index moved by more than 3.4%. In fact, there were 1,001. Convention would say an index change of more than 7% should happen only once every 300,000 years; in fact, that century saw 48 such days. Likewise, the official odds of the 1987 crash — a 29% collapse of the Dow in one day — were one in 10^{50}. Yet it happened. Mandelbrot's point, amplified by Taleb and others, is that these 'outliers,' far from aberrations, are at the heart of what makes a financial market work. To run a bank, trade stocks, rate bonds or manage an economy without this in mind is the height of irresponsibility.

But he has other messages worth heeding. Inherent in his multifractal models is that volatility clusters, and prices jump discontinuously. Troubles dont come singly and intermittently; they concentrate. Any experienced trader would agree: the biggest profits and losses happen in a few concentrated periods. The biggest names in speculation — from JP Morgan to George Soros — had long, successful careers but made their biggest fortunes in a few historic moments of sheer chaos.

Further, Mandelbrot argued, markets deceive. The oldest type of financial analyst is the chartist, who pores over market data in search of tradable patterns — a trend, a counter-trend, a 'break-out' point on which to place a bet and make a fortune. They usually fail, he said, because turbulence gives rise to spurious patterns — chance outcomes, that come and go like an eddy forming and dissipating in water. For similar reasons, market bubbles are inevitable; they appear to form, carry the unwary along, and then burst. Among investors, he wrote, "I perceive a tendency to grossly underestimate the ability of chance to generate extremely striking structures" in market data.

These have practical applications. The math underlying them has been picked up by some small hedge funds, and incorporated into trading models in New York, Paris and Zurich. (Mandelbrot himself never traded, and rejected offers to work in funds or, as he became famous, rent his name to one; it was curiosity, not greed, that motivated him.) But market regulators also listened to him; indeed, after our book was published, he met with several who were then struggling for new approaches to risk management. Mandelbrot's message to them was consistent: Your system is more fragile, and your models less useful, than you think. He was proved right in 2008 and its aftermath. Indeed, one of the tools he criticized, Value at Risk, was used by the Federal Reserve Board and the European Central Bank to 'stress test' bank portfolios.

His interest in markets was profound. The fundamental difference between the realms of physics and finance is money: "Money has properties which matter does not have. It can appear or disappear instantly, either by fiat of a central bank or by the effect of a sudden loss of value of a company," he said. "Expectation is pure vapor. It has no concrete reality. At a certain instant everyone assumes that a company is in very good shape; it is growing; and therefore I am going to pay more for it. But anticipation can change instantly. Anticipation is fundamental to finance, and does not exist in any other field of science.

"Nature didn't create the stock market. Man did over many centuries. My hunch was that this was the next frontier for the study of variability," he said. "My hope is not to change daily procedures at all, but to help move towards a better understanding, a better management, of markets. And to begin to see which parts of the system, with particular urgency, happen to be manageable."

Mr. Hudson is editor and CEO of Science—Business, a Brussels media company, and former managing editor of the Wall Street Journal Europe. In 2004, Mandelbrot and he co-authored The (Mis)Behavior of Markets. *He is now working on a biography of Mandelbrot.*

Bibliography

[1] N. Taleb, *The Black Swan*, Random House, New York, 2007.
[2] B. Mandelbrot, R. Hudson. *The (Mis)Behavior of Markets: A Fractal View of Risk, Ruin and Reward.* Basic Books, New York, 2004.
[3] L. Bachelier, *Theorie de la Speculation.* Annales Scientifiques de l'cole Normale Suprieure (iii) **17** (1900) 21-86.
[4] Mandelbrot and Hudson, *op. cit.*
[5] *Ibid.*

Chapter 12

Partition zeta functions, multifractal spectra, and tapestries of complex dimensions

Kate E. Ellis, Michel L. Lapidus[1]*, Michael C. Mackenzie and John A. Rock*

In memory of Benoit Mandelbrot, my long-time friend, mentor, and second father. We have shared love in beauty, poetry, history, harmony, and geometry; and so much more. May our last conversation, which turned out to also be your last words on this Earth, be engraved in my own memory forever. I miss you every day.

Michel Lapidus

[1]Partially supported by the US National Science Foundation (NSF) under the grants DMS-0707524 and DMS-1107750.

Abstract. For a Borel measure and a sequence of partitions on the unit interval, we define a multifractal spectrum based on coarse Hölder regularity. Specifically, the coarse Hölder regularity values attained by a given measure and with respect to a sequence of partitions generate a sequence of lengths (or rather, scales) which in turn define certain Dirichlet series, called the *partition zeta functions*. The abscissae of convergence of these functions define a multifractal spectrum whose concave envelope is the (geometric) Hausdorff multifractal spectrum which follows from a certain type of Moran construction. We discuss at some length the important special case of self-similar measures associated with weighted iterated function systems and, in particular, certain multinomial measures. Moreover, our multifractal spectrum is shown to extend to a *tapestry of complex dimensions* for a specific case of atomic measures.

12.1 Introduction

Multifractal analysis is the study of physical, mathematical, dynamical, probabilistic, and statistical concepts in which a whole range of fractals may arise from a single object. Such phenomena are often modeled by measures that have highly irregular concentrations of mass. These richly structured measures are called *multifractal measures*, or simply *multifractals*, and arise from situations such as, but certainly not limited to, rainfall distribution, turbulence, distribution of galaxies, spatial distribution of earthquakes, internet traffic modeling, and modeling of financial time series. See, for example, [1–8].

One setting for multifractal analysis that very much pertains to this paper is provided by number theory, specifically the study of N-ary (or base-N) expansions of real numbers, where N is an integer greater than 1. The set of numbers in the unit interval with N-ary expansions containing the digits $0, 1, \ldots, N-1$ in proportions given by a probability vector with N components generate a fractal set. For a fixed N, the collection of the various fractal sets constructed in this manner provides a multifractal decomposition of the unit interval. A tool used to study the structure of these fractal sets is the Hausdorff dimension and this tool plays an important role in our approach to multifractal analysis. In general, the collection of Hausdorff dimension values determines a *multifractal spectrum* which describes the multifractal decomposition of a set (or rather, of a mass distribution). See, for instance, [1–3, 9–18].

Our primary objective here is the determination of multifractal spectra as the abscissae of convergence values for specific collections of Dirichlet series. The abscissa of convergence of a Dirichlet series is analogous to the radius of convergence of a power series and plays an important role in this work and in the theory of complex dimensions of fractal strings. Our technique is motivated by the determination of the Minkowski dimension of fractal strings as abscissae of convergence of geometric zeta functions (which are Dirichlet series themselves, see [19–21] and Sec. 12.2). This determination allows for the definition of the generalization of Minkowski dimension called *complex dimensions* which are used, among other things, in expressions for counting functions and volume formulas in the study of the oscillatory phenomena of fractal strings.

In our setting, we take a measure supported on a subset of the unit interval and determine its one-parameter family of *partition zeta functions*, indexed by a countably infinite collection of coarse Hölder regularity values (which we call *regularity values*), and their abscissae of convergence. Regularity values are the exponents t for which a measure behaves locally like r^t, for small r, where r is a positive real number that determines scale. In particular, we show that these abscissae of convergence recover classical forms of the geometric and symbolic Hausdorff multifractal spectra in certain cases and Hausdorff dimensions of Besicovitch subsets of self-similar fractals in others.

This work, along with [22–24] and Sec. 13.3 of [21] (which is an exposition of some of the work in those references), marks the beginning of a new theory of complex dimensions for multifractals. In particular, this work greatly expands upon the results presented in [23] where the partition zeta functions and abscissa of convergence function for a generalized binomial measure supported on the Cantor set are developed and analyzed. Moreover, this paper provides significant strides toward the long-term goal stated (in a different but analogous manner) in [23] of developing a theory of oscillatory phenomena which are intrinsic to multifractal geometries. This theory would parallel that developed for fractal strings in [19–21] but would involve a whole family of partition zeta functions indexed by regularity and their complex dimensions.

Other works which examine (from a different perspective) multifractal measures similar to those examined in this paper are [1–3, 9–13, 15–18, 25–30]. A variety of other techniques in multifractal analysis can be found in [4–7, 14, 28, 31–39]. For the theory of complex fractal dimensions, one should consult the works of M. L. Lapidus and M. van Frankenhuijsen [19–21, 28, 40, 41] and their extensions with B. M. Hambly, H. Herichi, J. Lévy-Véhel, H. Lu, E. P. J. Pearse, J. A. Rock, and S. Winter, accordingly, in [22–24, 42–53].

The remainder of this chapter is organized as follows:

Section 12.2 provides a review of the relevant aspects of the theory of fractal strings and complex dimensions. In particular, we recall a connection between the counting function of the lengths of a fractal string and its complex dimensions from [19, 20], which motivates the definition of a suitable counting function in our multifractal setting given in Sec. 12.4.

Section 12.3 provides a review of the relevant aspects of multifractal analysis, in particular weighted iterated function systems, coarse Hölder regularity α, geometric and symbolic Hausdorff multifractal spectra, and Besicovitch subsets of Moran fractals. These notions will play an important role in our approach, particularly in Sec. 12.5.

Section 12.4 contains the definitions of our main objects of study: *α-lengths, partition zeta function, abscissa of convergence function, complex dimensions with parameter α, tapestry of complex of dimensions*, and *counting function of the α-lengths.*

Section 12.5 develops our main results regarding the partition zeta functions and abscissa of convergence functions of certain self-similar measures defined by weighted iterated function systems. In particular, connections with some of the well-known results found in [3, 9, 11, 13] and a recovery of the multifractal spectrum of the binomial measure on the unit interval (as described for example in [1–3]) are presented.

Section 12.6 develops the partition zeta functions, abscissa of convergence functions, tapestries of complex dimensions, and counting functions for a certain collection of multifractal atomic measures. In particular, exact explicit formulas for the counting functions of the associated α-lengths are given in terms of the underlying complex dimensions with regularity α. These examples are among the first steps toward a new theory of complex dimensions and oscillatory phenomena for multifractals.

Section 12.7 closes the chapter with a discussion of related works in progress and ideas for further development of multifractal analysis via zeta functions.

12.2 Fractal strings

A brief review of fractal strings, geometric zeta functions and complex dimensions (all of which are defined below) is in order. Results on fractal strings can be found in [22, 42, 46–49, 54–58] and results on geometric zeta functions and complex dimensions can be found in [19–24, 28, 40–53].

The primary references for the theory of complex dimensions of fractal strings are the monographs [19] and [20]. A significantly expanded second edition of [20] is forthcoming in [21], but for the convenience of the reader, we will mostly refer to [20] throughout the paper (except when required otherwise).

12.2.1 Fractal strings and Minkowski dimension

The primary example of a fractal string used throughout this work is the well-known Cantor string Ω_{CS} (the complement of the classical ternary Cantor set in the unit interval $[0, 1]$). The first several lengths of the Cantor string appear in Fig. 12.1 and the Cantor string is discussed in much more detail in 12.2.3 below.

Fractal strings are defined as follows:

Definition 12.2.1 A fractal string Ω is a bounded open subset of the real line.

As in [19, 20, 22], we distinguish a fractal string Ω from its sequence of lengths $\mathcal{L}$ (with multiplicities). The sequence $\mathcal{L} = \{\ell_i\}_{i=1}^{\infty}$ is the nonincreasing sequence of lengths of the disjoint open intervals (a_i, b_i), where $\Omega = \cup_{i=1}^{\infty}(a_i, b_i)$. More specifically, in this paper we follow the convention established in [19, 20] and generally do not consider the case where $\mathcal{L}$ comprises a finite collection of lengths. This is done in order to avoid discussion of trivial counterexamples to established results. We will, however, sometimes indicate what happens in the trivial case when $\mathcal{L}$ consists of finitely many lengths. Thus, throughout this paper, $\mathcal{L}$ typically comprises an infinite collection of lengths. We view $\mathcal{L}$ either as a decreasing sequence of positive *distinct* lengths denoted $\{l_n\}_{n=1}^{\infty}$ along with their multiplicities $\{m_n\}_{n=1}^{\infty}$, or as a nonincreasing sequence of (possibly equal) positive lengths denoted $\{\ell_i\}_{i=1}^{\infty}$ and repeated according to their multiplicities.

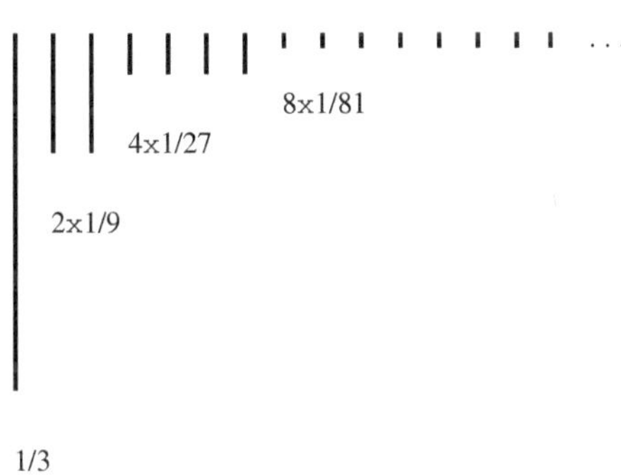

Figure 12.1 The lengths of the Cantor string.

A generalization of Minkowski dimension called *complex dimension* (defined in Sec. 12.2.2 below) is used to study the properties of certain fractal subsets of $\mathbb{R}$. For instance, the boundary of a fractal string Ω, denoted $\partial\Omega$, is often fractal and can be studied using complex dimensions. Throughout this text, a fractal string Ω is taken to be an open subset of the unit interval $[0,1]$ with $\mathcal{L}$ as its associated sequence of lengths.

The volume of the (inner) tubular neighborhood of radius ε of the boundary $\partial\Omega$ of a fractal string Ω is

$$V(\varepsilon) = |\,\{x \in \Omega \mid dist(x, \partial\Omega) < \varepsilon\}\,|,$$

where $|\cdot|$ denotes the Lebesgue measure (length). The Minkowski dimension of $\partial\Omega$, or simply of $\mathcal{L}$, is[2]

$$\dim_M(\partial\Omega) = D := \inf\left\{\alpha \geq 0 \;\middle|\; \limsup_{\varepsilon\to 0+} V(\varepsilon)\,\varepsilon^{\alpha-1} < \infty\right\}.$$

Note that one may refer directly to the Minkowski dimension of the sequence of lengths $\mathcal{L}$ because $V(\varepsilon)$ can be shown to only depend on $\mathcal{L}$ (see [20, 58]).

In [55], it is shown that if $F = \partial\Omega$ is the boundary of a bounded open set Ω, then $d - 1 \leq \dim_H(F) \leq \dim_M(F) \leq d$ where d is the Euclidean dimension of the ambient space, $\dim_H(F)$ is the Hausdorff dimension of F and $\dim_M(F)$ is the Minkowski dimension of F. We consider the case $d = 1$ in this paper, thus

$$0 \leq \dim_H(F) \leq \dim_M(F) \leq 1.$$

If F is self-similar and further satisfies the Open Set Condition (defined in Sec. 12.3), it is well known that $\dim_H(F) = \dim_M(F)$. (See, e.g., [59, 60] and Chap. 9 of [2].)

12.2.2 Complex dimensions and counting functions

The following equalities describe a relationship between the Minkowski dimension of a fractal string Ω (taken to be the Minkowski dimension of $\partial\Omega$) and the sum of each of its lengths with exponent $\gamma \in \mathbb{R}$. This relationship with suitably defined Dirichlet series (later called *geometric zeta functions* in [19, 20, 40]) was first observed in [56] using a key result

[2]This fractal dimension is also often called "box dimension" in the applied literature.

of A. S. Besicovitch and S. J. Taylor [61], and a direct proof can be found in [20, pp. 17–18] (and can also be found in [21]). We have[3]

$$\dim_M(\partial\Omega) = D = D_{\mathcal{L}} := \inf\left\{\gamma \in \mathbb{R} \,\middle|\, \sum_{i=1}^{\infty} \ell_i^{\gamma} < \infty\right\}. \tag{12.1}$$

Here, $D_{\mathcal{L}}$ can be considered to be the abscissa of convergence of the Dirichlet series $\sum_{i=1}^{\infty} \ell_i^s$, where $s \in \mathbb{C}$. This Dirichlet series is the *geometric zeta function* of $\mathcal{L}$; it is the function that has been generalized in [22–24] using notions from multifractal analysis and will in part motivate our proposed approach to multifractal zeta functions.

Definition 12.2.2 The geometric zeta function of a fractal string Ω with lengths $\mathcal{L}$ is

$$\zeta_{\mathcal{L}}(s) := \sum_{i=1}^{\infty} \ell_i^s = \sum_{n=1}^{\infty} m_n l_n^s, \tag{12.2}$$

where $\mathrm{Re}(s) > D_{\mathcal{L}}$.

To consider lengths $\ell_i = 0$, the convention $0^s = 0$ for all $s \in \mathbb{C}$ is used.

One can extend the notion of the Minkowksi dimension of a fractal string Ω to complex values by considering the poles of a meromorphic extension of $\zeta_{\mathcal{L}}$. In general, $\zeta_{\mathcal{L}}$ may not have a meromorphic extension to all of $\mathbb{C}$, yet one may consider suitable closed regions $W \subset \mathbb{C}$ where $\zeta_{\mathcal{L}}$ has a meromorphic extension, and collect the corresponding poles in these regions.

Assume that $\zeta_{\mathcal{L}}$ has a meromorphic extension to a connected open neighborhood of W and there is no pole of $\zeta_{\mathcal{L}}$ on ∂W. By a slight abuse of notation, $\zeta_{\mathcal{L}}$ denotes the geometric zeta function and its (necessarily unique) meromorphic extension to W.

Remark 12.2.1 More specifically, in [20, 21], the 'window' W is chosen to be the closed subset of $\mathbb{C}$ that is to the right of the 'screen' $S = \partial W$, defined as the graph (with the x and y axes interchanged) of a bounded and Lipschitz real-valued function on $(-\infty, D_{\mathcal{L}}]$; see Sec. 5.3 of [20].

---

[3] Strictly speaking, we must assume that $\mathcal{L}$ consists of infinitely many nonzero lengths; otherwise, $\dim_M(\partial\Omega) = \max\{0, D_{\mathcal{L}}\}$.

Definition 12.2.3 The set of (visible) complex dimensions of a fractal string Ω with lengths $\mathcal{L}$ is

$$\mathcal{D}_{\mathcal{L}}(W) := \{\omega \in W \mid \zeta_{\mathcal{L}} \text{ has a pole at } \omega\}. \tag{12.3}$$

Furthermore, if $W = \mathbb{C}$, then $\mathcal{D}_{\mathcal{L}} := \mathcal{D}_{\mathcal{L}}(\mathbb{C})$ is simply called the set of complex dimensions of $\mathcal{L}$.

The following proposition, which is a special case of [20, Theorem 5.10] (also found in [21]), uses the complex dimensions $\mathcal{D}_{\mathcal{L}}(W)$ of a fractal string in a formula for the *geometric counting function* of $\mathcal{L}$, denoted $N_{\mathcal{L}}(x)$ and defined by

$$N_{\mathcal{L}}(x) := \#\{i \geq 1 \mid \ell_i^{-1} \leq x\} = \sum_{n \geq 1 \mid l_n^{-1} \leq x} m_n,$$

where, as above, $\{\ell_i\}_{i=1}^{\infty}$ denotes the nonincreasing sequence of lengths of $\mathcal{L}$ repeated according to their multiplicities, whereas $\{l_n\}_{n=1}^{\infty}$ denotes the decreasing sequence of distinct lengths of $\mathcal{L}$ with associated multiplicities given by $\{m_n\}_{n=1}^{\infty}$.

Proposition 12.2.1 *Let Ω be a fractal string with lengths $\mathcal{L}$ such that $\mathcal{D}_{\mathcal{L}}(W)$ consists entirely of simple poles. Then, under certain mild growth conditions on $\zeta_{\mathcal{L}}$ (namely, if $\zeta_{\mathcal{L}}$ is languid of a suitable order, in the sense of* [20, 21]*), we have*

$$N_{\mathcal{L}}(x) = \sum_{\omega \in \mathcal{D}_{\mathcal{L}}(W)} \frac{x^{\omega}}{\omega} \operatorname{res}(\zeta_{\mathcal{L}}(s); \omega) + \{\zeta_{\mathcal{L}}(0)\} + R(x), \tag{12.4}$$

where $R(x)$ is an error term of small order and the term in braces is included only if $0 \in W \backslash \mathcal{D}_{\mathcal{L}}(W)$.

Remark 12.2.2 It is not necessary for the poles to be simple, but then the explicit formula for $N_{\mathcal{L}}$ is slightly more complicated to state; see [20, Sec. 6.2.1] for details.

Remark 12.2.3 If a fractal string Ω is *strongly languid*, then according to Theorem 5.14 of [20], Eq. (12.4) holds with no error term (i.e., $R(x) \equiv 0$). Examples of strongly languid fractal strings are self-similar strings (see [20, Chaps. 2 & 3]). In particular, the results for the counting functions to be presented in Sec. 12.6 follow from Theorem 5.14 of [20].

Before continuing to the next section on multifractal analysis, consider the following results on classic examples of fractal strings — the Cantor string and the Fibonacci string. The Cantor string plays an important role throughout this paper and the Fibonacci string is recovered in Example 12.5.4 below.

12.2.3 The Cantor string and the fibonacci string

The *Cantor string* is defined as the open set Ω_{CS} that is composed of all the deleted middle-third open intervals in the usual construction of the classical ternary Cantor set. Hence, its boundary $\partial\Omega_{CS}$ is simply the ternary Cantor set itself. An approximation of the Cantor string appears in Fig. 12.2.

The distinct lengths of the Cantor string are given by $l_n = 3^{-n}$ with multiplicity $m_n = 2^{n-1}$ for every positive integer n; see Fig. 12.1. Hence, for $\mathrm{Re}(s) > \log_3 2$,

$$\zeta_{\mathcal{L}}(s) = \zeta_{CS}(s) = \sum_{n=1}^{\infty} 2^{n-1} 3^{-ns} = \frac{3^{-s}}{1 - 2 \cdot 3^{-s}}. \tag{12.5}$$

Upon meromorphic continuation, we see that the last equality above holds for all $s \in \mathbb{C}$; hence, for $j := \sqrt{-1}$,

$$\mathcal{D}_{\mathcal{L}} = \mathcal{D}_{CS} = \left\{ \log_3 2 + \frac{2jz\pi}{\log 3} \,\middle|\, z \in \mathbb{Z} \right\} \tag{12.6}$$

and these poles are simple.

In order to illustrate Proposition 12.2.1 above, we give an exact formula for the counting function of the Cantor string, N_{CS}, in terms of the complex dimensions $\mathcal{D}_{CS}$ (see [20, Eq. (1.31), p.22]). For all $x > 1$, we have

$$N_{CS}(x) = \frac{1}{2\log 3} \sum_{z=-\infty}^{\infty} \frac{x^{D+jzp}}{D + jzp} - 1,$$

where $D = \log_3 2$ is the Minkowski dimension of the Cantor string (technically, of the Cantor set) and $p = 2\pi/\log 3$ is its oscillatory period.

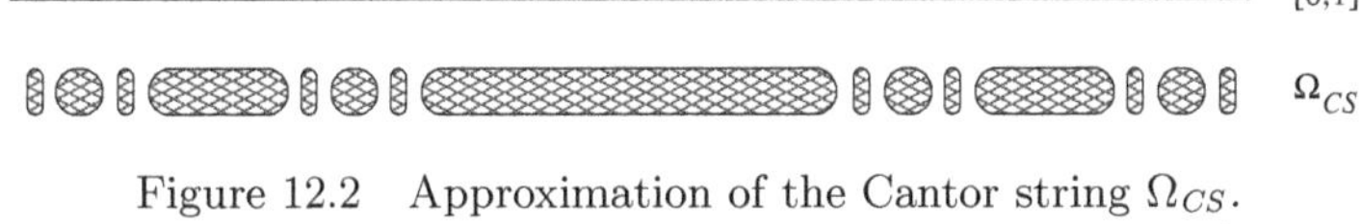

Figure 12.2 Approximation of the Cantor string Ω_{CS}.

This formula for $N_{CS}(x)$ is a special case of the explicit formula for the geometric counting function of general fractal strings provided by Theorems 5.10 and 5.14 in [20]. Note that in light of this formula, $N_{CS}+1$ can be written as the product of x^D and a multiplicatively periodic (or 'log-periodic') function of x.

Another example of a fractal string which is relevant to this paper is the Fibonacci string. The geometric zeta function of the Fibonacci string $\zeta_{\text{Fib}}(s)$ is recovered as a special case in Example 12.5.4. (See Sec. 2.3.2 of [20] for the development of the Fibonacci string.) The sequence of lengths for the Fibonacci string is

$$\mathcal{L}_{\text{Fib}} = \left\{2^{-n} \mid 2^{-n} \text{ has multiplicity } F_{n+1}, n \in \mathbb{N}\right\},$$

where $\mathbb{N}$ is the set of nonnegative integers and F_n denotes the nth Fibonacci number. (Recall that F_n is defined by the recurrence relation: $F_{n+1} = F_n + F_{n-1}$, and $F_0 = 0, F_1 = 1$.)

Furthermore, the geometric zeta function ζ_{Fib} of the Fibonacci string $\mathcal{L}_{\text{Fib}}$ is given by

$$\zeta_{\text{Fib}}(s) = \sum_{n=0}^{\infty} F_{n+1} 2^{-ns} = \frac{1}{1-2^{-s}-4^{-s}}. \tag{12.7}$$

The complex dimensions $\mathcal{D}_{\text{Fib}}$ can be found by solving the quadratic equation

$$2^{-2\omega} + 2^{-\omega} = 1, \qquad \omega \in \mathbb{C}.$$

Thus,

$$\mathcal{D}_{\text{Fib}} = \{D + jzp \mid z \in \mathbb{Z}\} \cup \{-D + j(z+1/2)p \mid z \in \mathbb{Z}\}, \tag{12.8}$$

where $\phi = (1+\sqrt{5})/2$ is the Golden Ratio, $D = \log_2 \phi$, and $p = 2\pi/\log 2$. In Sec. 2.3.2 of [20], these complex dimensions are used to determine a volume formula for the tubular neighborhood of the Fibonacci string.

We close this section by noting that both the Cantor string and the Fibonacci string are *self-similar* fractal strings, in the sense of [20, Chapters 2 and 3].

In the following section, we introduce some of the relevant theory and results on multifractal analysis currently available in the literature.

12.3 Multifractal analysis

Approaches to multifractal analysis which use multifractal measures closely related to those considered in this paper, and recalled below accordingly, can be found in [1, 2, 9–13, 15–18, 25–30, 62]. A variety of other techniques in multifractal analysis (incorporating wavelets, for instance) can be found in [4–7, 28, 31–39].[4] A common setting for multifractal analysis is that provided by self-similar measures defined by a probability vector and an Iterated Function System (IFS, or "map specified Moran construction" as in [11]) which satisfies the Open Set Condition (OSC, see [59, 60] and [2]). We construct such measures in this section by following the development found in [18, Sec. 1] and investigate the multifractal structure of these and other types of measures throughout this paper.

12.3.1 Iterated function systems and self-similar measures

Multifractal analysis of measures is the study of the fractal structure of the sets E_t of points $x \in E$ for which the measure $\mu(B(x,r))$ of the closed ball $B(x,r)$ with center x and radius r satisfies

$$\lim_{r\to 0^+} \frac{\log \mu(B(x,r))}{\log r} = t,$$

where $t \geq 0$ is *local Hölder regularity* and E is the support of μ. That is, from this traditional perspective, multifractal analysis is the study of the fractal geometry of the sets E_t where a Borel measure μ behaves locally like r^t.

The setting for multifractal analysis provided by a self-similar measure uniquely defined by an IFS which satisfies the OSC and a probability vector is developed as follows. For positive integers N and d and each $i \in \{1,\ldots,N\}$, let $S_i : \mathbb{R}^d \to \mathbb{R}^d$ be a contracting similarity with scaling ratio (or Lipschitz constant) $r_i \in (0,1)$. Let $\mathbf{r} = (r_1,\ldots,r_N)$, and let $\mathbf{p} = (p_1,\ldots,p_N)$ be a probability vector. The collection of contracting similarities $\{S_i\}_{i=1}^N$ is said to satisfy the OSC if there exists a nonempty, bounded, and open set $V \subset \mathbb{R}^d$ such that $S_i(V) \subset V$ and $S_i(V) \cap S_k(V) = \emptyset$ for all $i \neq k$ with $i,k \in \{1,\ldots,N\}$. We note that in this paper, as with many others on multifractal analysis, the collection of functions $\{S_i\}_{i=1}^N$ is

[4] The perspective adopted in all of these references is quite different, however, from the one adopted here, which consists in working with suitably defined partition (or multifractal) zeta functions.

assumed to satisfy the OSC. However, [18] and [63], for instance, do not require the OSC to be satisfied.

The multifractal measures for our setting are constructed as follows. Define the set E and the self-similar measure μ (supported on E) to be the unique nonempty compact subset of $\mathbb{R}^d$ and the unique Borel probability measure which satisfy, respectively, $E = \bigcup_{i=1}^N S_i(E)$ and $\mu = \sum_{i=1}^N p_i \mu \circ S_i^{-1}$ (see [60]). In particular, the measures considered in Proposition 12.3.1 and Sec. 12.5 below are defined in this manner. The Hausdorff dimension of the the support E is given by the solution of the Moran equation (see [59])

$$\sum_{i=1}^{N} r_i^s = 1 \qquad (s > 0). \tag{12.9}$$

12.3.2 Multifractal spectra

The multifractal spectra of Definitions 12.3.1 and 12.3.2 along with Proposition 12.3.1 below are presented as found in [18], as well as the corresponding references therein. See especially the work of R. Cawley and R. D. Mauldin in [11].

Definition 12.3.1 The geometric Hausdorff multifractal spectrum f_g of a Borel measure μ (on a Borel measurable subset of $\mathbb{R}^d$) supported on E is given by

$$f_g(t) := \dim_H(E_t), \tag{12.10}$$

where $t \geq 0$, $\dim_H$ is the Hausdorff dimension, and

$$E_t := \left\{ x \in E \,\middle|\, \lim_{r \to 0^+} \frac{\log \mu(B(x,r))}{\log r} = t \right\}. \tag{12.11}$$

The geometric Hausdorff multifractal spectrum f_g is difficult to compute for general self-similar measures. Thus, the symbolic multifractal spectrum f_s defined in terms of symbolic dynamics are often considered instead. This symbolic multifractal spectrum f_s, defined below, also serves as an analog to the approach to multifractal analysis developed in this paper.

For a nonnegative integer n and an integer $N \geq 2$, let

$$\Lambda^* := \{\mathbf{i} = i_1 \cdots i_n \mid k \in \mathbb{N}^*, i_k \in \{1, \ldots, N\}\}$$

and

$$\Lambda^{\mathbb{N}} := \{\mathbf{i} = i_1 i_2 \cdots \mid k \in \mathbb{N}^*, i_k \in \{1, \ldots, N\}\},$$

where $\mathbb{N} = \{0, 1, 2, \ldots\}$. For $\mathbf{i} \in \Lambda^{\mathbb{N}}$, let $\mathbf{i}|n = i_1 \cdots i_n$ be the truncation of $\mathbf{i}$ at the nth term. For $\mathbf{i} = i_1 \cdots i_n \in \Lambda^*$, we let $S_{\mathbf{i}} := S_{i_1} \circ \cdots \circ S_{i_n}$ and $E_{\mathbf{i}} := S_{\mathbf{i}}(E)$. Likewise, let $p_{\mathbf{i}} := p_{i_1} \cdots p_{i_n}$ and $r_{\mathbf{i}} := r_{i_1} \cdots r_{i_n}$. Finally, let $\pi : \Lambda^{\mathbb{N}} \to \mathbb{R}^d$ be defined by $\{\pi(\mathbf{i})\} := \cap_{n=0}^{\infty} E_{\mathbf{i}|n}$.

Definition 12.3.2 The symbolic Hausdorff multifractal spectrum f_s of a self-similar measure μ constructed as above is given by

$$f_s(t) := \dim_H \left(\pi \left\{ \mathbf{i} \in \Lambda^{\mathbb{N}} \,\middle|\, \lim_{n\to\infty} \frac{\log p_{\mathbf{i}|n}}{\log r_{\mathbf{i}|n}} = t \right\} \right) \tag{12.12}$$

for $t \geq 0$. (Here and henceforth, given $A \subset \mathbb{R}^d$, $\dim_H(A)$ denotes the Hausdorff dimension of A.)

The function $f_s(t)$ is usually easier to analyze than $f_g(t)$. Define $b : \mathbb{R} \to \mathbb{R}$ by

$$\sum_{i=1}^{N} p_i^q r_i^{b(q)} = 1, \tag{12.13}$$

and let $b^* : \mathbb{R} \to \mathbb{R} \cup \{-\infty\}$ be the Legendre transform of b. That is, for $t \in \mathbb{R}$,

$$b^*(t) := \inf_{q\in\mathbb{R}} (tq + b(q)). \tag{12.14}$$

When the OSC is satisfied, we have the following proposition (see, for instance, [11, 18, 25]).

Proposition 12.3.1 *Let μ be the unique self-similar measure on $\mathbb{R}^d$ defined, as above, by the IFS associated with $\{S_i\}_{i=1}^N$ which satisfies the Open Set Condition and weighted by the probability vector* $\mathbf{p}$. *Then*

$$f_g(-b'(q)) = f_s(-b'(q)) = b^*(-b'(q)), \tag{12.15}$$

where $q \in \mathbb{R}$ and b' is the derivative of b (assumed to exist here). Moreover, for the support E of the measure μ, we have

$$\dim_H(E) = b(0) = b^*(-b'(0)). \tag{12.16}$$

Other well-known properties of the function $f = f_g = f_s$ are described in the following section.

12.3.3 Properties of the multifractal spectrum

A full development of the properties of the multifractal spectrum $f = f_g = f_s$ of a self-similar measure μ described in this section, some of which are displayed in Fig. 12.3, can be found in [11, Sec. 1], specifically Fig. 1.3 therein.

A self-similar measure μ uniquely defined by $\{S_i\}_{i=1}^N$ and $\mathbf{p}$ has maximum and minimum regularity values $t_{\min}$ and $t_{\max}$ which define the domain of f. These values are given by

$$t_{\min} = \min_i \left\{\log_{r_i} p_i \mid i \in \{1, \ldots, N\}\right\}$$

and

$$t_{\max} = \max_i \left\{\log_{r_i} p_i \mid i \in \{1, \ldots, N\}\right\};$$

hence the domain of f (i.e., the values of t for which E_t is nonempty) is $[t_{\min}, t_{\max}]$. See Fig. 12.3.

Regardless of the values of $D_1 = f(t_{\min})$ and $D_2 = f(t_{\max})$, the slopes of f at the points $(t_{\min}, D_1)$ and $(t_{\max}, D_2)$ are infinite. The value $D_3 = t_1 = f(t_1)$ is the *information dimension* of μ. The value

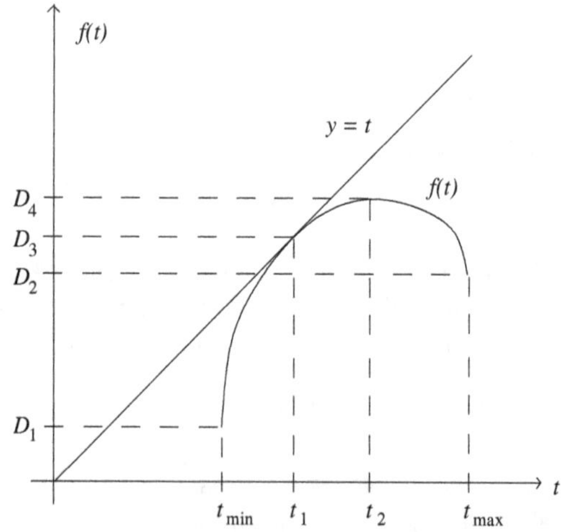

Figure 12.3 The graph of the multifractal spectrum $f(t) = f_g(t) = f_s(t)$ of a self-similar measure μ supported on a set defined by an IFS, where t is local Hölder regularity. The behavior of f is discussed in Sec. 12.3.3 and, for a specific case, is recovered in the context of the abscissa of convergence function and coarse Hölder regularity in Sec. 12.5.5.

$D_4 = f(t_2) = \max\{f(t) \mid t \in [t_{\min}, t_{\max}]\}$ is the Hausdorff dimension of μ. Thus, by Proposition 12.3.1 we have

$$D_4 = f(t_2) = \max\{f(t) \mid t \in [t_{\min}, t_{\max}]\} = \dim_H(E) = b^*(-b'(0)).$$

See Fig. 12.3.

If $p_i = r_i^D$ for all $i \in \{1, \ldots, N\}$, then $D = D_4 = f(D)$ and the domain of f is the singleton $\{D\}$. Excluding this case, f is concave and satisfies the following inequalities:

$$f(t_{\min}) = D_1 < f(t_1) = D_3 < f(t_2) = D_4$$

and

$$f(t_{\max}) = D_2 < D_4.$$

12.3.4 Besicovitch subsets of Moran fractals

Another setting of multifractal analysis that is important for our purposes is the one provided by the *Besicovitch subsets* of self-similar Moran fractals. (See [30, Sec. 1], for instance.)

For an IFS which satisfies the OSC on the unit interval $[0,1]$ with scaling ratios $\mathbf{r} = (r_1, \ldots, r_N)$ and a probability vector $\mathbf{q} = (q_1, \ldots, q_N)$ (that is, $\sum_{i=1}^N q_i = 1$ and $q_i \geq 0$ for $i \in \{1, \ldots, N\}$), the Besicovitch subset $E(\mathbf{q})$ of the self-similar set E (defined by the IFS) is defined by the coding mapping τ from $\{1, \ldots, N\}$ to E as follows:

$$E(\mathbf{q}) := \left\{ \tau(x) \in E \,\middle|\, \lim_{n\to\infty} \frac{1}{n} \sum_{k=1}^{n} \chi_i(x_k) = q_i, x \in \{1, \ldots, N\}^{\mathbb{N}^*} \right\}, \quad (12.17)$$

where $i \in \{1, \ldots, N\}$, χ_i is the characteristic function of the singleton $\{i\}$, $x = (x_k)_{k=1}^\infty$ with $x_k \in \{1, \ldots, N\}$, and $\mathbb{N}^*$ is the set of positive integers.

The well-known result in Proposition 12.3.2 below follows from the results of A. S. Besicovitch in [26] and ties the results of Theorem 12.5.1 to existing theory. (See, also, [11, 15, 30].)

Proposition 12.3.2 *For an IFS which satisfies the OSC on the unit interval $[0,1]$ with scaling ratios $\mathbf{r}$ and a probability vector $\mathbf{q}$, we have*

$$\dim_H(E(\mathbf{q})) = \frac{\sum_{i=1}^N q_i \log q_i}{\sum_{i=1}^N q_i \log r_i}. \quad (12.18)$$

Remark 12.3.1 In 1934, A. S. Besicovitch studied the unique nonterminating binary expansion of $x \in [0,1]$ (i.e., the case where $N = 2$ and $x_k \in \{0,1\}$) in [9] and proved that

$$\dim_H(E(\mathbf{q})) = \frac{-q_1 \log q_1 - q_2 \log q_2}{\log 2}.$$

In 1949, H. G. Eggleston generalized this result to N-ary expansions in [13] and found that

$$\dim_H(E(\mathbf{q})) = \frac{-\sum_{i=1}^{N} q_i \log q_i}{\log N}.$$

(Also see [3, Chap. 6].) In Sec. 12.5, we recover these results when $q_i \in \mathbb{Q} \cap [0,1]$ for $i \in \{1, \ldots, N\}$.

12.3.5 Coarse Hölder regularity

As a break from the mold defined by the results in this section up to this point, the multifractal structure investigated in upcoming sections of this paper is based on the notion of *coarse Hölder regularity* as defined in [21–24, 38], for example, and simply called *regularity* in this paper. Regularity is key to the development of the partition zeta functions defined in the next section.

Definition 12.3.3 For a given Borel measure μ with range in $[0,\infty]$ and an interval $U \subset [0,1]$ with positive Lebesgue measure denoted $|U|$, the regularity $A(U)$ of U is

$$A(U) := \frac{\log \mu(U)}{\log |U|}. \tag{12.19}$$

Equivalently, $A(U)$ is the exponent α that satisfies

$$|U|^{\alpha} = \mu(U).$$

Note that regularity can be considered for any interval, whether open, closed, or neither. To construct the partition zeta functions, intervals are gathered according to their regularity.

In general, regularity values α in the extended real numbers $[-\infty,\infty]$ may be considered. For infinite regularity values, we take

$$\alpha = \infty = A(U) \Leftrightarrow \mu(U) = 0 \quad \text{and} \quad |U| > 0,$$

and

$$\alpha = -\infty = A(U) \Leftrightarrow \mu(U) = \infty \quad \text{and} \quad |U| > 0.$$

However, in Sec. 12.5, we only consider finite regularity values and in Sec. 12.6, we consider $\alpha = \infty$ only briefly.

Fixing the regularity α when taking a measure μ and a sequence of partitions $\mathfrak{P}$ into consideration allows one to define the partition zeta functions, which is done in the next section.

12.4 Definitions

In this section, we define the main objects of study, in particular, the *partition zeta functions*. We consider self-similar measures which are supported on a subset of the unit interval $[0,1]$ and define the partition zeta functions in a manner which is similar to the way the geometric zeta functions are defined. (Compare Definition 12.2.2.) These definitions are similar in that both use a sequence of lengths to provide terms for certain Dirichlet series. However, unlike the geometric zeta functions, the lengths which define a partition zeta function are defined by the scales which stem from a sequence $\mathfrak{P}$ of partitions $\mathcal{P}_n$ (for $n \in \mathbb{N}^*$) of the unit interval and a fixed regularity value α (as defined in Sec. 12.3.5). To elaborate, we first define the appropriate sequence of lengths. Here and henceforth, we assume each partition $\mathcal{P}_n$ comprises a finite number of disjoint intervals with positive length.

Definition 12.4.1 For a Borel measure μ on the interval $[0,1]$ and a sequence of partitions $\mathfrak{P} = \{\mathcal{P}_n\}_{n=1}^{\infty}$ of $[0,1]$ with mesh tending to zero, the sequence of $\underline{\alpha\text{-lengths}}$ $\mathcal{L}^{\mu}_{\mathfrak{P}}(\alpha)$ corresponding to regularity $\alpha \in [-\infty, \infty]$ is given by

$$\mathcal{L}^{\mu}_{\mathfrak{P}}(\alpha) := \left\{ \ell \mid \ell = |P^i_n| \text{ and } A(P^i_n) = \alpha, \text{ where } n \in \mathbb{N}^*, P^i_n \in \mathcal{P}_n \right\}. \tag{12.20}$$

The α-lengths are essentially the distinct scales, along with their multiplicities, of the intervals at level n (for all $n \in \mathbb{N}^*$) in the partition $\mathcal{P}_n \in \mathfrak{P}$ which have regularity α. In turn, the α-lengths are used to define the partition zeta function as follows:

Definition 12.4.2 For a Borel measure μ on the interval $[0,1]$ and a sequence of partitions $\mathfrak{P} = \{\mathcal{P}_n\}_{n=1}^{\infty}$ with mesh tending to zero, the partition zeta function $\zeta^{\mu}_{\mathfrak{P}}(\alpha, s)$ corresponding to regularity $\alpha \in [-\infty, \infty]$ is given by

$$\zeta^{\mu}_{\mathfrak{P}}(\alpha, s) := \zeta_{\mathcal{L}}(s), \tag{12.21}$$

where $\mathcal{L} := \mathcal{L}^{\mu}_{\mathfrak{P}}(\alpha)$ and $\mathrm{Re}(s)$ is large enough. That is,

$$\zeta^{\mu}_{\mathfrak{P}}(\alpha, s) := \sum_{\ell \in \mathcal{L}^{\mu}_{\mathfrak{P}}(\alpha)} \ell^s = \sum_{n=1}^{\infty} \sum_{A(P^i_n)=\alpha} |P^i_n|^s, \tag{12.22}$$

where the inner sum is taken over the intervals P^i_n with regularity $A(P^i_n) = \alpha$ in the partition $\mathcal{P}_n$ for each $n \in \mathbb{N}^*$, and $\mathrm{Re}(s) > D_{\mathcal{L}}$, with $\mathcal{L} := \mathcal{L}^{\mu}_{\mathfrak{P}}(\alpha)$ as in Definition 12.4.1 above.

If there is no interval P^i_n such that $A(P^i_n) = \alpha_0$ for some regularity value α_0, then we set $\zeta^{\mu}_{\mathfrak{P}}(\alpha_0, \cdot)$ identically equal to zero and we refer to such regularity values as *trivial regularity values*. By extension, we will call *trivial* the regularity values α for which $\zeta^{\mu}_{\mathfrak{P}}(\alpha, \cdot)$ is an entire function. This is the case, for example, if $\mathcal{L}^{\mu}_{\mathfrak{P}}(\alpha)$ consists of at most finitely many nonzero α-lengths and is true iff we are in that situation provided $\zeta^{\mu}_{\mathfrak{P}}(\alpha, s)$ is assumed to have a meromorphic continuation to an open neigborhood of $[\mathrm{Re}(s) \geq 0]$; see Remark 12.4.2 below.

Remark 12.4.1 Unlike the case of the multifractal zeta functions from [21, 22, 24, 64], there is no particular partition zeta function which directly corresponds to the lengths of the fractal string of the complement in $[0, 1]$ of the support of a given measure. Still, there is a connection to the Hausdorff (and Minkowski) dimension of the support, as will be seen in Sec. 12.5. Also, for a given measure μ, there are at most countably many nontrivial regularity values attained with respect to a given sequence of partitions $\mathfrak{P}$. Hence, for a given measure and sequence of partitions, there are at most countably many partition zeta functions that are not entire (possibly in the broader context of Remark 12.5.6 below).

In order to connect our methods to those outlined in Sec. 12.3, with motivation provided by Definition 12.2.2 and Proposition 12.3.1, we consider the function $f^{\mu}_{\mathfrak{P}}$ on $[-\infty, \infty]$ which maps the regularity values α attained by a measure μ with respect to a sequence of partitions $\mathfrak{P}$ to the abscissa of convergence of the corresponding partition zeta function $\zeta^{\mu}_{\mathfrak{P}}(\alpha, \cdot)$.

Definition 12.4.3 Given a Borel measure μ on $[0,1]$ and a sequence of partitions $\mathfrak{P}$ with mesh tending to zero, the abscissa of convergence function $f_{\mathfrak{P}}^{\mu}(\alpha)$ is given by

$$f_{\mathfrak{P}}^{\mu}(\alpha) := \inf\left\{\gamma \in \mathbb{R} \mid \zeta_{\mathfrak{P}}^{\mu}(\alpha,\gamma) < \infty\right\}, \tag{12.23}$$

for $\alpha \in [-\infty,\infty]$. For a trivial regularity value α_0, we set $f_{\mathfrak{P}}^{\mu}(\alpha_0) = 0$.

That is, more precisely, in general $f_{\mathfrak{P}}^{\mu}(\alpha)$ is defined as the maximum of 0 and the abscissa of convergence (given by Eq. (12.23)) of the Dirichlet series defining $\zeta_{\mathfrak{P}}^{\mu}(\alpha,\cdot)$; so that $f_{\mathfrak{P}}^{\mu}(\alpha) \geq 0$ for all nontrivial regularity values $\alpha \in \mathbb{R}$ and when $\zeta_{\mathfrak{P}}^{\mu}(\alpha_0,\cdot)$ is entire for a given value of α_0 (i.e., when α_0 is trivial), $f_{\mathfrak{P}}^{\mu}(\alpha_0) = \max\{0,-\infty\} = 0$. (See Remark 12.4.2.) Accordingly, for a nontrivial regularity value α, $\{s \in \mathbb{C} \mid \mathrm{Re}(s) > f_{\mathfrak{P}}^{\mu}(\alpha)\}$ is the largest open right half-plane on which the Dirichlet series in Eq. (12.22) is absolutely convergent.

Remark 12.4.2 Note that if $\mathcal{L}_{\mathfrak{P}}^{\mu}(\alpha)$ consists of infinitely many nonzero α-lengths and $\zeta_{\mathfrak{P}}^{\mu}(\alpha,s)$ admits a meromorphic continuation to an open neigborhood of $[\mathrm{Re}(s) \geq 0]$, then by Eq. (12.22), $\zeta_{\mathfrak{P}}^{\mu}(\alpha,0) = \sum_{\ell\in\mathcal{L}_{\mathfrak{P}}^{\mu}(\alpha)} 1 = \infty$. Hence, by definition, the abscissa of convergence of $\zeta_{\mathfrak{P}}^{\mu}(\alpha,\cdot)$ is necessarily nonnegative in this case; in particular, the partition zeta function $\zeta_{\mathfrak{P}}^{\mu}(\alpha,\cdot)$ cannot be entire. Conversely, if $\mathcal{L}_{\mathfrak{P}}^{\mu}(\alpha)$ consists of finitely many nonzero α-lengths, then $\zeta_{\mathfrak{P}}^{\mu}(\alpha,\cdot)$ is clearly entire and hence, its abscissa of convergence is $-\infty$; from which it follows that $f_{\mathfrak{P}}^{\mu}(\alpha) = \max\{0,-\infty\} = 0$. This justifies, in particular, our terminology for trivial/nontrivial regularity values and the more precise definition of $f_{\mathfrak{P}}^{\mu}(\alpha)$ given in the text immediately following Definition 12.4.3. That is, regularity α is *trivial* iff $\mathcal{L}_{\mathfrak{P}}^{\mu}(\alpha)$ consists of finitely many nonzero lengths, or more generally, if $\zeta_{\mathfrak{P}}^{\mu}(\alpha,\cdot)$ is entire. Otherwise, α is *nontrivial*.

When the partition zeta function $\zeta_{\mathfrak{P}}^{\mu}(\alpha,\cdot)$ has a meromorphic continuation to a window $W_\alpha \subset \mathbb{C}$, we have the following definitions. Note that with a mild abuse of notation, $\zeta_{\mathfrak{P}}^{\mu}(\alpha,s)$ denotes the partition zeta function as well as its meromorphic continuation to W_α.

Definition 12.4.4 For a Borel measure μ on $[0,1]$, sequence $\mathfrak{P}$ of partitions of $[0,1]$ with mesh tending to zero, and regularity $\alpha \in [-\infty,\infty]$, the set of complex dimensions with parameter α, denoted by $\mathcal{D}_{\mathfrak{P}}^{\mu}(\alpha,W_\alpha)$, is given by

$$\mathcal{D}^{\mu}_{\mathfrak{P}}(\alpha, W_\alpha) := \{\omega \in W_\alpha \mid \zeta^{\mu}_{\mathcal{N}}(\alpha, s) \text{ has a pole at } \omega\}, \tag{12.24}$$

for an appropriate window W_α.

Gathering the sets $\mathcal{D}^{\mu}_{\mathfrak{P}}(\alpha, W_\alpha)$ by all finite regularity values α attained by μ with respect to $\mathfrak{P}$ yields the following collection.

Definition 12.4.5 For a Borel measure μ on $[0,1]$ and sequence $\mathfrak{P}$ of partitions of $[0,1]$ with mesh tending to zero, the tapestry of complex dimensions $\mathcal{T}^{\mu}_{\mathfrak{P}}$ with respect to the windows W_α is given by

$$\mathcal{T}^{\mu}_{\mathfrak{P}} := \{(\alpha, \omega) \mid \alpha \in (-\infty, \infty), \omega \in \mathcal{D}^{\mu}_{\mathfrak{P}}(\alpha, W_\alpha)\}. \tag{12.25}$$

Note that by definition, we have

$$\mathcal{T}^{\mu}_{\mathfrak{P}} \subset \mathbb{R} \times W_\alpha \subset \mathbb{R} \times \mathbb{C}.$$

In light of Definition 12.4.4 and Proposition 12.2.1, we define as follows the *counting function of the α-lengths* of a measure μ with respect to a sequence of partitions $\mathfrak{P}$ and the attained regularity values α.

Definition 12.4.6 For $\mathcal{L} := \mathcal{L}^{\mu}_{\mathfrak{P}}(\alpha)$, the counting function of the α-lengths of μ with respect to $\mathfrak{P}$ is

$$N^{\mu}_{\mathfrak{P}}(\alpha, x) := N_{\mathcal{L}}(x), \tag{12.26}$$

for $x > 0$.

Remark 12.4.3 Note that $N^{\mu}_{\mathfrak{P}}(\alpha, x)$ does not correspond to a fractal string per se, just a sequence of lengths (or rather, scales). In this setting, the corresponding explicit formulas for general fractal strings are discussed in [20, Chap. 5] (also [21]), and as mentioned in Sec. 12.2.3, can immediately be used to describe the multiscale behavior of μ with respect to $\mathfrak{P}$ and regularity α.

The following section investigates the properties of the partition zeta functions and the abscissa of convergence functions for self-similar measures and their natural sequences of partitions.

12.5 Partition zeta functions of self-similar measures

Now we develop and investigate the partition zeta functions of a self-similar measure μ on $[0,1]$ and its natural sequence of partitions of $[0,1]$, denoted $\mathfrak{P}$, where both μ and $\mathfrak{P}$ are defined by an IFS and a probability vector as in Sec. 12.3.1. The scaling ratios given by $\mathbf{r}$, along with the probability vector $\mathbf{p}$, completely determine the regularity values α which the measure μ attains with respect to $\mathfrak{P}$.

12.5.1 Self-similar measures and natural sequences of partitions

Consider an IFS with contracting similarities $\{S_i\}_{i=1}^N$ which satisfy the OSC and have scaling ratios $\mathbf{r} = (r_1, \ldots, r_N)$ such that $\sum_{i=1}^N r_i \leq 1$ and $r_i > 0$ for each $i \in \{1, \ldots, N\}$. (Here and thereafter, we have $N \geq 2$.) Without loss of generality, we may assume that $0 \in S_1([0,1])$ and $1 \in S_N([0,1])$. Furthermore, let $\mathbf{p} = (p_1, \ldots, p_N)$ be a probability vector; that is, $\sum_{i=1}^N p_i = 1$ and, without loss of generality in the setting of this section, $p_i > 0$ for all $i \in \{1, \ldots, N\}$.

At each stage $K \in \mathbf{N}^* = \{1, 2, \ldots\}$ of the recursive construction of the IFS, there are N^K distinct closed intervals with positive mass. These intervals, along with the distinct intervals which fill in the gaps between them (if any), constitute the partitions $\mathcal{P}_K$ for each $K \in \mathbb{N}^*$. In turn, $\{\mathcal{P}_K\}_{K=1}^\infty$ constitutes the *natural sequence of partitions* $\mathfrak{P}$. Each of the intervals in $\mathcal{P}_K$ with positive mass have their length and mass completely described by an ordered N-tuple of nonnegative integers $\mathbf{k} = (k_1, \ldots, k_N)$ such that $\sum_{i=1}^N k_i = K$. That is, an interval $P \in \mathcal{P}_K$ with positive mass has length $|P|$ of the form $r_1^{k_1} \cdots r_N^{k_N}$ and mass $\mu(P)$ of the form $p_1^{k_1} \cdots p_N^{k_N}$. See Fig. 12.4, where we use the notation introduced below.

Ultimately, $\mathbf{k} = (k_1, \ldots, k_N)$ and $\mathbf{r} = (r_1, \ldots, r_N)$ define the regularity value $\alpha(\mathbf{k})$ as below and, in turn, the Besicovitch subset $E(\mathbf{k}/K)$, defined by Eq. (12.17), of the Moran (self-similar) fractal $E = \text{supp}(\mu)$. See [11, 15, 18, 30] for the construction of Moran fractals and related results in more general settings.

Remark 12.5.1 Intervals without mass have regularity $\alpha = \infty$, but we do not investigate this case (except in Sec. 12.6, and there only briefly) since the resulting partition zeta functions are divergent everywhere. This is in

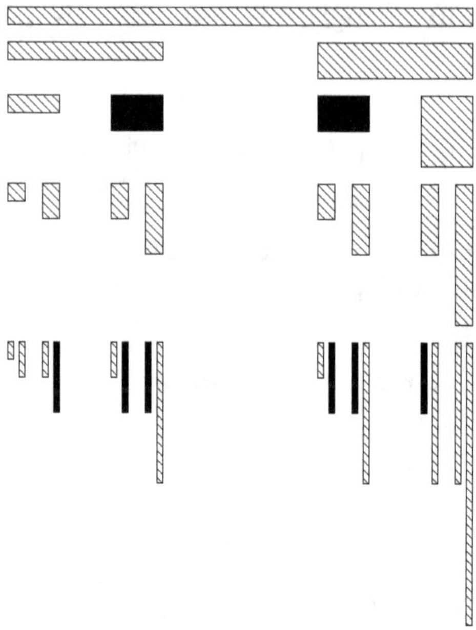

Figure 12.4 The solid black blocks correspond to intervals which have regularity $\alpha(1,2)$ stemming from the measure β (see Sec. 12.5.2) and its natural sequence of partitions.

stark contrast to the results obtained under the context of the multifractal zeta functions as described in [21–24, 64], where regularity $\alpha = \infty$ precisely recovers the geometric zeta function and $\alpha = -\infty$ yields the Hausdorff dimension of the boundary of a certain type of fractal string.

The breakdown of mass and length as above provides a complete description of the regularity values attained by a self-similar measure μ with respect to the natural sequence of partitions $\mathfrak{P}$. Specifically, $\mathbf{k} = (k_1, \ldots, k_N)$ corresponds to the regularity value $\alpha = \alpha(\mathbf{k})$ given by

$$\alpha(\mathbf{k}) = \frac{\log\left(p_1^{k_1} \cdots p_N^{k_N}\right)}{\log\left(r_1^{k_1} \cdots r_N^{k_N}\right)} = \frac{\sum_{i=1}^N k_i \log p_i}{\sum_{i=1}^N k_i \log r_i},$$

where the convention $0 \log 0 = 0$ is used. Note the similarity to the ratio of logarithms used in Eq. (12.12). Before stating and deriving our main results, we consider the partition zeta functions and abscissa of convergence function of a well-known binomial measure.

12.5.2 A multifractal measure on the Cantor set

Consider a binomial measure β supported on the classical ternary Cantor set defined by the weighted IFS given by $\mathbf{r} = (1/3, 1/3)$ with probabilities $\mathbf{p} = (p_1, p_2)$ such that $0 < p_1 < p_2 < 1$. At every stage $K \in \mathbb{N}^*$ of the IFS, there are 2^K intervals with positive mass. The number of intervals at stage K with regularity $\alpha(k_2, K) := \alpha(\mathbf{k})$ is given by the binomial coefficient

$$\binom{K}{k_1} = \binom{K}{k_2} = \frac{K!}{k_1!k_2!},$$

where the first equality holds since $k_1 = K - k_2$. See Fig. 12.4 for a depiction of the first five stages of the weighted IFS with $N = 2$, $\mathbf{r} = (1/3, 1/3)$, and $\mathbf{p} = (1/3, 2/3)$, resulting in the self-similar measure β and the natural sequence of partitions $\mathfrak{P}$.

The partition zeta function $\zeta_{\mathfrak{P}}^{\beta}(\alpha, s)$ with regularity $\alpha = \alpha(k_2, K)$ is given by

$$\zeta_{\mathfrak{P}}^{\beta}(\alpha, s) = \sum_{n=1}^{\infty} \binom{nK}{nk_2} 3^{-nKs}.$$

The abscissa of convergence function $f = f_{\mathfrak{P}}^{\beta}(\alpha)$ is given by

$$\begin{aligned} f_{\mathfrak{P}}^{\beta}(\alpha) &= -x \log_3 x - (1-x)\log_3(1-x) \\ &= -\left(\frac{1-\alpha}{\log_3 2}\right) \log_3 \left(\frac{1-\alpha}{\log_3 2}\right) \\ &\quad - \left(1 - \frac{1-\alpha}{\log_3 2}\right) \log_3 \left(1 - \frac{1-\alpha}{\log_3 2}\right), \end{aligned}$$

where $x = k_2/K = \frac{1-\alpha}{\log_3 2}$. (See Fig. 12.5.) The development of the partition zeta functions $\zeta_{\mathfrak{P}}^{\beta}(\alpha, s)$ and the abscissa of convergence function $f_{\mathfrak{P}}^{\beta}(\alpha)$ are provided below in more general settings.

12.5.3 First main result: distinct regularity

For a self-similar measure defined by an IFS and a probability vector, the task of collecting the intervals with identical regularity values from every partition in $\mathfrak{P}$ is relatively simple when the conditions of Theorem 12.5.1 are satisfied.

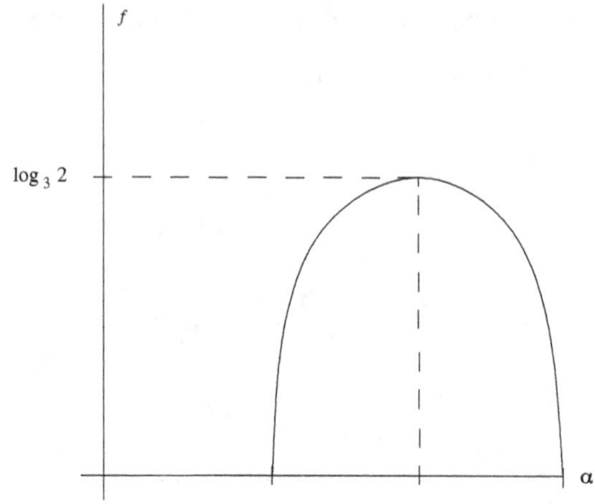

Figure 12.5 The graph of $f = f_{\mathfrak{P}}^{\beta}(\alpha)$ as a function of regularity α for the binomial measure β supported on the Cantor set.

Here and henceforth, we let

$$\binom{m}{m_1 \ldots m_N} := \frac{m!}{m_1! \cdots m_N!}$$

denote the multinomial coefficients, where $m_i \in \mathbb{N}$ for $i \in \{1, \ldots, N\}$ and $m := \sum_{i=1}^{N} m_i$.

Theorem 12.5.1 *Let μ be the self-similar measure and $\mathfrak{P}$ be the natural sequence of partitions defined by an IFS with scaling ratios $\mathbf{r} = (r_1, \ldots, r_N)$ which satisfies the OSC and is weighted by a probability vector $\mathbf{p}$ (as described in Sec. 12.5.1). Consider the following hypothesis:*

(***H***) *Suppose that for all $\mathbf{k} = (k_1, \ldots, k_N) \in \mathbb{N}^N$ where $\gcd(k_1, \ldots, k_N) = 1$ and $\mathbf{k} \neq \mathbf{0}$, we have that the regularity values $\alpha(\mathbf{k})$ are distinct. (See Remark 12.5.2 below.) That is, suppose $\alpha(z_1, \ldots, z_N) = \alpha(\mathbf{k})$ if and only if there exists $m \in \mathbb{N}^*$ such that $z_i = mk_i$ for all $i \in \{1, \ldots, N\}$.*

*Assume that hypothesis (**H**) holds. Then, for any $\mathbf{k} = (k_1, \ldots, k_N) \in \mathbb{N}^N$ with $\gcd(k_1, \ldots, k_N) = 1$, and letting $K := \sum_{i=1}^{N} k_i$, we have*

$$\zeta_{\mathfrak{P}}^{\mu}(\alpha(\mathbf{k}), s) = \sum_{n=1}^{\infty} \binom{nK}{nk_1 \cdots nk_N} (r_1^{k_1} \cdots r_N^{k_N})^{ns}. \tag{12.27}$$

Moreover, the abscissa of convergence $\sigma = f^{\mu}_{\mathfrak{P}}(\alpha(\mathbf{k}))$ *of the partition zeta function* $\zeta^{\mu}_{\mathfrak{P}}(\alpha(\mathbf{k}), s)$ *is given by*

$$f^{\mu}_{\mathfrak{P}}(\alpha(\mathbf{k})) = \frac{\sum_{i=1}^{N}(k_i/K)\log(k_i/K)}{\sum_{i=1}^{N}(k_i/K)\log r_i} = \dim_H(E(\mathbf{k}/K)), \tag{12.28}$$

where $E(\mathbf{k}/K)$ *is the Besicovitch subset of the Moran (self-similar) fractal* $E = supp(\mu)$ *defined by the scaling ratios* $\mathbf{r} = (r_1, \ldots, r_N)$ *and the probability vector* $\mathbf{k}/K = (k_1/K, \ldots, k_N/K)$*; see Eq. (12.17) in Sec. 12.3.4. Equivalently, and with use of the convention* $0^0 = 1$*, the abscissa of convergence* σ *is the unique real number satisfying the equation*

$$(r_1^{k_1} \cdots r_N^{k_N})^{\sigma} \frac{K^K}{k_1^{k_1} \cdots k_N^{k_N}} = 1; \tag{12.29}$$

in addition, $\sigma > 0$.

Proof Every interval from $\mathfrak{P}$ is taken into account since $\gcd(k_1, \ldots, k_N) = 1$. Specifically, for the given positive integers $(z_1, \ldots, z_N)$, there is an integer $m \in \mathbb{N}^*$ such that $mk_i = z_i$ for each i; hence the corresponding intervals have the same regularity since $\alpha(\mathbf{k}) = \alpha(n\mathbf{k})$ for all $n \in \mathbb{N}^*$. Thus, $\alpha = \alpha(\mathbf{k})$ is attained by μ in $\mathcal{P}_{nK}$ for each $n \in \mathbb{N}^*$ and the corresponding intervals contribute their lengths to the same partition zeta function. The coefficients $\binom{nK}{nk_1 \cdots nk_N}$ stem from the multinomial distribution of mass among the intervals in the partitions $\mathcal{P}_{nK}$.

To determine the abscissa of convergence function $f^{\mu}_{\mathfrak{P}}(\alpha(\mathbf{k}))$, an application of Stirling's formula and the n-th root test allows for the formulation of the abscissa of convergence, written σ for notational convenience, in terms of the N-tuples $\mathbf{k} = (k_1, \ldots, k_N)$ and $\mathbf{r} = (r_1, \ldots, r_N)$. Indeed, for a fixed real number s, Stirling's formula yields

$$\begin{aligned}
\kappa_{n,s} &:= \binom{nK}{nk_1 \cdots nk_N} (r_1^{k_1} \cdots r_N^{k_N})^{ns} \\
&= \frac{(nK)!}{(nk_1)! \cdots (nk_N)!} (r_1^{k_1} \cdots r_N^{k_N})^{ns} \\
&= \frac{(r_1^{k_1} \cdots r_N^{k_N})^{ns} K^{nK}}{k_1^{nk_1} \cdots k_N^{nk_N}} \cdot \frac{\sqrt{K}}{\sqrt{k_1}\sqrt{2\pi n k_2} \cdots \sqrt{2\pi n k_N}} (1 + \varepsilon_n),
\end{aligned}$$

where $\varepsilon_n \to 0$ as $n \to \infty$. Hence,

$$\kappa_{n,s}^{1/n} = \frac{(r_1^{k_1} \cdots r_N^{k_N})^s K^K}{k_1^{k_1} \cdots k_N^{k_N}}(1+\delta_n),$$

where $\delta_n \to 0$ as $n \to \infty$. Therefore, according to the n-th root test, the numerical series

$$\sum_{n=1}^{\infty} \kappa_{n,s} = \sum_{n=1}^{\infty} \binom{nK}{nk_1 \cdots nk_N} (r_1^{k_1} \cdots r_N^{k_N})^{ns}$$

converges for $s > \rho$ and diverges for $s < \rho$, where ρ is the unique real number such that

$$1 = \left(r_1^{k_1} \cdots r_N^{k_N}\right)^\rho \frac{K^K}{k_1^{k_1} \cdots k_N^{k_N}}.$$

Equivalently,

$$f_{\mathfrak{P}}^{\mu}(\alpha(\mathbf{k})) = \rho = \log_{r_1^{k_1} \cdots r_N^{k_N}} \left(\frac{k_1^{k_1} \cdots k_N^{k_N}}{K^K}\right) = \frac{\sum_{i=1}^N (k_i/K)\log(k_i/K)}{\sum_{i=1}^N (k_i/K)\log r_i}.$$

(The fact that ρ is well defined and $\rho > 0$ will be explained at the end of the proof.) By definition of the abscissa of convergence σ (see Definition 12.4.3), it follows that

$$f_{\mathfrak{P}}^{\mu}(\alpha(\mathbf{k})) = \sigma = \rho,$$

which in light of the previous expression for ρ, establishes part of Theorem 12.5.1.

Finally, by Proposition 12.3.2 applied to $\mathbf{r}$ and the probability vector $\mathbf{q} = \mathbf{k}/K$, we have

$$f_{\mathfrak{P}}^{\mu}(\alpha(\mathbf{k})) = \sigma = \frac{\sum_{i=1}^N (k_i/K)\log(k_i/K)}{\sum_{i=1}^N (k_i/K)\log r_i} = \dim_H(E(\mathbf{k}/K)),$$

where $E(\mathbf{k}/K)$ is the Besicovitch subset referred to in Theorem 12.5.1. (See, also, [11, 15, 30].) This last equality, combined with the fact that $\sigma = \rho$, enables us to conclude the proof of the main statement of Theorem 12.5.1.

As promised, we now supplement the proof by briefly explaining why ρ is well defined as the unique real solution of the equation $\varphi(s) = 1$, where

$$\varphi(s) := (r_1^{k_1} \cdots r_N^{k_N})^s \frac{K^K}{k_1^{k_1} \cdots k_N^{k_N}}.$$

First, note that φ is strictly decreasing on $(-\infty, \infty)$; indeed, $\varphi'(s) < 0$ since $0 < r_i < 1$ for $i = 1, \ldots, N$. Second, observe that $\varphi(0) = K^K/(k_1^{k_1} \cdots k_N^{k_N}) > 1$ since $K = \sum_{i=1}^N k_i$ and not all of the integers k_i are zero for $i = 1, \ldots, N$. (As noted above, the convention $0^0 = 1$ is used.) Hence, ρ is well defined and $\rho > 0$, as desired. □

Remark 12.5.2 We believe that the assumption (**H**) made in Theorem 12.5.1 is essentially superfluous. That is, we expect that the regularity values $\alpha(\mathbf{k})$ are either always distinct or can be broken down into distinct values with corresponding multiplicities, as is done (in a special case) in Lemma 12.5.1 below.

Remark 12.5.3 Theorem 12.5.1 applies to the results on the binomial measure β presented in Sec. 12.5.2. In that setting, $\alpha(k_2, K) = \log_r(p_1) + (k_2/K)\log_r(p_2/p_1)$ and these regularity values are distinct as long as $p_1 < p_2$. Indeed, using the substitution $x = k_2/K$, we have that $\alpha(k_2, K) = \alpha(x) = \log_r(p_1) + x\log_r(p_2/p_1)$ is a nonconstant linear function.

The condition of Theorem 12.5.1 requiring $\alpha(z_1, \ldots, z_N) = \alpha(\mathbf{k})$ if and only if there exists $m \in \mathbb{N}^*$ such that $z_i = mk_i$ for all $i \in \{1, \ldots, N\}$ is not a necessary condition. Indeed, the next section shows that this condition can be replaced with rational independence of the logarithm of the probability values when there is a single distinct scaling ratio used in the IFS. Moreover, the following corollary shows that, at least in the specific case of the binomial measure β_0 on the unit interval $[0, 1]$, we have $\hat{f}_{\mathfrak{P}}^{\beta_0}(t) = f_g(t) = f_s(t) = b^*(t)$, where $\hat{f}_{\mathfrak{P}}^{\beta_0}$ is the concave envelope of $f_{\mathfrak{P}}^{\beta_0}$ on the interval $[t_{\min}, t_{\max}]$. (See [9, 12, 13, 26, 30], as well as Figs. 12.3 and 12.5, but note that for β_0 we have $\mathbf{r} = (1/2, 1/2)$ and for β in Sec. 12.5.2 we have $\mathbf{r} = (1/3, 1/3)$.)

The following corollary also stems from the discussion that follows [23, Theorem 4.2] and the connection to the binomial measure on the unit interval (called β_0 below) discussed, for instance, in [1–3].

Corollary 12.5.1 *Consider the binomial measure β_0 defined by the similarities $S_1(x) = x/2$ and $S_2(x) = x/2 + 1/2$ with scaling ratios*

$\mathbf{r} = (1/2, 1/2)$ and the probability vector $\mathbf{p} = (1/3, 2/3)$. Then, for all $t \in [t_{\min}, t_{\max}]$, we have

$$\hat{f}_{\mathfrak{P}}^{\beta_0}(t) = f_g(t) = f_s(t) = b^*(t), \tag{12.30}$$

where $\hat{f}_{\mathfrak{P}}^{\beta_0}$ is the concave envelope of $f_{\mathfrak{P}}^{\beta_0}$ on the interval $[t_{\min}, t_{\max}]$. (Here, $[t_{\min}, t_{\max}] = [\log_2 3 - 1, \log_2 3]$.)

Proof By Theorem 12.5.1, we have

$$\begin{aligned} f_{\mathfrak{P}}^{\beta_0}(\alpha(\mathbf{k})) &= \frac{(k_1/K)\log(k_1/K) + (k_2/K)\log(k_2/K)}{(k_1/K)\log r_1 + (k_2/K)\log r_2} \\ &= -\frac{k_1}{K}\log_2\left(\frac{k_1}{K}\right) - \frac{k_2}{K}\log_2\left(\frac{k_2}{K}\right) \\ &= -\left(1 - \frac{k_2}{K}\right)\log_2\left(1 - \frac{k_2}{K}\right) - \frac{k_2}{K}\log_2\left(\frac{k_2}{K}\right). \end{aligned}$$

Further, we have

$$\alpha(\mathbf{k}) = \frac{\log(2^{k_2}/3^K)}{\log(1/2^K)} = \log_2 3 - \frac{k_2}{K}.$$

Since k_2 is a nonnegative integer and K is a positive integer where $k_2 \le K$, the maximum value of $\alpha(\mathbf{k})$ is $\log_2 3 = t_{\max}$ and the minimum value of $\alpha(\mathbf{k})$ is $\log_2 3 - 1 = t_{\min}$.

According to [1], we have

$$f_g(t) = -\frac{t_{\max} - t}{t_{\max} - t_{\min}}\log_2\left(\frac{t_{\max} - t}{t_{\max} - t_{\min}}\right) - \frac{t - t_{\min}}{t_{\max} - t_{\min}}\log_2\left(\frac{t - t_{\min}}{t_{\max} - t_{\min}}\right),$$

where $t \in [t_{\min}, t_{\max}] = [\log_2 3 - 1, \log_2 3]$.

For $t_1 = \log_2 3 - k_2/K = \alpha(\mathbf{k})$, we have $t_{\max} - t_{\min} = 1$, $t_{\max} - t_1 = k_2/K$, and $t_1 - t_{\min} = 1 - k_2/K$. Substitution yields

$$f_{\mathfrak{P}}^{\beta_0}(t_1) = f_g(t_1).$$

Note that the collection of all regularity values $t_1 = \log_2 3 - k_2/K = \alpha(\mathbf{k})$ is a dense subset of $[t_{\min}, t_{\max}] = [\log_2 3 - 1, \log_2 3]$; furthermore, recall that $f_g(t)$ is concave (see Sec. 12.3.3). Therefore, by Proposition 12.3.1 we have

$$\hat{f}_{\mathfrak{P}}^{\beta_0}(t) = f_g(t) = f_s(t) = b^*(t)$$

on $[t_{\min}, t_{\max}]$. □

Remark 12.5.4 Corollary 12.5.1 immediately holds in the slightly more general case where the components of the probability vector $\mathbf{p}$ are distinct (that is, $p_1 \neq p_2$). However, in this chapter we address only the case $\mathbf{p} = (1/3, 2/3)$, for clarity of exposition.

The next corollary is a simple consequence of Theorem 12.5.1 and the Moran equation.

Corollary 12.5.2 *Assume that the conditions of Theorem* 12.5.1 *hold and, additionally, that* $\mathbf{r} = (\lambda, \ldots, \lambda)$ *where* $0 < \lambda \leq 1/N$. *Then*

$$f_{\mathfrak{P}}^{\mu}(\alpha(1, \ldots, 1)) = \dim_H(supp(\mu)). \tag{12.31}$$

Proof In this case, $K = N$ and by Theorem 12.5.1 we have

$$f_{\mathfrak{P}}^{\mu}(\alpha(1, \ldots, 1)) = -\log_{\lambda} N.$$

The Moran equation (Eq. (12.9)) then becomes

$$\sum_{i=1}^{N} \lambda^s = N\lambda^s = 1,$$

which has the unique solution $s = -\log_{\lambda} N$ when $s > 0$. □

In light of Theorem 12.5.1, Theorem 12.5.2 (an analog of Theorem 12.5.1 in Sec. 12.5.4 below), Corollary 12.5.1, and Corollary 12.5.2, we make the following conjecture:

Conjecture 12.5.1 *For a self-similar measure* μ *on* $[0, 1]$ *and its natural sequence of partitions* $\mathfrak{P}$, *we have*

$$\hat{f}_{\mathfrak{P}}^{\mu}(t) = f_g(t) = f_s(t) = b^*(t), \tag{12.32}$$

for all $t \in [t_{\min}, t_{\max}]$, *where* $\hat{f}_{\mathfrak{P}}^{\mu}(t)$ *is the concave envelope of* $f_{\mathfrak{P}}^{\mu}(\alpha)$ *on* $[t_{\min}, t_{\max}]$.

Remark 12.5.5 Conjecture 12.5.1 would be proven in the case where the regularity values $\alpha(\mathbf{k})$ are distinct if, for instance, one could show that for all $\mathbf{k} = (k_1, \ldots, k_N) \in \mathbb{N}^N$ such that $\gcd(k_1, \ldots, k_N) = 1$,

$$E_{\alpha(\mathbf{k})} = E(\mathbf{k}/K),$$

where $E_{\alpha(\mathbf{k})}$ is given by Eq. (12.11) with $t = \alpha(\mathbf{k})$ and $E(\mathbf{k}/K)$ is given by Eq. (12.17) with $\mathbf{q} = \mathbf{k}/K$. See [15, Theorem A], [27, Proposition 5.1], and [11] for further evidence for the validity of Conjecture 12.5.1.

One condition of Theorem 12.5.1 that is required throughout the rest of this paper is that the components of $\mathbf{k} = (k_1, \ldots, k_N)$ satisfy $\gcd(k_1, \ldots, k_N) = 1$. This condition guarantees that every interval stemming from $\mathfrak{P}$ with regularity $\alpha(\mathbf{k})$ is taken into account in the corresponding partition zeta function. The next section develops and analyzes such partition zeta functions for a specific class of self-similar measures.

12.5.4 Second main result: two distinct probabilities and a single scaling ratio

Consider an IFS with N contracting similarities and a single scaling ratio (Lipschitz constant) r such that $0 < r = r_i \leq 1/N$, for all $i \in \{1, \ldots, N\}$. In contrast to what happens for the results in Sec. 12.5.3, the collection of finite regularity values attained by the corresponding measure μ on the natural sequence of partitions $\mathfrak{P}$ in this setting is determined solely by the probability vector $\mathbf{p}$.

Suppose there are $w \geq 2$ ($w \in \mathbb{N}^*$) distinct values among the components of $\mathbf{p} = (p_1, \ldots, p_N)$. (The case where $w = 1$ is a special case of Ex. 12.5.3 below.) Denote the distinct probabilities by $\mathbf{p'} = (p'_1, \ldots, p'_w)$, and denote the multiplicity of each of the numbers p'_i by c_i for $i = 1, \ldots, w$. In this setting, $\sum_{q=1}^{w} c_q = N$ and $p_1^{k_1} \ldots p_N^{k_N} = (p'_1)^{k'_1} \ldots (p'_w)^{k'_w}$, where $\sum_{i=1}^{N} k_i = \sum_{q=1}^{w} k'_q = K$. We continue to use the convention $0 \log 0 = 0$, and therefore, $0^0 = 1$.

Lemma 12.5.1 *With μ and $\mathfrak{P}$ as above, if $\gcd(k'_1, \ldots, k'_w) = 1$ and the numbers $\log_r p'_1, \ldots, \log_r p'_w$ are rationally independent, then the distinct regularity values attained by μ on $\mathfrak{P}$ are given by*

$$\alpha(k'_1, \ldots, k'_w) = \frac{1}{K} \log_r \left((p'_1)^{k'_1} \ldots (p'_w)^{k'_w} \right). \tag{12.33}$$

Moreover, for every $n \in \mathbb{N}^*$, the number of intervals P with regularity value

$$\alpha(nk'_1, \ldots, nk'_w) = \alpha(k'_1, \ldots, k'_w)$$

in the partition $\mathcal{P}_{nK}$ is

$$\binom{nK}{nk_1, \ldots, nk_N} = \binom{nK}{nk'_1, \ldots, nk'_w} c_1^{nk'_1} \cdots c_w^{nk'_w}.$$

Proof For every N-tuple of nonnegative integers $\mathbf{k}$, there exists an $n \in \mathbb{N}$ such that

$$\alpha(\mathbf{k}) = \alpha(nk'_1, \ldots, nk'_w).$$

The rational independence of the numbers $\log_r p'_i$ and the fact that

$$\log_{r^K}\left((p'_1)^{k'_1} \ldots (p'_w)^{k'_w}\right) = \frac{1}{K}\sum_{i=1}^{w} k'_i \log_r p'_i$$

imply that the regularity values $\alpha(k'_1, \ldots, k'_w)$ are indeed distinct. Moreover, we have

$$(p_1 + \cdots + p_N)^{nK} = (c_1 p'_1 + \cdots + c_w p'_w)^{nK},$$

which immediately yields

$$\binom{nK}{nk_1, \ldots, nk_N} = \binom{nK}{nk'_1, \ldots, nk'_w} c_1^{nk'_1} \cdots c_w^{nk'_w}. \qquad \square$$

Lemma 12.5.1 allows us to determine the corresponding partition zeta functions and abscissa of convergence function in Proposition 12.5.1, as we now explain.

Proposition 12.5.1 *If the conditions of Lemma* 12.5.1 *are satisfied, then*

$$\zeta^{\mu}_{\mathfrak{P}}(\alpha(\mathbf{k}), s) = \sum_{n=1}^{\infty} \binom{nK}{nk'_1, \ldots, nk'_w} c_1^{nk'_1} \cdots c_w^{nk'_w} r^{nKs} \tag{12.34}$$

and

$$f^{\mu}_{\mathfrak{P}}(\alpha(\mathbf{k})) = f^{\mu}_{\mathfrak{P}}(\alpha(k_1, \ldots, k_N)) = \log_{r^K}\left(\frac{(k'_1)^{k'_1} \cdots (k'_w)^{k'_w}}{c_1^{k'_1} \cdots c_w^{k'_w} K^K}\right). \tag{12.35}$$

Proof Equation (12.34) follows at once from Lemma 12.5.1. Next, much as in the proof of Theorem 12.5.1, it can be seen that an application of Stirling's formula and the n-th root test implies that the abscissa of convergence $\gamma := f^{\mu}_{\mathfrak{P}}(\alpha(\mathbf{k}))$ of $\zeta^{\mu}_{\mathfrak{P}}(\alpha(\mathbf{k}), \cdot)$ satisfies

$$1 = (r^K)^{\gamma} \frac{c_1^{k'_1} \cdots c_w^{k'_w} K^K}{(k'_1)^{k'_1} \cdots (k'_w)^{k'_w}}.$$

That is, $\gamma > 0$ is the unique real solution of the above equation. Hence,

$$f_{\mathfrak{P}}^{\mu}(\alpha(\mathbf{k})) = \gamma = \log_{r^K}\left(\frac{(k_1')^{k_1'}\cdots(k_w')^{k_w'}}{c_1^{k_1'}\cdots c_w^{k_w'}K^K}\right),$$

as desired. $\square$

In the special case of the conditions specified in Lemma 12.5.1 where $r_1 = \cdots = r_N = r$ and $w = 2$, the abscissa of convergence function $f_{\mathfrak{P}}^{\mu}(\alpha(\mathbf{k}))$ can easily be expressed as a function of the single variable $\alpha = \alpha(k_2', K) = \alpha(\mathbf{k})$. The results then mirror those described in [11, 18], among others, in the case of "map specified" Moran fractals (i.e., generated by an IFS) as popularized by [62].

Theorem 12.5.2 *Assuming the conditions of Lemma* 12.5.1 *are satisfied and, specifically,* $w = 2$ *and* $r_1 = \cdots = r_N = r$*, then we have*

$$\zeta_{\mathfrak{P}}^{\mu}(\alpha(\mathbf{k}), s) = \zeta_{\mathfrak{P}}^{\mu}(\alpha(k_2', K), s) = \sum_{n=1}^{\infty}\binom{nK}{nk_2'}(N-c_2)^{n(K-k_2')}c_2^{nk_2'}r^{nKs} \tag{12.36}$$

and

$$f_{\mathfrak{P}}^{\mu}(\alpha(\mathbf{k})) = f_{\mathfrak{P}}^{\mu}(\alpha(k_2', K)) = \log_{r^K}\left(\frac{(K-k_2')^{K-k_2'}(k_2')^{k_2'}}{(N-c_2)^{K-k_2'}c_2^{k_2'}K^K}\right). \tag{12.37}$$

(Recall that c_1 *and* c_2 *denote the multiplicities, respectively, of the two distinct values of* $p_1, \ldots, p_N$*.)*

Moreover, the concave envelope $\hat{f}_{\mathfrak{P}}^{\mu}$ *of* $f_{\mathfrak{P}}^{\mu}$ *has infinite slope at the extreme values of the attained regularity values.*

Lastly, we have

$$\max_{\alpha(\mathbf{k})}\{f_{\mathfrak{P}}^{\mu}(\alpha(\mathbf{k}))\} = f_{\mathfrak{P}}^{\mu}(\alpha(c_2/N)) = \dim_H(supp(\mu)). \tag{12.38}$$

Proof We have

$$\alpha(k_2', K) = \frac{\log\left((p_1')^{K-k_2'}(p_2')^{k_2'}\right)}{\log\left(r^K\right)} = \log_r(p_1') + \frac{k_2'}{K}\log_r\left(\frac{p_2'}{p_1'}\right).$$

Letting $x = k_2'/K$, we have

$$x = \frac{\alpha - \log_r(p_1')}{\log_r(p_2'/p_1')}.$$

This substitution allows one to express the abscissa of convergence function $g(x) := f_{\mathfrak{P}}^{\mu}(\alpha(k_2', K))$ in the following form:

$$\begin{aligned} g(x) = f_{\mathfrak{P}}^{\mu}(\alpha(k_2', K)) &= \log_{r^K}\left(\frac{(K-k_2')^{K-k_2'}(k_2')^{k_2'}}{(N-c_2)^{K-k_2'}c_2^{k_2'}K^K}\right) \\ &= x\log_r\left(\frac{x}{c_2}\right) + (1-x)\log_r\left(\frac{1-x}{N-c_2}\right). \end{aligned}$$

By temporarily allowing $x \in [0,1]$ and using a slight abuse of notation, we deduce that the first two derivatives of g are given by

$$g'(x) = \log_r\left(\frac{x}{1-x}\right) + \log_r\left(\frac{c_1}{c_2}\right), \qquad g''(x) = \frac{-1}{x(1-x)\log r}.$$

It immediately follows, assuming without loss of generality that $p_1' < p_2'$ and $x \in (0,1)$, that

$$\lim_{x\to 0^+} g(x) = -\log_r c_1, \quad \lim_{x\to 1^-} g(x) = -\log_r c_2,$$
$$\lim_{x\to 0^+} g'(x) = \infty, \quad \lim_{x\to 1^-} g'(x) = -\infty,$$

and

$$g''(x) < 0,$$

which implies that $\hat{f}_{\mathfrak{P}}^{\mu}$ is concave. A bit more calculus then shows that

$$\max_{\alpha(\mathbf{k})}\{f_{\mathfrak{P}}^{\mu}(\alpha(\mathbf{k}))\} = f_{\mathfrak{P}}^{\mu}(\alpha((c_1,c_2))) = -\log_r N = \dim_H(\mathrm{supp}(\mu)).$$

In particular, the last equality holds since $s = -\log_r N$ is the unique real-valued solution of the equation $Nr^{\sigma} = 1$. (See Proposition 12.3.1 and Eq. (12.13).) □

The following section discusses the way in which the results of Sec. 12.5.4 recover recent as well as classical results on self-similar measures.

12.5.5 Recovery of recent and classical results

Theorem 12.5.2 allows for the recovery of recent and classical results from the multifractal analysis of self-similar measures in the context provided by the partition zeta functions and the abscissa of convergence function, as we now discuss.

Example 12.5.1 (*The Trident Measure*) If $N = 3, \mathbf{r} = (1/5, 1/5, 1/5)$, and $\mathbf{p} = (1/5, 3/5, 1/5)$, then $w = 2, c_1 = 2$ and $c_2 = 1$; the resulting self-similar measure is called the *trident measure* ψ (see Fig. 12.6. The distinct regularity values attained by ψ on its natural sequence of partitions $\mathfrak{P}$ are

$$\alpha(\mathbf{k}) = \frac{\log(3^{nk_2'}/5^{nK})}{\log(1/5^{nK})} = 1 - \frac{k_2'}{K}\log_5 3;$$

furthermore, the partition zeta functions are

$$\zeta_{\mathfrak{P}}^{\psi}(\alpha(\mathbf{k}), s) = \sum_{n=1}^{\infty}\binom{nK}{nk_2'}2^{n(K-k_2')}5^{-nKs},$$

and the abscissa of convergence function is

$$f_{\mathfrak{P}}^{\psi}(\alpha(\mathbf{k})) = -x\log_5(x) - (1-x)\log_5(1-x) + (1-x)\log_5 2,$$

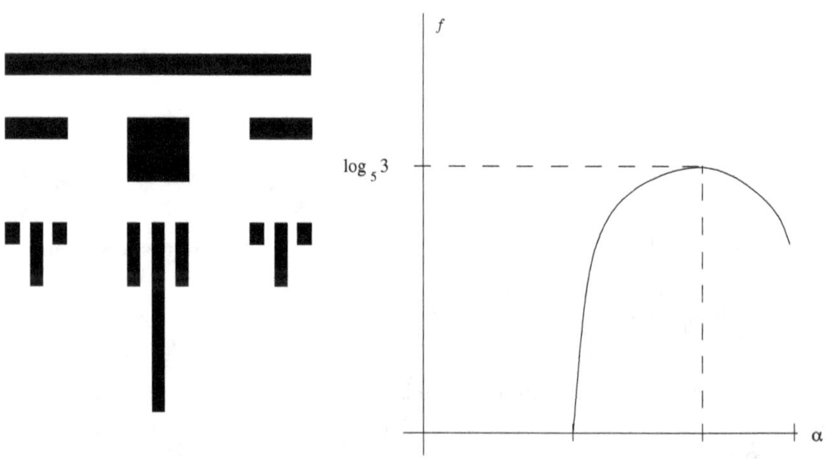

Figure 12.6 The first few stages in the construction of the trident measure ψ and its natural sequence of partitions $\mathfrak{P}$ (left), along with the graph of its abscissa of convergence function $f = f_{\mathfrak{P}}^{\psi}(\alpha)$ as a function of regularity α (right).

where $x = k_2'/K = \frac{1-\alpha}{\log_5 3}$ and $\alpha(\mathbf{k}) = \alpha$. Thus,

$$\begin{aligned} f_{\mathfrak{P}}^{\psi}(\alpha(\mathbf{k})) = f_{\mathfrak{P}}^{\psi}(\alpha) = &-\left(\frac{1-\alpha}{\log_5 3}\right)\log_5\left(\frac{1-\alpha}{\log_5 3}\right) \\ &-\left(1-\frac{1-\alpha}{\log_5 3}\right)\log_5\left(1-\frac{1-\alpha}{\log_5 3}\right) \\ &+\left(1-\frac{1-\alpha}{\log_5 3}\right)\log_5 2. \end{aligned}$$

See Fig. 12.6 for the graph of the concave envelope of $f_{\mathfrak{P}}^{\psi}$. Note that $f_{\mathfrak{P}}^{\psi}(\alpha)$ behaves as described in Sec. 12.3.3, as expected.

Example 12.5.2 (*Recovery of Classical Results*) Some of the results obtained by R. Cawley and R. D. Mauldin in [11], partially reproduced in Sec. 12.3.3, along with the classical results obtained by A. S. Besicovitch in [9] and H. G. Eggleston in [13] and described in Remark 12.3.1, can be recovered in the setting of this section. For instance, see Example 1.6 on page 205 of [11]. Setting $N = 4, \mathbf{r} = (r, r, r, r)$, and $\mathbf{p} = (p_1', p_1', p_2', p_2')$ (hence $w = 2$) yields the desired recovery of this example. Another classical example which can be recovered in our setting is the binomial measure β_0 supported on the unit interval, as noted in Corollary 12.5.1. Indeed, setting $N = 2, \mathbf{r} = (1/2, 1/2)$ and $\mathbf{p} = (1/3, 2/3)$ (hence $w = 2$) yields the desired recovery in this instance.

Example 12.5.3 (*Monofractal Measures*) If there exists $D > 0$ such that $r_i^D = p_i$ for all $i \in \{1, \ldots, N\}$, then the resulting measure ρ is the natural Hausdorff measure of the underlying generalized Cantor set as discussed on page 201 of [11] and the end of Sec. 12.3.3. More specifically, this Cantor set, which coincides with the support of ρ, is the self-similar set defined by the IFS $\{S_i\}_{i=1}^N$. The primitive of ρ is defined by $\int_0^x (d\rho) = \rho([0, x])$ and its graph, at least in the usual case of the ternary Cantor set when $N = 2$, $\mathbf{r} = (1/3, 1/3)$, and $\mathbf{p} = (1/2, 1/2)$, is the well-known Devil's staircase (see, for instance, [3, Chap. 6]). (In other words, the primitive of ρ is the Cantor–Lebesgue function.) Recall that this function is nondecreasing and continuous on $[0, 1]$ with zero derivative almost everywhere (on the Cantor string, in fact), yet its range is the full interval $[0, 1]$. See [20, Sec. 12.2] for an investigation of the Devil's staircase and a discussion of a new notion of fractality based on the distribution of complex dimensions of fractal strings. Fittingly, the only regularity values attained by ρ on the intervals from its natural sequence of partitions are ∞ and D. Also, $f_{\mathfrak{P}}^{\rho}(D) = D$ and according to Definition 12.4.3, this abscissa of convergence function is trivially equal to zero for all other finite regularity values since no length stemming from $\mathfrak{P}$ has finite regularity value different from D. (Note that in the special case where $\mu = \rho$, as above, we have $D = \log_3 2$.)

The following example is the result of work done by Scott Roby in the Multifractal Analysis research project at California State University, Stanislaus in December 2010 and January 2011. The weighted IFS associated with this example satisfies the OSC, but the resulting regularity values are not distinct in the sense of hypothesis (**H**) from Theorem 12.5.1, nor are the conditions of Lemma 12.5.1 satisfied. Nonetheless, the attained regularity values and the family of partition zeta functions are fully determined. As a special case, the geometric zeta function of the Fibonacci string $\zeta_{\mathrm{Fib}}(s)$ is recovered. (See Sec. 12.2.3 above for the development of the Fibonacci string and for a more thorough analysis, see Sec. 2.3.2 of [20].)

Example 12.5.4 (*Recovery of the Fibonacci String*) Consider a wei-ghted IFS that satisfies the OSC where $\mathbf{r} = (1/2, 1/4, 1/10)$ and $\mathbf{p} = (1/2, 1/4, 1/4)$. Then the regularity values attained by the resulting measure μ with respect to its natural sequence of partitions $\mathfrak{P}$ are given by

$$\alpha(\mathbf{k}) = \alpha(k_1, k_2, k_3) = \frac{\log 2^{-(k_1+2k_2+2k_3)}}{\log\left(2^{-(k_1+2k_2)}10^{-k_3}\right)},$$

where $k_1, k_2, k_3 \in \mathbb{N}$ and are not all zero. Note that the vectors $\mathbf{k}_1 = (1, 0, 0)$ and $\mathbf{k}_2 = (0, 1, 0)$ both yield regularity value 1. Hence, the collection of attained regularity values are not distinct in the sense of hypothesis (**H**) from Theorem 12.5.1. In order to distinguish the regularity values in this case, set $M := k_1 + 2k_2$ and suppose $k_3 \neq 0$. Define $\alpha(M, k_3)$ by

$$\alpha(M, k_3) := \alpha(\mathbf{k}) = 1 + \frac{\log 2 - \log 5}{\frac{M}{k_3}\log 2 + \log 10}.$$

The values of $\alpha(M, k_3)$ are distinct when $\gcd(M, k_3) = 1$, which we assume to be the case for the remainder of this example. For a given regularity value $\alpha(M, k_3)$ and positive integer n, the corresponding

α-lengths are

$$l_n(\alpha(M, k_3)) := (2^{-M} 10^{-k_3})^n,$$

with multiplicities given by

$$m_n(\alpha(M, k_3)) := \sum_{i=0}^{\lfloor \frac{nM}{2} \rfloor} \binom{\lfloor \frac{nM+2nk_3+1}{2} + i \rfloor}{nM - 2\lfloor \frac{nM}{2} \rfloor + 2i, \lfloor \frac{nM}{2} \rfloor - i, nk_3},$$

where $\lfloor \cdot \rfloor$ is the floor (i.e., integer part) function. That is, for $x \in \mathbb{R}$, $\lfloor x \rfloor$ is the greatest integer such that $\lfloor x \rfloor \leq x$. The partition zeta functions are then given by

$$\zeta_{\mathfrak{P}}^{\mu}(\alpha(M, k_3), s) = \sum_{n=1}^{\infty} m_n(\alpha(M, k_3))(l_n(\alpha(M, k_3)))^s.$$

In the case where $M = 1$ and $k_3 = 0$, so that $\alpha(1, 0) = 1$, we recover the geometric zeta function of the Fibonacci string up to an additive constant (or multiplication by a nowhere vanishing entire function in terms of the closed forms of these functions). Indeed, in this case the lengths are given by $l_n(1) = 2^{-n}$ and the multiplicities are given by

$$m_n(1) = \sum_{i=0}^{\lfloor \frac{n}{2} \rfloor} \binom{\lfloor \frac{n+1}{2} + i \rfloor}{n - 2\lfloor \frac{n}{2} \rfloor + 2i} = F_{n+1},$$

where F_{n+1} is the $(n+1)$th Fibonacci number. Refer to Sec. 12.2.3 above and to [20, Sec. 2.3.2] for a discussion of the Fibonacci string and its geometric zeta function ζ_{Fib}. The partition zeta function is therefore given by

$$\zeta_{\mathfrak{P}}^{\mu}(1, s) = \sum_{n=1}^{\infty} F_{n+1} 2^{-ns} = \zeta_{\text{Fib}}(s) - 1 = \frac{2^{-s} + 4^{-s}}{1 - 2^{-s} - 4^{-s}},$$

where ζ_{Fib} is given by Eq. (12.7). Thus, the corresponding complex dimensions, in both the classic sense and with respect to the parameter 1, are given by (cf. Eq. (12.8))

$$\mathcal{D}_{\text{Fib}} = \mathcal{D}_{\mathfrak{P}}^{\mu}(1) = \{D + jzp \mid z \in \mathbb{Z}\} \cup \{-D + j(z + 1/2)p \mid z \in \mathbb{Z}\},$$

where $\phi = (1 + \sqrt{5})/2$ is the Golden Ratio, $D = \log_2 \phi$, and $p = 2\pi / \log 2$.

The only other case in this example for which the partition zeta function and the complex dimensions are explicitly known is when $M = 0$ and

$k_3 = 1$. In this case, the partition zeta function is

$$\zeta_{\mathfrak{P}}^{\mu}(\alpha(0,1), s) = \sum_{n=1}^{\infty} 10^{-ns}.$$

The complex dimensions with respect to the regularity value $\alpha(0,1)$ are therefore given by

$$\mathcal{D}_{\mathfrak{P}}^{\mu}(\alpha(0,1)) = \{jzp \mid z \in \mathbb{Z}\},$$

where $p = 2\pi/\log 10$.

The remaining complex dimensions with respect to an arbitrary regularity value α, and even the corresponding abscissae of convergence, have yet to be determined.

Remark 12.5.6 The next step in the development of the theory of complex dimensions for self-similar measures is to determine $\mathcal{D}_{\mathfrak{P}}^{\mu}(\alpha, W_\alpha)$ and $\mathcal{T}_{\mathfrak{P}}^{\mu}$, in the general situation considered in Sec. 12.5 (or at least for interesting classes of examples, such as the multinomial multifractal measures). The work of D. Essouabri and the second author in [65] should provide a solid foundation for such a pursuit. It suggests, in particular, that the theory of complex fractal dimensions developed in [19] and [20] (or [21]) eventually should be extended to apply to zeta functions that are viewed as analytic functions on Riemann surfaces (rather than just on suitable domains of the complex plane $\mathbb{C}$ or of the Riemann sphere $\mathbb{C}^* = \mathbb{C} \cup \{\infty\}$). In the present situation, the classic Riemann surface associated with the logarithm (or the square root) would be required; see [65], which is motivated in part by the earlier, less general, results obtained in [23, 24] and described in [21, Sec. 13.3.6].

12.6 Partition zeta functions of atomic measures

In this section, we investigate the properties of certain atomic measures which are not self-similar in the sense of Sec. 12.5. Let σ_1 be given by

$$\sigma_1 = \sum_{i=1}^{\infty} 3^{-i}\delta_{3^{-i}}$$

and let $\Omega_1 = (0,1)\backslash\{3^{-i}\}_{i=1}^{\infty}$ be the fractal string determined by the complement in $[0,1]$ of the support of σ_1 (less the point 1). Let Ω_2 be the open

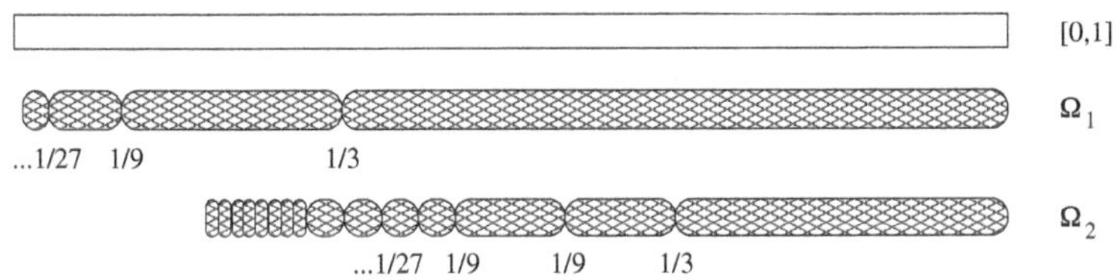

Figure 12.7 Approximation of the fractal strings Ω_1 and Ω_2.

subset of $[0, 1]$ obtained by placing disjoint open intervals with the lengths of the Cantor string $\mathcal{L}_{CS}$ end-to-end in nonincreasing order from right to left, with the single interval of length $1/3$ placed so that its right-endpoint is at 1 (see Fig. 12.7). Then, let σ_2 be the atomic measure supported on the left-endpoints of the fractal string Ω_2, where the left-endpoint of each distinct open interval has weight given by the length $l_n = 3^{-n}$ of said interval (see Fig. 12.7).

The sequence of distinct lengths $\mathcal{L}_1$ of the fractal string Ω_1 is given by $\mathcal{L}_1 = \{2 \cdot 3^{-n}\}_{n=1}^{\infty}$, where each length has multiplicity 1. Also, the sequence of distinct lengths $\mathcal{L}_2$ of the fractal string Ω_2 is exactly the same as the sequence of distinct lengths of the Cantor string. More specifically, $\mathcal{L}_2 = \mathcal{L}_{CS} = \{3^{-n}\}_{n=1}^{\infty}$, but where each length 3^{-n} has multiplicity 2^{n-1} (instead of 1). See Fig. 12.7 and Sec. 12.2.3.

In order to determine the corresponding partition zeta functions for σ_1 and σ_2, we must choose a suitable sequence of partitions. In the absence of naturally defined sequences of partitions for σ_1 and σ_2, throughout this section we take $\mathfrak{P}$ to be the sequence of partitions $\mathcal{P}_n$ of left-closed, right-open ternary intervals P_n^k of length 3^{-n} for $k \in \{1, \ldots, 3^n - 1\}$ and $P_n^{3^n} = [(3^n - 1)/3^n, 1]$, ordered from left to right by k. That is, for each $n \in \mathbb{N}^*$, the partition $\mathcal{P}_n$ is given by

$$\{[0, 1/3^n), [1/3^n, 2/3^n), \ldots, [(3^n - 2)/3^n, (3^n - 1)/3^n), [(3^n - 1)/3^n, 1]\}. \tag{12.39}$$

12.6.1 A full family of multifractal complex dimensions

As with the determination of the other partition zeta functions in this chapter, the most delicate part of the process in the case of the measures σ_1 and σ_2 and the sequence of partitions $\mathfrak{P}$ is to find and distinguish the nontrivial regularity values.

Lemma 12.6.1 *For the measure σ_1 and the sequence of partitions $\mathfrak{P}$ as given in Eq.* (12.39), *the distinct nontrivial regularity values have the following forms*: $1+\log_{3^n} 2, k_1/K$, *and* ∞, *where* $n, k_1, K \in \mathbb{N}^*, k_1 \leq K$, *and* $\gcd(k_1, K) = 1$.

Proof For each $n \in \mathbb{N}^*$, the leftmost interval P_n^1 of each partition $\mathcal{P}_n$ has regularity given by $A(P_n^1) = 1 + \log_{3^n} 2$. For $n, k_1, K \in \mathbb{N}^*, k_1 \leq K$, and $\gcd(k_1, K) = 1$, the intervals $P_{Kn}^{3^{k_1 n}}$ have regularity given by $A(P_{Kn}^{3^{k_1 n}}) = k_1/K$. No other interval P_n^k stemming from $\mathfrak{P}$ has mass, thus $A(P_n^k) = \infty$ for each of these intervals. See Fig. 12.8. $\square$

Remark 12.6.1 An immediate consequence of Lemma 12.6.1 is that the α-lengths of σ_1 for $\alpha = k_1/K$ are given by

$$\mathcal{L}_{\mathfrak{P}}^{\sigma_1}(k_1/K) = \{3^{-Kn} \mid 3^{-Kn} \text{ has multiplicity } 1, n \in \mathbb{N}^*\}; \tag{12.40}$$

see Fig. 12.8. The sequences $\mathcal{L}_{\mathfrak{P}}^{\sigma_1}(k_1/K)$ are strongly languid and trivially self-similar[5] (see Remark 12.2.3). This fact effects the forms of the explicit formulas for the counting functions of the α-lengths to be presented in Theorem 12.6.2.

Before stating and proving Theorem 12.6.2, we give the forms of the partition zeta functions, abscissa of convergence function, complex dimensions with respect to α, and tapestry of complex dimensions corresponding to the

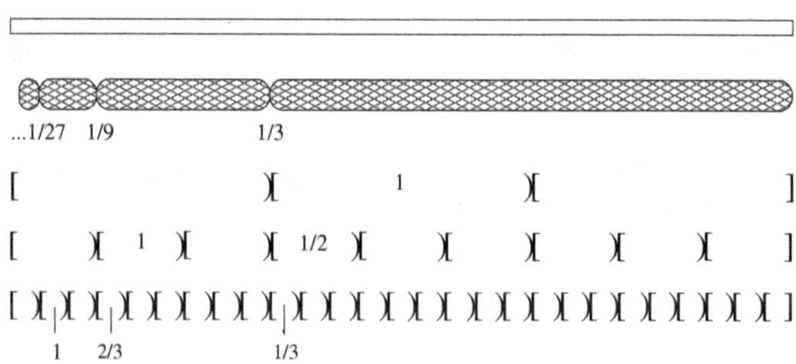

Figure 12.8 A breakdown of the regularity values attained by σ_1 with respect to the sequence of partitions $\mathfrak{P}$. The value inside each interval is its regularity. The leftmost blank intervals have regularity $\alpha = 1 + \log_{3^n} 2$ at each stage $n \in \mathbb{N}^*$. The remaining blank intervals have no mass and therefore have regularity $\alpha = \infty$.

[5]Strictly speaking, they are not self-similar since a single scaling ratio is involved and so the multiplicities are trivial.

nontrivial and finite regularity values α obtained by σ_1 with respect to $\mathfrak{P}$. At this point, the reader may wish to briefly review Sec. 12.4, specifically the definition of the complex dimensions with parameter α (Definition 12.4.4) and the definition of the tapestry of complex dimensions (Definition 12.4.5).

Theorem 12.6.1 *The partition zeta functions for the nontrivial and finite regularity values α obtained by σ_1 with respect to $\mathfrak{P}$ as in* Lemma 12.6.1 *are respectively given by*

$$\zeta_{\mathfrak{P}}^{\sigma_1}(1+\log_{3^n} 2, s) = 3^{-ns} \tag{12.41}$$

and

$$\zeta_{\mathfrak{P}}^{\sigma_1}(k_1/K, s) = \frac{3^{-Ks}}{1-3^{-Ks}}, \tag{12.42}$$

where $s \in \mathbb{C}, n, k_1, K \in \mathbb{N}^*, k_1 \leq K$, and $\gcd(k_1, K) = 1$.

Furthermore, the abscissa of convergence function is given by

$$f_{\mathfrak{P}}^{\sigma_1}(\alpha) = 0, \quad \textit{for all } \alpha \in (-\infty, \infty). \tag{12.43}$$

Moreover, for regularity $\alpha = k_1/K$, the complex dimensions with respect to α are given by

$$\mathcal{D}_{\mathfrak{P}}^{\sigma_1}(k_1/K, \mathbb{C}) = \left\{ \frac{2\pi j z}{K \log 3} \,\middle|\, z \in \mathbb{Z} \right\}. \tag{12.44}$$

Lastly, the tapestry of complex dimensions $\mathcal{T}_{\mathfrak{P}}^{\sigma_1}$ is given by

$$\mathcal{T}_{\mathfrak{P}}^{\sigma_1} = \left\{ (\alpha, \omega) | \alpha = \frac{k_1}{K}, \omega \in \mathcal{D}_{\mathfrak{P}}^{\sigma_1}(k_1/K, \mathbb{C}), \textit{ where } k_1 \leq K, k_1, K \in \mathbb{N}^* \right\}, \tag{12.45}$$

as portrayed in Fig. 12.9.

Proof Equations (12.41) and (12.42) follow immediately from Lemma 12.6.1 and Remark 12.6.1. Note that Eq. (12.42), explicitly

$$\zeta_{\mathfrak{P}}^{\sigma_1}(k_1/K, s) = \sum_{n=1}^{\infty} 3^{-Kns} = \frac{3^{-Ks}}{1-3^{-Ks}},$$

is first obtained by assuming that $\mathrm{Re}(s) > 0$, so that the geometric series involved converges. However, the end result clearly holds for all $s \in \mathbb{C}$, as can be seen upon meromorphic continuation. Hence, Eq. (12.42) holds for all $s \in \mathbb{C}$.

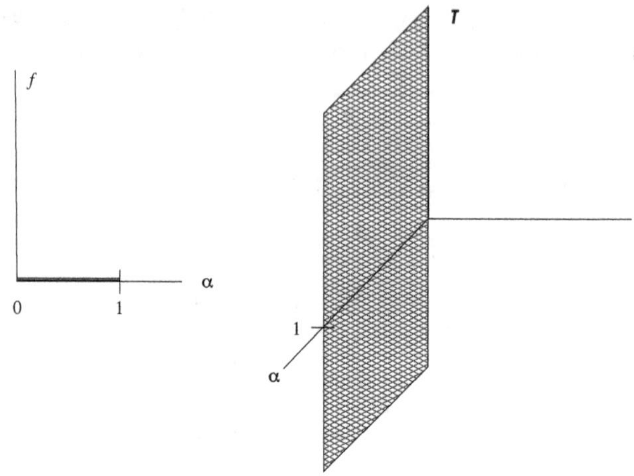

Figure 12.9 The abscissa of convergence function $f = f_{\mathfrak{P}}^{\sigma_1}(\alpha) = 0$ on $(0,1]$ (left) and tapestry of complex dimensions $\mathcal{T} = \mathcal{T}_{\mathfrak{P}}^{\sigma_1}$ of σ_1 with respect to $\mathfrak{P}$ (right).

Observe that the partition zeta functions with regularity $\alpha = 1+\log_{3^n} 2$ for all $n \in \mathbb{N}^*$ have no poles and hence no complex dimensions, so their abscissae of convergence are trivially equal to $-\infty$; and hence, according to Definition 12.4.3, we have $f_{\mathfrak{P}}^{\sigma_1}(\alpha) := \max\{0, -\infty\} = 0$ for all these regularity values α. For those partition zeta functions with regularity k_1/K, the abscissa of convergence is also zero since $s = 0$ is the unique real-valued solution to the equation $1 - 3^{-Ks} = 0$. Therefore, $f_{\mathfrak{P}}^{\sigma_1}(\alpha) = 0$ for all $\alpha \in (-\infty, \infty)$; see Fig. 12.9.

The expressions for the partitions zeta functions with regularity $\alpha = k_1/K$ have numerators which never vanish. Therefore, we deduce that the complex dimensions with parameter α are given by Eq. (12.44). In turn, in light of Definition 12.4.5, the complex dimensions with respect to α immediately yield the tapestry of complex dimensions given in Eq. (12.45). □

Remark 12.6.2 The only other regularity value attained by σ_1 with respect to $\mathfrak{P}$ is $\alpha = \infty$; see Fig. 12.8 along with Sec. 12.3.5. However, the intervals with such regularity are so numerous that the resulting partition zeta function $\zeta_{\mathfrak{P}}^{\sigma_1}(\infty, s)$ is divergent everywhere. A similar remark holds for the measure σ_2 discussed below in Sec. 12.6.2.

Remark 12.6.3 The complex dimensions $\mathcal{D}_{\mathfrak{P}}^{\sigma_1}(k_1/K, \mathbb{C})$ and the tapestry $\mathcal{T}_{\mathfrak{P}}^{\sigma_1}$ are exactly the same as those obtained for the measure $\sigma = \sigma_1$

in Corollary 13.55 and Remark 13.56 of [21, § 13.3.5] (which was written in conjunction with the fourth author of this paper and describes joint work of the second and fourth authors). However, those results are obtained in the context provided by *multifractal zeta functions*, which we discuss briefly in Sec. 12.7. The multifractal zeta functions are examined in [21–24] and are defined therein by a measure, a regularity value, and a sequence of scales (instead of a sequence of partitions). The multifractal structure of atomic measures similar to σ_1 and σ_2 are considered in [38], but not in the context of partition or multifractal zeta functions.

Next, we give an explicit formula, expressed in terms of the underlying complex dimensions $\mathcal{D}_{\mathfrak{P}}^{\sigma_1}(\alpha, \mathbb{C})$, for the counting functions of the α-lengths of σ_1 with respect to $\mathfrak{P}$, as given by $\mathcal{L}_{\mathfrak{P}}^{\sigma_1}(\alpha)$ in Remark 12.6.1 above.

Theorem 12.6.2 (*Exact pointwise formula for the α-lengths of σ_1*) *For each regularity value $\alpha = k_1/K$, with $k_1, K \in \mathbb{N}^*, k_1 \leq K$, and* $\gcd(k_1, K) = 1$ (*as in Lemma* 12.6.1), *the counting function of the α-lengths of σ_1 with respect to $\mathfrak{P}$ satisfies*

$$N_{\mathfrak{P}}^{\sigma_1}(\alpha, x) = N_{\mathfrak{P}}^{\sigma_1}(k_1/K, x) = M = \frac{1}{K \log 3} \sum_{\omega \in \mathcal{D}_\alpha} \frac{x^\omega}{\omega}, \qquad (12.46)$$

where the formula holds pointwise for every $x > 1$, with $M := \lfloor \log_{3^K} x \rfloor$ and $\mathcal{D}_\alpha := \mathcal{D}_{\mathfrak{P}}^{\sigma_1}(k_1/K, \mathbb{C})$ as given in Eq. (12.44) *of Theorem* 12.6.1.

Proof In this proof, we must assume that the reader has some familiarity with the theory developed in [20]. Fix a regularity value $\alpha = k_1/K$ as given by Lemma 12.6.1. In light of Eq. (12.40), the fact that $N_{\mathfrak{P}}^{\sigma_1}(k_1/K, x) = M = \lfloor \log_{3^K} x \rfloor$ is immediate.

Next, we justify the explicit formula 12.46 for $N_{\mathfrak{P}}^{\sigma_1}(k_1/K, x)$. This result follows from [20, Theorem 5.14], the pointwise explicit formula without error term, applied at level $k = 1$ (in the terminology of [20]) to the zeta function

$$\zeta_{\mathfrak{P}}^{\sigma_1}(\alpha, s) - \zeta_{\mathfrak{P}}^{\sigma_1}(k_1/K, s) = \frac{3^{-Ks}}{1 - 3^{-Ks}}, \quad s \in \mathbb{C},$$

viewed as the "geometric" (or rather "scaling") zeta function of the generalized fractal string associated with the α-lengths of σ_1 (see Definition 12.4.1). More specifically, since 0 is a pole of $\zeta_{\mathfrak{P}}^{\sigma_1}(\alpha, \cdot)$ and with our present notation,

[20, Theorem 5.14] yields for all $x > A$ (with $A := 1$, as explained below):

$$N_{\mathfrak{P}}^{\sigma_1}(\alpha, x) = \sum_{\omega \in \mathcal{D}_\alpha} \operatorname{res}\left(\frac{x^s}{s}\zeta_{\mathfrak{P}}^{\sigma_1}(\alpha, s); s = \omega\right)$$

$$= \sum_{\omega \in \mathcal{D}_\alpha} \frac{x^\omega}{\omega} \operatorname{res}\left(\zeta_{\mathfrak{P}}^{\sigma_1}(\alpha, s); s = \omega\right).$$

Hence (12.46) follows since $\zeta_{\mathfrak{P}}^{\sigma_1}(\alpha, s)$ and $\mathcal{D}_\alpha := \mathcal{D}_{\mathfrak{P}}^{\sigma_1}(\alpha, \mathbb{C})$ are given by Eqs. (12.42) and (12.44), respectively, and consequently, for $\omega = 2\pi jz/(K \log 3)$ (with $z \in \mathbb{Z}$), we have

$$\operatorname{res}\left(\zeta_{\mathfrak{P}}^{\sigma_1}(\alpha, s); s = \omega\right) = \operatorname{res}\left(\frac{3^{-Ks}}{1 - 3^{-Ks}}; s = \omega\right)$$

$$= \frac{3^{-K\omega}}{(K \log 3)3^{-K\omega}} = \frac{1}{K \log 3},$$

independently of $z \in \mathbb{Z}$.

Note that the aforementioned explicit formula of [20] can be applied here because an elementary computation (entirely analogous to the one performed on page 189 of [20, §6.4]) shows that $\zeta_{\mathfrak{P}}^{\sigma_1}(\alpha, \cdot)$ is strongly languid (in the sense of [20, Definition 5.3]) of order $\kappa = 0 < 1$ and with constant $A = 1$. (Here, we use the notation κ and A employed in [20]; see especially [20, Sec. 5.3].) More specifically, with $A := 1$, we clearly have

$$\left|\frac{3^{-Ks}}{1 - 3^{-Ks}}\right| = \left|\frac{1}{1 - 3^{-Ks}}\right| \leq 1 = (A^{-1})^{-|\mathrm{Re}(s)|},$$

as $\mathrm{Re}(s) \to -\infty$.[6] Also, we have $W = \mathbb{C}$ in this case. This concludes the proof of Theorem 12.6.2. □

Remark 12.6.4 We leave as an exercise the verification that given the simple form of the sequence of α-lengths obtained in Remark 12.6.1, it is possible to recover Eq. (12.46) by a direct computation (also involving a conditionally convergent Fourier series, but no longer using [20, Theorem 5.14]). (In more complicated situations, however, we would have to use the exact explicit formula in [20, Theorem 5.14], or its counterpart with error term given in [20, Theorem 5.10], or even more generally, their distributional

---

[6]Strictly speaking, in the above inequality, 1 should be replaced by η, for any given $\eta > 1$.

analogues obtained in [20, Sec. 5.4].) We note that the computation would then resemble the one carried out in a related context for the Cantor string in [20, Sec. 1.1.2]. Alos, it is useful to observe that Eq. (12.46) can be equivalently rewritten as follows:

$$N_{\mathfrak{P}}^{\sigma_1}(\alpha, x) = g(u), \quad \text{with } u := \lfloor \log_{3^K} x \rfloor, \tag{12.47}$$

where g is the 1-periodic function given by the (conditionally) convergent Fourier series

$$g(u) := \frac{1}{2\pi j} \sum_{z \in \mathbb{Z}} \frac{e^{2\pi j z u}}{z}, \quad u \in \mathbb{R}. \tag{12.48}$$

Observe that the lack of positive real part in the complex dimensions $\mathcal{D}_{\mathfrak{P}}^{\sigma_1}(k_1/K, \mathbb{C})$ stems from the unit multiplicity of each corresponding distinct α-length in $\mathcal{L}_{\mathfrak{P}}^{\sigma_1}(k_1/K)$. In the case of σ_2, however, the multiplicities of the distinct α-lengths are integer powers of 2. This results in a nonconstant linear multifractal spectrum for σ_2, as described in the next section.

12.6.2 A nonconstant linear multifractal spectrum

The determination of the distinct nontrivial regularity values in the case of σ_2 with respect to $\mathfrak{P}$ is actually easier than that of σ_1.

Lemma 12.6.2 *For the measure σ_2 and the sequence of partitions $\mathfrak{P}$ given at the beginning of Sec. 12.6, the distinct nontrivial regularity values have the following forms: $\frac{\log(3^{-k_1})}{\log(3^{-K})} = \frac{k_1}{K}$, and ∞, where $k_1, K \in \mathbb{N}^*, k_1 \leq K$, and $\gcd(k_1, K) = 1$. See Fig. 12.10.*

Proof Each interval P_{Kn}^k which contains two or more point-masses has regularity 1, since $\sigma_2(P_{Kn}^k) = |P_{Kn}^k|$ for all such intervals. Furthermore, each interval of length 3^{-nK} which contains a single point-mass of the form 3^{-nk_1}, with $n \in \mathbb{N}^*$, has regularity

$$\alpha = \frac{\log(3^{-k_1})}{\log(3^{-K})} = \frac{k_1}{K}.$$

Finally, observe that no other interval stemming from $\mathfrak{P}$ has mass and hence, $\alpha = \infty$ for each of these intervals; see Fig. 12.10. □

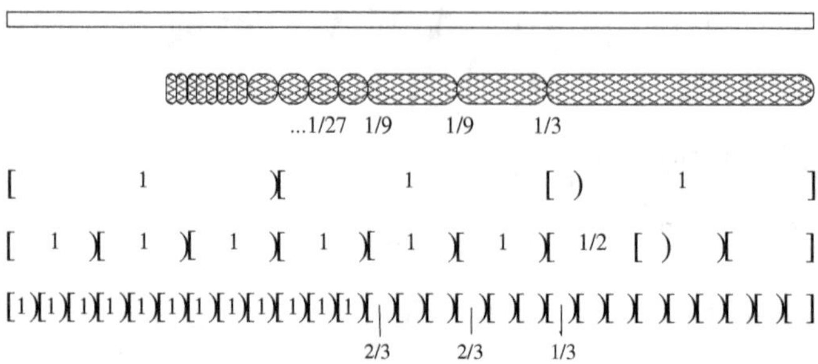

Figure 12.10 A breakdown of the regularity values attained by σ_2 with respect to the sequence of partitions $\mathfrak{P}$. The blank intervals have no mass and therefore have regularity $\alpha = \infty$.

Remark 12.6.5 The sequence of α-lengths $\mathcal{L}_{\mathfrak{P}}^{\sigma_2}(1)$ of σ_2 for $\alpha = k_1 = K = 1$ is given by

$$\mathcal{L}_{\mathfrak{P}}^{\sigma_2}(1) = \{3^{-n} \mid 3^{-n} \text{ has multiplicity } 3 \cdot 2^{n-1}, n \in \mathbb{N}^*\}. \tag{12.49}$$

Moreover, the sequence of α-lengths $\mathcal{L}_{\mathfrak{P}}^{\sigma_2}(k_1/K)$ of σ_2 for $\alpha = k_1/K$ with $k_1 < K$ is given by

$$\mathcal{L}_{\mathfrak{P}}^{\sigma_2}(k_1/K) = \{3^{-Kn} \mid 3^{-Kn} \text{ has multiplicity } 2^{k_1 n-1}, n \in \mathbb{N}^*\}. \tag{12.50}$$

As with the case of σ_1 above, the sequences $\mathcal{L}_{\mathfrak{P}}^{\sigma_2}(k_1/K)$ are also self-similar and strongly languid; see Remark 12.2.3. This effects the form of the counting function of the α-lengths presented in Theorem 12.6.4 below.

Before stating and proving Theorem 12.6.4, we give the forms of the partition zeta functions, abscissa of convergence function, complex dimensions with respect to α, and tapestry of complex dimensions corresponding to the nontrivial and finite regularity values α obtained by σ_2 with respect to $\mathfrak{P}$.

Theorem 12.6.3 *The partition zeta functions for the nontrivial and finite regularity values α obtained by σ_2 with respect to $\mathfrak{P}$ as in Lemma* 12.6.2 *are respectively given by*

$$\zeta_{\mathfrak{P}}^{\sigma_2}(1,s) = \frac{3 \cdot 3^{-s}}{1 - 2 \cdot 3^{-s}} \tag{12.51}$$

and

$$\zeta_{\mathfrak{P}}^{\sigma_2}(k_1/K,s) = \frac{2^{k_1-1} \cdot 3^{-Ks}}{1 - 2^{k_1} \cdot 3^{-Ks}}, \tag{12.52}$$

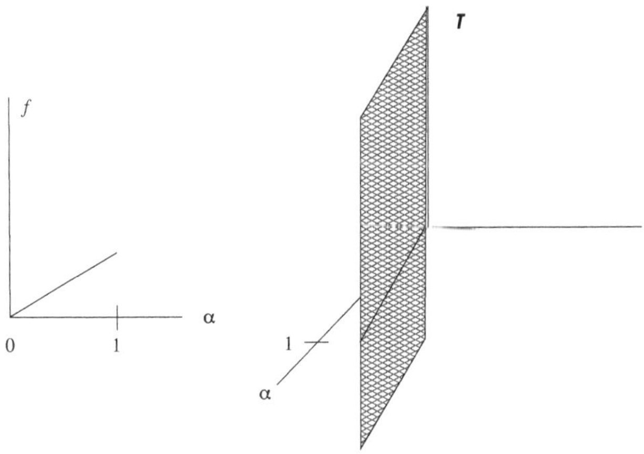

Figure 12.11 The abscissa of convergence function $f = f_{\mathfrak{P}}^{\sigma_2}(\alpha)$ (left) and tapestry $\mathcal{T} = \mathcal{T}_{\mathfrak{P}}^{\sigma_2}$ of complex dimensions of σ_2 with respect to $\mathfrak{P}$ (right).

where $s \in \mathbb{C}, k_1, K \in \mathbb{N}^*, k_1 < K$, *and* $\gcd(k_1, K) = 1$. (*Here and henceforth, when* $k_1 < K$, *we assume that* $\gcd(k_1, K) = 1$.)

Furthermore, the abscissa of convergence function is given by

$$f_{\mathfrak{P}}^{\sigma_2}(\alpha) = f_{\mathfrak{P}}^{\sigma_2}(k_1/K) = \frac{k_1}{K}\log_3 2, \tag{12.53}$$

where $\alpha = k_1/K$ *such that* $k_1 \leq K$ *with* $k_1, K \in \mathbb{N}^*$; *see Fig.* 12.11. (*Note that this equation also holds when* $\alpha = k_1 = K = 1$.)

Moreover, the set of complex dimensions $\mathcal{D}_{\mathfrak{P}}^{\sigma_2}(\alpha, \mathbb{C})$ *with respect to* $\alpha = k_1/K$, *where* $k_1 \leq K$ *with* $k_1, K \in \mathbb{N}^*$, *is*

$$\mathcal{D}_{\mathfrak{P}}^{\sigma_2}(k_1/K, \mathbb{C}) = \left\{ \frac{k_1}{K}\log_3 2 + \frac{2\pi j z}{K \log 3} \,\middle|\, z \in \mathbb{Z} \right\}. \tag{12.54}$$

(*Note that this equation still holds when* $\alpha = k_1 = K = 1$.)

Lastly, the tapestry of complex dimensions $\mathcal{T}_{\mathfrak{P}}^{\sigma_2}$ *is given by*

$$\mathcal{T}_{\mathfrak{P}}^{\sigma_2} = \left\{ (\alpha, \omega) | \alpha = \frac{k_1}{K}, \omega \in \mathcal{D}_{\mathfrak{P}}^{\sigma_2}(k_1/K, \mathbb{C}), \text{ where } k_1 \leq K, k_1, K \in \mathbb{N}^* \right\}, \tag{12.55}$$

as portrayed in Fig. 12.11.

Proof This proof is very similar to that of Theorem 12.6.1. Indeed, in light of Lemma 12.6.2, Eqs. (12.51) and (12.52) follow from Eqs. (12.49) and 12.50, respectively, first by summing a geometric series (for

$\mathrm{Re}(s) > \frac{k_1}{K}\log_3 2$) and then by meromorphically continuing the resulting expressions to all of $\mathbb{C}$. We deduce at once that the abscissa of convergence function is given by Eq. (12.53).

Finally, Eq. (12.54) (and then Eq. (12.55)) follows immediately from Eqs. (12.51) and (12.52). □

Remark 12.6.6 The concave envelope $\hat{f}_{\mathfrak{P}}^{\sigma_2}$ of $f_{\mathfrak{P}}^{\sigma_2}$ on $[0,1]$ is a nonconstant linear multifractal spectrum for the measure σ_2. Indeed, for every $t \in [0,1]$, we have

$$\hat{f}_{\mathfrak{P}}^{\sigma_2}(t) = t\log_3 2. \tag{12.56}$$

Note that, unlike for the results of Theorem 12.5.1, each nonzero value $\hat{f}_{\mathfrak{P}}^{\sigma_2}(t)$ cannot be equal to the Hausdorff dimension of some subset of the support of σ_2. Indeed, the support of σ_2 is the countable set $\partial\Omega_2$, which has Hausdorff dimension equal to zero.

Next, we close this section by giving an explicit formula, expressed in terms of the underlying complex dimensions $\mathcal{D}_{\mathfrak{P}}^{\sigma_2}(\alpha,\mathbb{C})$, for the counting functions of the α-lengths of σ_2 with respect to $\mathfrak{P}$, as given by $\mathcal{L}_{\mathfrak{P}}^{\sigma_2}(\alpha)$ in Remark 12.6.5.

Theorem 12.6.4 (*Exact pointwise formula for the α-lengths of σ_2*) *For each nontrivial regularity value of σ_2 with respect to $\mathfrak{P}$ given by Lemma* 12.6.2, *the counting function of the α-lengths is as follows:*

1. ($\alpha = k_1 = K = 1$). *For the regularity value $\alpha = 1$, we have*

$$N_{\mathfrak{P}}^{\sigma_2}(1,x) = 3\cdot(2^M - 1) = \frac{3}{2\log 3}\sum_{\omega\in\mathcal{D}_1}\frac{x^\omega}{\omega} - 3, \tag{12.57}$$

where this formula holds pointwise for every $x > 1$, with $M = \lfloor \log_3 x\rfloor$ and $\mathcal{D}_1 := \mathcal{D}_{\mathfrak{P}}^{\sigma_2}(1,\mathbb{C})$ as given in Eq. (12.54).

2. ($\alpha = k_1/K$). *For the regularity value $\alpha = k_1/K$ such that $k_1 < K$ with $k_1, K \in \mathbb{N}^*$ and* $\gcd(k_1,K) = 1$ (*as in Theorem* 12.6.3), *we have*

$$\begin{aligned} N_{\mathfrak{P}}^{\sigma_2}(\alpha,x) = N_{\mathfrak{P}}^{\sigma_2}(k_1/K,x) &= \frac{2^{k_1-1}(2^{k_1M}-1)}{2^{k_1}-1} \\ &= \frac{1}{2\log 3^K}\sum_{\omega\in\mathcal{D}_\alpha}\frac{x^\omega}{\omega} + \frac{2^{k_1-1}}{1-2^{k_1}}, \end{aligned} \tag{12.58}$$

where this formula holds pointwise for every $x > 1$, with $M := \lfloor \log_{3^K} x\rfloor$ (*as in Theorem* 12.6.2) *and $\mathcal{D}_\alpha := \mathcal{D}_{\mathfrak{P}}^{\sigma_2}(\alpha,\mathbb{C})$ as given in Eq.* (12.54).

Proof The proof parallels that of Theorem 12.6.2 and therefore follows from [20, Theorem 5.14] by showing (as on page 189 of [20, Sec. 6.4]) that for each given regularity α, $\zeta_{\mathfrak{P}}^{\sigma_2}(\alpha,\cdot)$ is strongly languid of order $\kappa = 0$ and with constant $A = A_\alpha = 1$. This last conclusion follows from the estimate

$$\left|\zeta_{\mathfrak{P}}^{\sigma_2}(\alpha,s)\right| \ll (A^{-1})^{-|\mathrm{Re}(s)|},$$

as $\mathrm{Re}(s) \to -\infty$, where $A = A_\alpha$ is given respectively by $A := 3^{-1}(3^{-1})^{-1} = 1$ when $\alpha = 1$ (as in Case **1**) and by $A := 3^{-K}(3^{-K})^{-1} = 1$ with $\alpha = k_1/K$ (as in Case **2**).

Moreover, since in either Case **1** or **2** of the theorem, 0 does not belong to $\mathcal{D}_\alpha$, the exact explicit formula of [20, Theorem 5.14] (applied at level $k = 1$) yields for all $x > A_\alpha = 1$:

$$\begin{aligned} N_{\mathfrak{P}}^{\sigma_2}(\alpha,x) &= \sum_{\omega\in\mathcal{D}_\alpha} \mathrm{res}\left(\frac{x^s}{s}\zeta_{\mathfrak{P}}^{\sigma_2}(\alpha,s); s=\omega\right) + \zeta_{\mathfrak{P}}^{\sigma_2}(\alpha,0) \\ &= \sum_{\omega\in\mathcal{D}_\alpha} \frac{x^\omega}{\omega}\mathrm{res}\left(\zeta_{\mathfrak{P}}^{\sigma_2}(\alpha,s); s=\omega\right) + \zeta_{\mathfrak{P}}^{\sigma_2}(\alpha,0). \end{aligned}$$

Consequently, the result follows since (in light of Theorem 12.6.3) an elementary computation shows that in Case **1** or **2**, respectively, we have for every $\omega \in \mathcal{D}_\alpha$:

$$\zeta_{\mathfrak{P}}^{\sigma_2}(\alpha,0) = -3, \; \mathrm{res}\big(\zeta_{\mathfrak{P}}^{\sigma_2}(\alpha,s); s=\omega\big) = \frac{3}{2\log 3},$$

while

$$\zeta_{\mathfrak{P}}^{\sigma_2}(\alpha,0) = \frac{2^{k_1-1}}{1-2^{k_1}}, \; \mathrm{res}\big(\zeta_{\mathfrak{P}}^{\sigma_2}(\alpha,s); s=\omega\big) = \frac{1}{2K\log 3}. \qquad \square$$

Remark 12.6.7 A comment completely analogous to the one made in Remark 12.6.4 (for the measure σ_1) applies to the measure σ_2.

In light of Theorem 12.6.3, we deduce at once the following result from Theorem 12.6.4:

Corollary 12.6.1 *The expression for the counting functions for the α-lengths can be rewritten as follows, in Case* **1** *($\alpha = 1$) or Case* **2**

($\alpha = k_1/K$, *with* $k_1 < K$ *and* $\gcd(k_1, K) = 1$) *of Theorem* 12.6.4:

$$N_{\mathfrak{P}}^{\sigma_2}(1,x) + 3 = \frac{3x^{\log_3 2}}{2\log 3} \sum_{z\in\mathbb{Z}} \frac{x^{\frac{2\pi jz}{\log 3}}}{\log_3 2 + \frac{2\pi jz}{\log 3}}$$

$$= x^{\log_3 2} G_1(\log_3 x)$$

$$= x^{f_{\mathfrak{P}}^{\sigma_2}(1)} G_1(\log_3 x), \tag{12.59}$$

and (*for* $\alpha = k_1/K$, *as in Case* **2**)

$$N_{\mathfrak{P}}^{\sigma_2}(\alpha,x) + \frac{2^{k_1-1}}{2^{k_1}-1} = \frac{x^{k_1 \log_{3^K} 2}}{2\log 3^K} \sum_{z\in\mathbb{Z}} \frac{x^{\frac{2\pi jz}{K\log 3}}}{k_1 \log_{3^K} 2 + \frac{2\pi jz}{K\log 3}}$$

$$= x^{k_1 \log_{3^K} 2} G_\alpha(\log_{3^K} x)$$

$$= x^{f_{\mathfrak{P}}^{\sigma_2}(\alpha)} G_\alpha(\log_{3^K} x), \tag{12.60}$$

where G_1 (*in Case* **1**) *or* G_α (*in Case* **2**) *is the* 1*-periodic function given by the conditionally convergent Fourier series*

$$G_1(u) := \frac{3}{2\log 3} \sum_{z\in\mathbb{Z}} \frac{e^{2\pi jzu}}{\log_3 2 + \frac{2\pi jz}{\log 3}}, \qquad u \in \mathbb{R}, \tag{12.61}$$

and (*for* $\alpha = k_1/K$, *as in Case* **2**)

$$G_\alpha(u) := \frac{1}{2\log 3^K} \sum_{z\in\mathbb{Z}} \frac{e^{2\pi jzu}}{\log_{3^K} 2 + \frac{2\pi jz}{\log 3^K}}, \qquad u \in \mathbb{R}. \tag{12.62}$$

Most of the key results of Sec. 12.6.2, namely, Lemma 12.6.2, Theorems 12.6.3 and 12.6.4, readily extend to cases involving slightly more general forms of the Cantor string. Such generalized results are stated here with less formality than in the previous sections and without proof since they follow those presented above, mutatis mutandis.

Example 12.6.1 (Generalization of the measure σ_2 Let $m \in \mathbb{N}$ such that $m \geq 2$ and let $\lambda = (2m-1)^{-1}$. Using this m and λ, construct the measure σ in the way σ_2 is constructed. The multiplicity of the initial intervals is $m-1$ and each of these intervals has length λ. (We note that $m = 2$ and hence $\lambda = 1/3$ in the following development would allow us to recover the results from Sec. 12.6.2.)

The sequence of partitions $\mathfrak{P}$ in this setting are partitions $\mathcal{P}_n$ which split the unit interval into disjoint subintervals of length λ^n. The nontrivial and finite regularity values α attained by σ with respect to $\mathfrak{P}$ are exactly the same as those attained in the case of σ_2. That is,

$$\alpha(k_1, K) = \frac{\log(\lambda^{k_1})}{\log(\lambda^K)} = \frac{k_1}{K},$$

and again we take $\alpha = k_1/K$ such that $k_1 \leq K$ with $k_1, K \in \mathbb{N}^*$ (and $\gcd(k_1, K) = 1$ when $k_1 < K$). The sequence of α-lengths $\mathcal{L}^{\sigma}_{\mathfrak{P}}(1)$ for $\alpha = k_1 = K = 1$ is given by

$$\mathcal{L}^{\sigma}_{\mathfrak{P}}(1) = \{\lambda^n \mid \lambda^n \text{ has multiplicity } \lambda^{-1}(m-1)m^{n-1}, n \in \mathbb{N}^*\}.$$

The sequence of α-lengths $\mathcal{L}^{\sigma}_{\mathfrak{P}}(k_1/K)$ for $\alpha = k_1/K$ with $k_1 < K$ is given by

$$\mathcal{L}^{\sigma}_{\mathfrak{P}}(k_1/K) = \{\lambda^{Kn} \mid \lambda^{Kn} \text{ has multiplicity } (m-1)m^{k_1 n-1}, n \in \mathbb{N}^*\}.$$

Hence, the partition zeta functions are given by

$$\zeta^{\sigma}_{\mathfrak{P}}(1, s) = \frac{(m-1)\lambda^{-1}}{m} \sum_{n=1}^{\infty} (m \cdot \lambda^s)^n = \frac{(m-1)\lambda^{s-1}}{1 - m\lambda^s}$$

and

$$\zeta^{\sigma}_{\mathfrak{P}}(k_1/K, s) = \frac{(m-1)}{m} \sum_{n=1}^{\infty} (m^{k_1} \cdot \lambda^{Ks})^n = \frac{(m-1)m^{k_1-1}\lambda^{Ks}}{1 - m^{k_1}\lambda^{Ks}},$$

where $s \in \mathbb{C}, k_1, K \in \mathbb{N}^*, k_1 < K$, and $\gcd(k_1, K) = 1$.

As a result, the abscissa of convergence function is given by

$$f^{\sigma}_{\mathfrak{P}}(\alpha) = f^{\sigma}_{\mathfrak{P}}(k_1/K) = \frac{k_1}{K} \log_{\lambda^{-1}} m,$$

where $\alpha = k_1/K$ such that $k_1 \leq K$ with $k_1, K \in \mathbb{N}^*$. Note that this equation also holds when $\alpha = k_1 = K = 1$. Furthermore, observe that $\log_{\lambda^{-1}} m > 0$ since $\lambda^{-1} = 2m - 1$ and $m \geq 2$.

Moreover, the set of complex dimensions $\mathcal{D}^{\sigma}_{\mathfrak{P}}(\alpha, \mathbb{C})$ with respect to $\alpha = k_1/K$, where $k_1 \leq K$ with $k_1, K \in \mathbb{N}^*$, is

$$\mathcal{D}^{\sigma}_{\mathfrak{P}}(k_1/K, \mathbb{C}) = \left\{ \frac{k_1}{K} \log_{\lambda^{-1}} m + \frac{2\pi j z}{K \log \lambda^{-1}} \,\middle|\, z \in \mathbb{Z} \right\}.$$

Note that this equation also holds when $\alpha = k_1 = K = 1$.

It follows that the tapestry of complex dimensions $\mathcal{T}_{\mathfrak{P}}^{\sigma}$ is given by

$$\mathcal{T}_{\mathfrak{P}}^{\sigma} = \left\{ (\alpha, \omega) | \alpha = \frac{k_1}{K}, \omega \in \mathcal{D}_{\mathfrak{P}}^{\sigma}(k_1/K, \mathbb{C}), \text{ where } k_1 \leq K, k_1, K \in \mathbb{N}^* \right\}.$$

Finally, we leave as an exercise obtaining the counterpart of Theorem 12.6.4 (the explicit formula for the counting function of the α-lengths), using the same line of reasoning as in the proof of that theorem.

The next section concludes the chapter with a brief description of natural questions and avenues of research provided by the approach to multifractal analysis via zeta functions adopted in this work.

12.7 Conclusion

The determination of the meromorphic continuation (or some other appropriate extension) of the partition zeta functions will be addressed in the near future. Work in this direction has already begun by the second author and D. Essouabri in [65]. Upon a suitable change of variable, the results of such an investigation will provide the poles, and hence complex dimensions, for this family of self-similar multifractal measures. This leads naturally to the search for an understanding of multifractal objects in more general settings, specifically those with non-multiplicative construction and properties. In the long term, motivated by [22–24] and the theory of complex dimensions in [19, 20] (also with consideration of the work done by J. Lévy Véhel and F. Mendivil in [37]), one may wish to investigate the physical or geometric oscillations of multiplicative and non-multiplicative multifractal objects in geometric, spectral and dynamical settings, as was done with fractal strings by way of their complex dimensions. (See, for example, [19–21, 40, 41], along with the relevant references mentioned in the introduction.) This work, [37, 65], as well as the exposition of some of the aspects of [23] given in [21, §13.3], should provide a nice foundation for such a theory of complex dimensions for multifractals.

A recent predecessor of this work is [22], by M. L. Lapidus, J. Lévy-Véhel and J. A. Rock, where certain Dirichlet series were introduced and used in order to study some geometric properties of fractal strings which are not accounted for in the theory developed in [19, 20]. The intent of the definition of the *multifractal zeta functions* from [22, 24], however, was to extend the techniques used in the theory of complex dimensions of fractal strings to multifractal analysis in some way. An elaboration on the difficulties of

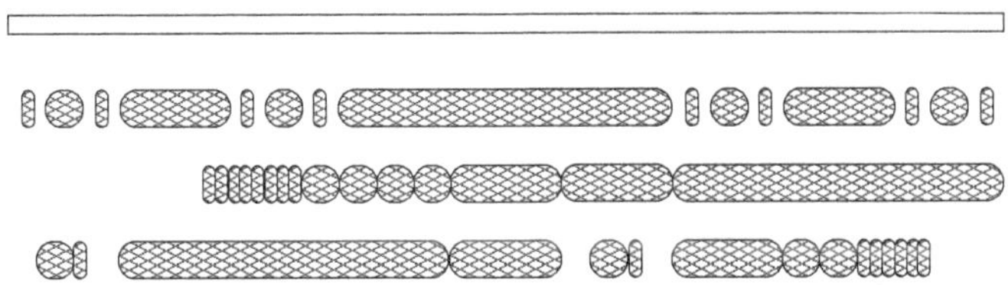

Figure 12.12 Approximation of three fractal strings with the same sequence of lengths $\mathcal{L}$, specifically the lengths of the Cantor string Ω_{CS}. The boundary of each fractal string has the same Minkowski dimension, but the Hausdorff dimensions differ.

using these multifractal zeta functions to this end is provided in [24], where the primary object study of this work, the *partition zeta function*, is first introduced.

In [22–24, 64], connections between the Hausdorff dimension of related fractal sets and the *topological zeta function* are established and examined. Specifically, in [64], building upon some examples in [22], certain collections of fractal strings Ω with identical sequence of lengths $\mathcal{L}$ are shown to have identical Minkowski dimension but varying Hausdorff dimension (see Fig. 12.12). The values for the Hausdorff dimension are computed, respectively, as the abscissa of convergence of the topological zeta function.

Another interesting extension of our results could lie within the investigation of self-similar measures constructed with weighted IFSs which do not satisfy the OSC. Such an investigation began with an example[7] developed by Scott Roby which was based in no small part upon the results pertaining to second-order identities from [63]. However, a full determination of the partition zeta functions in this setting has yet to be discovered.

Bibliography

[1] C. J. G. Evertsz, B. B. Mandelbrot, in [7], pp. 921–953.
[2] K. Falconer, *Fractal Geometry — Mathematical foundations and applications,* 2nd edn. (Wiley, Chichester, 2003).
[3] J. Feder, *Fractals* (Plenum, New York, 1988).
[4] B. B. Mandelbrot, *J. Fluid. Mech.* **62** (1974) 331.

[7] This example was presented by Scott Roby in the poster titled *Multifractal analysis of a measure when the open set condition is not satisfied* in the MAA Undergraduate Poster Session at the 2011 Joint Mathematics Meetings in New Orleans.

[5] B. B. Mandelbrot, *Multifractals and 1/f Noise: Wild self-affinity in physics* (Springer-Verlag, New York, 1999).

[6] G. Parisi, U. Frisch, in *Int. Sch. "Enrico Fermi", Course 88*, M. Ghil, ed. (North-Holland, Amsterdam, 1985), pp. 84–88.

[7] H.-O. Peitgen, H. Jürgens D. Saupe, *Chaos and Fractals: New frontiers of science* (Springer-Verlag, Berlin, 1992).

[8] M. R. Schroeder, *Fractals, Chaos, Power Laws: Minutes from an infinite paradise* (Freeman, New York, 1991).

[9] A. S. Besicovitch, *Math. Ann.* **110** (1934) 321.

[10] G. Brown, G. Michon J. Peyrière, *J. Statist. Phys.* **66** (1992) 775.

[11] R. Cawley R. D. Mauldin, *Adv. Math.* **92** (1992) 196.

[12] F. M. Dekking, W. X. Li, *Nonlinearity* **16** (2003) 1.

[13] H. G. Eggleston, *Quart. J. Math. Oxford* **20** (1949) 31.

[14] S. Graf, R. D. Mauldin, S. C. Williams, *Mem. Amer. Math. Soc., No. 381* **71** (1988) 1.

[15] W. Li, L. Olsen, Z. Wen, *Aequationes Math.* **77** (2009) 171.

[16] L. Olsen, *Adv. Math.* **116** (1996) 82.

[17] L. Olsen, in *Fractal Geometry and Stochastics II*, C. Bandt, S. Graf, M. Zähle, eds. (Birkhäuser, Basel, 2000), pp. 3–37.

[18] L. Olsen, *Math. Nachr.* **282** (2009) 1461.

[19] M. L. Lapidus, M. van Frankenhuijsen, *Fractal Geometry and Number Theory: Complex dimensions of fractal strings and zeros of zeta functions* (Birkhäuser, Boston, 2000).

[20] M. L. Lapidus, M. van Frankenhuijsen, *Fractal Geometry, Complex Dimensions and Zeta Functions: Geometry and spectra of fractal strings*, Research Monographs in Mathematics (Springer, New York, 2006).

[21] M. L. Lapidus, M. van Frankenhuijsen, *Fractal Geometry, Complex Dimensions and Zeta Functions: Geometry and spectra of fractal strings,* 2nd rev. and enl. edition (of the 2006 edition, [20]), Research Monographs in Mathematics (Springer, New York, 2013).

[22] M. L. Lapidus, J. Lévy-Véhel, J. A. Rock, *Lett. Math. Phys.* **88** (2009) 101.

[23] M. L. Lapidus, J. A. Rock, *Complex Variables and Elliptic Equations* **54** (2009) 545.

[24] J. A. Rock, *Ph.D. Dissertation*, Univ. Calif., Riverside, 2007.

[25] M. Arbeiter, N. Patzschke, *Math. Nachr.* **181** (1996) 5.

[26] A. S. Besicovitch, *Proc. Cambridge Philos. Soc.* **42** (1946) 1.

[27] D. J. Feng, K.-S. Lau, *J. Math. Pures Appl.* **92** (2000) 407.

[28] M. L. Lapidus, M. van Frankenhuijsen, eds., *Fractal Geometry and Applications: A Jubilee of Benoît Mandelbrot*, Proc. Sympos. Pure Math., Vol. 72, Parts 1 & 2 (Amer. Math. Soc., Providence, 2004).

[29] K. S. Lau, S. M. Ngai, *Studia Math.* **131** (1998) 225.

[30] W. Min, W. Yahao, *Chaos, Solitons, Fractals* **42** (2009) 2779.

[31] G. A. Edgar, R. D. Mauldin, *Proc. London Math. Soc.* **65** (1992) 604.

[32] R. S. Ellis, *Ann. Prob.* **12** (1984) 1.

[33] S. Jaffard, *SIAM J. Math. Anal.* **28** (1997) 994.

[34] S. Jaffard, *J. Math. Phys.* **38** (1998) 4129.
[35] S. Jaffard, *Probab. Theory Related Fields* **114** (1999) 207.
[36] S. Jaffard, Y. Meyer, *Mem. Amer. Math. Soc.*, No. 587, **123** (1996) 1.
[37] J. Lévy-Véhel, F.Mendivil, *Nonlinearity* **24** (2011) 259.
[38] J. Lévy-Véhel, C. Tricot, in *Fractal Geometry and Stochastics III*, C. Bandt, U. Mosco, M. Zähle, eds. (Birkhäuser, Basel, 2004), pp. 23–42.
[39] J. Lévy-Véhel, R. Vojak, *Adv. Appl. Math.* **20** (1998) 1.
[40] M. L. Lapidus, M. van Frankenhuijsen, in *Spectral Problems in Geometry and Arithmetic*, T. Branson, ed., Contemp. Math., Vol. 237 (Amer. Math. Soc., Providence, 1999), pp. 87–105.
[41] M. L. Lapidus, M. van Frankenhuijsen, *J. Experimental Math.* **12** (2003) 41.
[42] B. M. Hambly, M. L. Lapidus, *Trans. Amer. Math. Soc.* **358** (2006) 285.
[43] H. Herichi, M. L. Lapidus, *J. Phys. A: Math. Theor.* **45** (2012) 374005, 23 pp. Also: *Contemp. Math.*, Vol. 600 (Amer. Math. Soc., Providence, 2013), pp. 51–89.
[44] H. Herichi, M. L. Lapidus, *Fractal Strings, Quantized Number Theory and the Riemann Hypothesis: From infinitesimal shifts and spectral operators to phase transitions and universality*, research monograph, preprint, 2014 (approx. 200 pages).
[45] M. L. Lapidus, *In Search of the Riemann Zeros: Strings, fractal membranes and noncommutative spacetimes* (Amer. Math. Soc., Providence, 2008).
[46] M. L. Lapidus, H. Lu, *J. Fixed Point Theory Appl.* **3** (2008) 181.
[47] M. L. Lapidus, H. Lu, *p-Adic Numbers, Ultrametric Analysis and Applications* **1** (2009) 167.
[48] M. L. Lapidus, H. Lu, in *Advances in Non-Archimedean Analysis*, J. Araujo, B. Diarra, A. Escassut, eds., Contemp. Math., Vol. 551 (Amer. Math. Soc., Providence, 2011), pp. 163–206.
[49] M. L. Lapidus, H. Lu, M. van Frankenhuijsen, *Explicit tube formulas for p-adic fractal strings and nonarchimedean self-similar strings*, preprint, 2014. Also: *Contemp. Math.*, Vol. 600 (Amer. Math. Soc., Providence, 2013), pp. 185–203.
[50] M. L. Lapidus E. P. J. Pearse, *J. London Math. Soc.* **74** (2006) 397.
[51] M. L. Lapidus, E. P. J. Pearse, in *Analysis on Graphs and its Applications* P. Exner, *et al.*, eds., Proc. Sympos. Pure Math., Vol. 77 (Amer. Math. Soc., Providence, 2008), pp. 211–230.
[52] M. L. Lapidus, E. P. J. Pearse, *Acta Applicandae Mathematicae* **112** (2010) 91.
[53] M. L. Lapidus, E. P. J. Pearse, S. Winter, *Adv. Math.* **227** (2011) 1349.
[54] C. Q. He, M. L. Lapidus, *Mem. Amer. Math. Soc.*, No. 608, **127** (1997) 1.
[55] M. L. Lapidus, *Trans. Amer. Math. Soc.* **325** (1991) 465.
[56] M. L. Lapidus, in *Differential Equations and Mathematical Physics* C. Bennewitz, ed. (Academic Press, New York, 1992), pp. 151–182.

[57] M. L. Lapidus, in *Ordinary and Partial Differential Equations* B. D. Sleeman, R. J. Jarvis, eds., Research Notes in Math. Series, Vol. 289 (Longman Scientific and Technical, London, 1993), pp. 126–209.
[58] M. L. Lapidus, C. Pomerance, *Proc. London Math. Soc.* (*3*) **66** (1993) 41.
[59] P. A. P. Moran, *Math. Proc. Cambridge Philos. Soc.* **42** (1946) 15.
[60] J. E. Hutchinson, *Indiana Univ. Math. J.* **30** (1981) 713.
[61] A. S. Besicovitch. S. J. Taylor, *J. London Math. Soc.* **29** (1954) 449.
[62] M. Barnsley, S. Demko, *Proc. Roy. Soc. London Ser. A*, **399** (1985) 243.
[63] R. S. Strichartz, A. Taylor, T. Zhang, *J. Experimental Mathematics* **4** (1995) 101.
[64] S. Gibson, *Undergraduate research project*, Calif. State Univ., Stanislaus, 2009.
[65] D. Essouabri, M. L. Lapidus, J. A. Rock, *Analytic continuation of a class of multifractal zeta functions*, preprint, 2014.

Chapter 13

Benoit Mandelbrot, films, and me: a tribute

Nigel Lesmoir-Gordon

Benoit, pensive. *Photo credit: Nigel Lesmoir-Gordon.*

"The Mandelbrot set is one of the most beautiful and remarkable discoveries in the entire history of mathematics."

Sir Arthur C Clarke

Back in the early 1990s I was researching topics for possible science TV documentaries and reading Ian Stewart's book *Does God Play Dice?* [1] I came across the chapter he called The Gingerbread Man. It was about the Mandelbrot set. Reading that chapter I could see that here was the perfect subject for a film. The Mandelbrot set is a fractal (Fig. 13.1) and fractal geometry is the geometry of nature (Figs. 13.2 to 13.4). So here I had the stunning visual beauty of the set itself and the natural world to point my camera at. It was a science story I could tell in pictures. It wasn't the Big Bang or gravity or quantum mechanics, it was something you could actually see and which could be filmed directly without recourse to endless and expensive CGI.

I researched a lot more, wrote a treatment and submitted it to the UK's BBC, Discovery and Channel 4. And, as night follows day, the almost

Figure 13.1 Part of the Mandelbrot set. *Photo credit: William Rood.*

Figure 13.2 Fractals in a field. *Photo credit: Nigel Lesmoir-Gordon.*

Figure 13.3 Birds and fractals. *Photo credit: Gabriel Lesmoir-Gordon.*

Figure 13.4 Tropical fractals. *Photo credit: Gabriel Lesmoir-Gordon.*

predictable rejections followed. I fumed for a year, thrashing about for a way to finance the project.

There used to be a little shop in Ladbroke Grove, London called *Strange Attractions.* Sadly it's closed now, but at the time it was thriving, selling the products of the new mathematics — fractal geometry and chaos theory. Books, music, computer games, T-shirts, posters, coasters, cups and postcards. Inside, everywhere I looked, I saw the M set.

Chatting with the sales assistant, I found out that I knew the owner in London in the sixties. Amongst other things, Greg Sams pioneered the Vegiburger. Now it was fractals and chaos theory. Greg and I went out for coffee. We talked about old times, shared our excitement in the new maths and then I told him my problem.

Bingo! Two months ago, Greg told me, Fred Clarke had been driving Sir Arthur C. Clarke back to Arthur's apartment in Holland Park (Fig. 13.5), through Ladbroke Grove and past Greg's shop. Arthur C.'s sharp eye caught a glimpse of the colourful snapshot of the M set, swinging in the breeze on a big sign over Greg's shop door. Arthur had Fred reverse the car back to the shop and park. Greg was in.

Figure 13.5 Sir Arthur C. Clarke with the COLORS crew. L to R: Simon Gilbert Designer and Editor, Deepak Sethi Production Assistant, V. J. Veeraswami Production Manager, John Lamborn Director of Photography, Paul Sinclair Producer, Nigel Lesmoir-Gordon Director and Sir Arthur C. Clarke. *Photo credit: Nigel Lesmoir-Gordon.*

Arthur had just returned from Riyadh, where he wrote that he had "... the privilege of addressing the largest gathering of astronauts and cosmonauts ever assembled at one place (more than fifty, including Apollo 11's Buzz Aldrin and Mike Collins, and the first space-walker, Alexie Leonov). I decided to expand their horizons by introducing them to something really large! So, with astronaut Prince Sultan bin Salman bin Abdul Aziz in the chair, I delivered a lavishly illustrated lecture: The Colors of Infinity — Exploring the Fractal Universe."

Arthur had given Greg Sams permission to publish the speech. When I arrived Greg had just received the printer's proof. He gave me a copy of the booklet and Arthur's fax number in Colombo, Sri Lanka. Off I went to compose a persuasive letter.

I faxed my two-paged plea the next morning, went off to work, but forgot to inform my wife, Jenny, that I'd sent it, more or less expecting nothing to come back. So when Arthur C. Clarke phoned two hours later, Jenny was stunned and completely unprepared. Only her quick responses saved the situation as she took on board what was happening. She calmed down, took notes and came away with his agreement to a two-day shoot at his house in Colombo. And so, with finance raised from family and friends, the filming of *The Colors of Infinity* [2] began.

After we had finished the shooting in Sri Lanka and had returned to the UK, I approached Benoit to ask if I could come to America to interview him. He had the good sense to contact Arthur to ask him if I was the genuine article. Arthur was able to put his mind at rest and we set up a time and place to film Benoit.

We arrived at the Mandelbrot family home early in the morning. The crew were relaxed, but I was full of fear and trepidation and had to be hustled out of the van and up to the front door. My daughter, Daisy, was on the crew as production assistant. She gave me a warm and encouraging hug, which was just what I needed. With my knees shaking I pushed the door-bell. After a few seconds the door was opened by a smiling and gracious lady, Aliette Mandelbrot. She showed us into the sitting room where we sat and waited. Aliette offered us tea or coffee. While we waited the great man came in looking relaxed and comfortable. We jumped up and he shook our hands as I introduced the crew. He asked where I would like to do the shooting and offered to show me round the house. I said his sitting-room would be just fine.

While the crew set up the equipment, Benoit and I went to his study to discuss the shape that the interview would take. I had my questions prepared and we went through them. He was gracious and charming and quickly put me at ease. He was, in essence, happy with what I proposed, making the odd addition and amendment here and there. I commented uneasily that he looked a bit formal in his suit and tie and suggested that for the broad audience I was hoping to appeal to he should dress more casually. He laughed and said that he had a blue jacket he could wear and that he would drop the tie. I went back into the sitting-room. The camera and lights were set up and I checked the frame that the camera-man, John Lamborn, was proposing. It looked just fine to me and I said I would wait for final adjustments until Benoit was actually seated. When he came back into the room, instead of the bright blue jacket I was expecting, he was wearing a coat of subtle grey. I smiled to myself. Ah, well! It would do nicely. He took his seat and I sat down by the camera, clutching my notes and my list of questions. He looked good. Benoit was relaxed and I started to chill-out. I checked the framing, suggested a minor adjustment, which was taken on, and we were ready to roll. Lights, camera, action! Then everything popped. We had blown a fuse. Our electrician was dispatched to search for the source of the problem. He re-routed the lights and we started again. After 10 minutes the lights blew again. I started to get very agitated, but Benoit remained as cool as a cucumber. I was impressed.

We had another coffee break while the electrician did some serious re-wiring. Everyone stayed calm except for me. It was essential that I made a good job of the interview for the sake of our investors, the crew and most of all for Benoit. It was vital that I did not miss a trick or fail somehow to get this absolutely crucial interview just right. We managed to get started again and, save for a half hour lunch-break, worked on into the afternoon. Benoit was superb inspiring and lucid.

At the end of the session I was delighted and accepted the crew's generous congratulations.

The next day we flew to Atlanta to interview the Mathematician Michael Barnsley. This shoot went very well. Michael's enthusiasm for the topic was palpable. As he so clearly put it in the film:

> It took a long time for us to emerge and start to look out at the physical, observable universe — the totality of nature, and then say, 'my goodness me, we've got nothing to describe this with!' Clouds are not made with straight edges. Trees are not circles, they're not triangles. They are something very, very different indeed. There is a continual kind of a pattern that I can see as I look at the edge of a rising cumulus cloud, one of those very, very wrinkly, coruscated clouds, which have such fine structure in them. And you say, but there are no lines or circles there. The wonderful discovery has been that there's an extension of classical geometry — Euclidean Geometry — which is called fractal geometry.
>
> The discovery of fractal geometry changes completely the kind of patterns we can look for in nature. And that is really a fundamental change to the sort of things mathematicians and scientists can do. And that's got to have a big effect.
>
> No longer do you have to draw a straight line through your data to make science of it. You can actually draw a fractal curve through it or measure some fractal dimension of the data and do science! So the first application is in

> terms of a better description of the physical observable world. So there's a new branch of mathematics available to all scientists, and that application will stretch on through the centuries now as the primary tool for descriptive physical science."

We completed the filming for this production with an interview with Warwick University's Professor Ian Stewart and some bought-in footage of Sir Stephen Hawking. David Gilmour of Pink Floyd agreed to record the music for the film and I commissioned Bill Rood to produce the Mandelbrot set animations. With those elements in hand we had all the material that we needed. It took many months to complete the post-production, but when we were done I was truly delighted with the end product. We sent tapes off to the contributors and received warm, positive and encouraging responses from all of them. Benoit was generous with his praise and his expressions of gratitude.

We found a distributor for the film. It went on to sell in over 40 territories worldwide and has been sub-titled in 3 foreign languages. It was shown on 82 PBS stations in the USA.

When Benoit was recalling his research work at IBM for *The Colors of Infinity* he told me, "For me the first step with any difficult mathematical problem was to programme it, and see what it looked like. We started programming Julia sets of all kinds. It was extraordinary great fun! And in particular, at one point, we became interested in the Julia set of the simplest possible transformation, z goes to z squared plus c. ($z \to z^2 + c$). I made many, many pictures of it. The first ones were very rough. But the very rough pictures were not the answer. Each rough picture asked a question. So I made another picture, another picture. And after a few weeks we had this very strong, overwhelming impression that this was a kind of big bear we have encountered! This discovery was named after me. It is called the Mandelbrot set. I think the most important implication is that, from very simple formulas you can get very complicated results. It's fundamental from the viewpoint of the very basis of science. Because, what is science? We have all this mess around us. Things are totally incomprehensible. And then eventually we find simple laws, simple formulas. In a way, a very simple formula Newton's Law, which is just also a few symbols, can by hard work explain the motion of the planets around the sun and many, many other things to the 50th decimal! It's marvellous: a very simple formula explains all these very complicated things."

Figure 13.6 Benoit Mandelbrot from the film *Clouds are not Spheres. Photo credit: Nigel Lesmoir-Gordon.*

Given the success of *The Colors of Infinity*, I decided to try and make a documentary, which I had been planning, about Benoit's life and work (Fig. 13.6). I was calling it *Clouds are not Spheres* (Figs. 13.7 and 13.8) [3]. I raised some private funding and won an award from the Alfred P. Sloan foundation to produce the film. I spent two full days in Paris in an artist's studio where Benoit told me his life story. This interview formed the backbone of the production. Again Benoit was eloquent and enthusiastic, never losing his way. I completed this film with interviews from Nobel Laureate Professor Ivar Giaever; Yale University's Mathematics Professors Michael Frame, Peter Jones, Ronald Coifman; Warwick University's Emeritus Mathematics Professor Ian Stewart; Ralph Gomory, Heinz-Otto Peitgen; Physicists Carl Everstz, Martin Gutzwiller, Bernard Sapoval; Philosopher Pierre Roubinet; Composer Charles Wuorinen; and Economist Richard Olsen.

In the final scenes of *Clouds are not Spheres* Professor Ivar Giaever says, "There are people who think Mandelbrot should get the Nobel Prize, and

Figure 13.7 Mountains are not cones. *Photo credit: Nigel Lesmoir-Gordon.*

Figure 13.8 Neither spheres nor cones. *Photo credit: Gabriel Lesmoir-Gordon.*

I agree with that, but it's hard to find a field in which he should get it because there is no Nobel Prize in mathematics."

Following the completion and distribution of *Clouds are not Spheres* I produced another documentary, which I called *Is God a Number?* [4] This film focused on the work of Michael Barnsley, examining in depth

Photo credit: Gabriel Lesmoir-Gordon.

the power and the limits of mathematics to explain and describe the world we live in. *Is God a Number?* asks if we can compress and retrieve pictorial images in the interactive world, can we also use mathematics to describe the way our brains work and to help explain the mystery of consciousness?

Michael Barnsley, Sir Roger Penrose and Sir John Polkinghorne, three of the worlds most renowned scientists, use nature and exciting computer graphics to bring these ideas to the viewer in a fascinating review of current thinking.

Benoit, Michael Frame and I went on to make the educational DVD production called *Mandelbrot's World of Fractals*, which was commissioned by the National Science Foundation through Yale University.

This DVD concludes with Benoit addressing the camera:

> I've spent most of my life unpacking the ideas that became fractal geometry. This has been exciting and enjoyable, most times. But it also has been lonely. For years few shared my views. Yet the ghost of the idea of fractals continued to beguile me, so I kept looking through the long, dry years. So find the thing you love. It doesn't

> so much matter what it is. Find the thing you love and throw yourself into it. I found a new geometry; you'll find something else. Whatever you find will be yours. This search is what carried us from caves to the stars. Where will you take us?

The world we live in is not naturally smooth-edged and regularly shaped like the familiar cones (Fig. 13.9), circles, spheres and straight lines of Euclid's geometry. The real world is rough-edged, wrinkled, crinkled, and irregularly shaped. Its a wiggly world! Amazingly up until very recently we didnt have a geometry that described the natural world. But we do now. Benoit gave the world this new geometry, which puts him, as far as I am concerned, on a par with Euclid.

Fractals are aesthetically pleasing, frequently revealing exquisite beauty in the most subtle ways. Fractals are intimately connected to the concepts of beauty and elegance. Until recently, there was no word to describe the familiar shapes of nature, but now we can see fractals everywhere.

Apart from being a visionary and a maverick, Benoit was a kind and generous man, always patient and at pains to make his ideas clear and

Figure 13.9 Clouds are not spheres. *Photo credit: Nigel Lesmoir-Gordon.*

Figure 13.10 Farewell. *Photo credit: Gabriel Lesmoir-Gordon.*

understandable. He was also, in my humble opinion, a genius. The world is unquestionably a far lesser place without him.

At the end of the film *Clouds are not Spheres* Benoit closes with this:

> "In a certain sense I was extraordinarily fortunate to take the dream of a rebellious and ill-disciplined young man and follow it through (Fig. 13.10). The dream has brought me to many of the most fundamental issues of science. Some I improved, but certainly left very, very wide open and mysterious. This had been my hope as a young man. This hope has filled my whole life. In that sense I feel extremely fortunate."

The film *The Colors of Infinity* [2] has been shown in over 40 territories worldwide. *Clouds are not Spheres* [3] has been shown throughout the USA on PBS and in many other countries as has *Is God a Number?* [4] The DVD *Mandelbrot's World of Fractals* [5] has served its purpose well, informing teachers and pupils alike throughout the USA of the wonderful world of fractals.

I always felt highly privileged to have the opportunity to work with Benoit as I played my part in bringing his ground-breaking work to the attention of a broad and popular audience. Although I had a good grasp of the significance of fractal geometry and brought many years of experience and talent to the table I know that I was nonetheless merely a conduit for him. I appreciate now that my films served Benoit well in his efforts to communicate his discoveries and that he was most pleased with what I achieved for him. He always wanted to reach out and communicate warmly across as many platforms as he could and he was a great communicator. This was something he relished and took great pride and pleasure in doing. When you watch him in my films this much is obvious. My intention now is to raise the finances to make another film that will serve as a tribute to Benoit's brilliance and his achievements.

Bibliography

[1] I. Stewart, *Does God Play Dice?* (Basil Blackwell, New York, 1989).

[2] N. Lesmoir-Gordon, *The Colors of Infinity*, Films for the Humanities and Sciences (1995), 52 min.

[3] N. Lesmoir-Gordon, *Clouds are not Spheres*, Films for the Humanities and Sciences (2000), 53 min.

[4] N. Lesmoit-Gordon, *Is God a Number?*, Films for the Humanities and Sciences (1998), 53 min.

[5] N. Lesmoir-Gordon, *Mandelbrot's World of Fractals*, Gordon Films UK, Key Curriculum Press (2005), 63 min.

Chapter 14

Fractals and humor

Demetri Martin

Demetri Maritn prepared this palindromic poem as his project for Michael Frame's fractal geometry class at Yale. Notice the first, fourth, and seventh words in the second and next-to-second lines are palindromes, the first two and last two lines are palindromes, the middle line, "Be still if I fill its ebb" minus its last letter is a palindrome, and the entire poem is a palindrome.

Always appreciative of puzzles and wordplay, Benoit thought Demetri's work was amazing.

Dammit I'm mad

Dammit I'm mad
Evil is a deed as I live.
God, am I reviled?
I rise, my bed on a sun, I melt.
To be not one man emanating is sad. I piss.
Alas it is so late. Who stops to help? Man, it is hot.

I'm in it.
I tell.
I am not a devil.
I level "Mad Dog".

Ah, say burning is as a deified gulp
in my halo of a mired rum tin.
I erase many men. Oh, to be man, a sin.
Is evil in a clam? In a trap?
No. It is open.
On it I was stuck.

Rats peed on hope.
Elsewhere dips a web.
Be still if I fill its ebb.
Ew, a spider ... eh?
We sleep.

Oh no!
Deep, stark cuts saw it in one position.
Part animal, can I live? Sin is a name.
Both, one... my names are in it. Murder?
I'm a fool. A hymn I plug,
Deified as a sign in ruby ash — a Goddam level I lived at.

On mail let it in. I'm it.
Oh, sit in ample hot spots.
Oh, wet!
A loss it is alas (sip). I'd assign it a name.
Name not one bottle minus an ode by me:
"Sir, I deliver. I'm a dog."
Evil is a deed as I live.
Dammit I'm mad.

Chapter 15

Multifractal measures of time series: curvature surfaces of $f(\alpha)$ curves

William Martino and Michael Frame

15.1 Introduction

Benoit Mandelbrot's scientific legacy is immense. He combined his recognition that scaling is an important feature of nature with a handful of mostly 19th century counterexamples in analysis to produce fractal geometry. He showed that fractal dimension is the first reliable measure of roughness, he developed multifractal measures to study the complex patterns of turbulence, he showed the same techniques can and should be applied to financial markets, he discovered the Mandelbrot set and revitalized complex iteration, and on. Throughout his long career, one constant theme was the importance of the eye in science. However, he did not, as some incorrectly claim, simply look at pretty pictures. For example, every pretty picture of the Mandelbrot set suggests that the small copies surrounding the main body of the set are islands. Certainly, Benoit looked carefully at the pictures, but also he thought carefully. He saw similarities between magnifications of the Mandelbrot set and magnifications of some Julia sets, and he knew these Julia sets are connected, so he posed the question, later answered positively by A. Douady and J. Hubbard [1], "Is the Mandelbrot set connected?" The eye can help us ask good questions, but still we need to do the work.

This paper is a small step in this direction laid out by Benoit. From mostly financial data we construct a family of multifractal measures, look at the curves corresponding to these measures, and then compute and interpret a geometric property, the curvature of these curves. While we did not discuss the later stages of this project with Benoit — after all, even in the summer of 2010 he appeared to be in reasonable health, was preoccupied with his memoirs [2], and we would have plenty of time later to talk about this — he was familiar with early results, even reprinting some of Joseph Thornton's [3] pictures in *The (Mis)Behavior of Markets* [4]. We believe he would have been interested in the approach we present here.

15.2 Reading time series

Metric approaches to analyzing time series abound. See [5], for example. A different view, motivated in part by concern for the accuracy of measurements, is to coarse-grain, or bin, the data and then measure the likelihood that a point in bin i can be followed n time steps later by a point in bin j. Examples are [6–9].

In another direction, Jeffrey [10, 11] mapped DNA sequences to visual patterns using the four transformations

$$\begin{aligned} &T_1(x,y) = (x/2, y/2) \quad T_2(x,y) = (x/2, y/2) + (1/2, 0) \\ &T_3(x,y) = (x/2, y/2) + (0, 1/2) \quad T_4(x,y) = (x/2, y/2) + (1/2, 1/2) \end{aligned} \tag{15.1}$$

As iterated function system (IFS) rules, these generate the filled-in unit square. Jeffrey's method, an application of Stewart's [12] variation on Barnsley's chaos game [13], is this. Start with the point $(x_0, y_0) = (1/2, 1/2)$ and generate a sequence of points $(x_1, y_1), (x_2, y_2), \ldots,$ by $(x_{i+1}, y_{i+1}) = T_k(x_i, y_i)$, and the transformations T_k are determined by reading the DNA sequence: apply T_1 when an A is encountered, T_2 when T, T_3 when C, and

T_4 when G. Applying the transformations in the order determined by the DNA sequence is called *driving* the IFS by that sequence.

Note that the *time* part of time series is not necessary; the essential feature is a natural sequential ordering of the data. To get a sensible result the ordering must be natural. For example, some have suggested binning a text by the alphabetical ordering of the words. The arbitrariness of alphabetical ordering makes this approach nonsense.

On the left of Fig. 15.1 we see the IFS driven by the DNA sequence for the enzyme amylase. This sequence contains 3957 letters, so the driven IFS has that many points. Clearly, the square is not filled uniformly. How can we understand this nonuniformity?

In the left middle of Fig. 15.1 we have superimposed squares corresponding to compositions of two transformations (15.1), that is, to consecutive pairs in the amylase sequence. As shown on the left of Fig. 15.2, every point in the subsquare labeled A is the result of applying T_1, every point in the subsquare labeled T is the result of applying T_2, and so on. On the right of Fig. 15.2, every point in the subsquare labeled GC is the result of

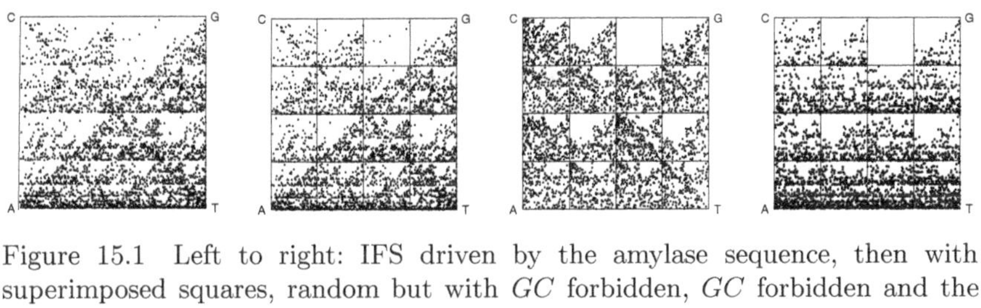

Figure 15.1 Left to right: IFS driven by the amylase sequence, then with superimposed squares, random but with GC forbidden, GC forbidden and the same probabilities as the amylase sequence.

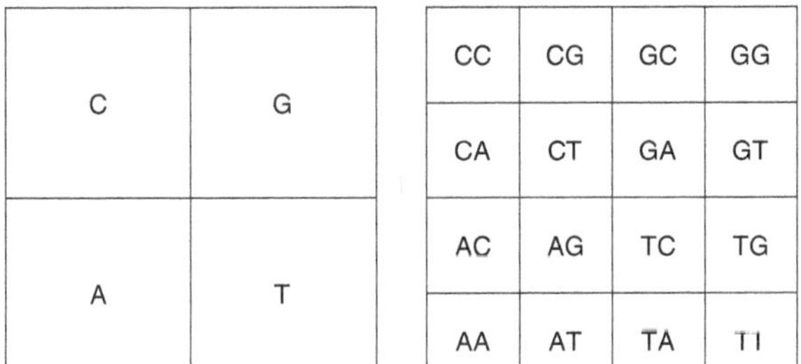

Figure 15.2 Length 1 (left) and length 2 (right) addresses for DNA driven IFS.

applying the composition $T_4 \circ T_3$. We say GC is the *length* 2 *address* of this subsquare. In the left middle of Fig. 15.1 we see this pair, GC, is rare in the amylase sequence; in the right middle of Fig. 15.1 we have driven the IFS with 3957 random choices of A, T, C, and G, subject to the restriction that G never follows C. Some, but not all, of the structure of the left driven IFS is replicated here.

Better surrogates can be made by adjusting the probabilities to reproduce the relative frequencies of the length 1 and 2 addresses. This is shown on the right of Fig. 15.1.

To quantify how well the surrogate-driven IFS matches the data-driven IFS, we need a way to measure the distance between the images. One approach is the Hutchinson metric [13], an extension of the familiar Hausdorff metric to measures, and applied to assign colors to images generated by IFS.

Rather than this, we use an extension of the correlation function developed by Kalla and Soban [14]. Given two sequences σ and τ of the same length N and drawn from $\{1, 2, 3, 4\}$, denote by $n_{jk}(\sigma)$ the number of times the pair jk occurs in σ. Then the length 2 correlation of the driven IFS is

$$\frac{N - (|n_{11}(\sigma) - n_{11}(\tau)| + |n_{12}(\sigma) - n_{12}(\tau)| + \cdots + |n_{44}(\sigma) - n_{44}(\tau)|)}{N}$$

If this correlation is 1, then the IFS driven by σ and τ produce the same occupancy of length 2 addresses; if this correlation is -1, the IFS driven by σ and τ occupy disjoint length 2 address spaces. Of course, using longer addresses gives a more detailed comparison of the driven IFS pictures, but this must be balanced against the issue of the confidence in occupancies of longer addresses derived from a finite data set.

Comparing time series by occupancies of addresses through length five was the basis for C. Noelle Thew's applied math senior thesis [15] at Yale.

These methods, and others, can be used to analyze numerical data by binning the data, mentioned earlier. Given a sequence $x_1, \ldots, x_N$, divide the range $\max(x_i) - \min(x_i)$ into four bins, B_1, B_2, B_3, and B_4. The order in which the T_j are applied is given by the order in which the x_i visit these bins. If $x_i \in B_j$, then the ith transformation applied is T_j. *Equal size* bins are those having the same length as intervals; *equal weight* or

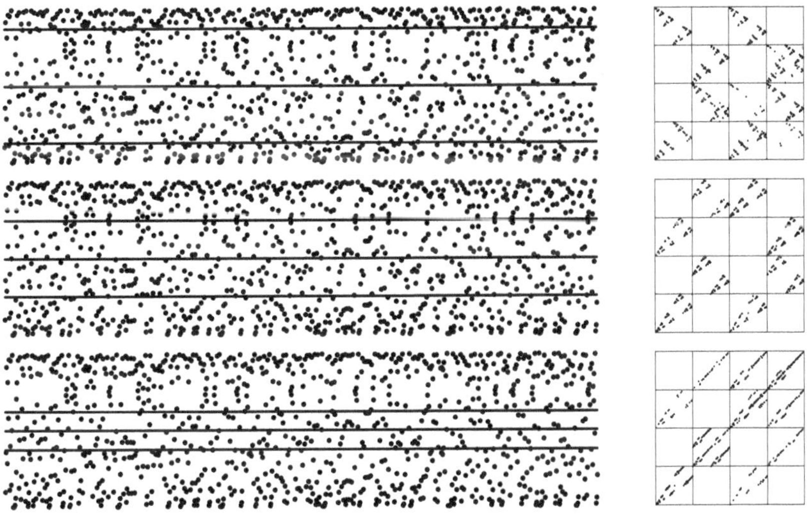

Figure 15.3 Right: a logistic map time series with bin boundaries indicated by horizontal lines. Right: the corresponding driven IFS.

maximum entropy bins have about the same number of data points in each bin.

Here we use bins of this form

$$\begin{aligned} B_1 &= (-\infty, \mu - \delta), \quad B_2 = [\mu - \delta, \mu), \\ B_3 &= [\mu, \mu + \epsilon), \quad B_4 = [\mu + \epsilon, +\infty) \end{aligned} \tag{15.2}$$

where μ is the median of the data values and δ and ϵ are adjustable parameters. Often, we take $\epsilon = \delta$.

For example, generate a time series by 1000 iterates of the logistic map, $x_{i+1} = 4x_i(1 - x_i)$. Then $\mu \approx 1/2$, and in Fig. 15.3 we see the time series, with bin boundaries indicated by horizontal lines. To the right of each time series is the driven IFS. Top to bottom, $\delta = \epsilon = 3/8$, $1/4$, and $1/8$.

Using a method described in Sec. 15.3, in [16] we found the $f(\alpha)$ curves for a collection of δs and assembled them into a surface. Properties of the surface gave information about the time series.

Here, we compute the curvature of the $f(\alpha)$ curves for a range of α values, and assemble these curvature graphs into a surface.

15.3 Four bin multifractals

For a time series divided into four bins we can approximate the transition probabilities by

$$p_{ij} = \text{Prob}(x_{k+1} \in B_i \text{ if } x_k \in B_j) \tag{15.3}$$

Then the standard formalism for computing the multifractal spectrum in this setting is

$$\rho[p_{ij}^q r_i^{\beta(q)}] = 1 \tag{15.4}$$

where $\rho[M]$ is the spectral radius of the matrix M, that is, the magnitude of the largest eigenvalue of M, and r_i is the contraction factor of the ith transformation. For the transformations (15.1), all $r_i = 1/2$, so Eq. (15.4) can be solved explicitly for $\beta(q)$:

$$\beta(q) = \log_2(\rho[p_{ij}^q]) \tag{15.5}$$

The local Hölder exponent is given by

$$\alpha = -\frac{d\beta}{dq} \tag{15.6}$$

and $f(\alpha)$ is defined by

$$f(\alpha) = \alpha \cdot q + \beta(q) \tag{15.7}$$

Before investigating real data, we demonstrate a test case for which we have a closed form expression for $f(\alpha)$ and so can compute the curvature $k(\alpha)$ by

$$k(\alpha) = \frac{\left|\frac{d^2 f}{d\alpha^2}\right|}{\left(1 + \left(\frac{df}{d\alpha}\right)^2\right)^{3/2}} \tag{15.8}$$

In Sec. 15.4 we sketch the binomial multifractal, for which we know an explicit expression for $f(\alpha)$, then in Sec. 15.4.1 we compute the curvature of these $f(\alpha)$ curves. In Sec. 15.4.2 we show a uniformly random time series divided into four bins produces an $f(\alpha)$ curve that is a vertical translation of the binomial multifractal curve, and so has the same curvature.

15.4 Binomial multifractals

For a starting example, suppose a time series of independent random numbers is divided into two bins, with probabilities p_1 and $p_2 = 1 - p_1$. The IFS driven by this series has 2×2 transition matrix $[p_{ij}]$, where p_{ij} is given by (15.3). Independence of successive time series entries gives

$$p_{ij} = p_i$$

Then the transition matrix for this driven IFS is

$$\begin{bmatrix} p_1 & p_1 \\ p_2 & p_2 \end{bmatrix}$$

By Eq. (15.5) we have

$$\beta(q) = \log_2 \left(\rho \begin{bmatrix} p_1^q & p_1^q \\ p_2^q & p_2^q \end{bmatrix} \right). \tag{15.9}$$

The eigenvalues of this matrix are 0 and $p_1^q + p_2^q$, so we obtain

$$\beta(q) = \log_2(p_1^q + p_2^q) \tag{15.10}$$

Next,

$$\alpha(q) = -\frac{d\beta}{dq} = -\frac{p_1^q \log_2(p_1) + p_2^q \log_2(p_2)}{p_1^q + p_2^q} \tag{15.11}$$

and $f(\alpha)$ is given by Eq. (15.7). Using Eqs. (15.10), (15.11), and (15.7), we obtain a parametric representation of points $(\alpha(q), f(\alpha(q)))$ on the $f(\alpha)$ curve. However, expressions of $df/d\alpha$ and $d^2f/d\alpha^2$ needed in Eq. (15.8) are most easily obtained if we have an explicit formulation of $f(\alpha)$ as a function of α.

In [18], Mandelbrot and Evertsz show that for binomial multifractals $f(\alpha)$ has this closed form expression

$$\begin{aligned} f(\alpha) = &-\frac{\alpha_{\max} - \alpha}{\alpha_{\max} - \alpha_{\min}} \log_2 \left(\frac{\alpha_{\max} - \alpha}{\alpha_{\max} - \alpha_{\min}} \right) \\ &-\frac{\alpha - \alpha_{\min}}{\alpha_{\max} - \alpha_{\min}} \log_2 \left(\frac{\alpha - \alpha_{\min}}{\alpha_{\max} - \alpha_{\min}} \right) \end{aligned} \tag{15.12}$$

where $\alpha_{\min} = -\log_2(\max\{p_1, p_2\})$ and $\alpha_{\max} = -\log_2(\min\{p_1, p_2\})$. A plot of this curve for $p_1 = .35$ is shown on the lower portion of Fig. 15.6.

15.4.1 Curvature of the binomial $f(\alpha)$ curve

The curvature of the binomial $f(\alpha)$ curves can be computed from Eq. (15.12). First,

$$\frac{df}{d\alpha} = \frac{\log_2\left(\frac{\alpha_{\max}-\alpha}{\alpha-\alpha_{\min}}\right)}{\alpha_{\max}-\alpha_{\min}} \tag{15.13}$$

Next,

$$\frac{d^2 f}{d\alpha^2} = \frac{1}{(\alpha-\alpha_{\min})(\alpha-\alpha_{\max})\ln(2)} \tag{15.14}$$

and using Eq. (15.8) we compute the curvature

$$k(\alpha) = \frac{(\alpha_{\max}-\alpha_{\min})^3}{\ln(2)(\alpha-\alpha_{\min})(\alpha_{\max}-\alpha)\left((\alpha_{\max}-\alpha_{\min})^2+\left(\log_2\left(\frac{\alpha_{\max}-\alpha}{\alpha-\alpha_{\min}}\right)\right)^2\right)^{3/2}} \tag{15.15}$$

In Fig. 15.4 we plot the $f(\alpha)$ surface (left) and corresponding curvature surface (right). These surface plots are made of polygons between consecutive $f(\alpha)$ curves and between consecutive $k(\alpha)$ curves.

For $\delta = 1/4$, the $f(\alpha)$ curve collapses to a single point, $(\alpha, f(\alpha)) = (2, 1)$. The most striking feature of the curvature plots is the divergence, truncated here. This occurs where $\alpha_{\min}$ approaches $\alpha_{\max}$, and in Sec. 15.8 we see that this narrowing of the range of α values causes the curvature to diverge, at least when $f(\alpha_{\min})$ and $f(\alpha_{\max})$ stay bounded below $\max\{f(\alpha)\}$. This

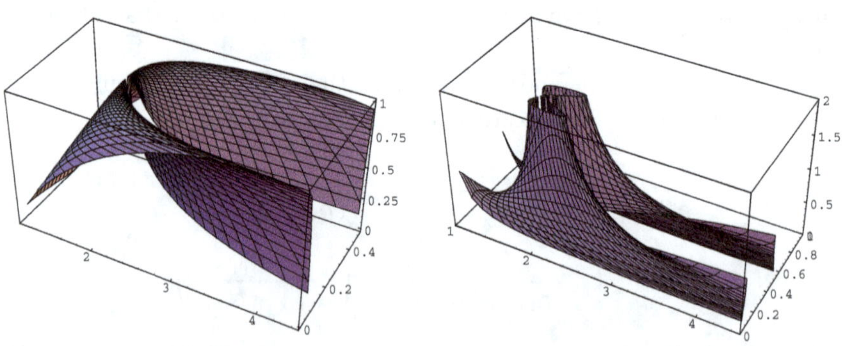

Figure 15.4 The $f(\alpha)$ and $k(\alpha)$ surfaces.

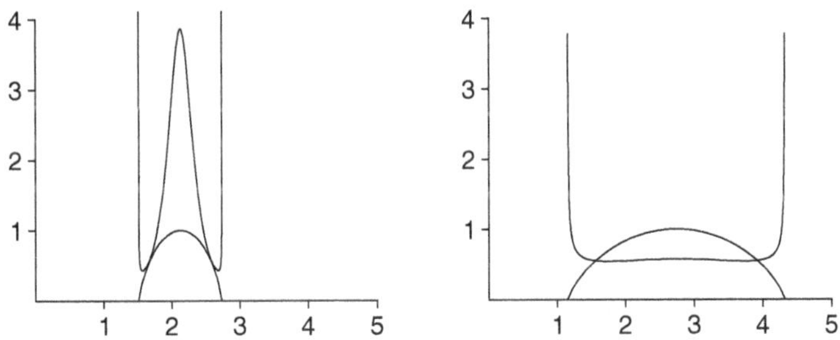

Figure 15.5 Two $f(\alpha)$ and $k(\alpha)$ curves.

curvature divergence does not interest us, so we truncate the curvature graphs.

Another feature we note now, in order to avoid later, is the increase in curvature near $\alpha_{\min}$ and $\alpha_{\max}$. This increase may seem surprising, but from the form of $k(\alpha)$ in Eq. (15.15) we see that $k(\alpha) \to \infty$ when $\alpha \to \alpha_{\min}$ and $\alpha \to \alpha_{\max}$. In Fig. 15.5 we see the graphs of the $f(\alpha)$ curve given by Eq. (15.12), and the $k(\alpha)$ curve given by Eq. (15.15), for $\delta = 0.15$ (left) and $\delta = 0.05$ (right). This curvature divergence at the extreme α values is the reason for restricting the range of α values for which we plot the curvature.

15.4.2 A four bin implementation of the random binomial multifractal

For a time series of random numbers uniformly distributed in $[0, 1]$ taking the bin boundaries as $1/2 - \delta$, $1/2$, and $1/2 + \delta$, the probabilities of being in each bin are

$$p_1 = 1/2 - \delta, \quad p_2 = \delta, \quad p_3 = \delta, \quad p_4 = 1/2 - \delta \tag{15.16}$$

Again, the independence of the random numbers gives $p_{ij} = p_i$. From Eq. (15.5) we have

$$\beta(q) = \log_2 \left(\rho \begin{bmatrix} p_1^q & p_1^q & p_1^q & p_1^q \\ p_2^q & p_2^q & p_2^q & p_2^q \\ p_3^q & p_3^q & p_3^q & p_3^q \\ p_4^q & p_4^q & p_4^q & p_4^q \end{bmatrix} \right)$$

Substituting in the values of p_i from Eq. (15.16), the matrix becomes

$$\begin{bmatrix} (1/2-\delta)^q & (1/2-\delta)^q & (1/2-\delta)^q & (1/2-\delta)^q \\ \delta^q & \delta^q & \delta^q & \delta^q \\ \delta^q & \delta^q & \delta^q & \delta^q \\ (1/2-\delta)^q & (1/2-\delta)^q & (1/2-\delta)^q & (1/2-\delta)^q \end{bmatrix} \tag{15.17}$$

This matrix has eigenvalues

$$0, \quad 0, \quad 0, \quad 2(\delta^q + (1/2-\delta)^q)$$

and so has spectral radius $\rho = 2(\delta^q + (1/2-\delta)^q)$, giving

$$\beta(q) = 1 + \log_2(\delta^q + (1/2-\delta)^q) = 1 + \log_2(p_1^q + p_2^q) \tag{15.18}$$

and

$$\alpha(q) = -\frac{d\beta}{dq} = -\frac{p_1^q \log_2(p_1) + p_2^q \log_2(p_2)}{p_1^q + p_2^q} \tag{15.19}$$

Note the equivalence of Eqs. (15.10) and (15.11) with Eqs. (15.18) and (15.19), with β of Eq. (15.18) equal to 1 plus the β of Eq. (15.10). Then from Eq. (15.7) we see the $f(\alpha)$ curve of the four bin random multifractal is that of the binomial multifractal shifted up by 1. The IFS driven by this time series has $f(\alpha)$ curve given by Eq. (15.12), where

$$\begin{aligned} \alpha_{\min} &= \min\{-\log_2(\delta), -\log_2(1/2-\delta)\} \\ \alpha_{\max} &= \max\{-\log_2(\delta), -\log_2(1/2-\delta)\} \end{aligned} \tag{15.20}$$

The derivation of Mandelbrot and Evertsz makes no use of the condition $p_1 + p_2 = 1$, which holds in the binomial multifractal of Sec. 15.4. For the construction of this section, $p_1 + p_2 = 1/2$, but that has no effect on the derivation of the formula (15.12).

Plotting the points

$$(\alpha(q), q \cdot \alpha(q) + \beta(q)) \tag{15.21}$$

using Eqs. (15.18) and (15.19) for $q = -10, \ldots, 10$, these points lie on the graph of $f(\alpha)$ given by Eq. (15.12), but shifted up by 1. Figure 15.6 shows the graph of the $f(\alpha)$ curve, the points (15.21), and those points shifted down by 1, for $\delta = 0.15$. The lower graph is the $f(\alpha)$ curve for the binomial multifractal; the lower points are the upper points shifted down by 1.

This shift makes sense because for the binomial multifractal, both $\alpha_{\min}$ and $\alpha_{\max}$ occur on single points, so $f(\alpha_{\min}) = f(\alpha_{\max}) = \dim(\text{point}) = 0$.

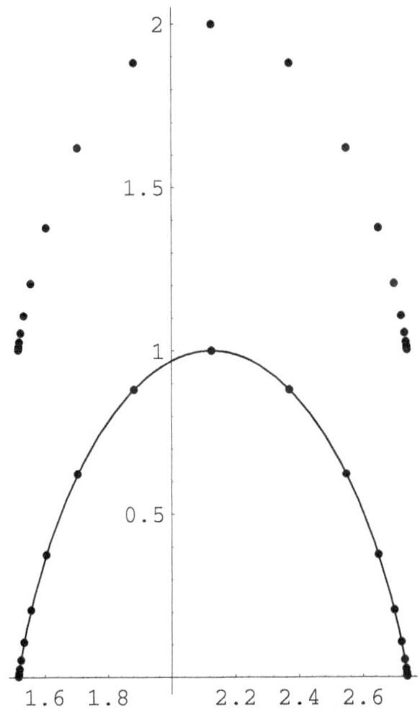

Figure 15.6 The $f(\alpha)$ curve, and points, for $\delta = 0.15$.

For IFS with all $r_i = r$, $\alpha_{\max}$ occurs on the set with minimum probablility, and $\alpha_{\min}$ on the set with maximum probability. Because $\delta < 1/4$ for this example, $p_2 = p_3 < p_1 = p_4$. Then $\alpha_{\max}$ occurs on the attractor of the IFS $\{T_2, T_3\}$, the line segment between $(1, 0)$ and $(0, 1)$. Similarly, $\alpha_{\min}$ occurs on the attractor of $\{T_1, T_4\}$. Consequently, $f(\alpha_{\min}) = f(\alpha_{\max}) = \dim(\text{line}) = 1$.

15.5 Finding the $f(\alpha)$ curvature from time series

In this section we describe the algorithm used to generate the $f(\alpha)$ curves and curvature plots from a time series. Although other choices are possible, we divide the time series values $\{x_1, x_2, \ldots, x_N\}$ into only four bins. This involves establishing three bin boundaries, $\beta_1 < \beta_2 < \beta_3$. The middle boundary, β_2, is the median of the x_i. Then β_1 is located a specified percent of $\beta_2 - \min\{x_i\}$ below β_2, and β_3 is located a specified percent of $\max\{x_i\} - \beta_2$ above β_2. The bins are

$$B_1 = [\min\{x_i\}, \beta_1), \quad B_2 = [\beta_1, \beta_2), \quad B_3 = [\beta_2, \beta_3), \quad B_4 = [\beta_3, \max\{x_i\}]$$

Given the bin boundaries, the transition probabilities p_{ij} are computed by Eq. (15.3). That is,

$$p_{ij} = \frac{\text{number of times } x_k \in B_j \text{ and } x_{k+1} \in B_i}{N-1}$$

Because we use an IFS with all scaling factors the same (all $r_i = 1/2$), Eq. (15.4) becomes

$$(1/2)^{\beta(q)} \rho[p_{ij}^q] = 1 \tag{15.22}$$

Solving Eq. (15.22) for β we find

$$\beta(q) = \log_2(\text{maximum } |\text{eigenvalues of } [p_{ij}^q]|)$$

The value of $\alpha(q)$ is approximated from Eq. (15.6):

$$\alpha(q) = -\frac{\beta(q)}{dq} = -\frac{1}{\ln(2) \cdot \rho(q)} \cdot \frac{\rho(q + (dq/2)) - \rho(q - (dq/2))}{dq} \tag{15.23}$$

we take $dq = 0.001$. Finally, $f(\alpha)$ is given by Eq. (15.7).

The points $(\alpha(q), f(\alpha(q))$ as q ranges from, say, -20 to 20 in steps of 0.1 give a parametric plot of the $f(\alpha)$ curve, but the points are spaced equally in q, not in α. If we are interested in just these curves [16], this is an adequate representation. To compute curvatures, we need first and second derivatives of f with respect to α; for this, points with equal q spacing are not adequate.

To address this issue, we set a collection of desired α values, α_i, and for each sweep through q values to find q_i minimizing $|\alpha_i - \alpha(q_i)|$. This is achieved by sequentially reducing the q increments, together with the range of q values searched, until q_i is found giving $|\alpha_i - \alpha(q_i)|$ to the prescribed accuracy.

First and second derivatives of f with respect to α are computed numerically, and the curvature is approximated by Eq. (15.8).

15.6 Building time series with prescribed memory

We conjecture that curvature of $f(\alpha)$ curves is related to the length of memory of the dynamical process generating the time series. In this section we show how to produce a time series with specified transition probabilities.

This method can be extended to produce time series with specified higher-length memories.

Recall p_{ij} is the probability of the transition $B_j \to B_i$, this ordering is chosen so subscripts correspond to compositions of IFS transformations: $p_{ij} > 0$ means the composition $T_i \circ T_j$ is allowed. Because a time series point in bin j must be followed by a time series point in *some* bin, for all j we have $p_{1j} + p_{2j} + p_{3j} + p_{4j} = 1$.

Given the time series point x_k, generating x_{k+1} requires two steps: determining in which bin x_{k+1} lies, and within that bin, finding an appropriate point. Suppose x_k lies in bin j. Select the bin, i, for x_{k+1} according to the probabilities p_{ij}. Selecting a point in a given bin can be done in many ways, but to simplify the calculations of Sec. 15.7, we take x_{k+1} uniformly randomly in bin i.

For example, given the transition matrix

$$\begin{bmatrix} .1 & .2 & .3 & .4 \\ .2 & .3 & .4 & .1 \\ .3 & .4 & .1 & .2 \\ .4 & .1 & .2 & .3 \end{bmatrix}$$

here are the actual transition matrices for implementations of this algorithm generating time series of 1000, 5000, 10000, and 20000 points.

$$\begin{bmatrix} .0878 & .2054 & .3418 & .4142 \\ .1870 & .2589 & .3600 & .0753 \\ .3282 & .4286 & .1418 & .2218 \\ .3969 & .1071 & .1563 & .2845 \end{bmatrix} \quad \begin{bmatrix} .1018 & .1942 & .2942 & .3717 \\ .2144 & .3238 & .3969 & .1056 \\ .2781 & .3678 & .1035 & .2167 \\ .4056 & .1141 & .2054 & .3052 \end{bmatrix}$$

$$\begin{bmatrix} .1067 & .2054 & .2874 & .3970 \\ .1969 & .2991 & .4045 & .1145 \\ .3019 & .4053 & .1000 & .1975 \\ .3945 & .0902 & .2076 & .2911 \end{bmatrix} \quad \begin{bmatrix} .1007 & .2057 & .3035 & .4029 \\ .1928 & .2932 & .3905 & .0953 \\ .3049 & .4033 & .1026 & .1970 \\ .4015 & .0978 & .2035 & .3046 \end{bmatrix}$$

Not surprisingly, longer time series give generally better approximations to the target transition matrix. There is some variability, smaller for longer time series, in transition matrix entries run with different random number generator seeds.

15.7 Transition matrix variations under moving bin boundaries

Assuming the time series points lie in the interval $[0, 1]$, a simple choice of bins is

$$B_1 = [0, 1/4), \quad B_2 = [1/4, 1/2), \quad B_3 = [1/2, 3/4), \quad B_4 = [3/4, 1] \tag{15.24}$$

Our use of families of $f(\alpha)$ curves to analyze time series is based on how changing the bin sizes affects the shape of the curve. In this section we use these bins

$$\begin{aligned} B_1 &= [0, 1/2 - \delta), \quad B_2 = [1/2 - \delta, 1/2), \\ B_3 &= [1/2, 1/2 + \epsilon), \quad B_4 = [1/2 + \epsilon, 1] \end{aligned} \tag{15.25}$$

where δ and ϵ lie in one of four regimes:

$$\begin{aligned} &\text{(a)} \quad 1/4 \le \delta \le 1/2, \quad 1/4 \le \epsilon \le 1/2 \\ &\text{(b)} \quad 1/4 \le \delta \le 1/2, \quad 0 \le \epsilon \le 1/4 \\ &\text{(c)} \quad 0 \le \delta \le 1/4, \quad 1/4 \le \epsilon \le 1/2 \\ &\text{(d)} \quad 0 \le \delta \le 1/4, \quad 0 \le \epsilon \le 1/4 \end{aligned}$$

Here we find how the transition matrix is modified for δ and ϵ in regime (a). Some care is needed, because changes in the domain and in the range affect the probabilities differently. Label the bins (15.24) as B_i^{old} and the bins (15.25) as B_i^{new}. Label the transition probabilities similarly:

$$p_{ij}^{\text{old}} = \Pr\left(B_j^{\text{old}} \to B_i^{\text{old}}\right) \quad \text{and} \quad p_{ij}^{\text{new}} = \Pr\left(B_j^{\text{new}} \to B_i^{\text{new}}\right)$$

Our goal is to express the p_{ij}^{new} in terms of δ, ϵ, and the p_{ij}^{old}.

First, to modify bins in the range we compare the sizes of new and old bins.

$$\begin{aligned} B_1^{\text{new}} &= (1/2 - \delta) B_1^{\text{old}} \\ B_2^{\text{new}} &= (\delta - 1/4) B_1^{\text{old}} \cup B_2^{\text{old}} \\ B_3^{\text{new}} &= B_3^{\text{old}} \cup (\epsilon - 1/4) B_4^{\text{old}} \\ B_4^{\text{new}} &= (1/2 - \epsilon) B_4^{\text{old}} \end{aligned}$$

where for example $(1/2-\delta)B_1^{\text{old}}$ means B_1^{old} scaled by $(1/2-\delta)$. Because our model time series has points uniformly distributed in each B_i^{old}, the probability that a point in B_1^{old} lies in B_1^{new} is the ratio of the lengths of these intervals, that is

$$\frac{|B_1^{\text{new}}|}{|B_1^{\text{old}}|} = \frac{1/2-\delta}{1/4} = 2-4\delta \tag{15.26}$$

Interpreting the modified domain bins is a bit more delicate. We illustrate this with two calculations.

$$\begin{aligned} p_{11}^{\text{new}} &= \Pr\big(B_1^{\text{new}} \to B_1^{\text{new}}\big) \\ &= \Pr\big((1/2-\delta)B_1^{\text{old}} \to (1/2-\delta)B_1^{\text{old}}\big) \\ &= \Pr\big(B_1^{\text{old}} \to (1/2-\delta)B_1^{\text{old}}\big) && (15.27) \\ &= \frac{1/2-\delta}{1/4}\Pr\big(B_1^{\text{old}} \to B_1^{\text{old}}\big) && (15.28) \\ &= (2-4\delta)p_{11}^{\text{old}} \end{aligned}$$

For Eq. (15.27) the relevant observations are $B_1^{\text{new}} \subseteq B_1^{\text{old}}$, and p_{11}^{old} is the probability that *every* point of B_1^{old} is followed by a point in B_1^{old}. That is, restricting a domain interval does not affect the transition probability.

As we see in Eq. (15.28), restricting a range interval does change the transition probability. Here we are asking for the probability that the next point occupies just a portion, B_1^{new}, of B_1^{old}. In Eq. (15.26) we see this scales the probability by the ratio of the lengths of the intervals.

Analyses of the other cases for regime (a), along with expressions for all the p_{ij}^{new} in terms of the p_{kl}^{old}, is given in the Appendix.

15.8 Interpreting the curvature of the $f(\alpha)$ curves

In an obvious fashion, we can think of $df/d\alpha$ and $d^2f/d\alpha^2$ as the velocity and acceleration of the dimension as a function of the Hölder exponent α. How are we to interpret the curvature of the $f(\alpha)$ curve? Some work has been done already. For example, in [20] the curvature of the $f(\alpha)$ curve at its maximum is used to distinguish true multifractals from signals exhibiting multifractal characteristics as a consequence of finite size artifacts.

Ideally, we are interested in interpretations of the curvature $k(\alpha)$ of the $f(\alpha)$ curve for all α, $\alpha_{\min} \le \alpha \le \alpha_{\max}$. As illustrated by Fig. 15.5, at least for binomial multifractals, the curvature diverges near the minimum and maximum values of α. The binomial is a simple multifractal, so if it exhibits these divergences, we expect others will as well. Moreover, our investigations of curvatures derived from real data appear to show these divergences. So our visual inspection of curvature will not be swamped by these divergences, we restrict the curvature plot to a subinterval of the α range determined by local minima of the curvature graphs.

Before we investigate what the curvature of $f(\alpha)$ curves can reveal, we recall some familiar data read from $f(\alpha)$ curves. See Chapter 11 of [21].

- The maximum height of the curve is the Hausdorff dimension of the support of the measure, and
- the point at which the $f(\alpha)$ curve intersects the diagonal line is the dimension of the measure.

In Fig. 15.7 we plot the $f(\alpha)$ curves from Fig. 15.5 on the left, the $k(\alpha)$ curves on the right. The $f(\alpha)$ curve for $\delta = 0.15$ has smaller α range; the wider α range is for $\delta = 0.05$. This difference is expected, because for $\delta = 0.05$ the data points are more concentrated in bins 1 and 4, while for $\delta = 0.15$ the data points are more evenly distributed among the bins. Near the middle of its α range, we see the $\delta = 0.05$ $f(\alpha)$ curve has almost constant curvature, while the $\delta = 0.15$ curve has much greater curvature. As we look across the α range, we see that except at the extreme values, the curvature of the $\delta = 0.05$ curve changes much more slowly than that of $\delta = 0.15$ curve.

Curvature of the $f(\alpha)$ curve gives a detailed picture of how the dimension of the iso-α sets accelerates along the curve. Visual inspection of

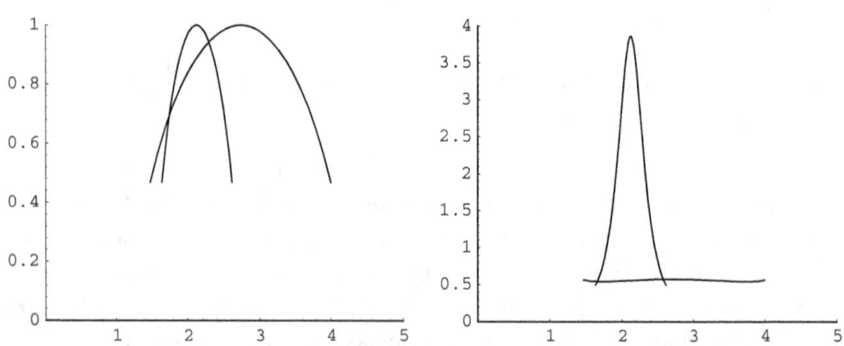

Figure 15.7 Left: the $f(\alpha)$ curves from Fig. 15.5. Right the $k(\alpha)$ curves from Fig. 15.5.

the $f(\alpha)$ curves sometimes can give a hint of coarse information about this acceleration, but not always, and never in much detail. Plotting the $k(\alpha)$ curve is more informative.

Motivated by Ruppeiner's Riemannian geometric model of thermodynamics [22–24] in which curvature of a metric he developed is related to correlation length, we conjecture that the curvature of the $f(\alpha)$ curve encodes information about the memory of the system generating the time series. We explore this conjecture with two experiments.

For the first experiment, we construct $f(\alpha)$ curves for a family of IFS using the transformations (15.1). Recall [25] a 1-IFS is specified by its 4×4 transition matrix $[p_{ij}]$, where the p_{ij} are pair composition probabilities $p_{ij} = \text{Prob}(T_i \circ T_j)$. Our family of examples are IFS specified by triple composition probabilities $p_{ijk} = \text{Prob}(T_i \circ T_j \circ T_k)$. These are called 2-IFS, and are determined by their $4 \times 4 \times 4$ transition array, which we represent as

$$[p_{ij1}][p_{ij2}][p_{ij3}][p_{ij4}]$$

Before presenting this family, we mention a caution about the relation of curvature to memory. In Fig. 15.8 we plot the $f(\alpha)$ and $\kappa(\alpha)$ curves for the IFS (15.1) with probabilities

$$p_1 = \delta, \quad p_2 = 1/4, \quad p_3 = 1/4, \quad p_4 = 1/2 - \delta$$

for $\delta = .05, .1$, and $.15$. Because $\min(p_i) = p_1$ occurs on the attractor of T_1, a point, $f(\alpha_{\max}) = \dim(\text{point}) = 0$, and similarly, $f(\alpha_{\min}) = 0$. Also,

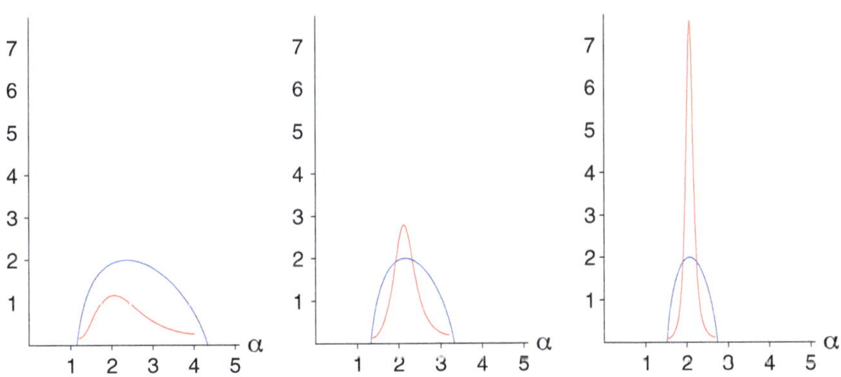

Figure 15.8 The $f(\alpha)$ and $k(\alpha)$ curves for the square IFS with $p_1 = \delta$, $p_2 = 1/4$, $p_3 = 1/4$, $p_4 = 1/2 - \delta$ for $\delta = .05, .1, .15$.

$\max(f(\alpha)) = 2$, the dimension of the attractor of the IFS $\{T_1, T_2, T_3, T_4\}$. When $\delta \to 1/4$, the multifractal collapses to the uniform distribution on the filled-in unit square, $\alpha_{\min} \to 2$ and $\alpha_{\max} \to 2$, so $\kappa(2) \to \infty$. That is, in this example, an IFS with no memory at all, the curvature can grow without bound. But this is a consequence of the range of α values going to 0. Any interpretation of curvature must be filtered through the effects of the range of α values.

The first 2-IFS of our family is

$$\text{(a) } [P][P][P][P]$$

where

$$[P] = \begin{bmatrix} 1/4 & 1/4 & 1/4 & \delta \\ 1/4 & 1/4 & 1/4 & 1/2-\delta \\ 1/4 & 1/4 & 1/4 & 1/4 \\ 1/4 & 1/4 & 1/4 & 1/4 \end{bmatrix}$$

Rather than vary δ, we set $\delta = 0.025$ and vary the matrices.

By Proposition 3.1 of [25], this 2-IFS is equivalent to the 1-IFS with transition matrix $[P]$. The $f(\alpha)$ and $k(\alpha)$ curves are plotted in Fig. 15.9.

In general, to compute the $f(\alpha)$ curve for a 2-IFS we use the higher block IFS method introduced in [25], based on the higher block shifts described in [26]. The transformations of this higher block IFS are the compositions $T_i \circ T_j$ given by the allowed pairs. The allowed transitions are $T_iT_j \to T_kT_l$ if $i = l$ and $klj = kij$ is an allowed triple. That is, the 2-IFS is equivalent to a 1-IFS having a larger set of transformations, and the probability of an allowed transition $T_iT_j \to T_kT_l$ is p_{kl}. The spectral radius, $f(\alpha)$ and $k(\alpha)$ curves are computed in the usual way.

Now take

$$[Q] = \begin{bmatrix} \delta & 1/4 & 1/4 & \delta \\ 1/2-\delta & 1/4 & 1/4 & 1/2-\delta \\ 1/4 & 1/4 & 1/4 & 1/4 \\ 1/4 & 1/4 & 1/4 & 1/4 \end{bmatrix}$$

In (b), (c), (d), and (e) of Fig. 15.9 we see the $f(\alpha)$ and $k(\alpha)$ curves for the 2-IFS with transition arrays

$$\text{(b) } [P][P][P][Q] \quad \text{(c) } [P][P][Q][Q] \quad \text{(d) } [P][Q][Q][Q] \quad \text{(e) } [Q][Q][Q][Q]$$

How can we interpret these graphs? The 2-IFS (a) is equivalent to a 1-IFS, so involves only one step of memory. The 2-IFS (b) is not equivalent to a

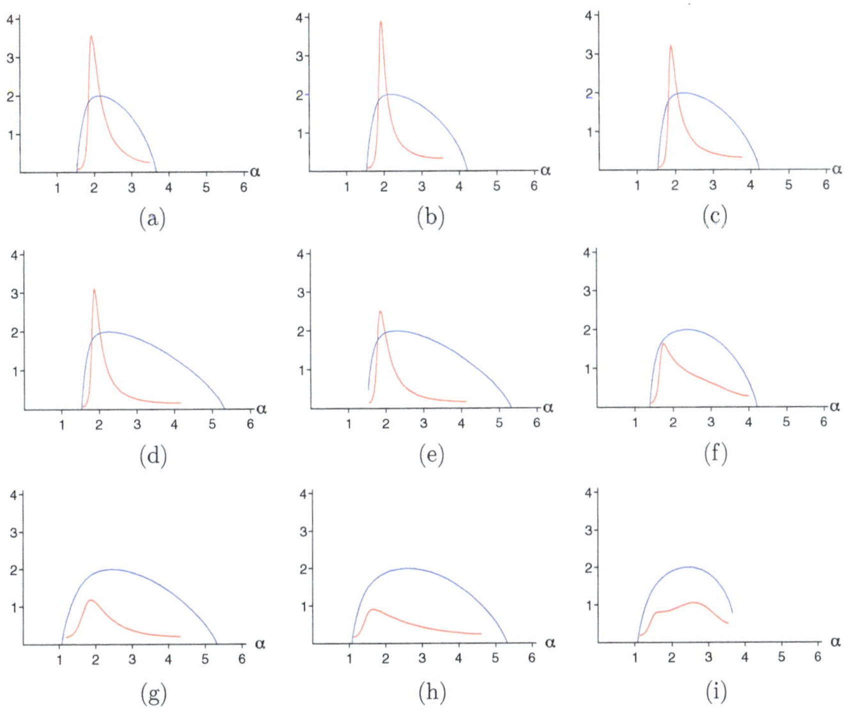

Figure 15.9 The $f(\alpha)$ and $k(\alpha)$ curves for 2-IFS (a) through (i).

1-IFS with transformations (15.1), and we note the curvature peak is higher and sharper. Continuing through graphs (c), (d), and (e), the curvature peak becomes lower. The 2-IFS (e), having the lowest curvature peak of those seen so far, is equivalent to the 1-IFS with transition matrix $[Q]$. Matrices P and Q involve placing δ and $1/2 - \delta$ in different locations, so combining P and Q in (b) involves more memory than do the homogeneous arrays (a) and (e). Because the curvature peak of (e) is lower than that of (a), it is plausible that adding more copies of Q in the memory array can lower the curvature peak. However, we have not yet investigated the possible effects of different placements of the factors in the transition array, so we have no expectation that these observations generalize.

The graph (f) of Fig. 15.9 has transition array $[P][Q][R][S]$, where

$$[R] = \begin{bmatrix} \delta & \delta & 1/4 & \delta \\ 1/2-\delta & 1/2-\delta & 1/4 & 1/2-\delta \\ 1/4 & 1/4 & 1/4 & 1/4 \\ 1/4 & 1/4 & 1/4 & 1/4 \end{bmatrix}$$

and

$$[S] = \begin{bmatrix} \delta & \delta & \delta & \delta \\ 1/2-\delta & 1/2-\delta & 1/2-\delta & 1/2-\delta \\ 1/4 & 1/4 & 1/4 & 1/4 \\ 1/4 & 1/4 & 1/4 & 1/4 \end{bmatrix}$$

The graph (g) has transition array $[R][R][R][R]$; the graph (h) has $[S][S][S][S]$. The graph (i) has transition array $[T][T][T][T]$, where

$$[T] = \begin{bmatrix} 1/4 & \delta & \delta & \delta \\ 1/4 & 1/2-\delta & 1/2-\delta & 1/2-\delta \\ 1/4 & 1/4 & 1/4 & 1/4 \\ 1/4 & 1/4 & 1/4 & 1/4 \end{bmatrix}$$

That is, T is just R with the column of 1/4s moved.

These examples show that as we progress from (b) to (e), the memory effects become more homogeneous, until at (e) we drop from a 2-IFS to a 1-IFS. Then from (e) to (h) the memory effects become even more homogeneous, with (h) reduced to a 0-IFS. This leads to the general conjecture that increasing memory requirements elevate the curvature peak, while decreasing memory requirements lower the peak, with obvious allowances made for the collapse of the $f(\alpha)$ curve to a single point.

However, the situation is more complicated than this: (g) and (i) have similar memory requirements. The only difference in their matrices is the location of the column of 1/4s. Yet their curvature graphs, including their maximum heights, differ,

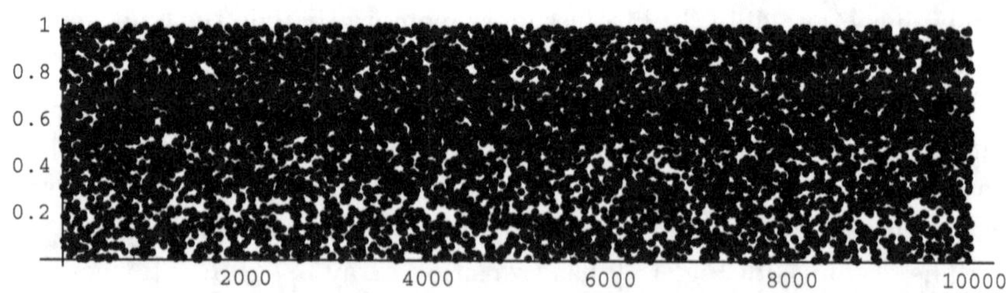

The second experiment uses this time series, constructed by the method of Sec. 15.6. The middle bin boundary is placed at the median, the other

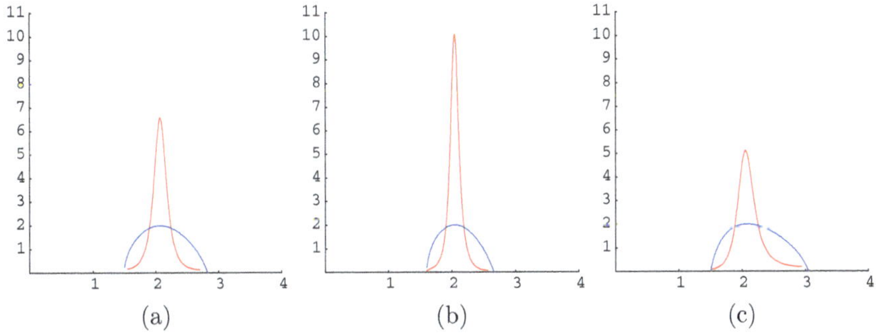

Figure 15.10 The $f(\alpha)$ and $k(\alpha)$ curves for this time series, δ increases left to right.

boundaries symmetrically above and below, by an amount δ, increasing from left to right in Fig. 15.10.

If a time series is determined just by the probabilities p_i of being in bin i, that is, of the previous state has no influence on the next, then each row of the transition matrix $[p_{ij}]$ is constant. So as a crude measure of the memory of the system, we sum the variances of each row of the measured transition matrices. Constant rows would give a sum of 0, motivating the notion that the lower the sum of variances, the shorter the memory. For these three examples, the variance sums are

(a) .0066 (b) .0057 (c) .0059

consistent with the idea that the higher the maximum curvature, the shorter the memory. Of course, this is just one example.

While the evidence investigated so far suggests some correlation between curvature and memory, several complications argue against a simple relation. This is only a first step.

15.9 Curvature surfaces of the data

To illustrate how curvatures can emphasize features not obvious in the surface of $f(\alpha)$ curves, we use time series of differences of successive daily closing prices for IBM, GE, and Google stock. The axes are arranged as shown here. Recall δ is the fraction of the range above and below the median where the first and third bin boundaries are placed.

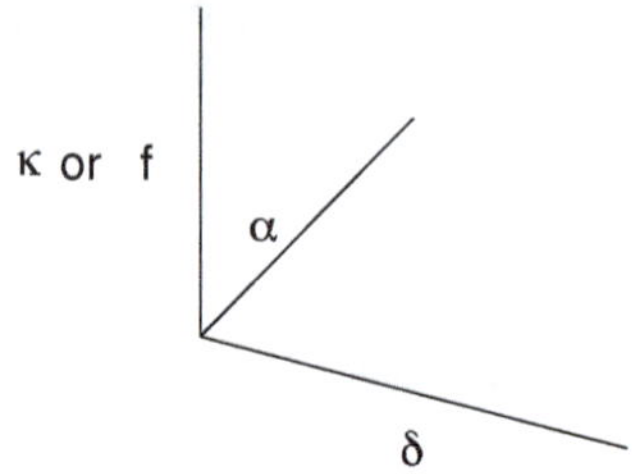

The surfaces of Fig. 15.11 are made of curves. For each δ sampled, we plot an $f(\alpha)$ curve and a $\kappa(\alpha)$ curve. The surfaces are made by interpolating polygons between adjacent curves.

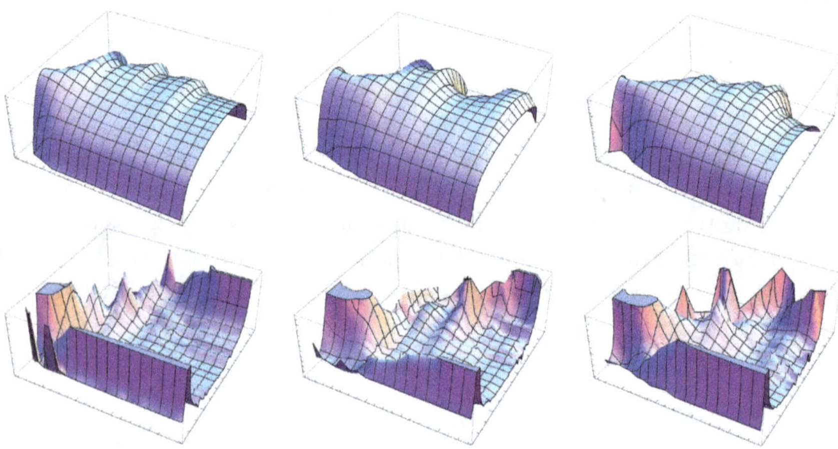

Figure 15.11 The $f(\alpha)$ and $k(\alpha)$ surfaces for IBM (left), GE (middle), and Google (right).

To avoid the curvature spikes at the extreme values of α, and where the range of α values dwindles, the $\kappa(\alpha)$ all have been truncated at a fixed height.

In [16] we interpreted some consequences of properties of $f(\alpha)$ surface plots. In the next section, we sample what the $\kappa(\alpha)$ surfaces can tell us.

15.10 Interpreting the curvature plots

The most obvious aspect of the $f(\alpha)$ surfaces is that, at least in major features, they look similar: hemicylinders with several waves superimposed. Common to the curvature surfaces is divergence near the left side of the graph. There, we see the range of α values decrease, so this divergence is due, at least in part, to the relation between maximum curvature and reduced α range mentioned in Sec. 15.8.

Of greatest interest is the trough in the middle of the curvature surfaces. For example, the two chains of peaks with ridge lines parallel to the α-axis in the GE surface mirror the two central waves in the GE $f(\alpha)$ surface. The right ridge has higher curvature, so if our conjecture is correct, using the corresponding δ will give a shorter memory representation of the data. The wider wave on the right of the GE surface has lower curvature, so bins for these δ will require a longer history to parse the data.

The IBM and Google curvature surfaces exhibit a complication: sharp, fairly high local peaks, local maxima among all nearby points in the (δ, α)-plane. Keeping with our conjecture, these peaks suggest that with a given δ, for some α values the dimension of the α-iso sets are sensitive to longer history than are the dimensions of other α-iso sets.

If true, this opens up the possibility that making predictions about different parts of an $f(\alpha)$ curve might require different amounts of memory, different length fragments of the time series. Of course, we never want to discard data, so the other interpretation is that using (all of) a given time series, we may be able to make more reliable predictions about some parts of an $f(\alpha)$ curve than we can about others.

The collection of small hills in the curvature surfaces suggests a complex interweaving of levels of memory dependence. Multifractals were developed in part because roughness can vary with location. By considering the curvature of $f(\alpha)$ curves, we may find that the depth of the time signal influencing part of a system's behavior can vary with the roughness of that behavior. Perhaps this variation can be organized on a higher level, a multifractal of multifractals.

A part of Benoit's legacy is the recognition that asking questions, making conjectures and guesses, is important part for how science advances. We are pleased to add our small contribution to this cause.

15.11 Conclusion

If curvature and length of memory do admit a manageable relationship, an application of this is to choose the bin boundaries to minimize the length of memory. This in turn would require knowing a smaller set of transition probabilities — say length three addresses rather than length five. Consequently, more reliable predictions can be made with smaller data sets, because estimating length five address occupancies requires more data than does estimating length three address occupancies.

Put another way, with a given data set we can be more confident length three address occupancies, and predictions based on those, than we can about length five address predictions.

It is also worth noting that curvature surface is drawn from the connection of individual line curvatures, each with the δ held constant. This course was chosen due to an intellectual obstacle: how to scale the δ axis to the $f(\alpha)$ axis. For closed form solutions that generate the curvature graphs, this relationship is not needed. But for the algorithmic methods required to generate surface curvatures (e.g. Voronoi-based curvature estimation) this relationship is fundamental.

Though this track of research was abandoned, with the line curvature method used in its stead, both authors believe the track to be an avenue worthy of further pursuit. The reason for this belief and for the aforementioned track's abandonment are almost entirely one in the same. In trying to match the scale of the δ axis to the $f(\alpha)$ an oddity was uncovered: the relationship between these axes seemed to be dependent on both of their values. The relationship was neither fixed nor, at first glance, at all obvious or continuous.

These predictions are only probabilisitc, of limited utility in the financial world where wild swings account for a substantial fraction of the big gains, and of the big losses. This limitation is familiar: often Benoit explained that his multifractal finance cartoons were not predictive. Rather, their principal virtue lay in warning that risk is underestimated by the most common trading models.

Benoit mentioned this point when talking about his multifractal cartoons to a group of high school and college teachers. A teacher asked if Benoit hadn't used his cartoons to make money on the stock market. Benoit thought a moment, then replied, "In public I never discuss four things: religion, politics, my portfolio, or sex." Silence from the audience, then the questioner continued, "I understand why you don't discuss religion or politics, and I think I get your point about portfolios. But really, has anyone ever asked you about sex?" With the slightest grin, Benoit replied, "No, but I keep hoping." The teachers burst into long, enthusiastic laughter. This example of Benoit's sense of humor is a good place to stop.

15.12 Appendix

Here we finish the calculations begun in Sec. 15.7. To understand the situation when B_i^{new} is not contained in B_i^{old}, we compute $p_{12}^{\text{new}} = \Pr(B_2^{\text{new}} \to B_1^{\text{new}})$. Because B_2^{new} consists of two portions, a part of B_1^{old} and all of B_2^{old}, we write p_{12}^{new} as the sum of two probabilities, weighted according to the fraction of B_2^{new} contained in B_1^{old} and in B_2^{old}. Again using ratios of lengths, these are

$$\frac{\delta - 1/4}{\delta} = \frac{4\delta - 1}{4\delta} \quad \text{and} \quad \frac{1/4}{\delta} = \frac{1}{4\delta}$$

Then we see

$$\begin{aligned}
p_{12}^{\text{new}} &= \frac{4\delta - 1}{4\delta}\Pr\big(B_1^{\text{old}} \to B_1^{\text{new}}\big) + \frac{1}{4\delta}\Pr\big(B_1^{\text{old}} \to B_1^{\text{new}}\big) \\
&= \frac{4\delta - 1}{4\delta}\Pr\big(B_1^{\text{old}} \to (1/2 - \delta)B_1^{\text{old}}\big) + \frac{1}{4\delta}\Pr\big(B_1^{\text{old}} \to (1/2 - \delta)B_1^{\text{old}}\big) \\
&= \frac{4\delta - 1}{4\delta}(2 - 4\delta)p_{11}^{\text{old}} + \frac{1}{4\delta}(2 - 4\delta)p_{12}^{\text{old}}
\end{aligned}$$

where the last line follows by applying the restricted range calculation.

The final complication occurs when both range and domain must be decomposed into pieces, illustrated by computing p_{22}^{new}.

$$\begin{aligned}
p_{22}^{\text{new}} &= \Pr\big(B_2^{\text{new}} \to B_2^{\text{new}}\big) \\
&= \frac{4\delta - 1}{4\delta}\Pr\big(B_1^{\text{old}} \to B_2^{\text{new}}\big) + \frac{1}{4\delta}\Pr\big(B_2^{\text{old}} \to B_2^{\text{new}}\big) \\
&= \frac{4\delta - 1}{4\delta}\big(\Pr\big(B_1^{\text{old}} \to (\delta - 1/4)B_1^{\text{old}}\big) + \Pr\big(B_1^{\text{old}} \to B_2^{\text{old}}\big)\big) \\
&\quad + \frac{1}{4}\big(\Pr\big(B_2^{\text{old}} \to (\delta - 1/4)B_1^{\text{old}}\big) + \Pr\big(B_2^{\text{old}} \to B_2^{\text{old}}\big)\big) \\
&= \frac{4\delta - 1}{4\delta}\big((4\delta - 1)p_{11}^{\text{old}} + p_{21}^{\text{old}}\big) + \frac{1}{4\delta}\big((4\delta - 1)p_{12}^{\text{old}} + p_{22}^{\text{old}}\big)
\end{aligned}$$

A bit of tedious algebra gives these results.

$$
\begin{aligned}
p_{11}^{\text{new}} &= p_{11}^{\text{old}}(2-4\delta) \\
p_{21}^{\text{new}} &= p_{21}^{\text{old}} + p_{11}^{\text{old}}(4\delta-1) \\
p_{31}^{\text{new}} &= p_{31}^{\text{old}} + p_{41}^{\text{old}}(4\epsilon-1) \\
p_{41}^{\text{new}} &= p_{41}^{\text{old}}(2-4\epsilon) \\
p_{12}^{\text{new}} &= p_{12}^{\text{old}}(2-4\delta)/4\delta + p_{11}^{\text{old}}(2-4\delta)(4\delta-1)/4\delta \\
p_{22}^{\text{new}} &= p_{22}^{\text{old}}/4\delta + (p_{12}^{\text{old}} + p_{21}^{\text{old}})(4\delta-1)/4\delta + p_{11}^{\text{old}}(4\delta-1)^2/4\delta \\
p_{32}^{\text{new}} &= p_{32}^{\text{old}}/4\delta + p_{42}^{\text{old}}(4\epsilon-1)/4\delta \\
&\qquad + p_{31}^{\text{old}}(4\delta-1)/4\delta + p_{41}^{\text{old}}(4\delta-1)(4\epsilon-1)/4\delta \\
p_{42}^{\text{new}} &= p_{42}^{\text{old}}(2-4\epsilon)/4\delta + p_{41}^{\text{old}}(2-4\epsilon)(4\delta-1)/4\delta \\
p_{13}^{\text{new}} &= p_{13}^{\text{old}}(2-4\delta)/4\epsilon + p_{14}^{\text{old}}(2-4\delta)(4\epsilon-1)/4\epsilon \\
p_{23}^{\text{new}} &= p_{23}^{\text{old}}/4\epsilon + p_{13}^{\text{old}}(4\delta-1)/4\epsilon \\
&\qquad + p_{24}^{\text{old}}(4\epsilon-1)(4\delta-1)/4\epsilon + p_{24}^{\text{old}}(4\epsilon-1)/4\epsilon \\
p_{33}^{\text{new}} &= p_{33}^{\text{old}}/4\epsilon + (p_{43}^{\text{old}} + p_{34}^{\text{old}})(4\epsilon-1)/4\epsilon + p_{44}^{\text{old}}(4\epsilon-1)^2/4\epsilon \\
p_{43}^{\text{new}} &= p_{43}^{\text{old}}(2-4\epsilon)/4\epsilon + p_{44}^{\text{old}}(2-4\epsilon)(4\epsilon-1)/4\epsilon \\
p_{14}^{\text{new}} &= p_{14}^{\text{old}}(2-4\delta) \\
p_{24}^{\text{new}} &= p_{24}^{\text{old}} + p_{14}^{\text{old}}(4\delta-1) \\
p_{34}^{\text{new}} &= p_{34}^{\text{old}} + p_{44}^{\text{old}}(4\epsilon-1) \\
p_{44}^{\text{new}} &= p_{44}^{\text{old}}(2-4\epsilon)
\end{aligned}
$$

Bibliography

[1] A. Douady, J. Hubbard, *Étude dynamique des polynômes complexes* Publ. Math. D'Orsay, 1er partie 84-02, 2me partie 85-04.

[2] B. Mandelbrot, *The Fractalist. Memoir of a Scientific Maverick*, Pantheon, New York, 2012.

[3] J. Thornton, "Fractal geometry course project," Yale University, 2002.

[4] B. Mandelbrot, R. Hudson, *The (Mis)Behavior of Markets. A Fractal View of Risk, Ruin, and Reward*, Basic Books, New York 2004.

[5] A. Weigend, N. Gershenfeld (eds.), *Time Series Prediction. Forecasting the Future and Understanding the Past*, Addison-Wesley, Reading, MA 1994.

[6] K. Mischaikow, M. Mrozek, J. Reiss, A. Szymczak, *Phys. Rev. Lett.* **82** (1999) 1144.

[7] K. Mischaikow, M. Mrozek, A. Szymczak, J. Reiss, *From time series to symbolic dynamics: an algebraic topological approach*, http://www.math.gatech.edu/ mischaik/papers/paperlist.html

[8] T. Suzuki, T. Ikeguchi, M. Suzuki, *Fractals* **14** (2006) 27.

[9] H. Teramoto, T. Komatsuzaki, *Chaos* **20** (2010) 037113.

[10] H. Jeffrey, *Nucleic Acids Research* **18** (1990) 2163.

[11] H. Jeffrey, *Computers and Graphics* **16** (1992) 25.

[12] I. Stewart, *Dynamics Newsletter* **3** (1989) 4.

[13] M. Barnsley, *Fractals Everywhere*, Academic Press, San Diego, 1993. (2nd ed).

[14] S. Kalla, N. Sobhan, "Fractal geometry course project," Yale University, 2000.

[15] N. Thew, "A dictionary of driven iterated function systems to characterize chaotic time series," senior thesis, Yale University, 2014.

[16] W. Martino, M. Frame, *Intl. J. Bif. Chaos* **20** (2010) 2453.

[17] W. Martino, M. Frame, *Multifractal measures of time series: the role of Gaussian curvature*, in preparation.

[18] C. Evertsz, B. Mandelbrot, *Multifractal measures*, Appendix B of [19].

[19] H.-O. Peitgen, H. Jürgens, D. Saupe, *Chaos and Fractals. New Frontiers of Science*, Springer-Verlag, New York, 1992.

[20] C. Godrèche, J. Luck, *J. Phys. A* **23** (1990) 3769.

[21] K. Falconer, *Techniques in Fractal Geometry*, John Wiley & Sons, Chichester, 1997.

[22] G. Ruppeiner, *Phys. Rev. A* **20** (1979) 1608.

[23] G. Ruppeiner, *Rev. Mod. Phys.* **67** (1995) 605.

[24] G. Ruppeiner, *Am. J. Phys.* **78** (2010) 1170.

[25] R. Bedient, M. Frame, K. Gross, J. Lanski, B. Sullivan, *Fractals* **18** (2010) 145.

[26] D. Lind, B. Marcus, *An Introduction to Symbolic Dynamics and Coding*, Cambridge Univ. Pr., Cambridge, 1995.

Chapter 16

Benoit Mandelbrot, educator

Nial Neger and Harlan J. Brothers

16.1 A natural teacher

Benoit Mandelbrot was remarkable not only for his broad mathematical insight, but also for the scope and depth of his interests. A conversation with him often would turn into an enlightening experience. His transparent love of knowledge might lead one through diverse subjects: the etymology of a word; a new piece of music he had heard; a piece of art he had seen in a new way; some practical wisdom in worldly matters; or a new development in fractal geometry. While such a torrent of information, taking the recursive form of nested footnotes, might at times place strenuous demands on an unsuspecting listener, to experience Benoit's encyclopedic mind in action was to join him in an educational adventure.

Clearly, Benoit had a childlike exuberance for learning and sharing this learning with others. He was a natural teacher who radiated his sense of wonder. He was especially interested in the education of students and

Figure 16.1 Benoit at the 2002 Fractal Geometry Workshop at Yale.

wanted very much to spark the joy of learning in them. This interest in education led to a collaboration with Michael Frame in editing *Fractals, Graphics, & Mathematics Education* [1]. The book is a collection of articles by high school and college teachers about their experiences in teaching fractal geometry. Its purpose was to pave the way for introducing fractals in the classroom and help inspire others to follow suit.

16.2 The fractal geometry workshops

For Benoit, the book was a starting point. He also wanted to speak with mathematics teachers and directly encourage them to begin teaching fractal geometry. He was not interested in having it taught as a separate entity, except perhaps as a college course. Rather, it seemed more natural, and practical, that it should blend into the existing mathematics curriculum.

To encourage the teaching of the subject in high schools and colleges, in the spring of 2000, he pushed to have a fractal geometry workshop at Yale. He enlisted Michael and Nial to organize and conduct the workshop. It was first offered in the summer of 2000 and ran Monday through Friday for five hours a day. Forty-nine high school and college teachers attended.

The content consisted of a basic introduction to fractals with a mixture of informal and technical information. Topics included: an introduction to fractals and iterated function systems (IFS); natural fractals and

Figure 16.2 Participants at the 2002 Fractal Geometry Workshop experimenting with the art technique of Decalcomania.

dimensions; the Mandelbrot set and Julia sets; cellular automata and fractal evolution; and random fractals and the stock market. The application and natural appearance of fractals were important themes. Discussion included: architectural fractals, fractal image compression, fractals in poetry, music, painting, and literature, fractals and mixing chemicals, fractal antennas, cascades and 1/f noise, fractals in history, diffusion-limited aggregation, bacterial growth in stressed environments, the large scale distribution of galaxies, and surrogates of the stock market.

There were two hours of activities in the morning, an hour for lunch and informal discussion, followed by two more hours of activities in the afternoon. Sessions comprised both computer based lectures and demonstrations along with hands-on labs in which participants often used custom software for directed exploration.

16.3 First meetings

How was it that we, the authors, found ourselves involved in this seminal project? Just as Benoit trusted his eye, he also trusted his instinct with regard to people. He was one of a rare breed of professionals who did not need to see a résumé to be certain that he wanted to work with someone.

So it was when Nial came to meet with Michael. Nial had taken a semester course in fractal geometry that Michael taught at Yale. He was captivated by the subject and wanted to discuss how he might continue his study of fractals. As Nial was waiting outside Michael's office, Benoit

Figure 16.3 Left to right: Harlan Brothers, Michael Frame, Benoit Mandelbrot, Nial Neger.

happened to pass by. "Hello," he said as he headed towards his office which was a few doors down from Michael's. A moment later, Benoit emerged from his office with an empty coffee cup and asked, "Are you one of Michael's students?" Nial replied that he was and had come to talk with Michael about how he could further his study of fractal geometry.

Benoit went down the hall, returned with a full coffee cup and said "When you have finished speaking with Michael would you please stop and see me. My office is right there. I might have some things for you to do." Here is Nial's description of what happened next.

> Unbelievable! He was going to talk to me about my working with fractals. The father of fractals himself. After talking with Michael, who described several projects he and Benoit were working on that he thought I might be interested in, I went to see Benoit. We spent close to forty minutes talking. He told me what he was doing. There were so many things I found it hard to keep track of them all. He asked about my work experience and seemed particularly interested when I said I had organized and run several workshops during the course of my teaching career. Aliette sat patiently and listened attentively, interjecting useful and insightful comments when appropriate. While we talked I noticed that he seemed to be actually interested in me. Why would a person of his stature be interested

> in someone he had just met? I came to find out he was simply interested in people. He also had a wonderful sense of humor and was very funny. At first I missed some of his quips because I couldn't believe that such a serious scholar could also be so humorous. At the end he said, 'So, would you be interested in working for me?' Would I! I had to restrain myself from jumping up and down in the office. That evening I tried to recall what had taken place and I wondered, 'Did it really happen or was it just a dream?' And so started a seven-year mathematical adventure.

From this chance meeting the fractal workshops at Yale began to take shape.

It was Harlan's mathematics professor, Miguel Garcia, who recommended him to Benoit. Benoit had called his former student Miguel looking for help in editing what was, at the time, planned to be the 20th anniversary edition of *The Fractal Geometry of Nature*, [2].

Harlan met with Benoit at his office at Yale. After greetings, Benoit sat down, his hands pointed in and resting on his legs. He began by saying, in his inimitable French accent, "So, Miguel tells me you are not the average cookie-cutter student..." Harlan shared some of his research and by the end of their meeting, Benoit had shared everything from the lesser known work of John Venn to the socio-political history of Budapest dating back to the 16th century, all of it fascinating. Harlan immediately began working on the book project and soon after as a teaching assistant at the workshops.

Benoit later suggested to Harlan that he pursue a rigorous exploration of the ways in which music can exhibit fractal structure. This led to a lecture and lab on fractal music that Harlan conducted at the 2004 workshop. Over the years, Benoit continued to share his overwhelming expertise, humor, and wisdom in practical matters. As with most natural teachers, his generosity of spirit and fundamental good nature provided a constant source of inspiration.

16.4 The NSF grant

We knew we wanted to run another workshop the following year but we were immediately confronted by a challenge. Many of the teachers were thrilled with the first workshop and wanted to do more the following year. When told that there were no such plans, only a repeated introduction the

Figure 16.4 Benoit with one of the workshop projects.

coming year, they were disappointed and loudly petitioned for a follow-up workshop. We thought we could do a three day advanced session, but it was not clear how it might be funded. Through Benoit's powers of persuasion, the mathematics department at Yale had generously funded the first workshop, but would they be willing to do it again and add three more days? While we were working for free, there was the expense of lunches, parking, materials for the labs, etc. Again, Benoit persuaded the department to fund us. However, it was clear that to continue to grow, we would need to apply for a grant.

The original plan resembled the second year's workshop: a five day introductory session for first-time participants and a three day session for those who had already attended. There would be money for lab materials, lunches, and parking, but it was a modest proposal. However, this time, serendipity played its part.

To our great good fortune, Benoit happened to be lecturing in Washington, D.C. In the audience were some administrators from the National Science Foundation. They were very interested when they heard his ideas on fractals and education. They suggested he apply for an NSF grant. But this was exactly what he was doing, and he described our proposal. Their response was to the effect, "Oh, this is too modest a proposal for such an important project. It might get overlooked if left as it is. You'll need to expand it." Their only concern whether it could succeed if both high school and college teachers participated together. Did they have anything in common to discuss? After two years, we knew we had a successful approach. The teachers clearly had much in common and gained substantially from working with each other.

Benoit returned with fresh enthusiasm for getting a grant and with new ideas for its content which were elaborated on by Michael and Nial. Happily, we found ourselves back at the drawing board to enlarge the grant proposal.

We expanded it to include provisions for posting a public web page containing the mathematical theory for the lectures; to write a lab manual and a set of classroom lesson plans; hire a professional software programmer; produce a DVD on fractal geometry; hire a teaching assistant; establish satellite fractal geometry programs; and pay for guest lecturers. We were also able to invite participants from an expanded geographical region by providing overnight accommodations for teachers along with a modest stipend. This became the foundation for the approved grant which ran for four years and engaged over 250 educators from around the country and several from abroad

While most of the lectures and labs were conducted by Michael and Nial, each summer there were two afternoon sessions where Benoit spoke about his latest research. He came early, often accompanied by his wife Aliette, to have lunch with the teachers. He would ask them questions about what they were doing, sign books for them, pose for pictures and be generally accessible. Teachers were often impressed that a person of his stature was so down to earth and sincere in engaging them on a one-to-one basis. Benoit's friendly, kind, and playful demeanor underscored the fun of education and gave participants the added gift of numerous stories and fond memories.

16.5 The website

Given that the primary purpose of the workshops was to have teachers take their experience back to classroom, great care was taken to present the material with a methodology that teachers could use when presenting the material to their students. Beginning with the first workshop in 2000, all notes for the lectures, the hands-on demonstrations, lab activities, and software were therefore placed online at a dedicated Yale web site and made accessible to the public.

The site, http://classes.yale.edu/fractals, is primarily authored, edited, and maintained by Michael. Starting with the original workshop notes, it began to include material from lectures he had given for his classes at Yale. The site has continued to grow steadily and is now a vast compendium of knowledge on the subject of fractal geometry. In addition to workshop and classroom notes, it includes workshop software; a broad cross-section of fundamental results and research; a lab manual (described in the following section); lesson plans; Mathematica code; Java software; and the *Panorama of Fractals and Their Uses* [3]. The Panorama is an enormous collection of information relating to the ubiquity of fractal structure in the universe. It includes applications and natural occurrences of fractals along with excerpts from articles written by prominent artists and scientists. In part due to Benoit's enthusiasm and vast network of contacts, the Panorama has attracted experts from wide variety of fields.

From the original series of lecture notes, the Yale website has now become the number one ranked site on the subject of fractal geometry.[1] It is a primary resource for research and education throughout the world.

16.6 The lab manual

As with most things that last any length of time, the grant components evolved over the course of the four years. The hands-on aspect of the workshop became more prominent and many lab activities were added.

A lab manual, *Kitchen Science Fractals* [4], now contains twenty-five labs that range from simple activities such as paper folding and finger painting to more complicated investigations concerning circle inversion, fractal basin boundaries, and driven IFS. All labs are meant to be done using inexpensive, easily obtained materials or free software designed for the labs.

Aside from providing a break from lectures, the labs proved to be valuable learning experiences for the teachers. All labs were done with at least two teachers working as partners. This created an *esprit de corps* among the participants, with many of them forming lasting bonds. Sometimes, labs (such as the Sierpinski Tetrahedron Project) incorporated tasks which were initially started by individuals but whose later stages required the individuals to work together in larger and larger groups to achieve the desired outcome.

---

[1] As ranked by Google for the term "fractal geometry," September 26, 2014.

16.7 The software

Software used in the workshops was written by Ginger Booth under the direction of Michael Frame and edited by Richard Bedient. It is Java- or Flash-based and was designed to be used in conjunction with the labs or as an investigative tool for researchers and the mathematically curious. The suite of programs is available to the public at both the Yale website and Ginger's site [5]. It consists of Deterministic IFS, Random IFS, Affine Transformation Calculator, Driven IFS, IFS with Memory, Random IFS with Memory, Box Counting Dimension, Mandelbrot and Julia Sets, 1-Dimensional Dynamics, Kelly Plots, Synchronized Oscillators, Cellular Automata, Circle Inversion Illustrator, Circle Inversion Limit Sets, L-Systems, Pascal's Triangle, Fractal Music Composer, and Fractal Statistics. The programs include theoretical overview, notes on usage, and built-in examples.

16.8 The lesson plans

The collection of classroom lesson plans was another area of focus for the workshops. Participants were encouraged to invent ways of integrating fractal geometry into their standard curricula. Nineteen lesson plans

Figure 16.5 Students involved in the Longfellow Sierpinski tetrahedron project, with teachers Bonnie Hole (left) and Janet Ortiz (center).

suitable for classroom use were developed, many of which can be found online [6]. They consists of teacher lesson guides, activities for the student and suggested excursions. All lesson plans cover topics commonly included in high school math curricula and assume no prior knowledge of fractals. They are interesting, self-contained modules meant to motivate students, particularly those who find mathematics challenging or, worse yet, boring. Many teachers have found them to be a valuable resource and have employed them with great success.

16.9 The DVD

Benoit thought that producing a DVD about fractals would be a wonderful way to make the basic ideas of fractal geometry accessible to high school teachers and students. Not only would it explain the fundamentals, but it would also offer examples of fractals from a variety of applications. It would additionally address some pedagogical issues for teachers.

Benoit knew a film maker who understood fractals and could do the topic justice, Nigel Lesmoir-Gordon, who had recently written a book on fractals [7], produced and directed the now classic 1995 documentary on fractal geometry, *The Colours Of Infinity*, narrated by Sir Arthur Clarke. Nigel was available and agreed to do the project. Some of Nigel's film experiences with Benoit are described in Chapter 13

The DVD comprises two parts. The first focuses on the fundamental ideas of fractals and the second on specific examples. As usual, Benoit was full of ideas and had numerous resources to draw on: computer generated landscapes from Dr. Ken Musgrave at Pandromeda; liver casts from Dr. Jean H. D. Fasel in Geneva; animal lung casts from Prof. R. W. Henry at the University of Tennessee; and casts of lungs and coronary arteries from Prof. Ewald R. Weibel at the University of Bern, to name a few. Benoit, of course, appeared in the DVD. He negotiated its marketing and it was released in 2005 as *Mandelbrot's World of Fractals* [8].

16.10 Satellite programs

As a part of the grant, we also established satellite programs at three different colleges. During the first year of the grant, three college teachers, Richard Bedient (Hamilton College), Michael Fisher (California State University at Fresno), and Michael Fraboni (Moravian College), attended the introductory week of the workshop not only to participate, but to

Figure 16.6 Detail of a Yale Sierpinski tetrahedron project.

observe the format and presentation. They also assisted in planning, editing, and presenting some of the material for the workshops. Over the following three years, introductory workshops similar to ours were given in the summer at Fresno State and Moravian College and an introductory course in fractal geometry was offered at Hamilton College.

With the expiration of the grant, the summer workshops ceased but the introductory course in fractals is still given at Hamilton College. Indeed, Richard recently conducted a well-received professional development workshop for the math department at Pittsfield High School, based on the materials from the Yale summer programs.

16.11 Conclusion

Benoit had an innate sense that fractal geometry could be used to both spark the interest of students, and inspire their teachers by pointing to the ubiquitous beauty of mathematics. Thanks to Benoit's vision, countless young minds around the globe have been introduced to the magic of mathematics through their exposure to fractal geometry. Our own experience with students attests to the fact that learning about fractals has given them a broader appreciation, not just of mathematics, but of the world around them.

Figure 16.7 Talking with MIT students.

We conclude with an excerpt from Benoit's memoirs [9]. A student had remarked to him:

> "You might have lived shortly after Newton. We can't believe that we could actually hear you discuss how part of our schoolwork had first come to your mind; to shake your hand would be a strange experience ... a big event."

Benoit reflected:

> Words from a charming young lady seemingly representing a group of college students who had packed a lecture I had just given. Of course, I was glad to shake that young lady's hand. Uncanny forms of flattery! Each lifted me to seventh heaven! Truly and deeply, each marked a very sweet day! Let me put it more strongly: it is occasions like that that make my life.

Bibliography

[1] M. Frame, B. Mandelbrot, *Fractals, Graphics, & Mathematics Education* (Math. Assoc. Amer., Washington, D.D., 2002).
[2] B. Mandelbrot, *The Fractal Geometry of Nature* (W. H. Freeman, San Francisco, 1982).
[3] M. Frame, B. Mandelbrot, *Fractal Geometry: A Panorama of Fractals and Their Uses*, http://classes.yale.edu/Fractals/Panorama/welcome.html.

[4] M. Frame, N. Neger, *Kitchen Science Fractals*, http://classes.yale.edu/fractals/Labs/Labs.html.
[5] G. Booth, *Fractals*, http://www.gingerbooth.com/pages/demosfractal.html.
[6] M. Frame, N. Neger, H. Brothers, *Fractal Geometry Lesson Plan Archive*, http://classes.yale.edu/Fractals/LP/LP.html.
[7] N. Lesmoir-Gordon, *Introducing Fractal Geometry* (Totem Books, Lanham, MD, 2000).
[8] N. Lesmoir-Gordon, B. Mandelbrot, M. Frame, *Mandelbrot's World of Fractals*, DVD, Key Curriculum Press, 2005.
[9] B. Mandelbrot, *The Fractalist. Memoir of a Scientific Maverick* (Pantheon, New York, 2012).

Chapter 17

The art of roughness

Emer O'Daly

"It is often said that 20th-century modern buildings are sterile, not built to human scale and, in fact, unnatural. The more I ponder such statements and their variants, the more elusive I find their meaning and the more I feel that a discussion of the logic and of the aesthetics of the notion of scale needs to be resumed." [1]

Benoit Mandelbrot

"Why is it that the silhouette of a storm-bent leafless tree against an evening sky in winter is perceived as beautiful, but the corresponding silhouette of any multipurpose university building is not, in spite of all efforts of the architect? The answer seems to me, even if somewhat speculative, to follow from the new insights into dynamical systems. Our feeling for beauty is inspired by the harmonious arrangement of order and disorder as it occurs in natural objects — in clouds, trees, mountain ranges, or snow crystals. The shapes of all these are dynamical systems jelled into physical forms, and particular combinations of order and disorder are typical of them." [2]

Physicist, Gert Eilenberger

Throughout my time studying and practising architecture, I have been deeply interested in understanding aesthetics from a scientific perspective. In university, I had read *The Fractal Geometry of Nature* and was drawn in by the visual basis of Mandelbrot's mathematics. At the time, I felt that there was some deep connection between fractal geometry and architectural aesthetics.

In 2009, I began a Masters in Architecture at Yale University and I studied fractal geometry under Dr. Michael Frame. This chapter is a result of my studies and the conversations I had throughout my time at Yale with two brilliant teachers — Michael Frame and architectural theorist Kent Bloomer.

17.1 Ornament in architecture

Fractal geometry brings a deeper understanding of the natural world. It may also deeply impact our understanding of the manmade world. Its potential contribution to the field of architecture was alluded to by Mandelbrot a number of times throughout his life. In *The Fractal Geometry of Nature* he writes that "in the context of architecture a Mies van der Rohe building is a scalebound throwback to Euclid, while a high period Beaux Arts building is rich in fractal aspects." [3] Benoit identified both the Palais Garnier [4] and the Eiffel Tower [5] as examples of scaling objects.

In an interview with the architect Paola Antonelli he gave in 2008, he said that "the idea that perfection is a cube is over." [6] Here he is referencing the architectural style of Modernism and its Euclidean shapes. Architectural forms of the Modernist style are characterized by simple, primary shapes devoid of ornament or extraneous detail, in contrast with the richly ornamented buildings of previous periods. Although the Modernist idiom has proliferated in the 20th century, there is little evidence that this style of architecture provides the best environment for mankind to inhabit. Surveys conducted by the AIA [7] (American Institute of Architects) and YouGov [8] in Britain indicate that the general public are disillusioned with Modernist buildings and, if asked for their opinion, are invariably drawn towards pre-twentieth century ornamented and Classical architecture. I believe fractal geometry may hold the key to understanding the aesthetic appeal of particular forms or styles of architecture.

So what did Mandelbrot mean by these references to the fractal aspects of Palais Garnier and the Eiffel Tower? What, architecturally speaking, does a high period Beaux Arts building such as the Palais Garnier possess, that a Mies van der Rohe building does not?

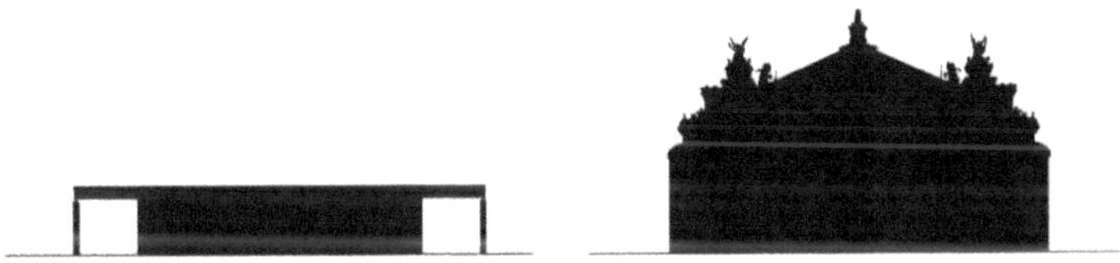

Figure 17.1 A comparison of the profiles of the Neue Nationalgalerie in Berlin by Miles van der Rohe (left) and the Palais Garnier in Paris by Charles Garnier (right).

Figure 17.2 A comparison of the level of detail or roughness in modernist architecture (left) and Beaux-Arts architecture (right).

The complaint often made against much modern architecture is that it is too "boxy." But the Parthenon — the quintessential Classical building — approximates a box. In fact, most buildings, regardless of architectural style, approximate Euclidean geometries at the large scale. The Pantheon in Rome can be approximated geometrically on the large scale by a box for the primary structure, a hemisphere for the roof, a triangle for the pediment, and cylinders for the columns. The fundamental difference between the geometry of Classical architecture and Modernist architecture lies at the small scales. If one visually compares Modern architecture with what came before, one of the most striking and basic differences lies in the profile or edges of these architectural styles. The outline of Modern architecture is flat, sharp and smooth, while the edges of traditional architecture have a more meandering or rough appearance. Taking much pre-twentieth century architecture, the striking visual difference lies at the scale of ornament.

The word ornament seems dated nowadays and is associated by many with applied decoration like cornices, friezes and fluting. However, prior to the 20th century, ornamentation was seen in many cultures as an essential property of both architecture and art. James Trilling calls it "the art we add to art." [9]

Ornament and decoration have been evident in most civilisations since the beginning of recorded history. For thousands of years ornament was seen as an essential property of architecture. Vitruvius, a Roman architect and writer in the first century A.D., saw ornament as architecture on the human scale. He traces the origin of temple ornament from its timber construction and gives directions, proportions, and rules for its design. The great Renaissance architect and writer, Leon Battista Alberti, in his treatise on architecture, *De Re Aedificatoria* (1450), devoted four of his ten books to ornament. He defines beauty as "that reasoned harmony of all the parts within a body," and that ornament may be defined "as a form of auxiliary light and complement to beauty." He states that beauty is an inherent quality of an object, whereas ornament is something that is attached or additional to the body of the object. While beauty and ornament were distinguished from each other, they were both seen as essential to the architectural whole.

In the 19th century, design theorist Christopher Dresser echoes this in his description of ornament as "that which, superadded to utility, renders the object more acceptable through bestowing upon it an amount of beauty which it would not otherwise possess." [10] Again, ornament is seen as additional to the body of art being ornamented and distinguished from the utility of the object.

Prior to the 20th century, theories on architecture see the act of ornamentation as a separate, but essential, element in the process of making architecture. It was seen as the culmination of the architectural life or expression of a structure. Its importance also lay in the fact that it was that part of architecture which existed at the human scale.

At the beginning of the 20th century, a revolution took place in architecture, which would eventually affect architecture worldwide today. Much received wisdom of traditional architectural forms, techniques, and methodologies was abandoned, and a new process replaced it which approached architecture from what was considered first principles. This new style of architecture — Modernism — rejected the heavily ornamented and detailed aesthetic that was typical of previous periods in favor of simple, Euclidean forms and smooth, flat surfaces.

Le Corbusier, one of the proponents of the Modernist movement, wrote at length on the beauty of Euclidean forms such as "cubes, cones, spheres, cylinders and pyramids," [11] although ironically, many of his best designs were spatially very complex. Ornament was demonised by Adolf Loos, an influential architect at the turn of the last century, in his essay "Ornament and Crime," in which he equated ornament with barbarism and uncivilized cultures: "The lower the cultural level of a people, the more extravagant it is with its ornament, its decoration." [12] He concluded that "the evolution of culture marches with the elimination of ornament from useful objects." [13] His ideas became one of the key formulations in the move away from historicism in Modern architecture. Ornament was seen as superfluous and inefficient, and to a culture enamored with the new Industrial age, it seemed to violate the beauty and efficiency of function. And with that, ornament disappeared from the architectural lexicon of the Western world, and has remained largely absent for the best part of the last hundred years.

Ornament may be described as the scale in architecture which most closely approximates forms in nature from the acanthus leaf to the human body. The figures and motifs of ornament may be organic or geometric in appearance and are generally characterized by their rhythmic arrangements. Ornament is composed of patterned figures which are syntactically distributed across surfaces, mediating between elements in a larger whole. Ornament is not synonymous with decoration and does not just encompass the detail or embellishment of an object; ornament is a visual language with its own figures, grammar, and syntax.

The repeatable figures or motifs in these patterns recur transhistorically and transculturally. In his seminal book, *Problems of Style*, art historian Alois Riegl, envisages the history of ornament as the endless reiteration of a very few fundamental motifs — the palmette, the rosette, the zigzag, the wavy line. [14] The spiral may also be described as one of these universal cultural icons or shapes. In *The Nature of Ornament*, architectural theorist Kent Bloomer writes that "... the spiral is the most generative figure in the history of ornament. Showing up in the ornament of all cultures, it speaks of virtual energy. It is also one of the most convenient shapes for visually articulating the actions of continuous dilation and contraction." [15]

Louis Sullivan, an American architect who devised an entirely new and personal system of ornament at the beginning of the 20th century, describes how the form of the ornament must emerge from the character of the architectural whole. He compares the design of an ornamental system to the growth patterns at different scales in nature: "It follows then, by the logic of growth, that a certain kind of ornament should appear on a certain kind of structure, just as a certain kind of leaf must appear on a certain kind of tree. An elm leaf would not 'look well' on a pine-tree — a pine-needle seems more 'in keeping'." [16]

Figure 17.3 Diagrams outlining three stages of ornamentation.

His protégée, Frank Lloyd Wright, elaborated on his philosophy of ornament and also developed a highly personal style. He believed ornament was not a matter of prettifying externals but should be organic with the structure it adorns. [17] Another way of stating this is that taken together, ornament and the object ornamented should possess a self-similar structure.

One of the primary functions of ornament is to mediate between different surfaces, volumes, materials, and spaces. It is common for ornament to fill the interstitial spaces in architecture — the gaps, the transition zones between elements, the height disparities between parts — and to bind an entire composition of architectural forms together. Throughout the history of architecture, ornament has often articulated how floor meets wall and wall meets ceiling; it can describe how elements and materials should converge and ornament has eased the transition between inside and outside.

Owen Jones, the influential design theorist of the 19th century, refers to this process in Proposition 7 in *The Grammar of Ornament.* "The general forms being first cared for, these should be subdivided and ornamented by general lines; the interstices may then be filled in with ornament, which may again be subdivided and enriched for closer inspection." [18]

There are also utilitarian reasons for ornamenting a building. The ornament often disguised the awkward moment where two materials would meet. We may illustrate this point by looking at the design of a door; or more specifically, where the void of a door meets the mass of a wall.

When designing how these two elements meet, for practical reasons, the architect or builder covers the joint to disguise the messiness of two different materials or systems meeting. In the case of a door, the covering of the joint is called the doorframe. The most basic, utilitarian form of a doorframe is a box that covers the edges of the door opening. However, when delineating the edges of a door, most cultures have gone far beyond this box. Georgian doors are renowned for their elaborate ornamentation. The Georgians crinkled the profile to create a variety of lengths at the edges of their openings and they are not alone in their celebration of an opening or passage from one space to another by this method.

This method of covering joints may be divided into three steps: firstly, with the creation of the utilitarian form; secondly, with the refining or "crinkling" of this form, and thirdly, with the addition of vegetal or geometric ornament. This process is not only performed on the architraves of the doors, but also the baseboards, cornices and window frames of their rooms, the eaves of the building where the roof meets the sky, the column capitals and the balustrades of the stairs. The distribution of these ornamental regions is driven by functional, utilitarian purposes, but the forms then produced and the refinement that takes place is driven by aesthetics.

In Fig. 17.4, these two photographs of the grand stair in the Palais Garnier in Paris also illustrate this process. In the photograph taken during the construction of the opera house (left, Fig. 17.4), the utilitarian forms can be seen taking shape and undergoing refinement. The second photograph (right, Fig. 17.4) reveals the transformation that takes place when the forms are overlaid with ornament. There is a proliferation of detail and form at the small scale that eases the transitional joints from column to capital and from door opening to wall and arch.

Profiling the edges and introducing detail at the small scale is not just limited to Classical European architecture. Traditional Chinese, Indonesian,

Figure 17.4 A comparison of the Palais Garnier during and after construction.

South American, Mughal and Indian cultures enriched their buildings with repeating forms and patterns at various scales, often over the entirety of the building. In the temple complex at Angkor Wat, the conical shape of the temple reappears at the scale of ornament, covering the surface with these repeating volumetric forms.

In *The Analysis of Beauty*, eighteenth-century British art theorist William Hogarth explores the rules of composition of form, in the design of a candlestick. Hogarth states that "the art of composing well is no more than the art of varying well." [19]

He goes on to explain this process of achieving better form:

1. "let every distance or length of divisions differ from the length of the socket, as also vary in their distances from each other,"
2. "let any two points, signifying distance, be placd farthest from any other two points, observing always that there should be one distance or part larger than all the rest; and you will readily see that variety could not be so complete without it."

He then warns against subdividing further because "it will want distinctness of form on a near view, and lose the effect of variety at a distance."

In Hogarth's description, he advocates the combination of a variety of lengths of "crinkle" in order to produce a well-composed form. He also advocates scale invariance, that is, self-similarity. He protests against the use of similar lengths (34), but instead indicates that various lengths at various scales produce a more pleasing effect (33). In this way, he is proposing that self-similarity at various scales, which introduce contrast in length or size in an overall form, is aesthetically appealing.

So, why do we go through this process of wrinkling, roughening, patterning and rusticating the surfaces of our structures? Why do we "crinkle" the profile of buildings as a method of beautification?

17.2 Ornament as fractal boundary

The most popular understanding of fractal geometry is the fact that fractals are self-similar; however, it may also be thought of as the field of mathematics that describes the geometry of edges of natural forms. Fractal geometry describes the mathematics of mediation in nature while in much pre-twentieth-century architecture, ornament mediates the edges of man-made forms.

In describing fractal shapes, Benoit Mandelbrot used the words, "grainy, hydralike, inbetween, pimply, pocky, ramified, seaweedy, strange, tangled, tortuous, wiggly, wispy, wrinkled." [20]

Similarly in architecture, these characteristics can also be used to describe the undulating surfaces and edges of much traditional ornamented architecture. When we see crinkled profiles and staggered edges in manmade forms, I believe we are seeing the influence of fractal geometry in our designs. Additionally, this fractal aspect to buildings usually becomes complete at the scale of ornament. When architects say that ornament "imports" or "super-adds" a new element into a building, along with the cultural significance and symbolic meaning it imbues, it also brings with it, quite literally, a new dimension. Ornamentation gives the edges of the form a fractal dimension. In this way, fractal dimension becomes a measureable property of an object or a space.

I believe that what Mandelbrot was articulating whenever he talked of the fractal nature of architecture was the fact that the craggy outlines of Beaux-Arts buildings lent the forms of those buildings a fractal dimension. Most forms in nature have a fractal edge, and therefore supplying such an edge to an object, gives it a more natural appearance.

In his paper on fractals and the classical orders, architect Andrew Crompton draws attention to the resemblance between the Doric entablature and the devil's staircase, a graph based on the Cantor set [21]. He points out that buildings can be seen as fractal in two ways. Firstly, a building can be classed as fractal if it repeats a motif or form over several orders of magnitude. Examples of this are Gothic and Romanesque architecture (Fig. 17.5, right). The second way in which a building may be fractal is by obeying a power law distribution in parts. What this means is that a building may be considered fractal if there is a exponentially larger quantity of elements or forms at the small scale, than at the large, regardless of the shape they take. Crompton concludes that "most old buildings are fractal in this sense because they have a lot of small-scale detail; in short, because they have mouldings." In the case of the Doric entablature both characteristics of a fractal are present; firstly, that there is a power law distribution between large and small scale, and secondly, that the forms (i.e. steps) remain similar at a number of scales.

What is salient here is the fact that fractals exhibit a power law distribution of shapes meaning that they possess a proliferation of detail at the small scale. Therefore, architecture can only possess fractal aspects if a

Figure 17.5 Jain Temple in Jaisalmer (left). The cathedral of Notre Dame at Bayeux (right).

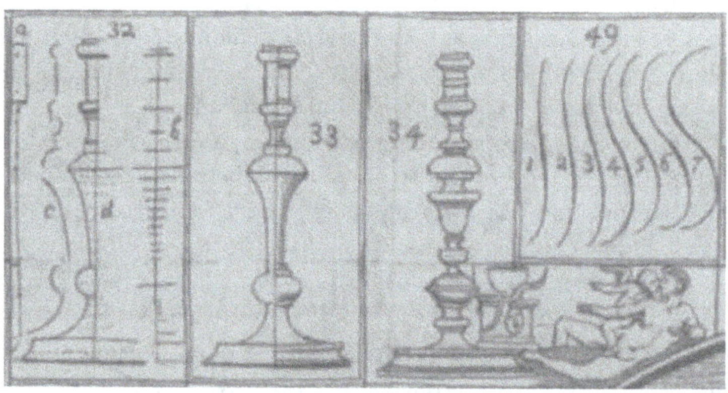

Figure 17.6 Hogarth's description of the amount of "crinkle" he advocates in the design of a candlestick.

cascade of detail appears at the small scale of the building. Ornament, in some form or other, is inevitable in a building with fractal aspects.

Louis Sullivan often wrote about the process of "awakening" an architectural form. He would begin with a Euclidean shape, often a pentagon or a circle, and through various symmetry operations, and elaborations of the form he would transform it into ornament that, in his words, "almost literally had life." [22] This method of developing an ornament may be seen in Sullivan's *A System of Architectural Ornament*, in a process he calls the "awakening of the pentagon." In a similar way, fractal geometry is itself a

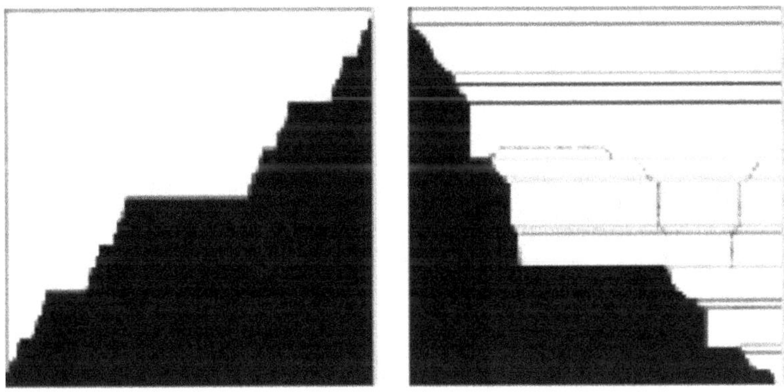

Figure 17.7 The devil's staircase (left) has a remarkable similarity to a Classical Doric Entablature (right).

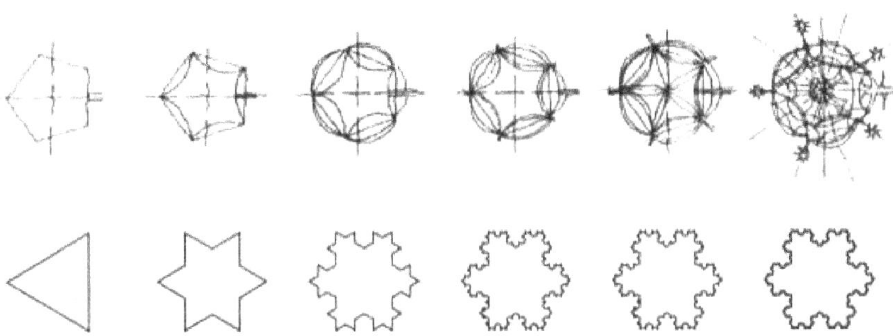

Figure 17.8 A comparison between successive iterates of the Koch snowflake and the development of ornament by Louis Sullivan exhibits remarkable similarities.

process of transformation from Euclidean to non-Euclidean forms. It could be said that Sullivan was talking about introducing a fractal dimension to his forms.

In Fig. 17.9, an image of part of the Mandelbrot set juxtaposed with a Rococo table from the 18th century, illustrates the similarities of these two formal vocabularies of architectural ornament and fractal mathematics. Appropriately, Mandelbrot refers to the forms that emerge in the Mandelbrot Set as, "extraordinary baroque decorations." [23]

In order to test the theory that the Rococo table has fractal aspect beyond the previous visual comparison with the Mandelbrot set, the number and radii of the circles required at various scales that approximated the

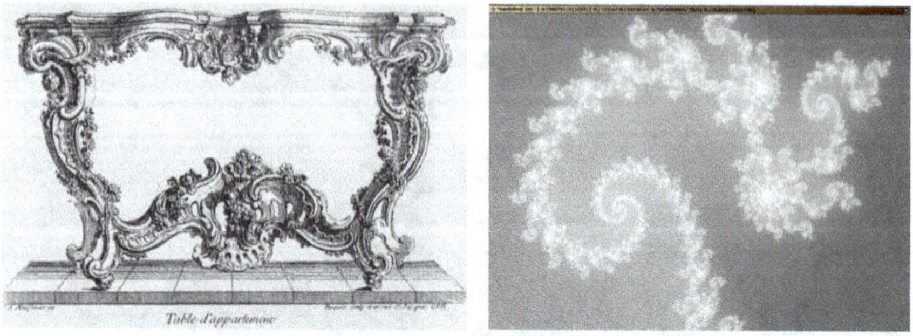

Figure 17.9 A comparison of an 18th century sketch of a Rococo table (left) and a portion of the Mandelbrot set (right).

r	Log$(1/r)$	$N(r)$	Log$(N(r))$
7.17	−0.85	2	0.3
3.8	−0.58	6	0.78
2.9	−0.46	7	0.85
1.6	−0.204	17	1.23
0.8	0.22	43	1.63

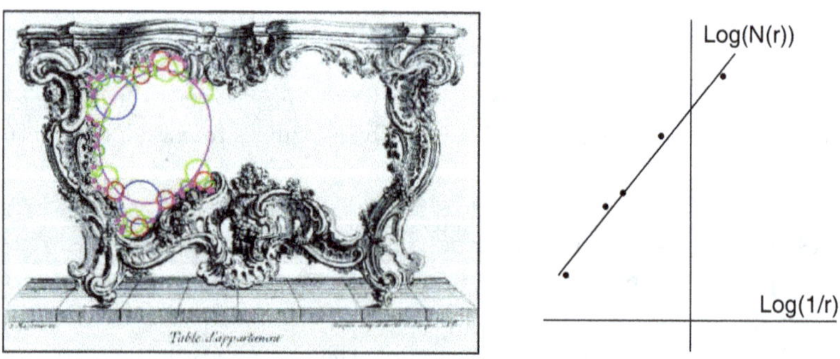

Figure 17.10 Estimating the dimension of an interior void edge.

meandering edge of the interior void of the table were calculated. The left side of Fig. 17.10 shows the circles used, the right side the log-log plot.

The results are then graphed, Log$(1/r)$ on the x-axis and Log$(N(r))$ on the y-axis. The points lie close ot a line, the slope of which is the fractal dimension of the edge of the interior void of the table. Linear regression gives

$$y = 1.23x + 1.41$$

Over the range of scales measured, the fractal dimension of the table interior is 1.23. This result indicates that the dimension as measured is larger than the dimension of a line.

In his book, *Fractal Geometry in Architecture and Design*, Carl Bovill uses the boxcounting method to ascribe fractal dimensions to facades of various buildings [24]. He concludes that those buildings with a cascade of scaling forms are the most aesthetically pleasing. He points out that Frank Lloyd Wright's architecture provides a good example of a progression of self-similar detail from the large to the small scale. The Robie house is calculated as displaying interesting form from a range of 12 feet down to less than 1 inch. He allows that the house has "self-similar textural characteristics" but only throughout a limited range of scales. However, he points out that we do not experience architecture by zooming our attention on small sections, but rather we experience architecture over time.

> We experience architecture by observing the overall profile of a building from a distance; as we approach closer, the patterns of window and siding come into attention; as we approach even closer, the details of doors and window frames come into attention, down to what the door knob is like. The process then continues inside the building. The fractal characteristic of an architectural composition presents itself in this progression of interesting detail as one approaches, enters and uses a building. [25]

Advances in evolutionary biology, environmental psychology, urban planning and computer sciences suggest that we are innately drawn to specific types of fractal environments. In his article "Fractal Architecture could be good for you," Yannick Joye describes how the empirical literature has shown that humans display a positive emotional affiliation with a specific set of natural elements and settings, namely vegetation (trees, plants and flowers) and savannas. [26]

However, when drawing parallels between fractals, nature, and aesthetics, I believe that complexity is only half the story.

While we are accustomed to seeing complex and swirling images of the Mandelbrot set, these images do not show the immense areas of

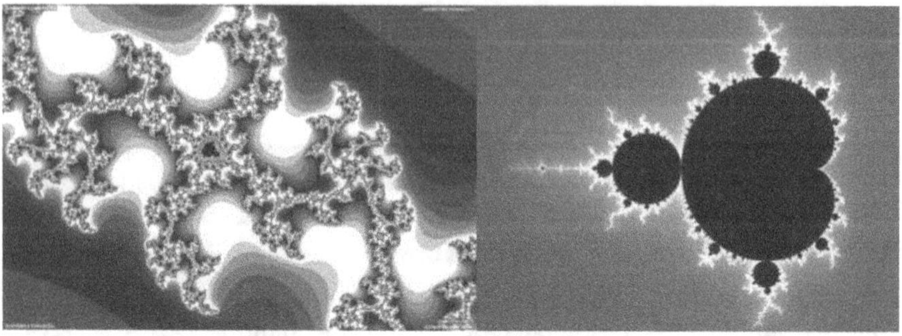

Figure 17.11 The complexity of the Mandelbrot set coundary (left) is in contrast to the relative simplicity of the entire set (right).

homogeneity that exist on either side of this intensely textured fractal boundary. Both the interior and exterior of the set are homogeneous. The boundary at which they meet is one of infinite complexity, but the overall shape of the Mandelbrot set appears very simple if one is at a sufficient distance to discount the meandering nature of the outline.

A coastline is often referred to as an example of a fractal in nature. We can also rephrase this to say that the joint between land and sea is fractal. By placing the emphasis on the context rather than the boundary condition, we can start to see a bigger picture. Fractal boundaries appear between areas of relative homogeneity, illustrating the fact that great complexity may arise out of great simplicity.

In nature, fractals assist in the metamorphosis from one state to another. For instance, the exterior surface of a cloud may be described as the transition zone between air moisture and its surroundings. The metamorphosis between one state and the next is achieved with a fractal boundary — the exterior of the cloud itself. Complexity only arises in conjunction with homogeneity or simplicity. Contrast is both fundamental and necessary.

In architecture, it is the ornament or detail which allows forms to metamorphose from one into the next. Ornament literally assists in the metamorphosis of floor to wall and from wall to ceiling. In the words of Kent Bloomer, "The principle strategy required to achieve an image of combination in ornament is to present a spectacle of transformation. In that spectacle, it may appear that one thing is turning into another and vice versa." [27] Henri Focillon, in *The Life of Forms in Art*, also identifies ornament as "the chosen home of metamorphoses." [28]

If ornament is equated with the complex boundary, then the opposite is also necessary — the simple and the homogeneous. In order for ornament to function, it needs objects to ornament — it needs the steady, smooth Euclidean forms in order to perform. Mandelbrot discussed the balance between homogeneity and detail in Renaissance portraiture: "... a typical Renaissance portrait depicts a seemingly obligatory juxtaposition of featureless areas (including draperies) and of finely detailed areas (including background landscapes that may often seem to lack any other purpose)." [29] It is not complexity or simplicity alone which creates a beautiful painting, but the contrast or interplay between the two. He points out that this is characteristic of a typical scaling in nature where, without any systematic bias, any portion selected for observation tends to include substantial featureless portions. Nature, and any scaling object, is not just composed of seething, swirling complexities. It is also composed of vast areas of homogeneity. The universe itself is predominately comprised of vast areas of nothing. But the areas that contain something — these are places of dense, layered life.

In the same way, good architecture often combines and contrasts the simple and the complex. In many styles of architecture, the overall form of a building may be approximated using simple, Euclidean shapes, but at the small scale, a more complex geometrical system is introduced. On approaching the Palais Garnier, for instance, it may appear initially as a regular Euclidean form, but as one nears the façade, one notices the elaborate ornamentation, the column capitals, the sinuous surfaces of the sculptured stone, and the experience becomes much more volumetric and geometrically complex. The overall shape is a box but the outline has a "fuzzy" quality, blurring the boundary and softening the box-like profile. In this way, a journey through a building becomes a process through which the simple and the complex interplay. Ultimately, in order to create aesthetically pleasing environments, we need to understand aesthetics. And much of what is aesthetically pleasing to people is that which is based on the laws of nature. Fractal geometry is a way to understand the fundamental rules of generation in nature which seem to favor a balance of interplay of complexity and simplicity. There is beauty in contrast.

Throughout the history of architecture, simplicity and complexity have been favored at different times. Rococo ornament represented a pinnacle of complexity in the Baroque style, and ultimately was rejected for the same reason. Currently in our architectural environments, we rely on a

Figure 17.12 Design for ornament using successive iterations of the Koch fractal to transform from smooth to roughly textured surfacres.

formal system that is heavily in favor of simplicity at both the large and small scale. When we eliminated ornament from our environments, we also eliminated much of the visual complexity that these forms lent our architecture. With the advancements in computer-aided design and digital fabrication techniques like 3D printing, we are in a position to explore much greater complexity in our designs. Alongside the smooth, glassy and homogeneous expanses we build today, we can also explore surfaces that are grainy and hydralike; inbetween, seaweedy and strange; tangled, tortuous, wiggly and wispy. Perhaps it is now time to reintroduce the intricate, the complex, the dense, layered life of ornament back into our architecture.

Bibliography

[1] B. Mandelbrot, *Leonardo* **14** (1981) 45.

[2] J. Gleick, *Chaos. Making a New Science*, Viking, New York, 1987, p. 117.

[3] B. Mandelbrot, *The Fractal Geometry of Nature*, W. H. Freeman, San Francisco, 1982, p. 24.

[4] B. Mandelbrot, http://www.ted.com/talks/benoit_mandelbrot-fractals-the-art-of-roughness.html

[5] http://seedmagazine.com/content/article/paola_antonelli_benoit_mandelbrot/

[6] http://seedmagazine.com/content/article/paola_antonelli_benoit_mandelbrot/

[7] American Institute of Architects, "About this Exhibit," FavoriteArchitecture.org website, 12/16/10

[8] E. Hopkirk, *Public Prefers Traditional Architecture,* suggests poll (bdonline.co.uk) 16 October, 2009.
[9] J. Trilling, *Ornament: a Modern Perspective,* Univ. of Washington Pr. 2003, p. 21.
[10] C. Dresser, *The Art of Decorative Design,* Amer. Life Found. 1977, p. 1.
[11] L. Corbusier, *Towards a New Architecture,* Architectural Pr., 1997, p. 29.
[12] A. Loos, "Das Luxusfuhrwerk," *Neue Freie Presse,* 3 Juky 1898; citations taken from *Sämtliche Schriften* Franz Glück, ed., Vienna 1962, 1, pg. 62–69 (p. 65).
[13] A. Loos, *Ornament and Crime,* Innsbruck, reprint Vienna, 1930, p. 80.
[14] A. Riegl, *Problems of Style,* Princeton Univ. Pr., 1992, p. xxi-xxii.
[15] K. Bloomer, *The Nature of Ornament,* W. W. Norton, New York, 2000, p. 71.
[16] L. Sullivan, *The Function of Ornament,* W. de Witt, ed., W. W. Norton, New York, 1986, p. 189.
[17] FLW
[18] O. Jones, *The Grammar of Ornament,* Bernard Quaritch, London, 1868, p. 5.
[19] W. Hogarth, *The Analysis of Beauty,* R. Paulson, ed., Yale Univ. Pr., 1997, p. 45.
[20] B. Mandelbrot, *The Fractal Geometry of Nature,* W. H. Freeman, 1982, p. 5.
[21] A. Crompton, *Environment and Planning B. Planning and Design* **29** (2002) 451.
[22] L. Sullivan, *The Function of Ornament,* W. de Witt, ed., W. W. Norton, New York, 1986, p. 187.
[23] B. Mandelbrot, http://www.ted.com/talks/benoit_mandelbrot-fractals-the-art-of-roughness.html
[24] C. Bovill, *Fractal Geometry in Architecture and Design,* Birkhäuser, Boston, 1996.
[25] C. Bovill, *Fractal Geometry in Architecture and Design,* Birkhäuser, Boston, 1996, p. 117.
[26] Y. Joye, *Nexus Network J.* **9** (2007) 318.
[27] K. Bloomer, *The Nature of Ornament,* W. W. Norton, New York, 2000. p. 27.
[28] K. Bloomer, *The Nature of Ornament,* W. W. Norton, New York, 2000. p. 27.
[29] B. Mandelbrot, *Leonardo* **14** (1981) 45.

Chapter 18

Long-range dependence of the two-dimensional Ising model at critical temperature

Vladas Pipiras and Murad S. Taqqu[1]

Abstract. The paper gives probabilists who are unfamiliar with the Ising model a coherent, integrated explanation of why the Ising model displays long-range dependence at critical temperature. The Ising model in two dimensions involves spins $\sigma_{j,k} = \pm 1$ located at every node (j, k) of the lattice, with nearest neighbor interactions. We shall focus on the covariances $\mathbb{E}\sigma_{0,0}\sigma_{0,N}$ and $\mathbb{E}\sigma_{0,0}\sigma_{N,N}$ between the spin at the origin and the spin located either on the axis at $(0, N)$ or located on the diagonal at (N, N), when the temperature equals a critical value. Using a recent formulation of the so-called "Szegö's theorem", we explain why these covariances decrease to zero like $N^{-1/4}$ as $N \to \infty$, thus at a slow enough rate so as to exhibit long-range dependence.

[1]Partially supported by the NSF grants DMS-0706786 and DMS-1007616 at Boston University, and the NSA grant H98230-13-1-0220 at the University of North Carolina, Chapel Hill.

18.1 Introduction

Benoit Mandelbrot's interests extended to Physics as well as to Mathematics. What attracted his attention was behavior, deterministic or stochastic, that can be described by power laws. In particular, he studied stationary phenomena whose covariances at large lags decrease to 0 as a power function, but so slowly that the covariances are not summable. Such behavior is now called *long-range dependence*, *long memory*, *strong dependence*, or $1/f$ *noise*. The goal of this paper is to show how it arises in the two-dimensional Ising model at critical temperature.

The literature on the Ising model is large, especially in Mathematical Physics. We shall assume that the reader has no experience with it. We will introduce the Ising model and develop the techniques necessary to achieve our goals. Many of these methods are scattered through the monograph of McCoy and Wu [1]. It is our aim to thread them together in this article.[2]

This is in large part an expository effort. One result we obtain is conjectured, in fact, in McCoy and Wu [1]. To prove the conjecture, we use a recently proven extension of the so-called strong Szegö's limit theorem. Our emphasis, however, is less on Szegö's theorem than on describing the various steps required to show that the two-dimensional Ising model at critical temperature displays long-range dependence.

We start with the notion of "long-range dependence" which is commonly defined for second-order stationary time series. Recall that a second-order stationary time series $X = \{X_n\}_{n\in\mathbb{Z}}$ has a constant mean $\mu = \mathbb{E}X_n$ and a covariance function

$$\gamma(n-m) = \mathbb{E}X_nX_m - \mathbb{E}X_n\mathbb{E}X_m = \mathbb{E}X_nX_m - \mu^2,$$

which depends only on the distance $|n-m|$. One can characterize the second order properties of the time series X by specifying its mean μ and covariance $\gamma(n)$, $n \in \mathbb{Z}$, or one can take a "spectral domain" perspective by focusing on the spectral density $f(w)$, $w \in [-\pi, \pi)$, of the time series, if it exists. The spectral density $f(w)$ is defined as

$$\int_{-\pi}^{\pi} e^{inw} f(w)dw = \gamma(n), \quad n \in \mathbb{Z},$$

that is, it is a function whose Fourier coefficients are the autocovariances $\gamma(n)$.

---

[2] We shall use to the extent possible the same notation as in McCoy and Wu [1].

We will say that a time series $X = \{X_n\}_{n\in\mathbb{Z}}$ exhibits *long-range dependence* if its covariance function satisfies[3,4]

$$\gamma(k) \sim c_1 k^{2d-1}, \quad \text{as } k \to \infty, \tag{18.1}$$

where c_1 is a positive constant and

$$0 < d < 1/2.$$

The higher d, the stronger the dependence. The bound $d < 1/2$ ensures that $\gamma(k) \to 0$, as $k \to \infty$. The bound $d > 0$ implies that the covariances are not summable, that is,

$$\sum_{k=-\infty}^{\infty} \gamma(k) = \infty. \tag{18.2}$$

Another consequence of the long-range dependence condition (18.1) is that the variance of the partial sums of X_n for $n = 1, \ldots, N$ does not grow like N as in the case of random walk, but grows faster than N. More precisely,

$$\mathrm{Var}\left(\sum_{n=1}^{N} X_n\right) = \sum_{j=1}^{N}\sum_{k=1}^{N} \gamma(j-k) \sim c_2 N^{2H}, \quad \text{as } N \to \infty,$$

where $c_2 = (d(2d+1))^{-1}c_1$ is a constant and

$$H = d + \frac{1}{2} \in \left(\frac{1}{2}, 1\right). \tag{18.3}$$

The higher the exponent H, the stronger the dependence. If, in fact, the X_n's were perfectly correlated, that is, $\gamma(k) = c_2$ for all k, then we would get $\mathrm{Var}(\sum_{n=1}^{N} X_n) = c_2 N^2$, that is, $H = 1$. We do not consider such an extreme case because we suppose that $\gamma(k) \to 0$ as $k \to \infty$.

[3] $a_n \sim b_n$ means $\lim_{n\to\infty} a_n/b_n = 1$.

[4] There are also more general definitions of long-range dependence, for example, $\gamma(k) - k^{2d-1}L(k)$, as $k \to \infty$, where L is a slowly varying function at infinity, such as a constant or a logarithm. There is also a definition involving the spectral density, namely, $f(w) \sim w^{-2d}L_1(w)$, as the frequency $w \to 0$, where L_1 is a slowly varying function at 0. That is, the spectral density blows up as the frequency tends to 0. These definitions are not always equivalent, however (see [2–4]). For us, here, the definition (18.1) will be sufficient.

The exponent H is often called the *Hurst exponent*, after Harold Edwin Hurst, the British hydrologist who studied the yearly changes in the level of the Nile through various centuries (Hurst [5]). Motivated by Hurst's studies, Benoit Mandelbrot also contributed to Hydrology by suggesting stochastic models to explain the fact that for some rivers, $H \neq 1/2$ (see [6–9]). For an overview, see Montanari [10].

The Ising model, which is the focus of this paper, involves a two-dimensional lattice with a random spin $\sigma_{j,k} = \pm 1$ at each node (j,k), with nearest neighbor interactions. We will study covariances

$$\mathbb{E}\sigma_{0,0}\sigma_{0,N} \quad \text{and} \quad \mathbb{E}\sigma_{0,0}\sigma_{N,N}$$

as $N \to \infty$. We will show that at the so-called "critical temperature" these covariances display long-range dependence with $d = 3/4$, or equivalently with $H = 7/8$. The result for $\mathbb{E}\sigma_{0,0}\sigma_{N,N}$ can be found in [1]. The one for $\mathbb{E}\sigma_{0,0}\sigma_{0,N}$ had been conjectured.

The idea is to introduce the notion of bonds in the Ising lattice, then replace this lattice by a larger one, called the counting lattice, to which a number of matrices are associated. The covariances of interest involve these matrices, their inverse and related determinants. Szegö's limit theorem describes the asymptotic behavior of these determinants.

The Ising model was first studied in the 1920's by Ernst Ising in his Ph.D. thesis, supervised by Wilhelm Lenz.[5] Ising considered the model in one dimension and found that it does not exhibit a phase transition. Although Ising conjectured that no phase transition occurs in higher dimension, further studies pointed to the contrary, culminating in the seminal work of Onsager [11] who provided an explicit calculation for the so-called "free energy" of the two-dimensional Ising model. The Ising model has since played a special role in Statistical Mechanics and other areas, as a source of new ideas and as one of few realistic and exactly solvable models. A nice account of the historical developments concerning the Ising model can be found in [12, 13].

The paper is structured as follows. In Sec. 18.2, we describe the Ising model and state the main theorem and a corollary. The proofs are developed in Secs. 18.3 and 18.4 where the notions of dimers and Pfaffian are introduced. Szegö's theorem and its extension are introduced in Sec. 18.5.

[5]The model is sometimes referred to as the Lenz–Ising model.

The proofs are completed in Sec. 18.6 by applying the extended version of Szegö's theorem. Section 18.7 contains some concluding remarks.

18.2 Model formulation and result

We consider the Ising model in two dimensions. The reason is that, in one dimension, the Ising model does not exhibit long-range dependence and, in three dimensions, the Ising model has not been solved exactly yet (in fact, this is still one of the famous open problems in Statistical Mechanics and, more generally, Physics — though see [14]). Thus, consider a two-dimensional square lattice (Fig. 18.1), referred to as the *Ising lattice*, with

$$\begin{aligned} 2L_v &: \text{ the number of rows,} \\ 2L_h &: \text{ the number of columns} \end{aligned}$$

("v" stands for "vertical", and "h" stands for "horizontal"). Denote a general point on the lattice as (j,k), $-L_v+1 \le j \le L_v$, $-L_h+1 \le k \le L_h$. Let

$$\sigma_{j,k} = \pm 1$$

be a variable associated with a lattice point (j,k).[6] We will consider below correlations between $\sigma_{0,0}$ and $\sigma_{0,N}$, and $\sigma_{0,0}$ and $\sigma_{N,N}$. In fact, we will first

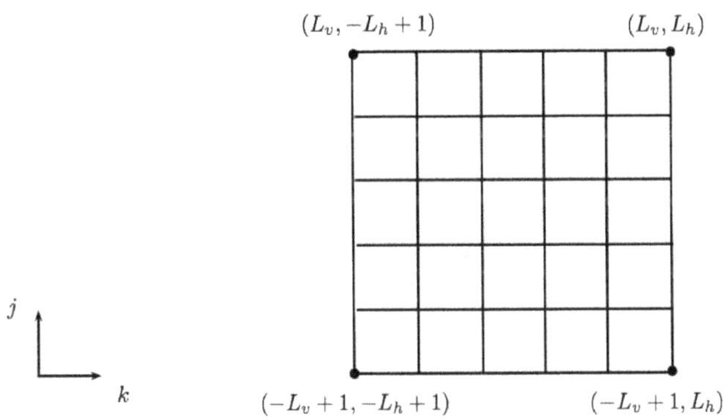

Figure 18.1 The Ising lattice.

[6]Note that j and k in (j,k) refer to row (vertical) and column (horizontal) directions, respectively. This is in accord with the matrix notation where (j,k) is used for a matrix element $a_{j,k}$. But this is the opposite of the (x,y) convention, where for example x refers to the horizontal direction.

let $L_v, L_h \to \infty$ (the so-called thermodynamic limit), and then investigate these correlations as $N \to \infty$.

In the Ising model, the collection of variables $\{\sigma_{j,k}, -L_v + 1 \le j \le L_v, -L_h + 1 \le k \le L_h\}$ is assumed to follow the joint distribution

$$\frac{1}{Z_{L_v,L_h}} e^{-\beta \mathcal{E}}, \tag{18.4}$$

where

$$\mathcal{E} = - \sum_{j=-L_v+1}^{L_v} \sum_{k=-L_h+1}^{L_h} (E^h \sigma_{j,k}\sigma_{j,k+1} + E^v \sigma_{j,k}\sigma_{j+1,k}) \tag{18.5}$$

with some positive constants $E^h, E^v > 0$, and where

$$Z_{L_v,L_h} = \sum_{\{\sigma_{j,k}\}} e^{-\beta \mathcal{E}} \tag{18.6}$$

is the normalization constant, where the sum is over all possible *configurations* (outcomes) $\{\sigma_{j,k}\}$. Since each $\sigma_{j,k}$ can take 2 values, there are $2^{(2L_v)(2L_h)}$ configurations. The parameter $\beta > 0$ is expressed as

$$\beta = \frac{1}{k_B T}, \tag{18.7}$$

where $k_B = 1.38 \times 10^{-23}$ is the Boltzmann's constant and $T > 0$.

The term $\mathcal{E}$ in (18.5) is referred to as *energy* (*Hamiltonian*) of a configuration $\{\sigma_{j,k}\}$. Note that the energy is determined only by interactions between neighboring sites, with the contributions E^h for horizontal interactions and E^v for vertical interactions. Note also the following effect of individual terms on the overall energy of the system. The energy is lower when neighboring sites align as $+1$ and $+1$, or as -1 and -1. Lower (negative) energy $\mathcal{E}$ translates into higher probability (18.4), that is, as expected in physical terms, the system favors configurations with lower energy.

The parameter $T > 0$ in (18.7) is referred to as *temperature* (and β as *inverse temperature*). When $T \to \infty$ (high temperature), note that $\beta \to 0$ and hence (18.4) approaches the uniform probability distribution on the lattice. In this case, the variables $\sigma_{j,k}$ tend to be independent and the system is in the disordered state. On the other hand, when $T \to 0$ (low temperature), note that $\beta \to \infty$ and the aligned configurations (all $+1$ or -1) are more likely. This is because the exponent in (18.4) will be large and positive. In this case, the system is in the ordered state.

Moving from the disordered to ordered state (large T to small T), one could expect that the system undergoes a *phase transition* at some *critical temperature* $T = T_c > 0$. We shall not define T_c in rigorous terms. Informally, the system exhibits very different characteristics for $T < T_c$ and $T > T_c$. The two-dimensional Ising model turns out to have such phase transition at $T_c > 0$.

Here are few other notes regarding the Ising model. The distribution (18.4) is also known as the *Boltzmann distribution* or the *Gibbs distribution.* The normalization factor

$$Z_{L_v,L_h} = Z_{L_v,L_h}(\beta) = \sum_{\{\sigma_{j,k}\}} e^{-\beta\mathcal{E}} \tag{18.8}$$

in (18.6) is called the *partition function.* In (18.8), the sum is over every value ± 1 of the variables $\sigma_{j,k}$ in the lattice. The energy $\mathcal{E}$ in (18.5) is often written more generally as

$$\mathcal{E} = - \sum_{j=-L_v+1}^{L_v} \sum_{k=-L_h+1}^{L_h} (E^h\sigma_{j,k}\sigma_{j,k+1} + E^v\sigma_{j,k}\sigma_{j+1,k} + H\sigma_{j,k}), \tag{18.9}$$

where the terms $H\sigma_{j,k}$ in (18.9) account for the so-called external magnetic field[7] (see the discussion below for the reason of using "magnetic"). With (18.5), we thus focus only on the situation of zero magnetic field $H = 0$. (In fact, the presence of the external magnetic field complicates the matters considerably and the Ising model with a magnetic field $H > 0$ has not been solved explicitly yet. See, for example, [15], pp. 277–280.)

Note also that the energy $\mathcal{E}$ in (18.5) involves the variables σ_{j,L_h+1} and $\sigma_{L_v+1,k}$ "outside" the lattice. What exactly are these variables? This is related to the so-called *boundary conditions.* The case

$$\sigma_{j,L_h+1} = \sigma_{j,-L_h+1}, \quad \sigma_{L_v+1,k} = \sigma_{-L_v+1,k} \tag{18.10}$$

is referred to as that of *periodic boundary* (toroidal or doughnut-shaped) conditions. The case

$$\sigma_{j,L_h+1} = 0, \quad \sigma_{L_v+1,k} = 0 \tag{18.11}$$

is known as that of *free boundary* conditions. It effectively corresponds to the situation where the terms involving σ_{j,L_h+1} and $\sigma_{L_v+1,k}$ are not present in (18.5).

[7]Do not confuse the magnetic field H in (18.9) with the Hurst parameter in (18.3).

Finally, one of the original applications of the Ising model is to the phenomenon of *ferromagnetism.* The variables $\sigma_{j,k}$ model a magnetic dipole of atoms of ferromagnetic material (e.g. iron). The variables $\sigma_{j,k}$ are referred to as *spins.* The critical temperature T_c is known as the Curie temperature. Below this temperature, ferromagnetic materials exhibit spontaneous magnetization.

We are interested here in the behavior of the correlation function of the spin at the origin and the spin at position (M, N), namely

$$\begin{aligned} C(M,N) &= \mathrm{Corr}(\sigma_{0,0}, \sigma_{M,N}) \\ &= \frac{\mathbb{E}\sigma_{0,0}\sigma_{M,N} - \mathbb{E}\sigma_{0,0}\mathbb{E}\sigma_{M,N}}{\sqrt{\mathbb{E}\sigma_{0,0}^2 - (\mathbb{E}\sigma_{0,0})^2}\sqrt{\mathbb{E}\sigma_{0,0}^2 - (\mathbb{E}\sigma_{0,0})^2}} \\ &= \mathbb{E}\sigma_{0,0}\sigma_{M,N}, \end{aligned}$$

since $\mathbb{E}\sigma_{j,k} = 0$ and $\mathbb{E}\sigma_{j,k}^2 = 1 = \sigma_{j,k}^2$. Thus correlations reduce to covariances. For the sake of simplicity and illustration, we shall only consider row and diagonal covariances

$$C(0,N) = \mathbb{E}\sigma_{0,0}\sigma_{0,N}, \quad C(N,N) = \mathbb{E}\sigma_{0,0}\sigma_{N,N}. \tag{18.12}$$

We shall also work in the case of free boundary conditions (18.11). Our goal is to show the following:

Theorem 18.2.1 *At the critical temperature $T = T_c$ characterized by*

$$1 = \sinh 2\beta E_h \sinh 2\beta E_v \tag{18.13}$$

and as $N \to \infty$,

$$\lim_{L_h\to\infty}\lim_{L_v\to\infty} \mathbb{E}\sigma_{0,0}\sigma_{0,N} \sim \left(\frac{1+\alpha_1}{1-\alpha_1}\right)^{1/4} AN^{-1/4}, \tag{18.14}$$

$$\lim_{L_h\to\infty}\lim_{L_v\to\infty} \mathbb{E}\sigma_{0,0}\sigma_{N,N} \sim AN^{-1/4}, \tag{18.15}$$

where $A = 0.6450024\ldots$ and

$$\alpha_1 = \frac{z_h(1-z_v)}{1+z_v}, \quad z_h = \tanh\beta E_h, \quad z_v = \tanh\beta E_v.$$

The asymptotic behavior of $\mathbb{E}\sigma_{0,0}\sigma_{N,N}$ in (18.15) is stated in [1], (4.43), p. 265, and proved but the asymptotic behavior of

$\mathbb{E}\sigma_{0,0}\sigma_{0,N}$ in (18.14), stated in Ref. [1], p. 267, is only conjectured ([1], p. 266).

Whereas, as expected, the asymptotic behavior of $\mathbb{E}\sigma_{0,0}\sigma_{N,N}$ is invariant under permutation of h and v, this is not the case for $\mathbb{E}\sigma_{0,0}\sigma_{0,N}$, for which

$$\frac{1+\alpha_1}{1-\alpha_1} = \frac{1+z_v+z_h-z_h z_v}{1+z_v-z_h+z_h z_v}. \tag{18.16}$$

Observe that in the special case $E_h = E_v = E$, this ratio simplifies and one has

Corollary 18.2.1 *If $E_h = E_v$, then at the critical temperature $T = T_c$, one has as $N \to \infty$,*

$$\lim_{L_h\to\infty} \lim_{L_v\to\infty} \mathbb{E}\sigma_{0,0}\sigma_{0,N} \sim 2^{1/8} A N^{-1/4}. \tag{18.17}$$

Since, at the critical temperature T_c, the covariances behave like $N^{-1/4}$ for large N, the Ising model exhibits long-range dependence. By identifying the exponent $2d-1$ with $-1/4$ and using $H = d + 1/2$, one gets that the Ising model exhibits long-range dependence with

$$d = \frac{3}{8} \quad \text{and} \quad H = \frac{7}{8}. \tag{18.18}$$

Theorem 18.2.1 and Corollary 18.2.1 are proved in Sec. 18.6.

18.3 Correlations, dimers and Pfaffians

We shall express the correlations $C(0,N)$ and $C(N,N)$ in terms of the determinants of $N \times N$ matrices in the so-called *thermodynamic limit* $L_v, L_h \to \infty$ first, and then analyze them as $N \to \infty$.

Observe that

$$\mathbb{E}\sigma_{0,0}\sigma_{0,N} = \frac{1}{Z_{L_v,L_h}} \sum_{\{\sigma_{j,k}\}} \sigma_{0,0}\sigma_{0,N} e^{-\beta\mathcal{E}}. \tag{18.19}$$

We shall analyze the two terms Z_{L_v,L_h} and $\sum_{\{\sigma_{j,k}\}} \sigma_{0,0}\sigma_{0,N} e^{-\beta\mathcal{E}}$ separately. The focus will be on the partition function Z_{L_v,L_h}, and similar arguments will be outlined for the second term. As for the diagonal correlation $\mathbb{E}\sigma_{0,0}\sigma_{N,N}$, we shall give its corresponding expression without proof.

18.3.1 Partition function Z_{L_v,L_h}

Observe that

$$Z_{L_v,L_h} = \sum_{\{\sigma_{j,k}\}} e^{-\beta\mathcal{E}} = \sum_{\{\sigma_{j,k}\}} e^{\beta\sum_{j=-L_v+1}^{L_v}\sum_{k=-L_h+1}^{L_h}(E_h\sigma_{j,k}\sigma_{j,k+1}+E_v\sigma_{j,k}\sigma_{j+1,k})}$$

$$= \sum_{\{\sigma_{j,k}\}} \prod_{j=-L_v+1}^{L_v} \prod_{k=-L_h+1}^{L_h} e^{\beta E_h\sigma_{j,k}\sigma_{j,k+1}} \prod_{j=-L_v+1}^{L_v} \prod_{k=-L_h+1}^{L_h} e^{\beta E_v\sigma_{j,k}\sigma_{j+1,k}}$$

$$= \sum_{\{\sigma_{j,k}\}} \prod_{j=-L_v+1}^{L_v} \prod_{k=-L_h+1}^{L_h} (\cosh\beta E_h + \sigma_{j,k}\sigma_{j,k+1}\sinh\beta E_h)$$

$$\times \prod_{j=-L_v+1}^{L_v} \prod_{k=-L_h+1}^{L_h} (\cosh\beta E_v + \sigma_{j,k}\sigma_{j+1,k}\sinh\beta E_v), \tag{18.20}$$

since $\cosh a = \frac{1}{2}(e^a + e^{-a})$, $\sinh a = \frac{1}{2}(e^a - e^{-a})$ and thus

$$e^{ax} = \cosh a + x\sinh a, \quad x = \pm 1.$$

Under the free boundary conditions, we get further that

$$Z_{L_v,L_h} = (\cosh\beta E_h)^{2L_v(2L_h-1)}(\cosh\beta E_v)^{(2L_v-1)2L_h}\widetilde{Z}_{L_v,L_h}, \tag{18.21}$$

where

$$\widetilde{Z}_{L_v,L_h} = \sum_{\{\sigma_{j,k}\}} \prod_{j=-L_v+1}^{L_v} \prod_{k=-L_h+1}^{L_h} (1 + \sigma_{j,k}\sigma_{j,k+1}z_h)$$

$$\times \prod_{j=-L_v+1}^{L_v} \prod_{k=-L_h+1}^{L_h} (1 + \sigma_{j,k}\sigma_{j+1,k}z_v) =: \sum_{\{\sigma_{j,k}\}} \widetilde{Z}_{L_v,L_h}(\{\sigma_{j,k}\}) \tag{18.22}$$

and

$$z_h = \frac{\sinh\beta E_h}{\cosh\beta E_h} = \tanh\beta E_h, \quad z_v = \tanh\beta E_v. \tag{18.23}$$

Consider now $\widetilde{Z}_{L_v,L_h}(\{\sigma_{j,k}\})$ given by (18.22). Expand all the products to express $\widetilde{Z}_{L_v,L_h}(\{\sigma_{j,k}\})$ as the sum over a number of terms as

$$\widetilde{Z}_{L_v,L_h}(\{\sigma_{j,k}\}) = 1 + \sigma_{-L_v+1,-L_h+1}\sigma_{-L_v+1,-L_h+2}z_h + \cdots + \sigma_{L_v-1,L_h}\sigma_{L_v,L_h}z_v + \cdots + \prod_{j=-L_v+1}^{L_v} \times \prod_{k=-L_h+1}^{L_h} \sigma_{j,k}\sigma_{j,k+1}z_h \prod_{j=-L_v+1}^{L_v}\prod_{k=-L_h+1}^{L_h} \sigma_{j,k}\sigma_{j+1,k}z_v. \tag{18.24}$$

Each term in the sum (18.24) can be thought as containing a factor for every pair of nearest-neighbor sites (j,k) and (j',k') on the lattice. This factor is either 1, when 1 from $1+\sigma_{j,k}\sigma_{j',k'}z$ is selected in the product, or $\sigma_{j,k}\sigma_{j',k'}z$, when $\sigma_{j,k}\sigma_{j',k'}z$ is selected from the product (z denotes z_h or z_v). For example, the first term in (18.24) occurs when each pair of nearest-neighbor sites contributes 1, and the last term in (18.24) occurs when each contributes $\sigma_{j,k}\sigma_{j',k'}z$. Each term in the sum (18.24) will therefore include either 1, $\sigma_{j,k}$, $\sigma^2_{j,k}$, $\sigma^3_{j,k}$ or $\sigma^4_{j,k}$ and hence one of the five quantities can be associated with each site (j,k). Note now that the terms in the sum $\widetilde{Z}_{L_v,L_h}(\{\sigma_{j,k}\})$ with sites having $\sigma_{j,k}$ or $\sigma^3_{j,k}$ will vanish in the total sum $\sum_{\{\sigma_{j,k}\}}$ in (18.22) because

$$\sum_{\sigma_{j,k}=\pm 1} \sigma_{j,k} = \sum_{\sigma_{j,k}=\pm 1} \sigma^3_{j,k} = 0.$$

Thus only

$$\text{the terms having } 1, \quad \sigma^2_{j,k} \equiv 1 \quad \text{or} \quad \sigma^4_{j,k} \equiv 1 \tag{18.25}$$

will not vanish in the total sum $\sum_{\{\sigma_{j,k}\}}$.

The presence of a term $\sigma_{j,k}\sigma_{j,k+1}z_h$ in (18.24) can be interpreted as indicating the presence of a horizontal bond z_h connecting the sites (j,k) and $(j,k+1)$. A corresponding interpretation holds for $\sigma_{j,k}\sigma_{j+1,k}z_v$. From a different perspective, when 1, $\sigma^2_{j,k}$ or $\sigma^4_{j,k}$ is associated with the site (j,k), then zero, two or four bonds traverse the site (j,k). Therefore, the terms

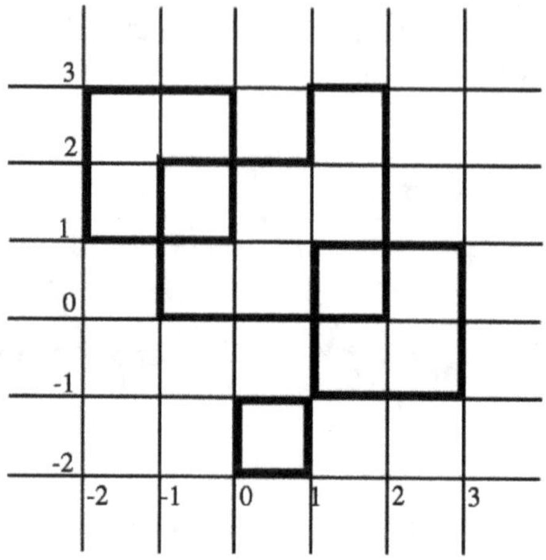

Figure 18.2 A figure (lines in bold) on the Ising lattice.

in the sum (18.24) where the conditions (18.25) hold can be regarded as depicting a figure on a lattice with the following properties:

(i) each bond between nearest neighbors may be used, at most, once;
(ii) an even number of bonds terminate at each site.

Figure 18.2 depicts one such figure. Observe that such figure would contribute a term

$$z_h^p z_v^q$$

to the total sum $\sum_{\{\sigma_{j,k}\}}$, where p is the number of horizontal bonds in the figure ($p = 16$ in Fig. 18.2) and q is the number of vertical bonds in the figure ($q = 16$ in Fig. 18.2). Statement (ii) above implies that p and q are even. Since there are

$$\sum_{\{\sigma_{j,k}\}} 1 = 2^{2L_v 2L_h}$$

different configurations, one gets that

$$\widetilde{Z}_{L_v,L_h} = 2^{2L_v 2L_h} \sum_{p,q} N_{p,q} z_h^p z_v^q =: 2^{2L_v 2L_h} g(z_h, z_v), \tag{18.26}$$

where $N_{p,q}$ is the number of figures on the lattice with the properties (*i*) and (*ii*) above, and where p, q are the numbers of horizontal and vertical bonds of the figure. By convention, $N_{0,0} = 1$. The next step is to compute the "generating function"

$$g(z_1, z_2) = \sum_{p,q} N_{p,q} z_1^p z_2^q \tag{18.27}$$

associated with counting of the figures as in Fig. 18.2. Observe that $g(z_1, z_2) \geq 0$ since p and q are even.

18.3.2 Dimers

There is an ingenious way to turn the problem of computing (18.27) into the so-called "problem of counting closest-packed dimer coverings" on a suitable lattice as explained in [1]. We will see later that the latter problem has a solution involving a matrix determinant which can then be studied more easily.

The first step is to replace the Ising lattice by a larger lattice, which we will call the *counting lattice*. More specifically, replace each site of the Ising lattice, as on the left side of Fig. 18.3, by a six-site cluster depicted on the right side of Fig. 18.3. Note that, as in the Ising lattice, the six-site cluster has four connecting lines. The difference is that one site is now replaced by six sites. The new, counting lattice has then $2K = 6(2L_v)(2L_h)$ sites and is formed by connecting six-site clusters and is depicted in Fig. 18.4.

In the next step, the idea is to replace each of eight possible Ising site configuration bonds with a suitable configuration of bonds on the counting lattice. The replacement is done according to Fig. 18.5. With this replacement, each figure counted in (18.27) is replaced by another figure in the counting lattice. In the Ising lattice, two adjacent sites may be either unconnected or connected by a bond. In the counting lattice,

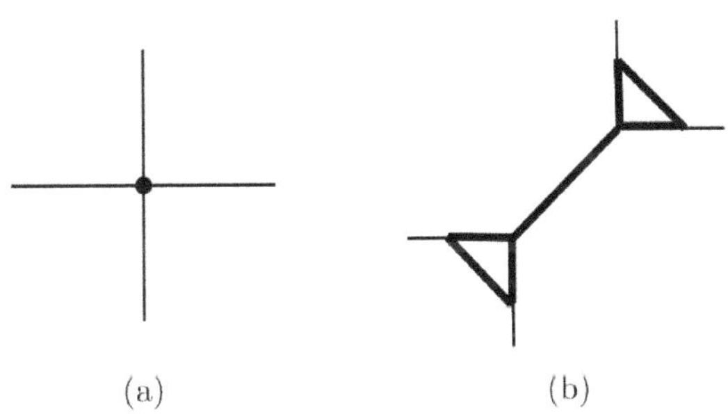

Figure 18.3 (a) A site of the Ising lattice. (b) A corresponding six-site cluster of the counting lattice.

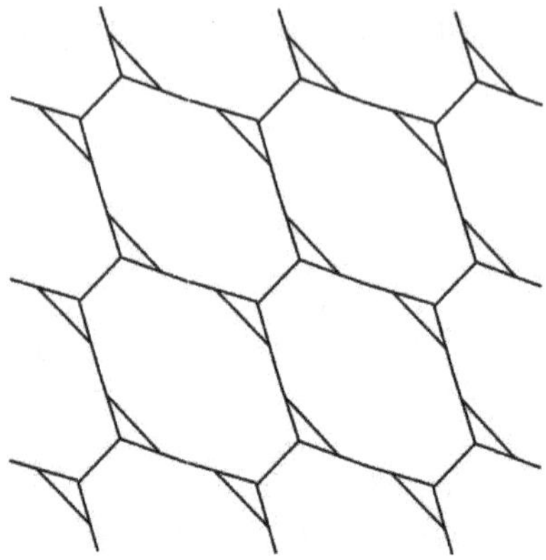

Figure 18.4 The counting lattice.

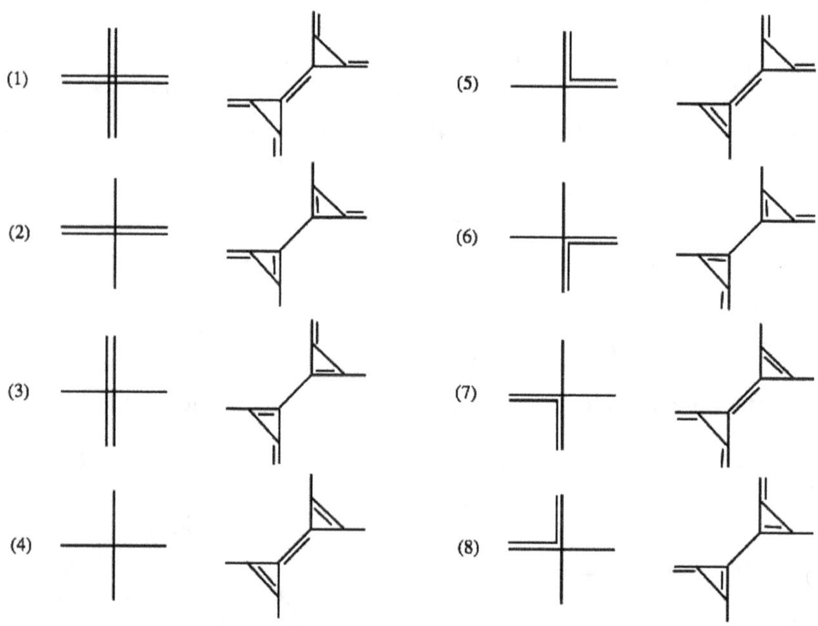

Figure 18.5 Replacement of eight possible Ising site configuration bonds by suitable configuration bonds on the counting lattice. The connecting bonds are indicated by *double* lines. Observe that a horizontal (resp. vertical) bond at an Ising site is associated with a horizontal (resp. vertical) bond connecting clusters.

each site has exactly one bond with one of its neighbors indicated by a double line in Fig. 18.5. Each such bond is called a *dimer*.[8] The figure one gets in the counting lattice is called a *closest-packed dimer configuration*.[9] Then, the problem of counting in (18.27) can be replaced by counting dimer configurations in the counting lattice.

To make this connection more precise, three classes of bonds need to be distinguished in the counting lattice: (1) horizontal bonds between clusters, (2) vertical bonds between clusters, and (3) bonds within a cluster. Let

$$G(z_1, z_2, z_3) = \sum_{p,q,r} N_{p,q,r} z_1^p z_2^q z_3^r \tag{18.28}$$

be the corresponding generating function, where $N_{p,q,r}$ is the number of dimer configurations on the counting lattice with p bonds of type (1), q bonds of type (2) and r bonds of type (3). In particular, one can think that bonds of types (1), (2) and (3) carry weights of sizes z_1, z_2 and z_3, respectively. The generating functions (18.27) and (18.28) then are related as follows:

$$g(z_1, z_2) = G(z_1, z_2, 1), \tag{18.29}$$

by assigning the weight $z_3 = 1$ to bonds within clusters. This is because a horizontal (resp. vertical) bond in the Ising lattice is associated with a horizontal (resp. vertical) bond between clusters (see Fig. 18.5).

The advantage of the formulation (18.28) or (18.29) is that it can be related to the determinant of a matrix. We shall do this in two steps. In the *first step*, we shall express (18.29) as

$$g(z_1, z_2) = G(z_1, z_2, 1) = \sum_p{}' \bar{b}_{p_1,p_2} \bar{b}_{p_3,p_4} \dots \bar{b}_{p_{2K-1},p_{2K}}, \tag{18.30}$$

where $\overline{B} = (\bar{b}_{pq})_{1 \le p,q \le 2K}$ is a suitable $2K \times 2K$ matrix (the bar does not refer here to the complex conjugate) and $\sum_p'$ is the sum over all

[8]In Chemistry, a dimer is a structure formed by two sub-units.

[9]One uses this term whenever each site has exactly one bond with one of its neighbors.

permutations $p_1, p_2 \ldots, p_{2K}$ of $1, 2, \ldots, 2K$ satisfying

$$p_{2m-1} < p_{2m}, \quad 1 \leq m < K, \quad p_{2m-1} < p_{2m+1}, \quad 1 \leq m < K-1. \qquad (18.31)$$

For example, if $2K = 4$, the sum is over the permutations 1234, 1324 and 1423 of 1234. Relation (18.31) is satisfied since these permutations are such that $p_1 < p_2$, $p_1 < p_3$ and $p_3 < p_4$, where, in the permutation 1324, for example, $p_1 = 1$, $p_2 = 3$, $p_3 = 2$, $p_4 = 4$. In the *second step*, the expression (18.30) will be written as the so-called *Pfaffian* of a matrix. As stated below, the Pfaffian of a matrix is the square root of its determinant. We shall now detail these two steps.

In the *first step*, the expression (18.30) and the permutations (18.31) above are naturally related to closest-packed dimer configurations on *any* lattice. Suppose a lattice (that is, *any* lattice) consists of $2K$ sites and enumerate these sites by $1, 2, \ldots, 2K$. Observe that there is a one-to-one correspondence between closest-packed dimer configurations on the lattice and permutations satisfying (18.31): the permutation

$$p_1 p_2 \,|\, p_3 p_4 \,|\, \cdots \,|\, p_{2K-1} p_{2K}$$

is associated with the closest-packed dimer configuration where dimers connect the sites p_1 and p_2, p_3 and p_4, $\ldots$, p_{2K-1} and p_{2K}. Here, note that a dimer can connect any two sites. For example, with a lattice of $2K = 4$ sites, there are 3 closest-packed dimer configurations associated with

$$12\,|\,34, \quad 13\,|\,24 \quad \text{and} \quad 14\,|\,23, \qquad (18.32)$$

as illustrated in Fig. 18.6. Suppose, in addition, that a dimer connecting sites p and q carries a weight $\bar{b}_{p,q}$. Then, we define the weight of

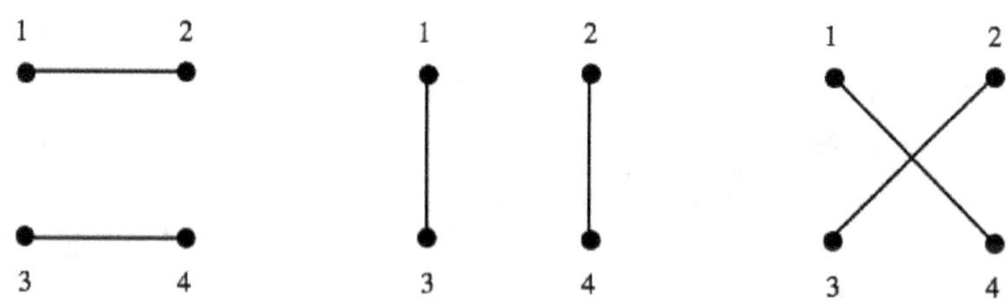

Figure 18.6 All possible closest-packed dimer configurations in a lattice of 4 sites.

the closest-packed dimer configuration associated with the permutation $p_1p_2 \,|\, p_3p_4 \,|\, \ldots \,|\, p_{2K-1}p_{2K}$ as

$$\bar{b}_{p_1,p_2}\bar{b}_{p_3,p_4}\ldots\bar{b}_{p_{2K-1},p_{2K}}.$$

Therefore,

$$\sum_p{}' \bar{b}_{p_1,p_2}\bar{b}_{p_3,p_4}\ldots\bar{b}_{p_{2K-1},p_{2K}}$$

is exactly the sum of the weights of all closest-packed dimer configurations.

With the latter fact in mind, we now turn to the relation (18.30). The function $G(z_1, z_2, 1)$ can then be written as

$$G(z_1, z_2, 1) = \sum_p{}' \bar{b}_{p_1,p_2}\bar{b}_{p_3,p_4}\ldots\bar{b}_{p_{2K-1},p_{2K}},$$

where

$$2K = 6(2L_v)(2L_h)$$

is the number of sites in the counting lattice, the sum $\sum_p{}'$ can be viewed as over all closest-packed dimer configurations, and the weights $\bar{b}_{p,q}$ are such that

$$\bar{b}_{pq} = \begin{cases} 1, & \text{if } p \text{ and } q \text{ are neighboring sites within the same six-site cluster,} \\ z_1, & \text{if sites } p \text{ and } q \text{ connect two six-site clusters in the vertical direction,} \\ z_2, & \text{if sites } p \text{ and } q \text{ connect two six-site clusters in the horizontal direction,} \\ 0, & \text{otherwise.} \end{cases} \tag{18.33}$$

The left side of Fig. 18.7 illustrates this. The weight 0 in $\bar{b}_{pq}$ ensures that only closest-packed dimer configurations with dimers connecting neighboring sites count for the function $G(z_1, z_2, 1)$. This means, in particular, that most of the elements $\bar{b}_{pq}$ are actually zero.

A matrix $\overline{B} = (\bar{b}_{pq})$ satisfying (18.33) can be constructed as follows. Consider first 6×6 matrices $\overline{B}(j, k; j', k')$ labeled by the Ising lattice sites (j, k) and (j', k'), where $-L_v + 1 \leq j,\ j' \leq L_v$, $-L_h + 1 \leq k,\ k' \leq L_h$.

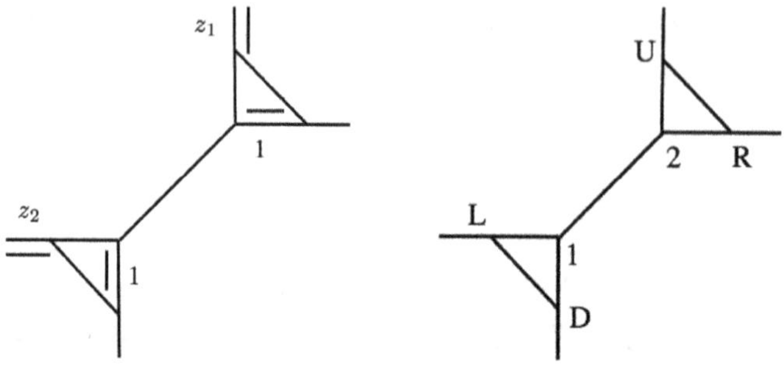

Figure 18.7 Left: The connecting bonds between sites are indicated by double lines. The weight b_{pq} connecting two sites p and q is indiacted. It is 1, z_1 or z_2 according to whether p and q are within the same six-site cluster, connect two six-site clusters in the vertical direction or connect two six-site clusters in the horizontal direction. All other bonds have weight 0. Right: Labels for the six sites of a cluster on the counting lattice.

They are zero matrices except in the following cases, where they are defined as:

$$\overline{B}(j,k;j,k) = \begin{matrix} \\ R \\ L \\ U \\ D \\ 1 \\ 2 \end{matrix} \begin{matrix} \begin{matrix} R & L & U & D & 1 & 2 \end{matrix} \\ \begin{pmatrix} 0 & 0 & 1 & 0 & 0 & 1 \\ 0 & 0 & 0 & 1 & 1 & 0 \\ 1 & 0 & 0 & 0 & 0 & 1 \\ 0 & 1 & 0 & 0 & 1 & 0 \\ 0 & 1 & 0 & 1 & 0 & 1 \\ 1 & 0 & 1 & 0 & 1 & 0 \end{pmatrix} \end{matrix},$$

for $-L_v + 1 \le j \le L_v$, $-L_h + 1 \le k \le L_h$,

$$\overline{B}(j,k;j,k+1) = \overline{B}(j,k+1;j,k)^T = \begin{matrix} \\ R \\ L \\ U \\ D \\ 1 \\ 2 \end{matrix} \begin{matrix} \begin{matrix} R & L & U & D & 1 & 2 \end{matrix} \\ \begin{pmatrix} 0 & z_1 & 0 & 0 & 0 & 0 \\ 0 & 0 & 0 & 0 & 0 & 0 \\ 0 & 0 & 0 & 0 & 0 & 0 \\ 0 & 0 & 0 & 0 & 0 & 0 \\ 0 & 0 & 0 & 0 & 0 & 0 \\ 0 & 0 & 0 & 0 & 0 & 0 \end{pmatrix} \end{matrix},$$

for $-L_v + 1 \le j \le L_v$, $-L_h + 1 \le k \le L_h - 1$,

$$\overline{B}(j,k;j+1,k) = \overline{B}(j+1,k;j,k)^T = \begin{matrix} & R & L & U & D & 1 & 2 \\ R & 0 & 0 & 0 & 0 & 0 & 0 \\ L & 0 & 0 & 0 & 0 & 0 & 0 \\ U & 0 & 0 & 0 & z_2 & 0 & 0 \\ D & 0 & 0 & 0 & 0 & 0 & 0 \\ 1 & 0 & 0 & 0 & 0 & 0 & 0 \\ 2 & 0 & 0 & 0 & 0 & 0 & 0 \end{matrix},$$

for $-L_v + 1 \le j \le L_v - 1$, $-L_h + 1 \le k \le L_h$. Each site (j,k) of the Ising lattice corresponds to a six-site cluster in the counting lattice. Think now of $\overline{B}(j,k;j',k')$ as containing weights $\bar{b}_{pq}$ between sites in six-site clusters denoted (j,k) and (j',k') (within the same cluster if $(j,k) = (j',k')$). The labels $R, L, U, D, 1$ and 2 correspond to the six sites in a cluster, depicted in the right side of Fig. 18.7. (R stands for "Right", L for "Left", U for "Up" and D for "Down".) Note that these matrices are exactly such that their elements $\bar{b}_{pq}$ satisfy (18.33). For example, within a cluster, a site R can only connect to U or 2 with the weight of 1.

Finally, the matrix $\overline{B} = (\bar{b}_{pq})_{1 \le p,q \le 2K}$ is made of the 6×6 blocks or submatrices $\overline{B}(j,k;j',k')$, $-L_v + 1 \le j,j' \le L_v$, $-L_h + 1 \le k,k' \le L_h$, we just defined. The exact placement of the blocks inside of the matrix $\overline{B}$ is as follows. Order the $d = (2L_v)(2L_h)$ sites (j,k) along the rows for increasing columns k, that is,

$$(-L_v + 1, -L_h + 1), \dots, (-L_v + 1, L_h),$$
$$(-L_v + 2, -L_h + 1), \dots, (-L_v + 2, L_h),$$
$$\cdots$$
$$(L_v, -L_h + 1), \dots, (L_v, L_h)$$

and renumber the sites $1 \le m \le d$. This renumbering associates then to each pair (j,k), (j',k') a pair (m,n), $1 \le m,n \le d$. In the matrix $\overline{B}$, its (m,n) block of size 6 is then defined as $\overline{B}(j,k;j',k')$.

18.3.3 Pfaffians

We now turn to the second step described in the discussion around (18.30), namely, to express (18.30) as the Pfaffian of a matrix. We first define the Pfaffian.

Definition 18.3.1 Consider a $2K \times 2K$ antisymmetric real-valued matrix $A = (a_{jk})_{1 \le j,k \le 2K}$, that is, with elements $a_{jk} = -a_{kj}$ and $a_{jj} = 0$. Its *Pfaffian* is defined as

$$\mathrm{Pf}A = \sum_p{}' \delta_p a_{p_1 p_2} a_{p_3 p_4} \cdots a_{p_{2K-1} p_{2K}}, \tag{18.34}$$

where, as in (18.31), $\sum_p'$ is the sum over all permutations $p_1, p_2 \ldots, p_{2K}$ of $1, 2, \ldots, 2K$ satisfying

$$p_{2m-1} < p_{2m}, \quad 1 \le m < K, \quad p_{2m-1} < p_{2m+1}, \quad 1 \le m < K-1 \tag{18.35}$$

and where δ_p, the parity of the permutation p, is 1 if the permutation p is made up of an even number of transpositions and -1 if p is made of an odd number of transpositions.

For example, if $2K = 4$, and

$$A = \begin{pmatrix} 0 & a_{12} & a_{13} & a_{14} \\ -a_{12} & 0 & a_{23} & a_{24} \\ -a_{13} & -a_{23} & 0 & a_{34} \\ -a_{14} & -a_{24} & -a_{34} & 0 \end{pmatrix},$$

then by (18.32),

$$\mathrm{Pf}A = a_{12}a_{34} - a_{13}a_{24} + a_{14}a_{23}.$$

This is because 1234, 1324, 1423 involve respectively 0, 1 and 2 transpositions.

The usefulness of the Pfaffian comes from the following formula: for an antisymmetric matrix A,

$$(\mathrm{Pf}A)^2 = \det(A) \tag{18.36}$$

(see, for example, pp. 47–51 in [1] for a proof). For example, if $2K = 2$ and

$$A = \begin{pmatrix} 0 & a_{12} \\ -a_{12} & 0 \end{pmatrix},$$

then one sees immediately that $(\text{Pf}A)^2 = \det(A)$ because $\text{Pf}A = a_{12}$ and $\det(A) = a_{12}^2$.

The relation (18.30) is not exactly the Pfaffian of the matrix $\overline{B}$ in that it does not include the parity factor δ_p. In fact, the parity factor can be introduced by suitably altering the signs of the elements of the matrix $\overline{B}$. We shall describe but not prove the assignment for doing so. See, for example, pp. 51–67 in [1] for more details.

For the sum (18.30) to have the parity factor δ_p, the blocks of the matrix $\overline{B}$ have to be replaced by the blocks:

$$\overline{A}(j,k;j,k) = \begin{array}{c} \\ R \\ L \\ U \\ D \\ 1 \\ 2 \end{array} \begin{array}{c} \begin{array}{cccccc} R & L & U & D & 1 & 2 \end{array} \\ \begin{pmatrix} 0 & 0 & -1 & 0 & 0 & 1 \\ 0 & 0 & 0 & -1 & 1 & 0 \\ 1 & 0 & 0 & 0 & 0 & -1 \\ 0 & 1 & 0 & 0 & -1 & 0 \\ 0 & -1 & 0 & 1 & 0 & 1 \\ -1 & 0 & 1 & 0 & -1 & 0 \end{pmatrix} \end{array}, \tag{18.37}$$

for $-L_v + 1 \le j \le L_v, \; -L_h + 1 \le k \le L_h$,

$$\overline{A}(j,k;j,k+1) = -\overline{A}(j,k+1;j,k)^T = \begin{array}{c} \\ R \\ L \\ U \\ D \\ 1 \\ 2 \end{array} \begin{array}{c} \begin{array}{cccccc} R & L & U & D & 1 & 2 \end{array} \\ \begin{pmatrix} 0 & z_1 & 0 & 0 & 0 & 0 \\ 0 & 0 & 0 & 0 & 0 & 0 \\ 0 & 0 & 0 & 0 & 0 & 0 \\ 0 & 0 & 0 & 0 & 0 & 0 \\ 0 & 0 & 0 & 0 & 0 & 0 \\ 0 & 0 & 0 & 0 & 0 & 0 \end{pmatrix} \end{array}, \tag{18.38}$$

for $-L_v + 1 \le j \le L_v, \; -L_h + 1 \le k \le L_h - 1$,

$$\overline{A}(j,k;j+1,k) = -\overline{A}(j+1,k;j,k)^T = \begin{array}{c} \\ R \\ L \\ U \\ D \\ 1 \\ 2 \end{array} \begin{array}{c} \begin{array}{cccccc} R & L & U & D & 1 & 2 \end{array} \\ \begin{pmatrix} 0 & 0 & 0 & 0 & 0 & 0 \\ 0 & 0 & 0 & 0 & 0 & 0 \\ 0 & 0 & 0 & z_2 & 0 & 0 \\ 0 & 0 & 0 & 0 & 0 & 0 \\ 0 & 0 & 0 & 0 & 0 & 0 \\ 0 & 0 & 0 & 0 & 0 & 0 \end{pmatrix} \end{array}, \tag{18.39}$$

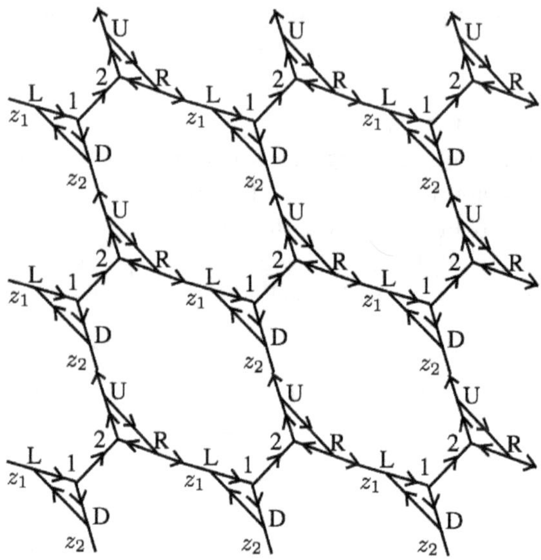

Figure 18.8 The counting lattice with directions between sites.

for $-L_v + 1 \le j \le L_v - 1$, $-L_h + 1 \le k \le L_h$. Denote the corresponding matrix by $\overline{A}$ and its elements by $\overline{a}_{p,q}$. (The $2K \times 2K$ matrix $\overline{A}$ is defined in the same way as $\overline{B}$ but using blocks $\overline{A}(j,k;j',k')$ instead of $\overline{B}(j,k;j',k')$.) Note that the only difference between the elements of $\overline{A}$ and $\overline{B}$ is that some of the entries have different signs. It is convenient to think of the sign assignment graphically as depicted in Fig. 18.8. Each bond between sites of the counting lattice now not only carries a weight (z_1, z_2, 1 or 0) but also a direction. The sign of the elements of the matrix blocks $\overline{A}(j,k;j',k')$ now corresponds to this direction. For example, the sign is positive if the connecting bond is in the forward direction, as indicated by the arrows in Fig. 18.9. Thus, in the $\overline{A}(j,k;j,k)$ block, the (U,R) entry is 1 because it corresponds to the a rrow $U \to R$ in Fig. 18.9, but the $(U,2)$ entry is -1 because $U \to 2$ is not in the direction of the arrow. Thus, one can show ([1], pp. 81, 51–58) that

$$g(z_1,z_2) = \sum_p{}' \delta_p \overline{a}_{p_1p_2}\overline{a}_{p_3p_4}\ldots\overline{a}_{p_{6d-1}p_{6d}} = \mathrm{Pf}\,\overline{A} = (\det(\overline{A}))^{1/2}, \quad (18.40)$$

where the sum is as in (18.30), δ_p is the parity factor and $\mathrm{Pf}\overline{A}$ is the Pfaffian in Definition 18.3.1. The sign at the square root in (18.40) is positive since $g(z_1,z_2)$ is defined in (18.27) with even exponents.

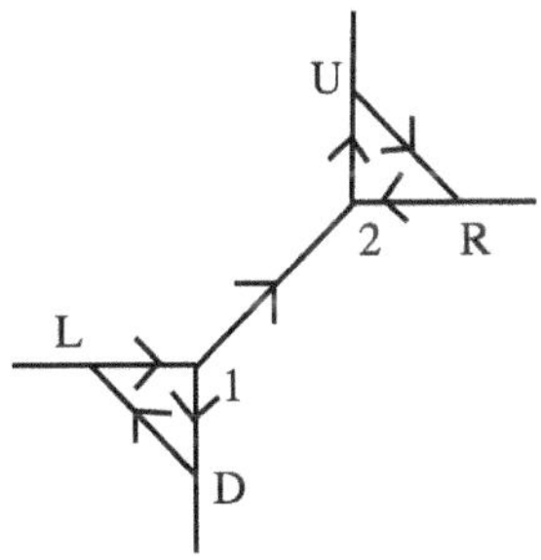

Figure 18.9 Directions in the six-site cluster of the counting lattice.

Let c_R, c_L, c_U, c_D, c_1, c_2 denote the columns of any 6×6 block $\overline{A}(j,k;j',k')$, and let us perform the following operations:

$$c_R - c_1, \quad c_U + c_1, \quad c_L + c_2, \quad c_D - c_2.$$

After these operations, the block $\overline{A}(j,k;j,k)$ becomes

$$\begin{array}{c} \\ R \\ L \\ U \\ D \\ 1 \\ 2 \end{array} \begin{array}{c} \begin{array}{cccccc} R & L & U & D & 1 & 2 \end{array} \\ \begin{pmatrix} 0 & 1 & -1 & -1 & 0 & 1 \\ -1 & 0 & 1 & -1 & 1 & 0 \\ 1 & -1 & 0 & 1 & 0 & -1 \\ 1 & 1 & -1 & 0 & -1 & 0 \\ 0 & 0 & 0 & 0 & 0 & 1 \\ 0 & 0 & 0 & 0 & -1 & 0 \end{pmatrix} \end{array} \tag{18.41}$$

and the other blocks $\overline{A}(j,k;j',k')$ remain the same. Note that the last two rows of the block (18.41) are zero except the 2×2 submatrix

$$\begin{pmatrix} 0 & 1 \\ -1 & 0 \end{pmatrix} \tag{18.42}$$

in the bottom-right corner. Since the determinant of the submatrix (18.42) equals 1, the rows and columns of the matrix $\overline{A}$ corresponding to the submatrix can be eliminated without affecting the determinant of $\overline{A}$. We thus have

$$\det(\overline{A}) = \det(A), \tag{18.43}$$

where

$$A(j,k;j,k) = \begin{array}{c} \\ R \\ L \\ U \\ D \end{array} \begin{array}{c} \begin{array}{cccc} R & L & U & D \end{array} \\ \begin{pmatrix} 0 & 1 & -1 & -1 \\ -1 & 0 & 1 & -1 \\ 1 & -1 & 0 & 1 \\ 1 & 1 & -1 & 0 \end{pmatrix} \end{array} \tag{18.44}$$

and all other $A(j,k;j',k')$ are identical to $\overline{A}(j,k;j',k')$ with the rows and columns labeled 1 and 2 removed. Thus, we also have

$$g(z_1, z_2) = (\det(A))^{1/2}, \tag{18.45}$$

where A is now a $2K \times 2K$ matrix with

$$2K = 4(2L_v)(2L_h).$$

Combining (18.21), (18.26) and (18.45), we obtain that

$$\begin{aligned} Z_{L_v,L_h} &= (\cosh \beta E_h)^{2L_v(2L_h-1)} \\ &\quad \times (\cosh \beta E_v)^{(2L_v-1)2L_h} 2^{2L_v 2L_h} (\det(A))^{1/2}, \end{aligned} \tag{18.46}$$

where z_1, z_2 in A are replaced by z_h and z_v.

The right-hand side of (18.46) provides an expression for the denominator (partition function) of (18.19). One can derive similarly an expression for the numerator of (18.19). Proceeding as in (18.20) and (18.21), we have

$$\begin{aligned} \sum_{\{\sigma_{j,k}\}} \sigma_{0,0}\sigma_{0,N} e^{-\beta \mathcal{E}} &= (\cosh \beta E_h)^{2L_v(2L_h-1)} (\cosh \beta E_v)^{(2L_v-1)2L_h} \\ &\quad \times \sum_{\{\sigma_{j,k}\}} \sigma_{0,0}\sigma_{0,N} \prod_{j=-L_v+1}^{L_v} \prod_{k=-L_h+1}^{L_h} (1 + \sigma_{j,k}\sigma_{j,k+1} z_h) \\ &\quad \times \prod_{j=-L_v+1}^{L_v} \prod_{k=-L_h+1}^{L_h} (1 + \sigma_{j,k}\sigma_{j+1,k} z_v). \end{aligned} \tag{18.47}$$

Since $\sigma_{i,j}^2 = 1$, we can write

$$\sigma_{0,0}\sigma_{0,N} = (\sigma_{0,0}\sigma_{0,1})(\sigma_{0,1}\sigma_{0,2}) \cdots (\sigma_{0,N-1}\sigma_{0,N})$$

and

$$\sigma_{0,k}\sigma_{0,k+1}(1 + \sigma_{0,k}\sigma_{0,k+1} z_h) = z_h(1 + \sigma_{0,k}\sigma_{0,k+1} z_h^{-1}),$$

so that

$$\sigma_{0,0}\sigma_{0,N}\prod_{k=0}^{N-1}(1+\sigma_{0,k}\sigma_{0,k+1}z_h) = z_h^N\prod_{k=0}^{N-1}(1+\sigma_{0,k}\sigma_{0,k+1}z_h^{-1}).$$

Then the relation (18.47) can be expressed as

$$\begin{aligned}\sum_{\{\sigma_{j,k}\}}\sigma_{0,0}\sigma_{0,N}e^{-\beta\mathcal{E}} &= (\cosh\beta E_h)^{2L_v(2L_h-1)}(\cosh\beta E_v)^{(2L_v-1)2L_h}\,z_h^N\\ &\quad\times\sum_{\{\sigma_{j,k}\}}\prod_{k=0}^{N-1}(1+\sigma_{0,k}\sigma_{0,k+1}z_h^{-1})\\ &\quad\times\prod_{j=-L_v+1}^{L_v}\ \prod_{k=-L_h+1}^{L_h}{}'\ (1+\sigma_{j,k}\sigma_{j,k+1}z_h)\\ &\quad\times\prod_{j=-L_v+1}^{L_v}\ \prod_{k=-L_h+1}^{L_h}(1+\sigma_{j,k}\sigma_{j+1,k}z_v),\end{aligned}\tag{18.48}$$

where $\prod_{k=-L_h+1}^{L_h}{}'$ means that the terms with $j=0$, $k=0,\ldots,N-1$, are omitted. The last expression is similar to (18.21) and (18.22) except for the factor z_h^N and for the fact that the weights z_h of the bonds on the horizontal line connecting $(0,0)$ and $(0,N)$ are now replaced by z_h^{-1}. This is also reflected in the fact that the term $\prod_{k=0}^{N-1}(1+\sigma_{0,k}\sigma_{0,k+1}z_h^{-1})$ now replaces the corresponding term $\prod_{k=0}^{N-1}(1+\sigma_{0,k}\sigma_{0,k+1}z_h)$ in (18.22). One can then argue in the same way as for Z_{L_v,L_h}. Since Z_{L_v,L_h} is expressible as (18.46), by combining the expressions for the numerator and denominator of (18.19), we obtain that

$$\mathbb{E}\sigma_{0,0}\sigma_{0,N} = \frac{1}{Z_{L_v,L_h}}\sum_{\{\sigma_{j,k}\}}\sigma_{0,0}\sigma_{0,N}e^{-\beta\mathcal{E}} = z_h^N\left(\frac{\det(A')}{\det(A)}\right)^{1/2},\tag{18.49}$$

where

$$A' = A+\delta$$

or equivalently $\delta = A' - A$ with

$$\delta(0,k;0,k+1) = -\delta(0,k+1;0,k)^T = \begin{array}{c} \\ R\\ L\\ U\\ D\end{array}\!\!\begin{array}{c}\begin{array}{cccc} R & L & U & D\end{array}\\ \begin{pmatrix} 0 & z_h^{-1}-z_h & 0 & 0\\ 0 & 0 & 0 & 0\\ 0 & 0 & 0 & 0\\ 0 & 0 & 0 & 0\end{pmatrix}\end{array}\tag{18.50}$$

if $0 \le k \le N-1$ and zero otherwise. Observe that the presence of δ affects only the "horizontal" edges $(0,k) \to (0,k+1)$ and $(0,k+1) \to (0,k)$ of the Ising lattice.

The matrix δ, which has dimensions $2K \times 2K$ with $2K = 4(2L_v)(2L_h)$, is zero everywhere except on its $2N$ columns and $2N$ rows. Let y be that $2N \times 2N$ submatrix of δ where it does not vanish. We have from (18.50) that

$$y = \begin{pmatrix} y_{RR} & y_{RL} \\ y_{LR} & y_{LL} \end{pmatrix}, \tag{18.51}$$

where

$$y_{RR} = \begin{array}{cc} & \begin{array}{cccc} 00 & 01 & \dots & 0N-1 \\ R & R & \dots & R \end{array} \\ \begin{array}{cc} 00 & R \\ 01 & R \\ \vdots & \vdots \\ 0N-1 & R \end{array} & \begin{pmatrix} 0 & 0 & \dots & 0 \\ 0 & 0 & \dots & 0 \\ \vdots & \vdots & \ddots & \vdots \\ 0 & 0 & \dots & 0 \end{pmatrix} \end{array},$$

$$y_{LR} = \begin{array}{cc} & \begin{array}{cccc} 00 & 01 & \dots & 0N-1 \\ R & R & \dots & R \end{array} \\ \begin{array}{cc} 01 & L \\ 02 & L \\ \vdots & \vdots \\ 0N & L \end{array} & \begin{pmatrix} -(z_h^{-1} - z_h) & 0 & \dots & 0 \\ 0 & -(z_h^{-1} - z_h) & \dots & 0 \\ \vdots & \vdots & \ddots & \vdots \\ 0 & 0 & \dots & -(z_h^{-1} - z_h) \end{pmatrix} \end{array},$$

$$y_{RL} = \begin{array}{cc} & \begin{array}{cccc} 01 & 02 & \dots & 0N \\ L & L & \dots & L \end{array} \\ \begin{array}{cc} 00 & R \\ 01 & R \\ \vdots & \vdots \\ 0N-1 & R \end{array} & \begin{pmatrix} z_h^{-1} - z_h & 0 & \dots & 0 \\ 0 & z_h^{-1} - z_h & \dots & 0 \\ \vdots & \vdots & \ddots & \vdots \\ 0 & 0 & \dots & z_h^{-1} - z_h \end{pmatrix} \end{array},$$

$$y_{LL} = \begin{array}{cc} & \begin{array}{cccc} 01 & 02 & \dots & 0N \\ L & L & \dots & L \end{array} \\ \begin{array}{cc} 01 & L \\ 02 & L \\ \vdots & \vdots \\ 0N & L \end{array} & \begin{pmatrix} 0 & 0 & \dots & 0 \\ 0 & 0 & \dots & 0 \\ \vdots & \vdots & \ddots & \vdots \\ 0 & 0 & \dots & 0 \end{pmatrix} \end{array}.$$

In the expressions above, following δ in (18.50), $00, 01, \dots, 0N-1, 01, \dots, 0N$ refer to the Ising lattice sites with non-zero weights, and R, L refer to the site in the cluster of the counting lattice.

Let also Q be the $2N \times 2N$ submatrix of A^{-1} in this same subspace as y. Then, in view of (18.49),

$$\begin{aligned}(\mathbb{E}\sigma_{0,0}\sigma_{0,N})^2 &= z_h^{2N}\frac{\det(A+\delta)}{\det(A)} = z_h^{2N}\det(A^{-1})\det(A+\delta) \\ &= z_h^{2N}\det(I + A^{-1}\delta) = z_h^{2N}\det(I+Qy) \\ &= z_h^{2N}\det(y)\det(y^{-1}+Q) \\ &= z_h^{2N}(z_h^{-1} - z_h)^{2N}\det(y^{-1}+Q) = (1-z_h^2)^{2N}\det(y^{-1}+Q),\end{aligned} \tag{18.52}$$

so that

$$(\mathbb{E}\sigma_{0,0}\sigma_{0,N})^2 = \det((1-z_h^2)(y^{-1}+Q)). \tag{18.53}$$

18.4 Computation of the inverse

To evaluate $\det(y^{-1}+Q)$ in (18.53), we need an expression for the inverse A^{-1} of A and, more specifically, for the elements of the $2N \times 2N$ submatrix Q of A^{-1} in the same subspace as y. Then, we have

$$Q = \begin{pmatrix} Q_{RR} & Q_{RL} \\ Q_{LR} & Q_{LL} \end{pmatrix}, \tag{18.54}$$

where

$$Q_{RR} = \begin{pmatrix} 0 & \dots & A^{-1}(0,0;0,N-1)_{RR} \\ A^{-1}(0,1;0,0)_{RR} & \dots & A^{-1}(0,1;0,N-1)_{RR} \\ \vdots & \ddots & \vdots \\ A^{-1}(0,N-1;0,0)_{RR} & \dots & 0 \end{pmatrix},$$

$$Q_{LR} = \begin{pmatrix} A^{-1}(0,1;0,0)_{LR} & \dots & A^{-1}(0,1;0,N-1)_{LR} \\ A^{-1}(0,2;0,0)_{LR} & \dots & A^{-1}(0,2;0,N-1)_{LR} \\ \vdots & \ddots & \vdots \\ A^{-1}(0,N;0,0)_{LR} & \dots & A^{-1}(0,N;0,N-1)_{LR} \end{pmatrix},$$

$$Q_{RL} = \begin{pmatrix} A^{-1}(0,0;0,1)_{RL} & \dots & A^{-1}(0,0;0,N)_{RL} \\ A^{-1}(0,1;0,1)_{RL} & \dots & A^{-1}(0,1;0,N)_{RL} \\ \vdots & \ddots & \vdots \\ A^{-1}(0,N-1;0,1)_{RL} & \dots & A^{-1}(0,N-1;0,N)_{RL} \end{pmatrix},$$

$$Q_{LL} = \begin{pmatrix} 0 & \dots & A^{-1}(0,1;0,N)_{LL} \\ A^{-1}(0,2;0,1)_{LL} & \dots & A^{-1}(0,2;0,N)_{LL} \\ \vdots & \ddots & \vdots \\ A^{-1}(0,N;0,1)_{LL} & \dots & 0 \end{pmatrix},$$

and where zeroes are on the diagonal since A^{-1} and hence Q is antisymmetric. The entries $A^{-1}(\cdot,\cdot,\cdot,\cdot)_{..}$ are place holders for the corresponding entries of the matrix A^{-1} and will need to be evaluated.

Recall that the matrix A is made up of blocks $A(j,k;j',k')$ given by (18.44) if $j'=j, k'=k$, and otherwise, by (18.38) and (18.39) with the rows and columns labeled 1 and 2 removed. For practical purpose, it is convenient to replace A by the following matrix $\widetilde{A}$ which, as indicated below, has a cyclic structure,

$$\begin{aligned} \widetilde{A} = I_{2L_v} \otimes I_{2L_h} \otimes \begin{pmatrix} 0 & 1 & -1 & -1 \\ -1 & 0 & 1 & -1 \\ 1 & -1 & 0 & 1 \\ 1 & 1 & -1 & 0 \end{pmatrix} \\ + I_{2L_v} \otimes \widetilde{H}_{2L_h} \otimes \begin{pmatrix} 0 & z_h & 0 & 0 \\ 0 & 0 & 0 & 0 \\ 0 & 0 & 0 & 0 \\ 0 & 0 & 0 & 0 \end{pmatrix} + I_{2L_v} \otimes \widetilde{H}^T_{2L_h} \otimes \begin{pmatrix} 0 & 0 & 0 & 0 \\ -z_h & 0 & 0 & 0 \\ 0 & 0 & 0 & 0 \\ 0 & 0 & 0 & 0 \end{pmatrix} \\ + \widetilde{H}_{2L_v} \otimes I_{2L_h} \otimes \begin{pmatrix} 0 & 0 & 0 & 0 \\ 0 & 0 & 0 & 0 \\ 0 & 0 & 0 & z_v \\ 0 & 0 & 0 & 0 \end{pmatrix} + \widetilde{H}^T_{2L_v} \otimes I_{2L_h} \otimes \begin{pmatrix} 0 & 0 & 0 & 0 \\ 0 & 0 & 0 & 0 \\ 0 & 0 & 0 & 0 \\ 0 & 0 & -z_v & 0 \end{pmatrix}, \end{aligned} \tag{18.55}$$

where $\otimes$ indicates the Kronecker product and where

$$\widetilde{H}_n = \begin{pmatrix} 0 & 1 & 0 & \dots & 0 \\ 0 & 0 & 1 & \dots & 0 \\ \vdots & \vdots & \vdots & \ddots & \vdots \\ 0 & 0 & 0 & \dots & 1 \\ -1 & 0 & 0 & \dots & 0 \end{pmatrix}$$

is an $n \times n$ matrix with a near-cyclic structure and corresponding eigenvalues $e^{i\theta}$, $\theta = \pi(2\ell - 1)/n$, $\ell = 1, \dots, n$ ([1], p. 84). The matrix $\widetilde{A}$ differs from A only at the rows and columns corresponding to the boundary positions (j, k), that is, when $j = -L_h + 1, L_h$ or $k = -L_v + 1, L_v$. Indeed, a row and a column of A associated with a position (j, k) away from the boundary contain 5 non-zero blocks $A(j, k; j, k)$, $A(j, k; j, k + 1)$, $A(j, k + 1; j, k)$, $A(j, k; j + 1, k)$ and $A(j + 1, k; j, k)$. These blocks are captured by the respective 5 terms in the sum (18.55).

We will consider below the so-called thermodynamic limit of $L_h, L_v \to \infty$. Since the difference between A and $\widetilde{A}$ is only at the columns and rows corresponding to the (j, k) which are at boundary positions, in the thermodynamic limit, the elements of A^{-1} away from these columns and rows will be those of $\widetilde{A}^{-1}$. Since the elements in (18.54) are away from the boundary, we will thus suppose without loss of generality that A is actually given by (18.55).

The advantage of working with A given by (18.55) is that it now has a cyclic structure which makes it easier to compute its inverse. One can show ([1], pp. 183, 148–150) that the entries of its inverse are

$$A^{-1}(j, k; j', k') = \frac{1}{2L_h} \sum_{\theta} e^{i\theta(k-k')} \Big(B^{-1}(\theta) \Big)_{j,j'}, \tag{18.56}$$

where the sum is over $\theta = \pi(2\ell - 1)/2L_h$, $\ell = 1, \dots, 2L_h$, and where the $4(2L_v) \times 4(2L_v)$ matrix $B(\theta)$ is given by

$$B_{j,j}(\theta) = \begin{array}{c} \\ R \\ L \\ U \\ D \end{array} \begin{array}{c} \begin{array}{cccc} R & L & U & D \end{array} \\ \begin{pmatrix} 0 & 1 + z_h e^{i\theta} & -1 & -1 \\ -1 - z_h e^{-i\theta} & 0 & 1 & -1 \\ 1 & -1 & 0 & -1 \\ 1 & 1 & -1 & 0 \end{pmatrix} \end{array}, \tag{18.57}$$

for $-L_v+1 \le j \le L_v$,

$$B_{j,j+1}(\theta) = -B_{j+1,j}(\theta)^T = \begin{pmatrix} 0 & 0 & 0 & 0 \\ 0 & 0 & 0 & 0 \\ 0 & 0 & 0 & z_v \\ 0 & 0 & 0 & 0 \end{pmatrix}, \tag{18.58}$$

for $-L_v+1 \le j \le L_v$, and all other matrix elements are zero. A matrix $(B^{-1}(\theta))_{j,j'}$ in (18.56) has dimension 4×4. Labeling its rows and columns by R, L, U and D as in (18.57), denote its elements as $(B^{-1}(\theta))_{jl,j'l'}$ with the labels $l, l' = R, L, D, U$. One can further show ([1], p. 185, formulas (2.26)) that, as $L_v \to \infty$ (thermodynamic limit),

$$\begin{aligned}(B^{-1}(\theta))_{jR,j'R} &= -(B^{-1}(\theta))_{jL,j'L} \\ &= \begin{cases} -\alpha^{-|j'-j|}(1-z_h^2)^{-1}, & \text{if } j \neq j', \\ 0, & \text{if } j = j', \end{cases}\end{aligned} \tag{18.59}$$

and that

$$\begin{aligned}(B^{-1}(\theta))_{jR,j'L} &= -(B^{-1}(\theta))^*_{jL,j'R} \\ &= z_h(1-z_h^2)^{-1}e^{i\theta}\delta_{j,j'} - \alpha^{-|j'-j|}(1-z_h^2)^{-1}e^{i\theta} \\ &\quad \times \left(\frac{(1-\alpha_1 e^{i\theta})(1-\alpha_2 e^{-i\theta})}{(1-\alpha_1 e^{-i\theta})(1-\alpha_2 e^{i\theta})}\right)^{1/2},\end{aligned} \tag{18.60}$$

where * stands for the Hermitian transpose, $\delta_{j,j'}$ is here the Kronecker delta (1 if $j = j'$, 0 if $j \neq j'$),

$$\alpha_1 = \frac{z_h(1-z_v)}{1+z_v}, \quad \alpha_2 = \frac{z_h^{-1}(1-z_v)}{1+z_v}, \tag{18.61}$$

$$\begin{aligned}\alpha = &\frac{1}{2z_v(1-z_h^2)} \\ &\times ((1+z_h^2)(1+z_v^2) - z_h(1-z_v^2)(e^{i\theta}+e^{-i\theta}) \\ &+ (1-z_v^2)|1-\alpha_1 e^{i\theta}||1-\alpha_2^{-1}e^{i\theta}|).\end{aligned}$$

(See formula (3.16) on p. 121, and formula (3.3) on p. 87 in [1] for a definition of α.) The square root in (18.60) is such that it is positive at $\theta = \pi$.

When $j = j' = 0$, these formulas (18.59) and (18.60) simplify to

$$(B^{-1}(\theta))_{0R,0R} = -(B^{-1}(\theta))_{0L,0L} = -(1-z_h^2)^{-1} \tag{18.62}$$

and

$$(B^{-1}(\theta))_{0R,0L} = -(B^{-1}(\theta))^*_{0L,0R}$$
$$= \frac{1}{1-z_h^2}\left(z_h e^{i\theta} - e^{i\theta}\left(\frac{(1-\alpha_1 e^{i\theta})(1-\alpha_2 e^{-i\theta})}{(1-\alpha_1 e^{-i\theta})(1-\alpha_2 e^{i\theta})}\right)^{1/2}\right). \tag{18.63}$$

Letting $L_v \to \infty$, $L_h \to \infty$ (thermodynamic limit) in (18.56), we get the 4×4 matrices

$$\lim_{L_h\to\infty}\lim_{L_v\to\infty} A^{-1}(0,k;0,k') = \frac{1}{2\pi}\int_0^{2\pi} e^{i\theta(k-k')}(B^{-1}(\theta))_{0,0}d\theta. \tag{18.64}$$

For example, each matrix has R, L entries

$$\lim_{L_h\to\infty}\lim_{L_v\to\infty} A^{-1}(0,k;0,k')_{RL} = \frac{1}{2\pi}\int_0^{2\pi} e^{i\theta(k-k')}(B^{-1}(\theta))_{0R,0L}d\theta.$$

Substituting (18.62)–(18.63) into (18.64), we obtain for the entries of the $2N \times 2N$ matrix Q in (18.54) that

$$\lim_{L_h\to\infty}\lim_{L_v\to\infty} A^{-1}(0,k;0,k')_{RR}$$
$$= \lim_{L_h\to\infty}\lim_{L_v\to\infty} A^{-1}(0,k;0,k')_{LL} = 0 \tag{18.65}$$

and

$$\lim_{L_h\to\infty}\lim_{L_v\to\infty} A^{-1}(0,k;0,k')_{RL}$$
$$= z_h(1-z_h^2)^{-1}\delta_{k-k'+1,0} - (1-z_h^2)^{-1}(2\pi)^{-1}$$
$$\times \int_0^{2\pi} e^{i(k-k'+1)\theta}\left(\frac{(1-\alpha_1 e^{i\theta})(1-\alpha_2 e^{-i\theta})}{(1-\alpha_1 e^{-i\theta})(1-\alpha_2 e^{i\theta})}\right)^{1/2} d\theta$$
$$= \frac{1}{1-z_h^2}\left(z_h\delta_{k'-k-1,0} - a_{k'-k-1}\right), \tag{18.66}$$

after setting

$$a_n = \frac{1}{2\pi}\int_0^{2\pi} d\theta\, e^{-in\theta}\phi(\theta), \tag{18.67}$$

$$\phi(\theta) = \left(\frac{(1-\alpha_1 e^{i\theta})(1-\alpha_2 e^{-i\theta})}{(1-\alpha_1 e^{-i\theta})(1-\alpha_2 e^{i\theta})} \right)^{1/2} \tag{18.68}$$

and using $\int_0^{2\pi} e^{i\theta(k-k'+1)} d\theta = 2\pi\delta_{k-k'+1,0} = 2\pi\delta_{k'-k-1,0}$. One gets similarly

$$\begin{aligned}
&\lim_{L_h\to\infty} \lim_{L_v\to\infty} A^{-1}(0,k;0,k')_{LR} \\
&\quad = -z_h(1-z_h^2)^{-1}\delta_{k-k'-1,0} + (1-z_h^2)^{-1}(2\pi)^{-1} \\
&\qquad \times \int_0^{2\pi} e^{i(k-k'-1)\theta} \left(\frac{(1-\alpha_1 e^{-i\theta})(1-\alpha_2 e^{i\theta})}{(1-\alpha_1 e^{i\theta})(1-\alpha_2 e^{-i\theta})} \right)^{1/2} d\theta \\
&\quad = \frac{1}{1-z_h^2}(-z_h\delta_{k-k'-1,0} + a_{k-k'-1}).
\end{aligned} \tag{18.69}$$

We now want to substitute (18.65), (18.66) and (18.69) into the expression (18.53). Observe from (18.54) and (18.65)–(18.69) that the $2N \times 2N$ matrix Q becomes, in the thermodynamic limit,

$$\lim_{L_h\to\infty} \lim_{L_v\to\infty} Q$$

$$= \begin{pmatrix}
0 & 0 & \cdots & 0 & \frac{z_h-a_0}{1-z_h^2} & \frac{-a_1}{1-z_h^2} & \cdots & \frac{-a_{N-1}}{1-z_h^2} \\
0 & 0 & \cdots & 0 & \frac{-a_{-1}}{1-z_h^2} & \frac{z_h-a_0}{1-z_h^2} & \cdots & \frac{-a_{N-2}}{1-z_h^2} \\
\vdots & \vdots & \ddots & \vdots & \vdots & \vdots & \ddots & \vdots \\
0 & 0 & \cdots & 0 & \frac{-a_{-N+1}}{1-z_h^2} & \frac{-a_{-N+2}}{1-z_h^2} & \cdots & \frac{z_h-a_0}{1-z_h^2} \\
\frac{-z_h+a_0}{1-z_h^2} & \frac{a_{-1}}{1-z_h^2} & \cdots & \frac{a_{-N+1}}{1-z_h^2} & 0 & 0 & \cdots & 0 \\
\frac{a_1}{1-z_h^2} & \frac{-z_h+a_0}{1-z_h^2} & \cdots & \frac{a_{-N+2}}{1-z_h^2} & 0 & 0 & \cdots & 0 \\
\vdots & \vdots & \ddots & \vdots & \vdots & \vdots & \ddots & \vdots \\
\frac{a_{N-1}}{1-z_h^2} & \frac{a_{N-2}}{1-z_h^2} & \cdots & \frac{-z_h+a_0}{1-z_h^2} & 0 & 0 & \cdots & 0
\end{pmatrix}.$$

Note from (18.51) that

$$y^{-1} = \begin{pmatrix} 0 & 0 & \dots & 0 & \frac{-z_h}{1-z_h^2} & 0 & \dots & 0 \\ 0 & 0 & \dots & 0 & 0 & \frac{-z_h}{1-z_h^2} & \dots & 0 \\ \vdots & \vdots & \ddots & \vdots & \vdots & \vdots & \ddots & \vdots \\ 0 & 0 & \dots & 0 & 0 & 0 & \dots & \frac{-z_h}{1-z_h^2} \\ \frac{z_h}{1-z_h^2} & 0 & \dots & 0 & 0 & 0 & \dots & 0 \\ 0 & \frac{z_h}{1-z_h^2} & \dots & 0 & 0 & 0 & \dots & 0 \\ \vdots & \vdots & \ddots & \vdots & \vdots & \vdots & \ddots & \vdots \\ 0 & 0 & \dots & \frac{z_h}{1-z_h^2} & 0 & 0 & \dots & 0 \end{pmatrix}.$$

Then,

$$(1 - z_h^2)(y^{-1} + Q)$$
$$= \begin{pmatrix} 0 & 0 & \dots & 0 & -a_0 & -a_1 & \dots & -a_{N-1} \\ 0 & 0 & \dots & 0 & -a_{-1} & -a_0 & \dots & -a_{N-2} \\ \vdots & \vdots & \ddots & \vdots & \vdots & \vdots & \ddots & \vdots \\ 0 & 0 & \dots & 0 & -a_{-N+1} & -a_{-N+2} & \dots & -a_0 \\ a_0 & a_{-1} & \dots & a_{-N+1} & 0 & 0 & \dots & 0 \\ a_1 & a_0 & \dots & a_{-N+2} & 0 & 0 & \dots & 0 \\ \vdots & \vdots & \ddots & \vdots & \vdots & \vdots & \ddots & \vdots \\ a_{N-1} & a_{N-2} & \dots & a_0 & 0 & 0 & \dots & 0 \end{pmatrix}$$

or

$$(1 - z_h^2)(y^{-1} + Q) = \begin{pmatrix} 0 & -B^T \\ B & 0 \end{pmatrix},$$

where $B = (a_{j-k})_{1\leq j,k\leq N}$ and 0 is a $N \times N$ zero matrix. By using the formula $\det((A\ B; C\ D)) = \det(AD - CB)$ if $AB = BA$ for $N \times N$ blocks A, B, C and D, we deduce that

$$\det((1 - z_h^2)(y^{-1} + Q)) = (\det(B))^2.$$

Using the expression (18.53) for $(\mathbb{E}\sigma_{0,0}\sigma_{0,N})^2$, this leads to the following expression which now involves the determinant of an $N \times N$ matrix:

$$S_N := \lim_{L_h\to\infty} \lim_{L_v\to\infty} \mathbb{E}\sigma_{0,0}\sigma_{0,N}$$

$$= \det \begin{pmatrix} a_0 & a_{-1} & a_{-2} & \cdots & a_{-N+1} \\ a_1 & a_0 & a_{-1} & \cdots & a_{-N+2} \\ a_2 & a_1 & a_0 & \cdots & a_{-N+2} \\ \vdots & \vdots & \vdots & \ddots & \vdots \\ a_{N-1} & a_{N-2} & a_{N-3} & \cdots & a_0 \end{pmatrix}. \tag{18.70}$$

Similarly, one can show ([1], p. 199, formula (3.31)) that

$$\widetilde{S}_N := \lim_{L_h\to\infty} \lim_{L_v\to\infty} \mathbb{E}\sigma_{0,0}\sigma_{N,N}$$

$$= \det \begin{pmatrix} \widetilde{a}_0 & \widetilde{a}_{-1} & \widetilde{a}_{-2} & \cdots & \widetilde{a}_{-N+1} \\ \widetilde{a}_1 & \widetilde{a}_0 & \widetilde{a}_{-1} & \cdots & \widetilde{a}_{-N+2} \\ \widetilde{a}_2 & \widetilde{a}_1 & \widetilde{a}_0 & \cdots & \widetilde{a}_{-N+2} \\ \vdots & \vdots & \vdots & \ddots & \vdots \\ \widetilde{a}_{N-1} & \widetilde{a}_{N-2} & \widetilde{a}_{N-3} & \cdots & \widetilde{a}_0 \end{pmatrix}, \tag{18.71}$$

where $\widetilde{a}_n$ is defined as a_n in (18.67) but using $\widetilde{\phi}$ instead of ϕ, defined by

$$\widetilde{\phi}(\theta) = \left(\frac{\sinh 2\beta E_h \sinh 2\beta E_v - e^{-i\theta}}{\sinh 2\beta E_h \sinh 2\beta E_v - e^{i\theta}}\right)^{1/2}. \tag{18.72}$$

18.5 The strong Szegö limit theorem

We are now interested in the behavior of the determinants S_N and $\widetilde{S}_N$ in (18.70) and (18.71), resp., as $N \to \infty$ at the critical temperature $T = T_c$, which is such that

$$1 = \sinh 2\beta E_h \sinh 2\beta E_v, \quad \beta = 1/T_c. \tag{18.73}$$

The determinants S_N and $\widetilde{S}_N$ are defined in terms of functions $\phi(\theta)$ and $\widetilde{\phi}(\theta)$ in (18.68) and (18.72). We will need expressions of these functions at the critical temperature.

Lemma 18.5.1 *At the critical temperature T_c, $\phi(\theta)$ in (18.68) becomes*

$$\phi(\theta) = \left(\frac{1-\alpha_1 e^{i\theta}}{1-\alpha_1 e^{-i\theta}}\right)^{1/2} e^{i(-1/2)(\theta-\pi)} \tag{18.74}$$

and $\widetilde{\phi}(\theta)$ in (18.72) becomes

$$\widetilde{\phi}(\theta) = e^{i(-1/2)(\theta-\pi)}. \tag{18.75}$$

Proof Consider first $\widetilde{\phi}(\theta)$ in (18.72), which becomes by (18.73),

$$\widetilde{\phi}(\theta) = \left(\frac{1-e^{-i\theta}}{1-e^{i\theta}}\right)^{1/2} = \left(-e^{-i\theta}\right)^{1/2} = ie^{-i\theta/2} = e^{i(-1/2)(\theta-\pi)}. \tag{18.76}$$

Turning to the function $\phi(\theta)$ in (18.68), recall that the parameters α_1 and α_2 were defined in (18.61), and z_h and z_v in (18.23). We first show that if (18.73) holds, then $\alpha_2 = 1$. To do so, use the relations $\cosh^2 x - \sinh^2 x = 1$ and $2(\sinh x)(\cosh x) = \sinh 2x$. Then, by using $(1-z_k^2)/(2z_k) = (\sinh 2\beta E_k)^{-1}$ for $k = h, v$, the relation (18.73) can be expressed as $(1-z_h^2)(1-z_v^2) = 4z_h z_v$. This is the same as $(1-z_h z_v)^2 = (z_h+z_v)^2$ or $1-z_h z_v = z_h+z_v$. The latter yields $1 = z_h^{-1}(1-z_v)/(1+z_v) = \alpha_2$. This now implies that

$$\alpha_1 = z_h^2 \alpha_2 = z_h^2 < 1. \tag{18.77}$$

Hence, by (18.76), $\phi(\theta)$ in (18.68) becomes (18.74). □

To obtain the asymptotic behavior of the determinants S_N and $\widetilde{S}_N$ as $N \to \infty$, we will apply the so-called "strong Szegö limit theorem". This theorem concerns the limiting behavior, as $N \to \infty$, of the $N \times N$ Toeplitz determinant

$$D_N = \det \begin{pmatrix} c_0 & c_{-1} & \cdots & c_{-N+1} \\ c_1 & c_0 & \cdots & c_{-N+2} \\ \vdots & \vdots & \ddots & \vdots \\ c_{N-1} & c_{N-2} & \cdots & c_0 \end{pmatrix}, \tag{18.78}$$

where the entries c_n are the Fourier coefficients of a function C on the unit circle, that is,

$$c_n = \frac{1}{2\pi}\int_0^{2\pi} d\theta\, e^{-in\theta} C(e^{i\theta}).$$

Thus, $C(e^{i\theta}) = \sum_{n=-\infty}^{\infty} c_n e^{-i\theta n}$ is the discrete Fourier transform of the sequence $\{c_n\}$. In the case of interest here, $C(e^{i\theta})$ is chosen to be either $\phi(\theta)$ or $\widetilde{\phi}(\theta)$.

The classical strong Szegö limit theorem supposes that $C(e^{i\theta})$ is "smooth" and allows one to conclude that

$$D_N \sim E(C)\mu^N, \tag{18.79}$$

as $N \to \infty$, where μ and $E(C)$ are two constants given by

$$\mu = \frac{1}{2\pi}\int_0^{2\pi} d\theta \, \ln C(e^{i\theta}), \quad E(C) = \exp\left\{\sum_{n=1}^{\infty} n(\ln C)_n(\ln C)_{-n}\right\},$$

and $(\ln C)_n$ denotes the nth Fourier coefficient of the function $\ln C(e^{i\theta})$. See [16]. A typical assumption for (18.79) to hold is that $C(\xi)$ be continuous on the unit circle $|\xi| = 1$. Note that this is not the case with the functions $\widetilde{\phi}(\theta)$ and $\phi(\theta)$ in (18.75) and (18.74). For example, note that $\widetilde{\phi}(0) = ie^{-i0} = i$ and $\widetilde{\phi}(2\pi) = ie^{-i\pi} = -i$. In the case when $C(\xi)$ is not continuous, the function $C(e^{i\theta})$ and the determinant D_N are said to have a *singularity*.

Though conjectured in the past by Fisher and Hartwig [17], the asymptotic behavior of determinants with singularities has been established only recently in Bootcher and Silbermann [18], Ehrhardt and Silbermann [19], Deift *et al.* [20]. We state next a corollary of Theorem 2.5 in [19] which will be sufficient for our purposes.[10] Suppose the function $C(e^{i\theta})$ can be expressed as

$$C(e^{i\theta}) = b(e^{i\theta})t_\beta(e^{i\theta}), \tag{18.80}$$

where

$$t_\beta(e^{i\theta}) = e^{i\beta(\theta-\pi)}, \quad \beta \in \mathbb{R}.$$

Note that the function $t_\beta(e^{i\theta})$ satisfies $t_\beta(e^{i0}) = e^{-i\beta\pi}$, $t_\beta(e^{i2\pi}) = e^{i\beta\pi}$, and hence is a discontinuous function on the unit circle (unless β is an integer). The function $b(e^{i\theta})$ will be assumed to be continuous on the unit circle. We

[10]The result we use follows the statement of Theorem 2.5 in [19].

will also need the Wiener-Hopf factorization of b, namely,

$$b(e^{i\theta}) = b_+(e^{i\theta})g(b)b_-(e^{i\theta}),$$

where the factors are defined as

$$b_\pm(e^{i\theta}) = \exp\left\{\sum_{n=1}^{\infty} e^{\pm i\theta n}(\ln b)_{\pm n}\right\}, \quad g(b) = \exp((\ln b)_0) \tag{18.81}$$

and where $(\ln b)_n$ denotes the nth Fourier coefficient of the function $\ln b$.

Theorem 18.5.1 *(Strong Szegö limit theorem for determinants with singularity). Suppose that the function $C(e^{i\theta})$ can be expressed as* (18.80) *where $b(e^{i\theta})$ is infinitely differentiable on the unit circle. Then, the determinant D_N in* (18.78) *satisfies, as $N \to \infty$,*

$$D_N \sim E\, g(b)^N N^{-\beta^2}, \tag{18.82}$$

where $g(b)$ is defined in (18.81), *and*

$$E = E(b)b_+(1)^{\beta}b_-(1)^{-\beta}G(1+\beta)G(1-\beta). \tag{18.83}$$

In (18.83), *$b_+(1)$, $b_-(1)$ are defined in* (18.81),

$$E(b) = \exp\left(\sum_{n=1}^{\infty} n(\ln b)_n(\ln b)_{-n}\right)$$

and

$$G(1+z) = (2\pi)^{z/2}e^{-(z+1)z/2-\gamma z^2/2}\prod_{n=1}^{\infty}\left\{\left(1+\frac{z}{n}\right)^n e^{-z+z^2/2n}\right\} \tag{18.84}$$

stands for the so-called Barnes G-function (γ is the Euler's constant).

18.6 Long-range dependence at critical temperature

Applying Theorem 18.5.1 to the determinants S_n and $\widetilde{S}_N$ in (18.70) and (18.71) with the respective functions $\phi(\theta)$ and $\widetilde{\phi}(\theta)$ in (18.68) and (18.72) at the critical temperature yields the following result which concludes the proof of Theorem 18.2.1.

Theorem 18.6.1 *Let S_n and $\widetilde{S}_N$ be the determinants in* (18.70) *and* (18.71). *At the critical temperature* (18.73), *as $N \to \infty$,*

$$S_N = \lim_{L_h\to\infty} \lim_{L_v\to\infty} \mathbb{E}\sigma_{0,0}\sigma_{0,N} \sim \left(\frac{1+\alpha_1}{1-\alpha_1}\right)^{1/4} AN^{-1/4}, \tag{18.85}$$

$$S_N = \lim_{L_h\to\infty} \lim_{L_v\to\infty} \mathbb{E}\sigma_{0,0}\sigma_{N,N} \sim AN^{-1/4}, \tag{18.86}$$

where α_1 is defined in (18.61),

$$A = G(1/2)G(3/2) = 0.6450024\ldots$$

and G is the Barnes G-function in (18.84).

Proof Consider first the determinant S_N with the function $\phi(\theta)$ in (18.74). We will apply Theorem 18.5.1. Note that $\phi(\theta)$ can be written as in (18.80),

$$\phi(\theta) = b(e^{i\theta})t_\beta(e^{i\theta}),$$

where

$$b(e^{i\theta}) = \left(\frac{1-\alpha_1 e^{i\theta}}{1-\alpha_1 e^{-i\theta}}\right)^{1/2}, \quad t_\beta(e^{i\theta}) = e^{i\beta(\theta-\pi)}, \quad \beta = -1/2. \tag{18.87}$$

The factors (18.81) of b in its Wiener-Hopf factorization are

$$\begin{aligned} b_+(e^{i\theta}) &= (1-\alpha_1 e^{i\theta})^{1/2}, \\ b_-(e^{i\theta}) &= \frac{1}{(1-\alpha_1 e^{-i\theta})^{1/2}}, \quad g(b) = 1. \end{aligned} \tag{18.88}$$

In particular,

$$b_+(1) = (1-\alpha_1)^{1/2}, \quad b_-(1) = 1/(1-\alpha_1)^{1/2}$$

and hence $b_+(1)^{-1/2}b_-(1)^{1/2} = (1-\alpha_1)^{-1/2}$. For $b(e^{i\theta})$ in (18.87), we have

$$\begin{aligned} \ln b(e^{i\theta}) &= \frac{1}{2}\ln(1-\alpha_1 e^{i\theta}) - \frac{1}{2}\ln(1-\alpha_1 e^{-i\theta}) \\ &= -\frac{1}{2}\sum_{k=1}^{\infty}\frac{\alpha_1^k e^{i\theta k}}{k} + \frac{1}{2}\sum_{k=1}^{\infty}\frac{\alpha_1^k e^{-i\theta k}}{k} \end{aligned}$$

and hence the Fourier coefficients of the function $\ln b$ are given by

$$(\ln b)_n = \frac{1}{2\pi}\int_0^{2\pi} d\theta e^{-in\theta}\ln b(e^{i\theta}) = \begin{cases} -\dfrac{\alpha_1^n}{2n}, & \text{if } n > 0, \\ 0, & \text{if } n = 0, \\ \dfrac{\alpha_1^{|n|}}{2|n|}, & \text{if } n < 0, \end{cases} \tag{18.89}$$

so that

$$\begin{aligned} E(b) &= \exp\left(\sum_{n=1}^{\infty} n(\ln b)_n(\ln b)_{-n}\right) \\ &= \exp\left(-\frac{1}{4}\sum_{n=1}^{\infty}\frac{(\alpha_1^2)^n}{n}\right) \\ &= \exp\left(\frac{1}{4}\ln(1-\alpha_1^2)\right) = (1-\alpha_1^2)^{1/4}. \end{aligned} \tag{18.90}$$

In view of (18.90), (18.83) and (18.88), we deduce (18.85). The result (18.86) for the determinant $\widetilde{S}_N$ can be obtained similarly. □

Remark. The asymptotic behavior (18.86) of $\widetilde{S}_N$ was obtained in [1] directly without using the strong Szegö limit theorem as follows. In view of (18.75), the elements $\widetilde{a}_n$ of $\widetilde{S}_n$ can be evaluated as

$$\widetilde{a}_n = \frac{1}{2\pi}\int_0^{2\pi} d\theta\, e^{-in\theta} ie^{-i\theta/2} = \frac{2}{\pi(2n+1)}.$$

Hence, by (18.71),

$$\widetilde{S}_N = (2\pi^{-1})^N \det\left(\frac{1}{2m-2n+1}\right)_{0\le m,n\le N-1}. \tag{18.91}$$

The latter determinant is the special case of the so-called Cauchy determinant,

$$\widetilde{D}_N = \det\left(\frac{1}{\mu_m+\nu_n}\right)_{0\le m,n\le N-1}, \tag{18.92}$$

where $\mu_m = 2m+1$ and $\nu_n = -2n$. It is known (for example, [1], p. 261) that

$$\widetilde{D}_N = \frac{\prod_{0\le m<n\le N-1}(\mu_m-\mu_n)(\nu_m-\nu_n)}{\prod_{m=0}^{N-1}\prod_{n=0}^{N-1}(\mu_m+\nu_n)}. \tag{18.93}$$

Applying (18.93) to (18.91) allows one to evaluate the determinant $\widetilde{S}_N$ directly, and to deduce the asymptotic behavior (18.86). See [1], pp. 261–264. In particular, one can show that the constant $A = G(1/2)G(3/2)$ in Theorem 18.6.1 can also be written as

$$A = e^{-\gamma/4+\overline{A}-1/4}$$

where

$$\gamma = \lim_{N\to\infty}\left(\sum_{l=1}^{N} l^{-1} - \ln N\right) = 0.5772157\ldots$$

is the Euler's constant, and

$$\overline{A} = \sum_{l=1}^{\infty} l\left(\ln\left(1-\frac{1}{4l^2}\right)+\frac{1}{4l^2}\right).$$

The constant A is numerically about $A = 0.6450024\ldots$ as stated in Theorem 18.6.1.

Proof of Corollary 18.2.1: Suppose $E_h = E_v = E$ and set $c = \cosh\beta E$, $s = \sinh\beta E$, $c_2 = \cosh 2\beta E$ and $s_2 = \sinh 2\beta E$. We have

$$\frac{1+\alpha_1}{1-\alpha_1} = \frac{c^2-s^2+2sc}{c^2+s^2} = \frac{1+s_2}{\sqrt{1+s_2^2}},$$

where the last equality follows from the relations $c^2 - s^2 = 1$, $2cs = s_2$ and $c^2 + s^2 = c_2 = \sqrt{1+s_2^2}$. But (18.13) yields $s_2 = 1$, and therefore

$$\frac{1+\alpha_1}{1-\alpha_1} = \sqrt{2}. \qquad \square$$

18.7 Conclusion

Benoit Mandelbrot was interested in any phenomenon exhibiting power laws, be it in Mathematics, Physics, Finance, Geology or Hydrology. He passed this interest to his student Murad Taqqu — the second author of this paper, — who in turn passed it to his own student, Vladas Pipiras — the first author.

Power laws often occur at critical junctures, where there is a "phase transition". This is a typical situation in Physics. We have focused here on the Ising model in two dimensions and shown that at critical temperature, the correlations between the spin at the origin and one at "distance" N decreases like $N^{-1/4}$. New research on the Ising model — as well as on other lattice and growth models — has recently explored their behavior when the edges of the lattice become infinitely small. These scaling limits are generally characterized by conformal invariance, and involve the Schramm-Loewner Evolution (SLE), the Conformal Loop Ensemble (CLE) and related probabilistic objects (see [21–23]). For example, the *interface* between the +1 and −1 spins in the Ising model at critical temperature has long been conjectured to converge to the SLE model with parameter $\kappa = 16/3$. This is studied rigorously by Smirnov [24]. SLE techniques were used by Lawler *et al.* [25] to show that the Hausdorff dimension of the *frontier* of planar Brownian motion is equal to 4/3. This celebrated result was conjectured by Mandelbrot [26].

Acknowledgments

We would like to thank Clément Hongler for his comments.

Bibliography

[1] B. M. McCoy, T. T. Wu. *The Two-Dimensional Ising Model* (Harvard Univ. Pr., Cambridge, 1973).

[2] M. S. Taqqu, *Theory and Applications of Long-Range Dependence*, P. Doukhan, G. Oppenheim, M. S. Taqqu, eds. (Birkhäuser, Boston, 2003), pp. 5–38.

[3] G. Samorodnitsky, *Found. Trends Stoch. Syst.* **1** (2006) 163.

[4] V. Pipiras, M. S. Taqqu, *Long-Range Dependence and Self-Similarity*, In Preparation.

[5] H. E. Hurst, *Trans. Amer. Soc. Civil Eng.* **116** (1951) 770.

[6] B. B. Mandelbrot, *C. R. Acad. Sci. Paris* **260** (1965) 3274.

[7] B. B. Mandelbrot, J. W. Van Ness, *SIAM Review* **10** (1968) 422.

[8] B. B. Mandelbrot, J. R. Wallis, *Water Resources Research* **4** (1968) 909.

[9] B. B. Mandelbrot, J. R. Wallis, *Water Resources Research* **5** (1969) 321.

[10] A. Montanari, *Theory and Applications of Long-Range Dependence*, P. Doukhan, G. Oppenheim, M. S. Taqqu, eds. (Birkhäuser, Boston, 2003), pp. 461–472.

[11] L. Onsager, *Phys. Rev.* **65** (1944) 117.

[12] M. Niss, *Archive Hist. Exact Sci.* **59** (2005) 267.

[13] M. Niss, *Archive Hist. Exact Sci.* **63** (2009) 243.

[14] B. Cipra, *Science* **288** (2000) 1561.

[15] B. M. McCoy, *Advanced Statistical Mechanics* (Oxford Univ. Pr., Oxford, 2010).

[16] G. Szegö, *Comm. Sém. Math. Univ. Lund [Medd. Lunds Univ. Mat. Sem.]* **1952**(Tome Supplementaire) (1952) 228.

[17] M. E. Fisher, R. E. Hartwig, in *Advances in Chemical Physics: Stochastic Processes in Chemical Physics, Vol. 15*, K. E. Shuler, ed. (John Wiley & Sons, Hoboken, 1968), pp. 333–353.

[18] A. Böttcher, B. Silbermann, *J. Functional Analysis* **63** (1985) 178.
[19] T. Ehrhardt, B. Silbermann, *J. Functional Analysis* **148** (1997) 229.
[20] P. Deift, A. Its, I. Krasovsky, *Ann. of Math.* **174** (2011) 1243.
[21] O. Schramm, *Israel Journal of Mathematics* **118** (2000) 221.
[22] W. Werner, in *Lectures on probability theory and statistics*, *Lecture Notes in Math.*, Vol. 1840 (Springer, Berlin, 2004), pp. 107–195.
[23] G. F. Lawler, *Conformally Invariant Processes in the Plane*, (Amer. Math. Soc., Providence, (2005).
[24] S. Smirnov, *Ann. of Math.* **172** (2010) 1435.
[25] G. F. Lawler, O. Schramm, W. Werner, *Mathematical Research Letters* **8** (2001) 401.
[26] B. B. Mandelbrot, *The Fractal Geometry of Nature* (W. H. Freeman, San Francisco, 1982).

Chapter 19

Benoit Mandelbrot, W. H. Freeman, and the launch of *The Fractal Geometry of Nature*

Peter Renz

Chance plays a role in the affairs of men. The importance of chance was Mandelbrot's theme. These comments, delivered at Yale's memorial meeting for Benoit, show some of the uncertainties surrounding his key publications. It is Benoit we are celebrating, and I am sure that had he been a classicist or philologist a similar celebration would be being held, with other participants, but with similar stories and tributes. At this memorial, Ralph Gomory spoke of Benoit's courage in pursuing his insights. His enthusiasm and good will were important as well. How did his seminal books come to be published? Here is the story.

Phil Morrison ended his November 1975 *Scientific American* review of Benoit Mandelbrot's *Les Objets Fractal* [1] saying that it was "a good bet to become a classic of its deep yet elementary genre." Adding, "Some American publisher ought to get it into English soon." At the time Phil was the book review editor for *Scientific American* and I was mathematics editor at W.H. Freeman and Company, the magazine's book-publishing affiliate.

Phil sent me a copy of his review ahead of publication. His final sentence seemed more than a friendly suggestion though less than a direct order. The review suggested breadth of Mandelbrot's scope and the appeal of his methods. I knew the recursively constructed counter-examples of analysis and topology and had studied measure dimensions in a course taught by A.S. Besicovitch, so the ideas of self-similarity and of cascade constructions were familiar.

I remember particularly that when Besicovitch worked out the Hausdorff dimension of the Cantor set, he cautioned us not to try to think of measure dimension in direct geometric terms. But Mandelbrot showed us that direct geometric vision was exactly what we should apply in this case. Furthermore, Mandelbrot showed that these fractal dimensions have physical and biological meanings as well.

I was won over by Mandelbrot's suggested application — with their sensible inner and outer cut-offs. I got in touch with Benoit, who sent me copies of his *How Long Is the Coast of Britain?* [2] and *Noah, Joseph* [3] papers and told me that he was at work on a much-expanded English version of his book: more strongly visual and more technical than the French version. The papers mentioned, together with Phil Morrison's review, gave me a clear idea of the excitement that was to come.

Addison-Wesley and other publishers were vying for the book. The division of labor between Author and Publisher that held in 1975 was upset for Benoit's book because the text and illustrations were to be produced at IBM using a computer-driven typesetter. This meant that our production people would have to deal with whatever Mandelbrot provided. Production departments were wary of such innovations. Computers were rare and expensive. Output came via chain-drive or daisy-wheel printers, and high-resolution graphics were a specialty product. Only an author at IBM or Bell Labs might have access to such technology, and authors could not be relied upon to understand the requirements of professional book publishing.

This was a complex project requiring special author and production arrangements. Benoit was sensibly asking for a larger royalty. He also insisted on greater control than Authors usually were allowed. Our negotiations were delicate and vigorous.

Freeman had certain advantages: its connection to *Scientific American* gave links to an important audience. Its motto was "our books are our salesmen," and Acquisitions, Production, and Marketing were designed to sign, develop, and promote books that would make their way without a sales force, which Freeman did not have then. Freeman also had an ace designer in Bob Ishikawa. I spoke to Bob recently about these books and he told me that they were the most interesting, important, and difficult projects he worked on. IBM's printer produced beautiful, high-resolution images on photographic paper, but its typographic capabilities were limited, and the images were done at whatever size was convenient — with no regard to what size the images needed to be to fit on the printed page. (These were physical images, not resizable digital ones.)

The text was one long column of type that Bob cut with an Exacto knife, waxed, and ironed down to put the book together. Bob spent long hours resizing the images on a photostat machine and waxing them in place. Along with Benoit and his colleagues, Bob is responsible for the visual effectiveness of *Fractals: Form, Chance, and Dimension* (1977) [4] and for that of *The Fractal Geometry of Nature* (1982) [5]. Benoit appreciated Bob's ability and stayed in touch with him over the years, asking for advice about other projects. Bob and I helped to land Benoit's book for Freeman, and, once landed, we helped it realize its potential.

It helped that Freeman accepted most of Benoit's terms. But those concessions led to a further bump in the road to publication. Early in the summer of 1976 both parties had come to terms. The Mandelbrots were visiting in La Jolla and in August they would drive north to San Francisco, visit Freeman, and the agreement would be signed.

I spent the two weeks just before this event on the Telot Glacier in the Coast Range of British Columbia, completely out of touch. I flew back on the weekend expecting to get together with Mandelbrot on Tuesday or Wednesday and sign the agreement, but when I arrived I was summoned to the office of Richard Warrington, the president of Freeman. It turned out that in my absence Warrington had decided that Benoit and his book were more trouble than they were worth. The deal was off.

I objected, but Warrington had the final say. I called Benoit and gave him the news, saying that it was not my doing. I expected the publishing opposition to sign the project quickly, but this did not happen. Benoit and I remained in touch via a curious back channel.

Martin Gardner was working on a column on fractals and was in regular contact with Benoit. I spoke to Martin by phone often and through Martin,

Benoit knew of my continued interest. By chance I called Martin when Benoit was visiting, and Martin put Benoit on the line. This tenuous connection kept Freeman's trashed chances for doing the book alive and sealed our friendship.

Luck favored Freeman when Art Bartlett, the biology editor, went to a FASEB meeting early that fall. There he ran into Lori Henline, advanced books editor for Addison-Wesley. She congratulated Art on Freeman's managing "the signing of the century." Art asked, "What book?" and she replied "Mandelbrot." When Art returned to San Francisco with this story, Warrington allowed that he might have made a mistake. He agreed that I should approach Benoit and see whether a deal might still be struck.

The Contract was signed on October 23, 1976. *Fractals: Form, Chance, and Dimension* [4] was published in 1977. The typeset material and illustrations from IBM were in order, and Bob Ishikawa did a beautiful job with the book.

Fractals: Form, Chance, and Dimension [4] seemed clear, useful, and true from the start. However, it was twenty plus years later, when Benoit asked me to look at material he was developing for a second edition of *The Fractal Geometry of Nature* [5] in 2003, that I appreciated what a bold and all-encompassing book this was.

When some new, intriguing fractal observation came to my attention, I would call Benoit to discuss it. Bill Gosper mentioned that he had discovered, to his surprise, that when you carry out the analogue of the Koch snowflake construction starting with a solid regular tetrahedron you get a cube as the limit solid, not some highly complex fractal shape. See Fig. 19.1 and the left image of Fig. 19.3. I mentioned this to Benoit, who pointed out that this fact was on page 139 of [5] in the section "Secrets of the Koch pyramid," though the discussion there is more complicated because Benoit considered the evolution of the surface rather than the solid.

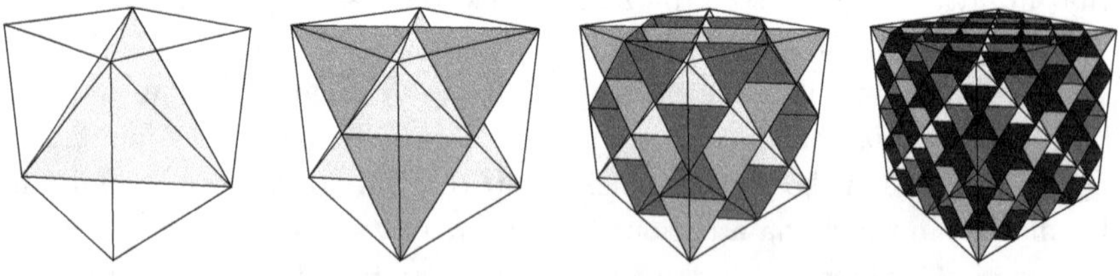

Figure 19.1 Iterates of the Koch pyramid, converging to a cube.

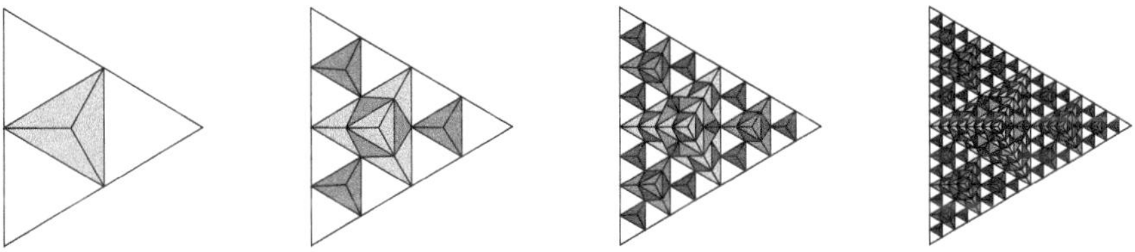

Figure 19.2 Paper model of a stage of constructing the Koch pyramid, outside (left) and insdie (right).

Figure 19.3 Views of a paper model of an early stage of building the Koch pyramid.

Fig. 19.2 and the right image of Fig. 19.3 shows chambers off of chambers inside the Koch pyramid.

Gosper was guided to his observation by computer graphics, while Benoit's analysis was by pure geometric reasoning. Scott Kim had thought about this tetrahedral process and has a nice way of looking at it. At the second state, you have a *stella octangula*, the second image of Fig. 19.1, which Scott thinks of as being inscribed in a cube. (The edges of this cube are included in the images of Fig. 19.1.) This arrangement is then repeated in a lattice of cubes that fill all of space. The voids are regular octahedra. Each successive stage partially fills in each of these octahedral voids, leaving one half-scale regular octahedral at each of the six vertices of the original void and reducing the volume of the voids by a factor of 3/4. Passing to the limit, the original cube is completely filled. Scott gives us a third way of reaching the conclusion Benoit set down long ago in his book and that Bill rediscovered by computer.

In my editorial travels I asked many people what they made of Benoit's first book. Generally they were positive. I found that most had skimmed the book for what interested them or seemed useful rather than reading it systematically. Almost none paid serious attention to Benoit's definition of a fractal as an object whose measure-theoretic dimension exceeds its topological dimension. Rightly so, since Benoit himself later set this aside. It was the free use of exact, approximate, or statistical self-similarity that spoke to readers.

I remember that Michael Aizenman told me that Benoit's book had led him to construct an elegant counter-example to a conjecture in fluid dynamics. For Aizenman, it was the images in Benoit's book, not the formulas and words, that led to insight.

Fractals: Form, Chance, and Dimension [4] and, later, *The Fractal Geometry of Nature* [5] are initial chapters of an ongoing story, whose breadth and depth are such that no single book could now aspire to tell it all. These books teem with suggestions for uses of a new geometry, whose full character was yet to be discovered. Their success is evidenced by those who worked on mapping this realm: Hutchinson, Douady, Hubbard among the earlier ones. Equally, we look to the meteorologists, geologists and geographers, ecologists and biologists, physicists and engineers who have been applying fractal ideas in order to see how useful Mandelbrot's suggestions have been.

Chapter XII of [5] is titled "Of Men and Ideas." It gives the history of fractal ideas and is a window into Benoit's thinking. He was concerned with the human and historical side of science as well as with the technical side. In the history of this field, the spread of fractal ideas was aided by developments in computing and computer graphics. Benoit's courage and commitment in following his ideas was crucial to the eventual success of the field.

I miss particularly our long and wide-ranging conversations, especially those touching on the many interesting colleagues and friends he had and his thoughts about historical and cultural matters: French academics; people he knew, from his uncle Szolem Mandelbrojt to Claude Levi-Strauss; mutual friends such as Mark and Kitty Kac and Shizuo Kakutani — yes, and some mathematics too, including, not long ago, the Kakeya Problem.

What a treasure trove he has left us all; how many have seized his ideas, made them their own and built on them; how I will miss his unique insights. It was a privilege to have had a hand in the success of these two books and to have known Benoit.

Bibliography

[1] B. Mandelbrot, *Les Objets Fractals: Forme, Hasard et dimension*, Flammarion, Paris, 1975.
[2] B. Mandelbrot, *Science* **156** (1967) 636.
[3] B. Mandelbrot, J. Wallis, *Water Resources Research* **4** (1968) 909.
[4] B. Mandelbrot, *Fractals: Form, Chance, and Dimension*, W. H. Freeman, San Francisco, 1977.
[5] B. Mandelbrot, *The Fractal Geometry of Nature*, W. H. Freeman, New York, 1982.

Chapter 20

Math and physics: Lévy flights and drives

Michael Shlesinger

Abstract. Benoit Mandelbrot had a profound influence in the physics community by connecting scaling and geometry. One example is the Lévy flight. In this memorial issue, in honor of Benoit, I review some personal history and ideas on the mathematics of Lévy flights and then introduce Lévy drives for applications to physical problems.

20.1 Introduction

I first met Benoit way back in 1972 when he presented the physics colloquium at the University of Rochester where I was a graduate student. He was invited by my Ph.D. thesis advisor, Elliott Montroll, a renowned physicist and a former Director of General Sciences at IBM in the early 1960s. This is the only time I remember that Montroll invited a colloquium speaker and certainly the only time he recommended his students to attend the colloquium. The talk was unique. Benoit showed slides of dots, in what we call today a fractal pattern. The dots represented the end points of Lévy flight jumps, but how the pictures were generated was never mentioned. Benoit also displayed pictures of stars that seemed to be similar to the fractal pictures. He would note the dimension of the pictures, saying something like this is 1.7 dimensional. The term fractal was not yet introduced. Hands would be raised with question of how could something not be an integer dimension. Benoit asked the audience to use their intuition to judge dimension without defining it. There was general dissatisfaction in the audience with this approach to star cluster and galaxy structure, but I was excited and all fired up to try to understand the lecture.

20.2 Lévy flights

After the talk I asked Montroll what was going on. He drew a Lévy flight on the board showing clusters of jumps on a small scale followed eventually by an order of magnitude larger jump repeated over several scales. He showed about four different scales of jumps in his drawing and mentioned the connection to Paul Lévy's work, saying he didn't quite understand what it meant, but as a beautiful piece of mathematics it would surely one day find its way into physics. I wrote down, almost immediately, in trying to understand the Lévy flight on my terms, the following equation that allows jumps, for a $1D$ random walk, on all scales b^j $(0 < j < \infty)$, but each larger scale b^j occurring with a lower probability proportional to c^{-j}, $b > 1$ and $c > 1$

$$p(x) = \frac{c-1}{2c} \sum_{j=0}^{\infty} \frac{1}{c^j} (\delta_{x,b^j} + \delta_{x,b^{-j}}) \tag{20.1}$$

Although, I did this in 1972 and wrote the structure function $\hat{p}(k)$ as

$$\hat{p}(k) \equiv \sum_{x=-\infty}^{\infty} e^{ikx} p(x) = \frac{c-1}{c} \sum_{j=0}^{\infty} c^{-j} \cos(kb^j) \tag{20.2}$$

the complete analysis of this random walk [1] waited until 1981 when Barry Hughes joined Montroll and myself at the U of Maryland. Hughes pointed out that $\hat{p}(k)$ is the Weierstrass function that is everywhere continuous, but nowhere differentiable. One can write

$$\hat{p}(k) = \frac{1}{c}\hat{p}(bk) + \frac{c-1}{c} \cos(k) \tag{20.3}$$

And by substitution see that k^β is a solution to the homogeneous part of the equation when $\beta = \ln(c)/\ln(b)$. When $b^2 > c$, then $\beta < 2$ and $\langle x^2 \rangle = \infty$

and there is no characteristic scale to the jump sizes. By writing the cosine in terms of its own inverse Mellin transform, $\hat{p}(k)$ in Eq. (20.2)) can now be written now as a contour integral

$$\begin{aligned}\hat{p}(k) &= \frac{1}{2\pi i}\sum_{j=0}^{\infty}\oint \frac{\Gamma(s)\cos(\pi s/2)}{c^j b^{sj}|k|^s}ds \\ &= \frac{1}{2\pi i}\oint \frac{\Gamma(s)\cos(\pi s/2)}{1-c^{-1}b^{-s}}|k|^{-s}ds \qquad (20.4)\end{aligned}$$

While the full series expansion can be calculated, we only remark here that the simple pole in the denominator gives a $|k|^{\beta}$ term with $\beta = \ln(c)/\ln(b)$. We called this the Weierstrass random walk in honor of the Weierstrass function appearance in the Fourier transform of the jump probabilities. Benoit particularly liked this work where the fractal dimension appeared explicitly as the log of the number of sub-clusters in a cluster divided by the log of the size ratio between a cluster and sub-cluster.

I next saw Benoit in 1976 when I picked him up together with Mark Kac at the Rochester airport and drove them to Montroll's house for a party the day before the Montroll 60th birthday symposium. At the symposium Benoit showed the amazing self-affine pictures that he with Richard Voss created of realistic landscapes, including the moon. At this time the word fractal was in play.

My Ph.D. thesis work followed the model of Harvey Scher and Elliott Montroll on the transport of electrons in an amorphous semiconductor film used in the xerographic process [2]. The behavior was non-standard and at first Montroll posited a Lévy flight for the electrons, but Harvey Scher using the Montroll-Weiss [3] continuous time random walk model showed the data could be explained if charge (actually holes) became intermittently trapped with a random release time showing no characteristic time scale. A time sequence of jump times looked like the points in a randomized Cantor set. We solved several aspects of this problem, including the charge mobility behavior as a function of sample size and the external electric field strength and the nonstationary current decay regimes [2, 4, 5]. Harvey Scher called this dispersive transport, but at a discussion at the Montroll celebration, Benoit said it was a case of *fractal time* and he pointed us to his paper on errors in telephone transmissions [6]. By the way, this is a great technical paper, but overshadowed by Benoit's later geometric works on Julia and Mandelbrot sets.

I had run into the Cantor Set in a real analysis course as an undergrad. It was presented as a pathology (a weird set all of whose points are limit points, but the set is totally disconnected). The Cantor example came with a warning that don't let intuition fool you, math must be done in a rigorous way, and things like Cantor Sets are counter-intuitive. In the same vein, in this course theorems were stated under the condition that functions were of bounded variation. I only later realized that this meant fractal curves were excluded. It was Benoit's tireless insistence that brought fractal sets and curves into mainstream mathematics and physics, instead of relegating them to a corner as oddities to be avoided through the careful wording of theorems.

Benoit was widely read and had a vast store of knowledge and a quick and sharp mind. He also had sitzfleich. He would sit in the first row and sit through every talk at our fractals meetings absorbing new results and reminding speakers of old results. Sometimes the mathematicians and the physicists did not see eye to eye about what was new, which is perhaps an older argument between pure and applied math. I like the analogy of a great chess match whose sequence of moves is thrilling and novel, but the game is played under the rules of chess. Was something new accomplished or was the game just one example out of many that follow from the rules of chess? Benoit's intuition, insight, and fortitude created the field of fractals. Although many of the pieces, including the Hausdorff dimension, Julia sets, Kock curves, etc., were in place, it took Benoit's contributions to create the field of fractals. It is now commonplace to find the use of fractals in many fields of science and technology.

In 1982 I organized a conference at the National Bureau of Standards on random walks, including Lévy flights. I invited Benoit as the after dinner speaker and he was a big hit. We decided to organize the first US conference on fractals. I was doing all the work and as the time of the meeting was approaching I asked Benoit what he would like to contribute. He answered immediately, "I'm inviting the press" and Ivars Peterson from *Science News* did come and write up a report about the meeting and he continued to write pieces about fractals. A few months later we had a second meeting at Les Houches in France, where I was listed as an organizer, but did nothing in way of organizing.

While the Lévy flight can be readily generalized to higher dimensions in Cartesian coordinates, it is perhaps more instructive to consider N-dimensional spherical coordinates where the angular dependence is uniform and the radial dependence is Lévy distributed. In this case, we have for the jump probability

$$p(\vec{x}) = \frac{\Gamma(N/2)}{2\pi^{N/2}|x|^{N-1}} \frac{c-1}{c} \sum_{j=0}^{\infty} c^{-j} \delta(|x| - b^j) \tag{20.5}$$

and

$$\hat{p}(\vec{k}) = \frac{\Gamma(N/2)}{|kb^j/2|^{N/2-1}} \frac{c-1}{c} \sum_{j=0}^{\infty} c^{-j} J_{N/2-1}(|k|b^j) \tag{20.6}$$

where J is a Bessel function. We see that this generalizes the Weierstrass function from a lacunary cosine series to a lacunary Bessel function series where in $1D$ the cosine is now recognized as an order $1/2$ spherical Bessel function.

20.3 Lévy drives

The Lévy flight only considers the start and end points of jumps and produces a sequence of dots. These are like airplane flights with only the takeoff and landing points shown. If one drove a car between the points the path would look like the connected lines in Fig. 20.1, so I call the driving process a Lévy drive. Of course one drives with a velocity so there is the need to introduce velocity into the Lévy drive and this makes the process more physical than a Lévy flight where the movement between dots is instantaneous. As one can drive with different speeds depending on the road conditions, one can study Lévy drives with different velocities for different segments. For example, one may go at slow speeds in a residential area and high speeds on a highway over long distances. We will see this type of behavior in studying turbulent motions with low velocities in small vortices and large velocities in large vortices, first we consider in Sec. 20.4 a diffusion type model with the jumps occurring with a velocity V.

20.4 Bachelier revisited as a telegrapher drive

In the standard Montroll-Weiss continuous time random walk, one has a probability $p(x)$ for jumping a distance x, and a probability density $\Psi(t)$ for waiting a time t between jumps. For the Lévy drive, one needs to couple x and t to relate space and time, i.e. to incorporate a velocity. One

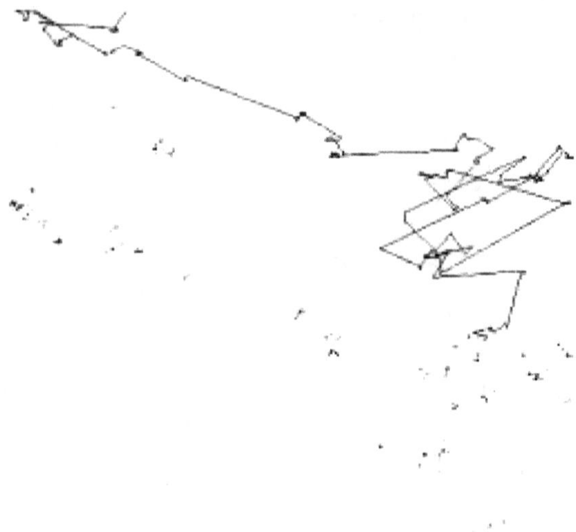

Figure 20.1 The dots are the points visited by a Lévy flight random walk. They form a fractal pattern and in a 1972 lecture Mandelbrot posited their similarity to star clusters. The solid lines shifted above the dots are the trajectories of the random walk whose turning points are the Lévy flight points. I call these lines a Lévy drive as this would be the straight line path if driving beteen the Lévy flight points.

introduces [7]

$$\Psi(x,t) = p(x|t)\psi(t) \tag{20.7}$$

that takes into account the probability of a jump of distance x given that the jump is of duration t, times the probability density that the jump is of duration t. A simple case is constant velocity and jumps with a finite mean square displacement, i.e. let us choose,

$$p(x|t) = \delta(|x| - Vt) \tag{20.8}$$

$$\psi(t) = \lambda e^{-\lambda t} \tag{20.9}$$

The mean trajectory duration is $1/\lambda$ so the mean jump length is V/λ. Without going through all of the detailed calculation, one finds [8] for $Vt < |x|$ that the probability for being at site x at time t is given approximately by

$$P(x,t) \sim \exp(-\lambda t/2) I_0 \left(\frac{\lambda}{2} \sqrt{t^2 - \frac{t^2}{V^2}} \right) \Theta(Vt - |x|) \tag{20.10}$$

Where I_0 is a modified Bessel function and the Heavyside Θ function sets $P(x,t) = 0$ for distances that cannot be reached in a time t. The equation has a Gaussian type behavior for $|x| < Vt$ with an effective diffusion constant of V^2/λ, further and it can be shown that $P(x,t)$ satisfies the telegraphers equation rather than the diffusion equation. The diffusion equation is unphysical in that it has a finite probability that a random walker can be very far away from its origin even at very short times, while the telegraphers equation does not allow the walker to get beyond a distance x that is greater than Vt. The means that the diffusion has a propagating front at $|x| = Vt$.

Benoit was at the revival of the appreciation of Bachelier's work on stock market prices. Bachelier included the velocity of change of prices and derived a diffusion equation for the price of a bond or stock. Somewhat later Lévy objected to with Bachelier's approach because of the mathematical inconsistency in that a velocity is not defined for diffusion. What Bachelier should have derived for his model with changes occurring with a finite velocity is the telegraphers equation.

20.5 Supradiffusion [9]

For a random walk with a constant velocity V for each jump, the duration T of the jump translates into a displacement $x = VT$. Choosing the probability waiting time density to have the asymptotic behavior,

$$\psi(t) \approx t^{-1-\beta} \tag{20.11}$$

gives rise to Lévy drive jumps without a characteristic size, when β is less than unity, yielding the following asymptotic time results for the mean squared displacement of a symmetric random walk in $1D$

$$\langle x^2(t)\rangle \approx \begin{cases} t^2 & 0 < \beta < 1 \\ t^2/\ln(t) & \beta = 1 \\ t^{3-\beta} & 1 < \beta < 2 \\ t\ln(t) & \beta = 2 \\ t & \beta > 2 \end{cases} \tag{20.12}$$

More complicated processes can be treated including a case with velocity that depends on the jump distance. Let us choose the algebraic $\Psi(t)$ in

Eq. (20.11) with

$$p(R|t) = \delta(R - V(R)t) \tag{20.13}$$

And $V(R)$ varying as $R^{1/3}$. This is the velocity-distance relationship for Kolmogorov inertial range turbulence where larger vortices have more energy than smaller ones, so longer trajectories would have larger velocities. This leads to the following asymptotic behavior for the mean squared displacement

$$\langle R^2(t)\rangle \approx \begin{cases} t^3 & \beta \le 1/3 \\ t^{2+3(1-\beta)/2} & 1/3 \le \beta \le 5/3 \\ t & 5/3 \le \beta \end{cases} \tag{20.14}$$

The t^3 case is called Richardson's Law for turbulent diffusion. One can study more complicated situations, including relativistic turbulence in a gravitational field [8].

Even deterministic motion can appear to be Lévy-like, as seen in Fig. 20.2. The Zaslavsky map in Eq. (20.15) for a kicked nonlinear oscillator shows Lévy type properties in the (u, v) phase space [10]

$$\begin{aligned} u_{n+1} &= (u_n + K\sin(v_n))\cos(2\pi/q) + v_n \sin(2\pi/q) \\ v_{n+1} &= -(u_n + K\sin(v_n))\sin(2\pi/q) + v_n \cos(2\pi/q) \end{aligned} \tag{20.15}$$

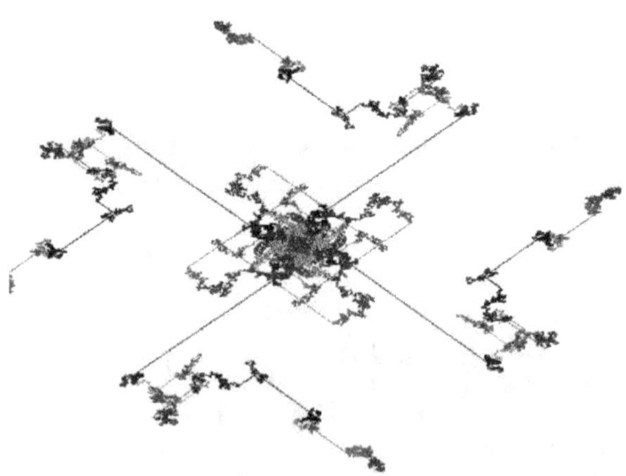

Figure 20.2 The trajectory of a four-fold $q = 4$ symmetric Zaslavsky map. It was surprising to me that the deterministic trajectories of a Hamiltonian nonlinear map can resemble the stochastic trajectory of a Lévy drive.

The equation only has trigonometric terms, but a typical trajectory has the Lévy drive signature and a fractal dimension that depending on the value of K and the symmetry on the value of q. There is no apparent pre-selected fractal dimension exponent in the equation.

Lévy drive behaviors appear in many areas including, Josephson junction phase rotations [11], polymers motions [12], Couette fluid flow [13], animal movements [14], the movement of money [15] and people [16] and price changes in the stock market [17]. The best place to be introduced to all aspects of fractals remains the works of Benoit where his insight, genius, esthetics, and enthusiasm combine into a transcendental experience for the dedicated reader of [18].

Bibliography

[1] B. Hughes, M. Shlesinger, E. Montroll, *Proc. Nat. Acad. Sci. USA* **87** (1981) 3287.

[2] E. Montroll, H. Scher, *J. Stat. Phys.* **9** (1074) 101.

[3] E. Montroll, G. Weiss, *J. Math. Phys.* **6** (1965) 167.

[4] M. Shlesinger, *J. Stat. Phys.* **10** (1974) 421.

[5] H. Scher, E. Montroll, *Phys. Rev.* **B12**, (1975) 2455.

[6] J. Berger, B. Mandelbrot, *IBM J. Res. Dev.* **7** (1963) 224.

[7] M. Shlesinger, J. Klafter, J. Wong, *J. Stat. Phys.* **27** (1982) 111.

[8] M. Shlesinger, in *Processes with Long-Range Correlations*, G. Rangarajan, M. Ding, eds. (1993), pp. 139–147.

[9] J. Klafter, M. Shlesinger, G. Zumofen, *Phys. Today* **49** (1982) 33.

[10] G. M. Zaslavsky, M. Yu. Zakharov, R. Z. Sagdeev, D. A. Usikov, and A. A. Chernikov, *Sov. Phys. JETP* **64** (1986) 294.

[11] T. Geisel, J. Nierwetberg, and A. Zacherl, *Phys. Rev. Lett.* **54** (1985) 616.

[12] A. Ott, J. P. Bouchaud, D. Langevin, and W. Urbach, *Phys. Rev. Lett.* **65** (1990) 2201.

[13] T. H. Solomon, E. R. Weeks, and H. Swinney, *Phys. Rev. Lett.* **71** (1993) 3975.

[14] N. E. Humphries *et al.*, *Nature* **465** (2010) 1066.

[15] D. Brockman, L. Hufnagel, and T. Geisel, *Nature* **439** (2005) 462.

[16] C. Song, T. Koren, P. Wang, and A. L. Barabasi, *Nature Physics* **6** (2010) 818.

[17] R.N. Mantegna, *Physica A* **179** (1991) 232.

[18] B. Mandelbrot, *The Fractal Geometry of Nature* (W. H. Freeman, San Francisco, 1982).

Chapter 21

Benoit Mandelbrot in finance

Christian Walter

21.1 Introduction

Stock market graphs representing changes in the prices of securities over a period of time appear as irregular forms that seem to be reproduced and repeated in all scales of analysis, rising periods follow periods of fall. However, the rises are broken up with intermediate falling phases and, falls are interspersed with partial rises and this continues almost indefinitely until the quotations of securities reach the natural limit of the scale.

This entanglement of rising and falling waves of repetitive patterns in all the scales was discovered in the 1930s by Ralph Elliott, to whom this idea occurred while observing the ebb and flow of tides on the sands of a seashore. From this, he formulated a financial symbolization known as "stock market waves" or "Elliott's waves," which he broke up into huge tides, normal waves and wavelets, and also "tsunami," from the name given in Japan to huge waves arising due to earthquake. The theory called "Elliott's waves" presents a deterministic *fractal description* of the stock market from self-similar geometric figures that we find on all scales of observation and compiles a toolbox in the form of graphic analysis of stock market fluctuations used by certain market professionals technical analysts.

Elliott's figures propose a calibration of rise and fall variations from a Pythagorician numerology based on the usage of golden ratio and Fibonacci sequence, which are predictions strongly tinged with subjectivity, in so far as detection and positioning of waves depend on the graphic analyst's view of the market which he examines. For the lack of an appropriate mathematical tool, this conceptualization of stock market variations was confirmed, as alchemy before chemistry, in the pre-scientific field until the emergence of fractals.

The fractals of Benoit Mandelbrot, though developed in a radically different approach, fit in this understanding of stock market variations and present, as common point with Elliott's waves, the fact of finding how to untangle the inextricable interlacing of stock market fluctuations in all the scales. Using stock market language, do we find ourselves in fall correction of a rising phase or in a fall period contradicted by a temporary rise? This represents adequate conceptualization allowing the translation of intuitions of graphic analysts in rigorous mathematical representations.

However, the adventure of fractals in finance does not have a smooth history (it's a fractal history). It rather refers to an eventful progression of Mandelbrot's assumptions through the evolution of finance theory over 50 years, from 1960 until today, which stirred up a vehement controversy on modeling in infinite variance or infinite memory. The connecting thread of Mandelbrot's works, followed by others — including his contradictors — was the research of scaling laws on stock market fluctuations, irrespective of whether this research followed the direction of scaling invariance, or pure fractal approach of markets, as proposed by Mandelbrot, or according to that of an instrumentation of multiscaling analysis of markets, such as that corresponding to mixed processes or of ARCH type that emerged in the 1980s, or to the changing system in the 1990s.

The starting point of this controversy was the existence of leptokurtic (or non-Gaussian) distributions on stock market variations. This distributional anomaly in relation to the Brownian hypothesis of traditional financial modeling led Mandelbrot to propose α-stable distributions in 1962 to Paul Lévy's infinite variance by replacing Gaussian for modeling of periodic returns.

However, very soon, this new hypothesis provoked a relatively fierce controversy on the existence of variance and other new candidate processes appeared, all the more easily while scaling invariance of α-stable laws, a cardinal property of Mandelbrot's fractal hypothesis, did not appear, or only with difficulty, experimentally validated. The attempt to resolve leptokurtic problems by conservation of IID hypothesis and the proposal of α-stable distributions did not solve all the anomalies, since a new type of anomalies, or scaling anomalies appeared.

Therefore, the theoretical research is interested in modelling leptokurticity in other possible ways and we turn towards the second pivot of financial modelling of independence of successive returns, which was equally challenged. Hence, we looked in different forms of dependence between returns (linear and then non-linear dependence) for the cause of leptokurticity observed. This is the second time of empirical investigations. After highlighting the absence of short memory on returns, the research is turned towards the detection of long memory on returns. This attempt does not succeed either. Then, the focus is shifted the process of volatilities with the formalization of short memory on volatilities, i.e. an approach that led to the trend of ARCH modeling and then by highlighting long memory on volatilities (or long dependence), i.e. a trend that led to the rediscovery of scaling laws in finance.

Finally, the fractal hypothesis was validated on the generating process of stock market volatilities. Today, the long memory of volatilities (i.e., a hyperbolic law on correlations among volatilities) has become an acquired fact on financial markets and financial modeling seeks to reconcile the absence of memory on returns and the presence of long memory on volatilities.

21.2 The Noah and Joseph effects and the non-Gaussian and non-Brownian issues of the financial theory

Let $(S(t), t \geq 0)$ denote the price process of a given asset. The basic model in finance is the exponential Brownian motion, i.e. the stochastic process defined by

$$S(t) = S(0)\exp(\mu t + \sigma W(t)) \tag{21.1}$$

with $W(t)$ standard Brownian motion, that is, $W(0) = 0$, $\mathbb{E}[W_1] = 0$ and $\mathbb{E}[W_1^2] = 1$. This is the standard model of price variations in finance. It was introduced by Samuelson in 1965 and is sometimes called in the early financial literature "geometric Brownian motion" [1]. We introduce the notation

$$X(t) = \ln S(t) - \ln S(0) \tag{21.2}$$

as the continuously compounded return of the asset. According to the standard model (21.1),

$$X(t) = \mu t + \sigma W(t) \tag{21.3}$$

is the Brownian motion introduced by Bachelier in 1900 [2, 3]. The distribution of $X(t)$ is the normal (Gaussian) one. In other words, with the standard model, the marginal distributions of returns have to be Gaussian.

It is important to clearly distinguish between the normality assumption and the continuity assumption in (21.3). The two assumptions are confounded in the standard model because of the properties of Brownian motion, but they refer to different conceptual frameworks. The *normality* of asset returns has played a central role in finance for a large number of *static* models, beginning with the Markowitz' standard portfolio theory (1952) and the Sharpe's Capital Asset Pricing Model (1964) and more recently with the Value at Risk computations in the Basle framework. On the other hand, the *continuity* of trajectories has played a central role in the *dynamic* models built on the building blocks of the Black, Scholes and Merton's standard option pricing theory (1973) and the default risk probabilities computed with the Vasicek's yield curve model (1977).

It is now well documented that both the normality and continuity assumptions are contradicted by the data: the normal distribution is a poor model for the marginal returns, and all the trajectories exhibit a lot of jumps. Tackling these challenges has represented the aims pursued by Benoit Mandelbrot[1]:

> In science, all important ideas need names and stories to fix them in the memory. It occurred to me that the market's first wild trait, abrupt change or discontinuity, is prefigured in the Bible tale of Noah. As *Genesis* relates, in

[1]A short biography of Benoit Mandelbrot can be found in the Richard Hudson's prelude to [4], pgs. xv–xxiv : "Introducing a Maverick in Science". Other existing short biography is "A maverick's apprenticeship", by Benoit Mandelbrot himself (available on website). Most complete is Benoit's memoirs [5].

> Noah's six hundredth year God ordered the Great Flood to purify a wicked world. [...] The flood came and went — catastrophic but transient. Market crashes are like that: at times, even a great bank or brokerage house can seem like a little boat in a big storm.
>
> The market's second wild trait — almost cycles — is prefigured in the story of Joseph. Pharaoh dreamed that seven fat cattle were feeding in the meadows, when seven lean kine rose out of the Nile and ate them. [...] Joseph, a Hebrew slave, called the dreams prophetic: Seven years of famine would follow seven years of prosperity. [...] Of course, this is not a regular or predictable pattern. But the appearance of one is strong. Behind it is the influence of long-range dependence in an otherwise random process — or, put another way, a long-term memory through which the past continues to influence the random fluctuations of the present.
>
> I called these two distinct forms of wild behavior the *Noah effect* and the *Joseph effect*. They are two aspects of one reality[2]".

We present below the three main models of Benoit Mandelbrot and their legacy. The next section presents the description of the Noah effect by the α-stable motion (1962). Section 21.4 shows the relation between Joseph effect and fractional Brownian motion (1965). Section 21.5 presents the fundamental intuition of the time-changed Brownian motion (1967) for disentangling the Noah and Joseph effects.

21.3 The first model of Mandelbrot (1962): α-stable motion with paretian tails

Adopting a static standpoint, the issue of non normality is exhibited by the so-called leptokurtic phenomenon (following the terminology introduced in [6]) which appeared very early in financial literature. To solve this puzzle, the idea of Mandelbrot can be summarized as follows: let us abandon both the continuity and normality assumptions, but keep the assumption that the process has independent increments. Hence, in 1962, Mandelbrot replaced

[2] [4] pgs. 200–201.

the standard model (21.3) by a new model called α-stable Lévy motion ([7], followed by a more detailed description [8]), where the distributions have fat or pareto-type tails, and in fact, infinite variance. This precisely allows to understand the Mandelbrot's intuition: the search for paretian tails in finance. "The power of power laws" ([4], pg. 13) is a tale of fat tails:

> Examine price records more closely, and you typically find a different kind of distribution than the bell curve: the tails do not become imperceptible but follow a "power law". These are *common in nature.* [...] In economics, one classic power law was discovered by italian economist Vilfredo Pareto a century ago. It describes the distribution of income in the upper reaches of society. That power laws *concentrates* much more of a society's wealth among the very fews; a bell curve would be more equitable, scattering incomes more evenly around an average. Now we reach *one of my main findings.* A power law also applies to positive or negative price movements[3] of many financial instruments. It leaves room for many more big price swings than would the bell curve. And it fits the data for many price series. *I provided the first evidence in a 1962 research report.* (our italics).

The legacy of paretian approach The sequel of the story is now well known.[4] After a period of strong arguments on the pros and cons of the infinite variance hypothesis, the 1962 model was abandoned because of the lack of second moment, considered as too strong for financial uses. However the ideas didn't disappear. On one hand, the theory of extreme values, the rebirth of which dates back to 1993, kept the intuition of paretian tails but subsumed it through the generalized Pareto distribution which includes many cases of fat tails and not only the α-stable ones [10]. On the other hand, the general Lèvy processes which were considered in the 1990s maintained the discontinuity feature but without the infinite variance characteristic of the α-stable motion [11, 12]. There is now a large body of research on these topics and *financial modelling under Non Gaussian*

[3]Exactly the two sides of the Lévy measure defined in (21.13).

[4]See [9] for a comprehensive approach of the story of the α-stable attempts for solving the leptokurtic character of returns.

distributions[5] is becoming a new consensual approach in the financial literature.

21.4 The second model of Mandelbrot (1965): fractional brownian motion with aperiodic cycles

An interesting feature of the standard model defined in (21.3) is that the process $X(t)$ is *self-similar* when the drift μ is zero. We say that $X(t)$ is self-similar with exponent 1/2 (hereafter 1/2-ss) if

$$X(at) \stackrel{d}{=} a^{1/2} X(t) \tag{21.4}$$

where the notation $\stackrel{d}{=}$ defines equality in distributions. The presence of self-similarity in classical finance stems from the fact that Brownian motion is a 1/2-ss process. The second Mandelbrot's idea was to generalize the Eq. (21.4) and to propose representing the fluctuations of returns by a self-similar process with exponent $H \in (0,1]$. Hence the relation

$$X(at) \stackrel{d}{=} a^{H} X(t) \tag{21.5}$$

defines the fractional Brownian motion[6] (FBM) introduced in 1965. This process can display what is called a "long-range dependence" or *long memory*, which is of great interest in modelling the complex dependency of economic variables, termed by Granger, "the typical spectral shape of an economic variable" [16]. In fact, the intuition behind this modification is the notion of long memory, termed "Joseph effect" by Mandelbrot, in reference to the slow and aperiodic cycles evoked by the biblical Joseph with regard to fluctuations of the harvest. This led Mandelbrot to rediscover Hurst's law [17] by introducing the concept of fractional Brownian motion (FBM) which is a self-similar process with exponent H and whose increments admit as autocovariance function

$$C_H(h) = \frac{1}{2}[|h+1|^{2H} - 2|h|^{2H} + |h-1|^{2H}] \tag{21.6}$$

With the FBM, the span of interdependence between the increments can be said to be infinite. There is a power law in the correlation as there

[5] According to the title of [13].

[6] [14], followed by a more detailed description [15].

was a power law in the returns (paretian tails) in the 1962 model. Thus, power laws, albeit with a different interpretation, are involved in both cases.

The legacy of long memory approach The search for long memory failed for returns but was successfully achieved for volatility. In the first case, the long memory properties of returns were hard to find, either using the Hurst R/S analysis [18, 19] or with the attempt of ARFIMA modelling [16, 20]. But it appeared that long memory properties exist in volatility. This was at the origin of the FIGARCH models of volatility [21] which led to an important stream of academic studies on volatility processes and volatility forecast with long memory.

21.5 The third model of Mandelbrot (1967): time changed Brownian motion with stochastic clock

It was difficult, however, to reconcile both fat paretian tails (non-normal returns) and long aperiodic cycles (volatility correlations) in the same model and to isolate the two effects. In order to disentangle the Noah and Joseph contributions, Mandelbrot's third idea in 1967 was to consider the price variations in a different time grid, related to market activity (or business activity) and not to physical time [22]. That means to replace in the standard model (21.3) the physical time t by a social economically relevant time $\Theta(t)$ to design a new process $\xi(t) = X(\Theta(t))$ defined in physical time as

$$\xi(t) = \mu\Theta(t) + \sigma W(\Theta(t)) \tag{21.7}$$

In fact, what is called *time-changed Brownian motion* can simultaneously address these two issues. It is easy to see that the choice of the new clock $\Theta(t)$ determines the new model $\xi(t)$ in physical time. By replacing the usual clock of an independent Brownian motion by a stochastic one, Mandelbrot and Taylor paved the way for a new interpretation of the non-normality of returns. The intuition behind it is the following, quoted from page 39 of [23].

> The key step is to introduce an auxiliary quantity called *trading time*. The term is self-explanatory and embodies two observations. While price changes over fixed clock

> time intervals are long-tailed, price changes between successive transactions stay near-Gaussian over sometimes long period between discontinuities. Following variations in the trading volume, the time interval between successive transactions vary greatly. This suggests that trading time is related to volume.'

In the setup of the 1967 model, $W(t)$ was Brownian motion and hence had independent increments, but this constraining hypothesis was relaxed later on. For instance, if one replaces $W(t)$ by $W_H(t)$ which is the FBM defined above, then $\xi(t)$ is called a *multifractal process*. From a general standpoint, the resulting process $\xi(t)$ is a time changed process, i.e. a process using a stochastic clock in social time.

The legacy of Brownian subordination After the first investigation of Clark in 1973 [24], the idea that return distribution should be normal when measured per unit of relevant clock was rediscovered in the end of the 1990s by several authors [25–27] who generalize Clark's results to be able to build a general framework for time-change modelling by using Lévy processes. The models based on Brownian subordination are now of a great interest and two recent models (variance gamma process and normal inverse Gaussian process [28, 29]) represent a good illustration of the wide applicability of this approach.

21.6 Appendix: a tale of fat tails

A Lévy process $(X(t), t \geq 0)$ is a process with stationnary (identically distributed) and independent increments (hereafter i.i.d.). It follows that $X(t)$ is infinitely divisible. The characteristic function of the random variable X is $\Phi_X(u) = \mathbb{E}[\exp(iuX)]$. If $X(t)$ is a Lévy process, then $\Phi_{X_t} = \exp(\Psi_{X_t}(u))$ where $\Psi_{X_t}(u)$ is the characteristic exponent of $X(t)$. The characteristic exponent of the return process $X(t)$ in the case of the standard model (21.3) is

$$\Psi_{X_t}(u) = t\left(i\mu u - \frac{1}{2}\sigma^2 u^2\right) \tag{21.8}$$

An interesting feature of the Lévy processes is that a Lévy process is completely determined by any of its one-dimensional marginal distributions: $\Psi_{X_t}(u) = t\,\Psi_{X_1}(u)$ and it suffices to define $\Psi_{X_t}(u)$ for $t = 1$.

In 1962, Mandelbrot replaced the standard model (21.8) by a new model termed the α-stable Lévy motion [7, 8] and defined by

$$\Psi_{X_t}(u) = t\left(i\mu u - \gamma^\alpha |u|^\alpha \left(1 - i\beta \frac{u}{|u|} \operatorname{tg} \frac{\pi\alpha}{2}\right)\right) \tag{21.9}$$

The α-stable Lévy motion is another special case of the general Lévy processes, with both activity and variation infinite. For an heuristic understanding of the Mandelbrot's idea and the choice he made of this type of Lévy process, one intoduces the Lévy-Khinthine representation

$$\Psi_{X_t}(u) = t\left(i\mu u - \frac{1}{2}\sigma^2 u^2\right) + t\int_{R^*} \psi(u,x)\,\nu(dx) \tag{21.10}$$

with

$$\psi(u,x) = e^{iux} - 1 - iux\,\mathbf{1}_{|x|<1}(x)$$

With this representation, the characteristic exponent (21.9) of the α-stable Lévy motion [30] is

$$\Psi_{X_t}(u) = t\left(i\mu u + \int_{-\infty}^{0} \psi(u,x)\,\frac{C_-}{|x|^{1+\alpha}}dx + \int_{0}^{+\infty} \psi(u,x)\,\frac{C_+}{x^{1+\alpha}}dx\right) \tag{21.11}$$

From (21.10) and (21.11), it follows that the Lévy measure of the α-stable motion is

$$\nu(dx) = \frac{C_-}{|x|^{1+\alpha}}\mathbf{1}_{(-\infty,0)}(x)dx + \frac{C_+}{x^{1+\alpha}}\mathbf{1}_{(0,+\infty)}(x)dx \tag{21.12}$$

or in a more intuitive form

$$\nu(x) = \begin{cases} \dfrac{C_-}{|x|^{1+\alpha}} & \text{if } x < 0 \\ \dfrac{C_+}{x^{1+\alpha}} & \text{if } x > 0 \end{cases} \tag{21.13}$$

This last form (21.13) clearly exhibits a power paretian law in the Lévy measure. This precisely allows to understand the Mandelbrot's intuition: the search for paretian tails in finance. A tale of fat tails, he named "the power of power laws" (page 13 of [4]).

Bibliography

[1] P. Samuelson, *Industrial Management Review* **6** (1965) 13.
[2] J. Courtault, Y. Kabanov, *Louis Bachelier. Aux origines de la finance mathèmatique* (Besançon, Presses universitaires Franc-comtoises, 2002).
[3] C. Walter, Pages 111–163 of [2].
[4] B. Mandelbrot, R. Hudson, *The (mis)behavior of markets: A fractal view of risk, ruin, and reward* (Basic Books, New York, 2004).
[5] B. Mandelbrot, *The Fractalist. Memoir of a Scientific Maverick* (Pantheon, New York, 2012).
[6] C. Walter, *Finance* **23** (2002) 15.
[7] B. Mandelbrot, *Compte-rendus Acad. Sci. Paris* **254** (1962) 3968.
[8] B. Mandelbrot, *J. of Business* **36** (1963) 394.
[9] C. Walter, Pages 437–464 in P. Abry, P. Gonçalvés, J. Lévy Véhel, eds., *Scaling, Fractals and Wavelets* (Wiley, New York, 2009).
[10] P. Embrechts, C. Klüppelberg, T. Mikosch, *Modelling Extremal Events for Insurance and Finance* (Springer-Verlag, Berlin, 1997).
[11] R. Cont, P. Tankov, *Financial Modelling with Jump Processes* (Chapman and Hall, London, 2004).
[12] O. Le Courtois, C. Walter, *Extreme Financial Risks and Asset Allocation* (Imperial College Press, London, 2014).
[13] E. Jondeau, S.-H. Poon, M. Rockinger, *Financial Modeling Under Non-Gaussian Distributions* (Springer-Verlag, New York, 2007).
[14] B. Mandelbrot, *Comptes-rendus Acad. Sci. Paris* **260** (1965) 3274, pp. 422–437.
[15] B. Mandelbrot, J. van Ness, *SIAM Review* **10** (1968) 422.
[16] C. Granger, *Econometrica* **34** (1966) 150.
[17] H. Hurst, *Trans. Amer. Soc. Civil Engineers* **116** (1951) 770.
[18] A. Lo, *Econometrica* **59** (1991) 1279.
[19] J. Beran, *Statistics for Long Memory Processes* (Chapman and Hall, London, 1994).
[20] J. Hosking, *Water Resources Research* **20** (1981) 1898.
[21] R. Baillie, T. Bollerslev, H. Mikkelsen H, *J. Econometrics* **74** (1996) 3.
[22] B. Mandelbrot, H. Taylor, *Operations Research* **15** (1967) 1057.
[23] B. Mandelbrot, *Fractals and Scaling in Finance. Discontinuity, Concentration, Risk* (Springer, New York, 1997).
[24] P. Clark, *Econometrica* **41** (1973) 135.
[25] T. Ané, H. Geman, *J. Finance* **55** (2000) 2259.
[26] H. Geman, D. Madan, M. Yor, *Mathematical Finance* **11** (2001) 79.
[27] P. Carr, L. Wu, *J. Financial Economics* **71** (2004) 113.
[28] D. Madan, P. Carr, E. Chang, *European Finance Rev.* **2** (1998) 79.
[29] O. Barndoff-Nielsen, *Finance and Stochastics* **2** (1998) 41.
[30] G. Samorodnistky, M. Taqqu, *Stable Non-Gaussian Random Processes* (Chapman and Hall, London, 1994).

Chapter 22

How Benoit Mandelbrot changed my thinking about biological form

Ewald R. Weibel

22.1 Introduction

The evolution of science progresses in leaps when new paradigms or new tools come into play — some call this "scientific revolutions" [1]. One such leap was initiated when Benoit Mandelbrot showed that the "Geometry of Nature" can not be satisfactorily described by Euclidian geometry and introduced "Fractal Geometry" in his seminal book *Fractals. Form, Chance and Dimension* published in 1977 [2]. This new paradigm has fundamentally changed my approach to understanding biological form, as I will explain.

However, I had already been induced to a first important change in my thinking about anatomical structures in 1959 when I took up a new job at Columbia University's Cardiopulmonary Laboratory at Bellevue

Hospital in New York. I had for a few years worked in an anatomy institute studying some enigmatic structural features of the pulmonary circulation by means of the classical tools of microscopy. Now I was working in a laboratory of physiology, headed by the two Nobel Laureates André F. Cournand and Dickinson W. Richards who had been honored in 1956 for their revolutionary work on cardiac and pulmonary physiology by pioneering cardiac catheterization. In their invitation to this job they had asked me "to do anything on the structure of the lung of interest for physiology," which meant that my research should be useful in some way. In fact, I was expected to shed some light into the black box that the lung constituted with respect to how gases, oxygen and carbon dioxide, are exchanged between air and blood in the deep parts of the lung, not easily accessible for physiological probing. Interacting as a morphologist with the physiologists of the laboratory, particularly with the biomathematician Domingo M. Gomez, it became soon clear that what was needed was accurate information on the critical dimensions of internal lung structure, such as the extent of the contact area between the air in the small air chambers called alveoli and the blood in the dense capillary network contained in the alveolar walls. This was needed for the conception of physico-mathematcal models describing the functional processes of gas exchange as a function of structural design [3]. To approach these questions called for measurements on samples of the complex hierarchy of the structures making up the lung's gas exchange apparatus, what we call lung morphometry.

We had, in this endeavor, to solve a number of methodological problems:

(1) there was a problem of size: the microscopically small gas exchange units made up a very large organ of several liters in the human, so the accuracy would depend on appropriate sampling strategies;
(2) for microscopy we depended on studying thin sections that essentially presented a 2-dimensional image, but we were interested in the 3-dimensional structures: we had to find ways to estimate the 3-D alveolar surface area from measurements on the 2-D outline of the alveolar borders seen in the microscope, for example (Fig. 22.1).

This new project therefore required a completely new approach to morphological studies, and this required new tools, solving at once the statistical sampling problem and the limitations of measuring 3D structures on 2D sections by the microscope. The solution came by the development of stereological methods that happened to occur simultaneously around 1960

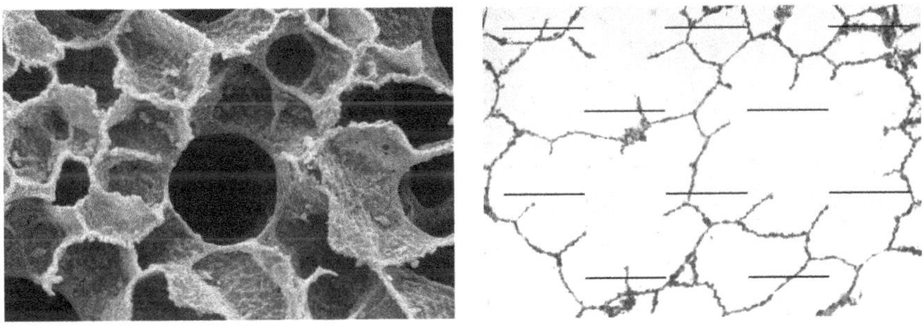

(a) (b)

Figure 22.1 (a) Low power scanning electron micrograph of human lung showing spatial arrangement of alveoli around alveolar duct in cross-section. (b) Light micrograph of thin section at similar magnification with test lines for estimating surface area.

in various fields, from mineralogy and metallurgy to biology. Stereology is a set of methods based on the mathematical principles of integral geometry and geometrical probability, the most compact definition being "a set of methods for sampling structure with geometric probes." The development of these methods asked for the collaboration between mathematicians and microscopists, and this was strongly fostered by the foundation of the International Society for Stereology in 1961 that brought together workers in these different fields.

The basis for these methods was, however, already there for many years, partly for centuries. One example relates to the method for estimating 3D surfaces on sections. The most appropriate, indeed even the only accurate and efficient method, is that of probing the surface with "needles" of defined length and counting the number of times these needles intersect the surface, a practical solution being to overlay a section with a set of test lines of defined total length L and to count the number of times I these lines intersect the alveolar surface (Fig. 22.1b): the surface area S per unit volume of lung V is then found as $S/V = 2 \cdot I/L$, a formula derived several times independently [4]. That I/L is proportional to S/V is intuitively plausible, but the proportionality factor 2 must be derived from considerations of the geometric probability that a random line hits a spatial surface, and this has its root in the so-called "Buffon needle problem" as follows.

In 1777, the great French naturalist Buffon formulated the first problem of geometrical probability when he asked the French Academy of

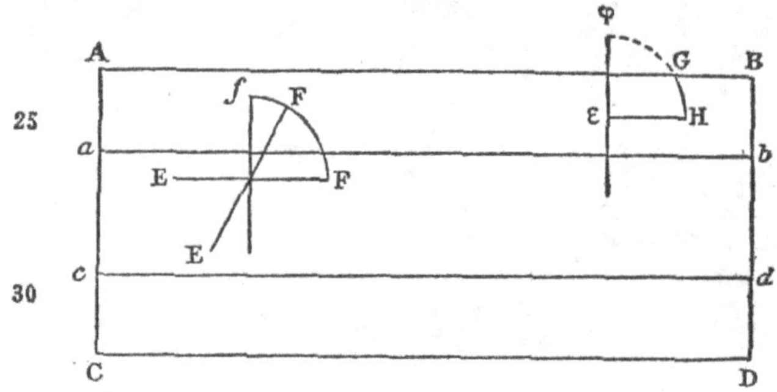

Figure 22.2 Diagram of the Buffon needle problem [5].

Sciences [5]: what is the probability P that a randomly tossed needle of length l intersects a set of parallel lines spaced at an equal distance d, e.g. the seam lines of a parquet floor? Buffon solved the problem himself (Fig. 22.2): considering the chances to have the needle at different orientations (angles) and different distances from the lines he derived $P = (2/\pi) \cdot (l/d)$ where $2/\pi$ considers the random orientation between lines and needle in this 2-dimensional situation [5]; the equivalent factor in the 3-dimensional case is 2 [4]. The further developments of this principle have led to stereological methods, for example the method to estimate the surface area by probing the tissue with needles, as stated above [4].

22.2 Enters Benoit Mandelbrot

The Buffon needle problem was arguably the first realization of "Form, Chance and Dimension". And with this I have introduced the occasion that brought about my first encounter with Benoit Mandelbrot and this had a very great influence on my way of thinking about natural structures. We met in 1977 in Paris at a symposium on "Geometric probability and biological structures" that the Australian mathematician Roger Miles and I had organized together with Jean Serra to commemorate 200 years of the Buffon needle problem outlined above [6]. The topic was related to the theoretical foundations of stereology, but it was, in a way, also related to "form, chance and dimension". It happened that Benoit Mandelbrot had just published in 1977 his seminal book *Fractals: Form, Chance and Dimension* [2]. He was invited as a keynote speaker (Fig. 22.3) and it became immediately evident that this marked a leap in the understanding of the geometry of

Figure 22.3 Benoit Mandelbrot and participants of the Buffon symposium of 1977 beneath the statue of Georges-Louis Leclerc Comte de Buffon at the Jardin des Plantes in Paris.

life. Mandelbrot's concept of fractal geometry gained significant influence on the way we now describe the geometric design of living systems. But it also had a significant impact on the further development of stereology, since these methods are the eminent tools with which to characterize structures quantitatively, in geometrical terms.

At the Buffon symposium Mandelbrot talked about "the fractal geometry of trees and other natural phenomena" [7] among which were the structure of natural boundaries which are never simple, and the systematic structure of trees that abound in nature, in animals for example in the form of blood vessels and airways. By presenting these fundamental concepts Mandelbrot opened the eyes of biological morphologists for unpredicted complexities in the structure of internal organs. I will, in the following give but a few indications to how the concept of fractal geometry has changed our views of biological design, and I will concentrate on one organ system, the lung.

22.3 How Mandelbrot's concepts affected the way we look at lung structure

22.3.1 How accurate is the estimate of surface area really?

Many biological processes such as the exchange of substances or chemical reactions, take place at interfaces between different compartments of the cell or the body. The quantitative description of structure-function relationships therefore often depends on the measurement of the surface area of such membranes. This is for example the case in the lung where oxygen is transferred from the air to the blood across a large surface area. In 1963, I had used the light microscope to estimate this surface and found it to measure $80\,m^2$ [8]. In the 1970s we used the electron microscope for the same measurement and obtained a value of $130\,m^2$ [9]. We first did not know whether this was due to better preparations in the new study or to some measuring error or artifact in one or the other of the studies. But in a systematic study of estimates of alveolar surface area at increasing magnifications we became convinced that this was simply an effect of differences in microscopic resolution. This study was presented in 1976 at the International Stereology Congress [10]. When one year later, at the Buffon symposium, Mandelbrot showed on the example of Richardson's problem of the indeterminate length of the coast of Britain, that the length of a boundary depends on the yardstick used to obtain the measurement, and that this was related to its fractal dimension [2], this pointed the way on how to resolve these paradoxical results of lung morphometry. Figure 22.4 shows electron micrographs at the magnifications used for obtaining surface estimates by the line intersection counting method; comparing these micrographs with those shown in Fig. 22.1 it becomes quite evident that the electron microscope reveals many more details of surface structure so that it is not surprising that the surface area estimate will increase with resolution. This is related to the fact that the lung's internal surface appears like a space-filling fractal surface whose dimension is estimated at 2.2 [11]. However, we must realize that such real structural surfaces, defined by cell membranes, are not true fractals as they are truncated. So how much of this surface is really available for gas exchange cannot be answered from fractal theory but must consider various boundary conditions, such as the effects of surface tension at the air-tissue interface — this is however beyond this essay.

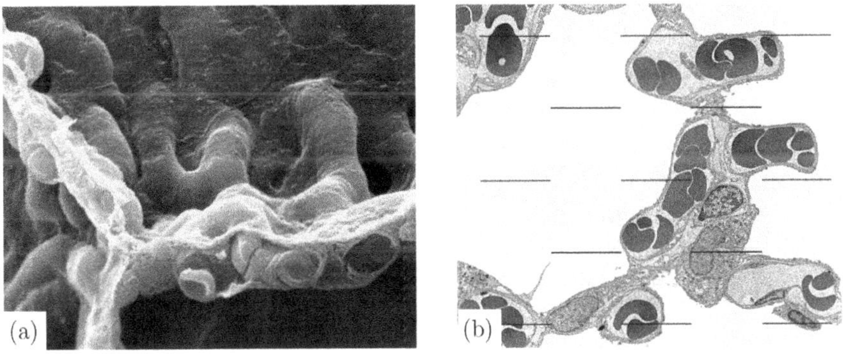

Figure 22.4 (a) Scanning electron micrograph of alveolar septum with capillaries in uman lung. (b) Thin section transmission electron micrograph from same lung with test lines for estimating surface area.

A very similar problem could then also be solved the same way as a result of Benoit Mandelbrot's presentation. Stereologic studies on liver cells had also presented controversial results because the surface area of the endoplasmic reticulum membrane, the site of drug metabolism and of protein synthesis, had been estimated at $6\,m^2/cm^3$ by Loud [12] whereas we had obtained a value of $11\,m^2/cm^3$ [13]. And yet, the methods used were the same, with one exception: we obtained our measurements at $90,000\ \times$ magnification of the electron microscope whereas the other group had used a much lower magnification. In a systematic study of liver cell structure by stereology using different electron microscopic magnifications Paumgartner *et al.* [14] found the estimates of the surface of cellular membranes to increase with increasing magnification or decreasing resolution scale (Fig. 22.5) concluding that the fractal dimension of the endoplasmic reticulum of liver cells was about 2.7 which fully explained the differences in surface measurement obtained at different microscopic resolutions, i.e. with yardsticks of different length. Similar results were obtained for other membranes such that the fractal dimension of inner mitochondrial membranes was estimated at 2.54 (Fig. 22.5).

On a personal note these studies have given me periodic opportunities to get together with Benoit Mandelbrot. One of the two collaborators on the liver cell study [14], Gabriele Losa, made the topic of fractals in cell biology his pet research topic and he initiated the International Symposia on Fractals in Biology and Medicine that took place for the first time in Ascona in 1993 with Benoit Mandelbrot as keynote speaker [15]. At the fourth of

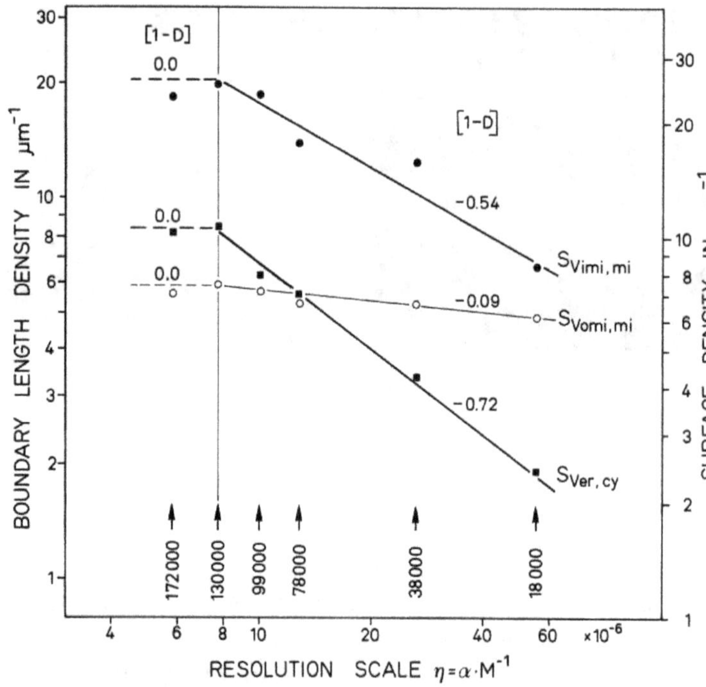

Figure 22.5 Measured surface density of endoplasmic reticulum and inner mitochondrial membranes increase with increasing magnification. The slope of the log-log regressions are related to the fractal dimension [14].

these memorable symposia, regularly attended by Benoit and Aliette, I had the honor to address the audience with a Tribute to Benoit Mandelbrot on his 80th Birthday [16]. Clearly, these were not our only encounters, I remember particularly our invitation to a symposium in Zermatt on Chaos and Creativity where he presented a fractal formula for the Matterhorn, and we skidded over the snowfields on Gornergrat.

22.3.2 The lung's airway tree: optimal function and fractals

To ensure gas exchange in the alveoli oxygen must be efficiently supplied and the waste gas carbon dioxide eliminated from the lung, the essential function of ventilation. To achieve this the 400 million alveoli are arranged around the last few generations of an airway tree that branches over 23 generations on average (Fig. 22.6) and by designing the airway tree in

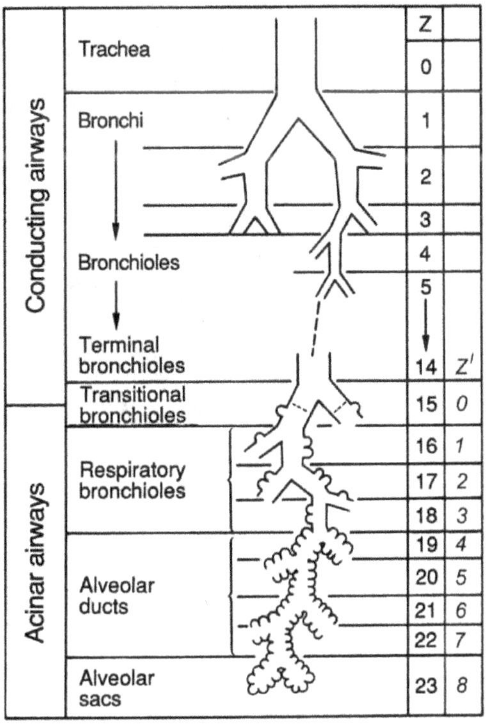

Figure 22.6 Hierarchical model of human airways for idealized symmetric tree. From [3].

such a way that all parts of the gas exchanging lung tissue are reached and this with minimal energy expenditure. Figure 22.7 shows that the human airway system forms a tree whose branches become progressively shorter and smaller in diameter as branching progresses.

The reduction of airway diameter and length at each branch point by a constant factor is of functional significance, both in blood vessels and in airways. It was proposed on theoretical grounds by W.R. Hess [17] and C.D. Murray [18] that the dissipation of energy due to the flow of blood or air in a branched tube system can be minimized if the diameter of the two daughter-branches d_1 and d_2 are related to the diameter d_o of the parent branch as $d_o^3 = d_1^3 + d_2^3$. If we consider a simplified symmetric branching tree, where the two daughter-branches have equal diameter and length, then the diameter of the daughter branch is reduced with respect to d_o by a factor $2^{-1/3}$.

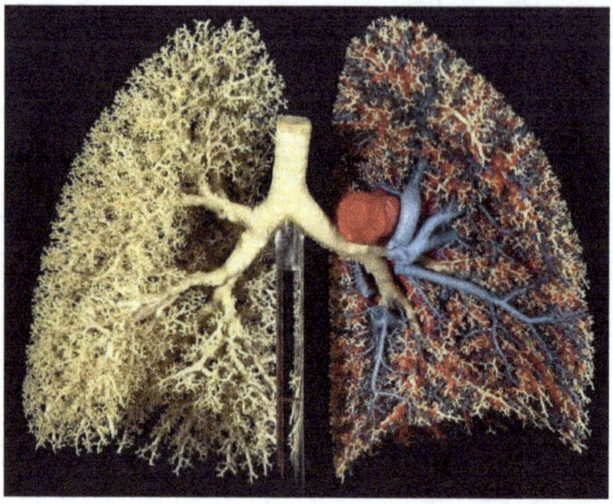

Figure 22.7 Plastic cast of human lung with the bronchial tree shown in yellow, the pulmonary arteries in red, and the veins in blue in the left lung.

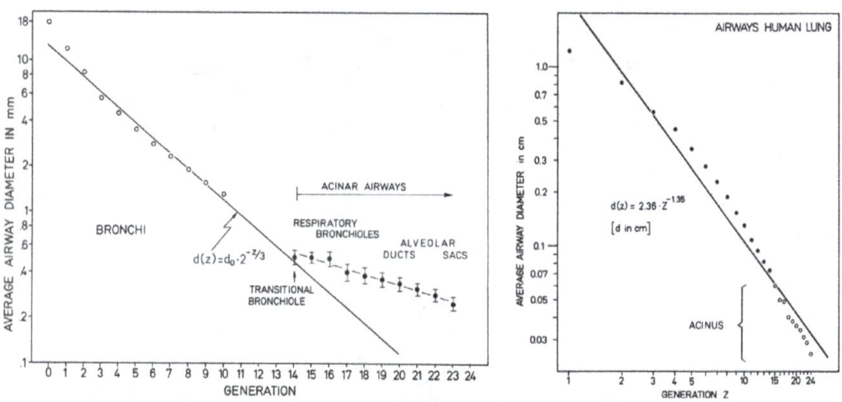

(a) (b)

Figure 22.8 (a) Semi-log plot of average airway diameter in human lungs against generations, z, of branching. After [19]. (b) Double-logarithmic plot of the same data. After [11].

Does this law apply in the airways of the human lung? In 1962 Domingo Gomez and I had analyzed the dimensions of the human bronchial tree [19]. We found that it branches over 23 dichotomous generations. When the average diameters of the airways were plotted semi-logarithmically against the generations (Fig. 22.8a) we observed that the data points lie closely around a straight line down to generation 14, the last generation of so-called

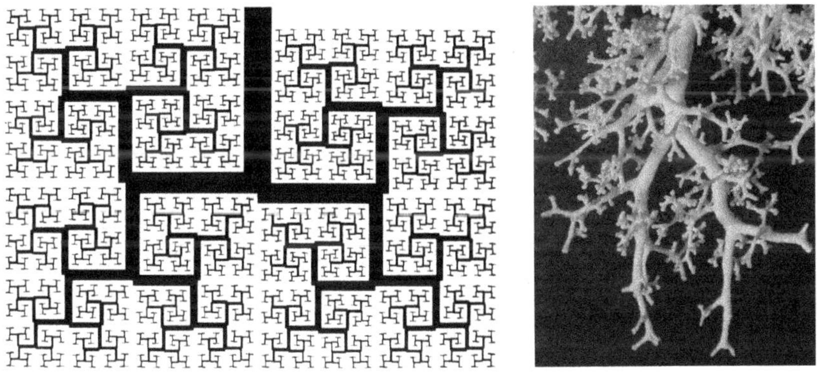

Figure 22.9 (a) Koch tree model of airways (from [2]) compared to (b) cast of human airway tree.

conducting airways, and the slope of this line was $2^{-z/3}$. The average diameter of airways in each generation can thus be predicted from the Hess-Murray law. The airways of generations 15–23 follow a different rule as explained below.

In introducing fractals at the 1977 Buffon symposium Benoit Mandelbrot presented the Koch tree model (Fig. 22.9a), a self-similar space-filling fractal based on dichotomous branching whereby the length and diameter of the daughter-branches is reduced by a constant factor from one generation to the next. Even though the airway tree of the human lung shows considerable irregularity the principle of a systematic reduction of airway size seems similar to the Koch tree model (Fig. 22.9b). In presenting this model in 1977 Mandelbrot remarked that, "the lung can be self-similar and it is" — and he based this conclusion on our findings of 1962 [19]. If we now plot the data of Fig. 22.8a on a log-log scale (Fig. 22.8b) it is evident that the data points follow a scaling function with slope −1.36. With this approach it was later demonstrated by a systematic analysis that the airway tree in different species shows a common fractal structure, in spite of some gross differences in airway morphology [20]. In the context of fractal geometry the reduction factor depends on the fractal dimension of the branching tree so that the correct formula is

$$d_1 = d_o \cdot 2^{-1/D}.$$

In the case of the Hess-Murray law $D = 3$ because the tree is considered to be space-filling.

Benoit Mandelbrot has thus shown us that the conducting airways of the human lung are designed as a self-similar and space-filling fractal tree: the airways reach into all corners of the lung's space with similar distances of all tips from the origin of the airways in the trachea — the Koch tree of Fig. 22.9a is a reasonable idealized model of airway morphology. But there is a point of particular interest related to the fact that the conducting airways end at about 15 generations, followed by 8 generations of gas exchanging airways where a conducting air duct is enwrapped by a sleeve of gas exchanging alveoli (Fig. 22.6); in these peripheral airways oxygen reaches the gas exchanging surface predominantly by diffusion in the air phase and this mode of transport is favored by a slightly larger diameter than that predicted by the Hess-Murray law, as seen to be the case in Fig. 22.9a. A fractal Koch tree becomes truly space-filling only when the branching goes on to infinity, but the real airways are truncated at generation 23, on average: the space between the air channels becomes filled by alveoli, a feature that Benoit Mandelbrot has called the "alveolar inner cutoff" [21]. The fact that also the length of these last generations is greater than predicted by the similarity principle led Mandelbrot to conjecture that "much of the program for the lung's design is imposed by the properties of space and need not be separately encoded" genetically [21].

But it turns out that the real lung may not completely comply to the theoretical principles that seem to govern design. It has indeed been suggested that "an optimal bronchial tree may be dangerous" [22] because the reduction factor $2^{-1/3} \approx 0.79$ is critical with respect to determining the airway resistance that varies inversely with the fourth power of the airway diameter. Thus, a small reduction of this factor would cause the airway resistance to increase very dramatically in the smaller peripheral bronchioles. This could lead to catastrophic situations, for example in asthma, a pathological condition characterized by progressive narrowing of small airways. A closer analysis of the data in Fig. 22.8a shows that the reduction factor corresponds about to the critical value of 0.79 in generation 6 but then slowly increases to about 0.9 in the 16th generation [22]. The average factor for small airways is therefore about 0.85 and this means that the flow resistance decreases in the small airways, a feature that can be considered a safety factor in a functionally critical region [22].

22.3.3 Fractal pairs: matching ventillation and perfusion by fractal desgin

It has long been realized that vascular trees should be designed on fractal principles in order to be efficient functional systems [2, 21]. In the lung, the situation is particularly interesting because the arteries and veins must ensure supply of blood to the capillaries in the gas exchange units matched to ventilation of the airspaces. Part of this match is set up by design: the arteries branch in parallel with the bronchial tree and penetrate into the acini along the airways whereas the veins are located between airway units and collect the blood from the periphery of the acini (Fig. 22.10).

There have been a large number of studies attempting to characterize the fractal geometry of the blood vascular system. One of them looked in great details at the pulmonary arteries [23]. These authors analyzed the pulmonary arterial trees according to a Strahler ordering system, i.e. ordering the branches "upwards" starting with the most peripheral branch rather than "downwards" assuming dichotomy as in our airway study. They therefore derived a branching ratio of 3.36 rather than 2. They then estimated the fractal dimension of the arterial tree by observing the changes in the diameter of arteries with successive orders calculating the fractal dimension from the ratio of the log of the total cross section to the log of

Figure 22.10 Cast of airways and blood vessels in human lung. The bronchial tree (yellow) is accompanied by arteries (red), whereas the veins (blue) lie in between the broncho-arterial units.

the average cross section at each order: they found a fractal dimension of 2.71 for pulmonary arteries and 2.64 for pulmonary veins. This is interesting because Mandelbrot had predicted already in 1977 [2, 21] that the fractal dimension of an arterial tree should be smaller than 3 and he predicted 2.7 as a characteristic value!

This has important consequences on the functional effects of arterial dimensions. With respect to the bronchial tree we have discussed the importance of the size reduction (or homothety) factor which in a dichotomous tree we have seen to be $h = 2^{-z/3}$. In a dichotomous tree 2 is the branching ratio (by definition) and the diameter decreases with increasing generation z so that the exponent is negative. With a Strahler ordering system, as used by Huang *et al.* [23] the smallest branches are numbered 1; higher order branches have an increasingly larger diameter. The ratio of the number of branches in one generation to that in the next is called the branching ratio that is estimated to be 3.36 in this study. Accordingly, the homothety factor h is the ratio of the mean diameter in order n to that in order $n-1$. When one then introduces the value of $R = 3.36$ into the Hess-Murray law $h = R^{1/D}$ we obtain $h = 0.77$ which is *smaller* than the ideal value of 0.79. This means that vascular resistance *increases* progressively towards the smaller branches of the arterial tree, a well-known fact since the largest resistance is found in arterioles, the last branches of the arterial tree before the capillaries. This is now in contrast to the condition in the airways where we had found an average value of $h = 0.85$ which means that the resistance to ventilatory air flow becomes lower in peripheral airways [22].

22.3.4 Fractal design of the acinar airways for efficient gas exchange

The airways in the last 8 generations constitute what is called an acinus: the ventilatory unit of the gas exchanger (Fig. 22.11a). They are designed to serve two functions: they must (1) allow oxygen to be transferred from the air to the capillary blood and (2) ensure that oxygen can reach along the air ducts into the deepest corners of the lung. The first function requires a very large surface to be wrapped around the airways in form of alveoli (Fig. 22.11b), a surface that has fractal properties as discussed above [11].

This surface is densely perfused with blood through a dense capillary network (Fig. 22.11c), and a very thin air-blood tissue barrier affords this surface a high permeability for oxygen. The second function is promoted

Figure 22.11 The structure of airways and blood vessels leading into the gas exchange region of the lung. (a) The last branches of the conducting bronchioles lead into the acinar ducts which are surrounded by alveoli. (b) Cast of airways (yellow) with alveoli surrounding the duct and arteries (red) penetrating into the acinus along the airways. (c) Terminal blood vessels leading into the capillary networks in the walls of alveoli around an alveolar duct in cross-section.

by arranging the alveoli around a system of branching channels, the alveolar ducts, whose diameter decreases very gently towards the periphery (Fig. 22.8a) thus reducing flow resistance and facilitating diffusion of oxygen in the air phase. With this arrangement oxygen is taken up from the air all along the acinar airways thus establishing an oxygen partial pressure gradient. The problem is, however, to make sure that oxygen can reach to the last alveoli and this depends on an adequate balance between permeability of the alveolar surface and diffusivity of oxygen within the air channels; and it also depends on the size of the acini that determines the total length of the channels along which oxygen must diffuse in air [24]. Efficient conditions are achieved if the size of the acinus is matched to the ratio of membrane permeability and oxygen diffusivity: the rule is: "Smaller is better but not too small" and it turns out that the size of acini in mammalian lungs is such as to ensure efficient gas exchange [24]. Acini are fractal structures in terms of their design for gas exchange, and they have just the right size.

22.4 The "fractal leap" in morphological sciences

After Benoit Mandelbrot's presentation of a new "Geometry of Nature" the morphological sciences could no longer be the same. A leap had occurred particularly in biology when we came to realize that the evident but partly obscure complexities of biological structures had an explanation in the design principles of fractal geometry, and that this could even be part of the blueprint for the development of organs, complementing genetic programming. I have shown this change in our understanding on the particular case of the lung which is, indeed, permeated by fractals at all levels and on all axes, airways and blood vessels as well as the gas exchanging microstructures. This was one of the revelations that my encounter with Benoit Mandelbrot has brought forth. Another one was the realization that what we see and measure in the microscope may not be the whole story, that our measurements, even if done with the most sophisticated stereological methods, may not be as accurate as we have claimed them to be as long as we do not precisely define the conditions, specifically the resolution, at which the measurements were done. And all this is indeed of great functional significance.

Bibliography

[1] T. S. Kuhn, *The structure of scientific revolutions*, 2nd edn. (Univ. of Chicago Pr., Chicago, 1970).

[2] B. B. Mandelbrot, *Fractals: Form, Chance, and Dimension* (W. H. Freeman, San Francisco, 1977).

[3] E. R. Weibel, *Swiss Med. Wkly.* **139** (2009) 375.

[4] E. R. Weibel, *Stereological Methods*, (Academic Pr., New York, 1979).

[5] G. Buffon, *Essai d'arithmétique morale*, (Acad. Sci., Paris, 1777).

[6] R. E. Miles, J. Serra, *Geometrical Probability and Biological Structures*, (Springer Verlag, Berlin, 1978).

[7] B. B. Mandelbrot, in [6] pp. 235–250.

[8] E. R. Weibel, *Morphometry of the Human Lung* (Springer Verlag, Berlin, 1963).

[9] P. Gehr, M. Bachofen, E. R. Weibel. *Respir. Physiol.* **32** (1978) 121.

[10] H. J. Keller, H. P. Friedli, P. Gehr, M. Bachofen, E. R. Weibel, in *Proceedings Fourth International Congress Stereology, Gaithersburg 1975* (National Bureau of Standards, 1975), pp. 409–410.

[11] E. R. Weibel, *Am. J. Physiol.* **261** (1991) L361.

[12] A. V. Loud, *J. Cell Biol.* **37** (1968) 27.

[13] E. R. Weibel, W. Stäubli, H. R. Gnägi, F. A. Hess, *J. Cell Biol.* **42** (1969) 68.

[14] D. Paumgartner, G. Losa, E. R. Weibel, *J Microsc.* **121** (1981) 51.

[15] T. F. Nonnenmacher, G. Losa, E. R. Weibel, eds. *Fractals in Biology and Medicine*, Vol. 1 (Birkhäuser, Basel, 1994).

[16] E. R. Weibel, in *Fractals in Biology and Medicine. Vol. 4*, eds. G. Losa, D. Merlini, T. F. Nonnenmacher, E. R. Weibel (Birkhäuser, Basel, 2005), pp. 3–16.

[17] W. R. Hess, *Archiv für Anatomie und Physiologie. Physiologische Abteilung*, 1914.

[18] C. D. Murray, *J. Gen. Physiol.* **9** (1926) 835.

[19] E. R. Weibel, D. M. Gomez, *Science* **137** (1962) 577.

[20] B. J. West, V. Barghava, A. L. Goldberger, *J. Appl. Physiol.* **60** (1086) 1089.

[21] B. B. Mandelbrot, *The Fractal Geometry of Nature* (Freeman, San Francisco, 1983).

[22] B. Mauroy, M. Filoche, E. R. Weibel, B. Sapoval, *Nature* **427** (2004) 633.

[23] W. Huang, R. T. Yen, M. McLaurine, G. Bledsoe, *J. Appl. Physiol.* **81** (1996) 2123.

[24] B. Sapoval, M. Filoche, E. R. Weibel, *Proc. Natl. Acad. Sci. USA* **99** (2002) 10411.

Chapter 23

Entropic origin of allometry relations

Bruce J. West and Damien West

Abstract. The theoretical allometry relation (AR) between the size of a network Y and a property of the network X is of the form $X = aY^b$ and has been known for nearly two centuries. The allometry coefficient a and allometry exponent b have been fit by various data sets over that time. The ubiquity of ARs in biology, sociology, ecology and indeed in virtually all the other science disciplines entreats science to find the origin of ARs. Data analysis indicates that the empirical AR is obtained with the replacements $X \rightarrow \langle X \rangle$ and $Y \rightarrow \langle Y \rangle$ and the brackets denote an average over an ensemble of realizations of the network. It has been shown that the empirical AR cannot usually be derived from the theoretical one by simple averaging due to the fractal statistics of the fluctuations. Consequently we hypothesize that a possible origin of AR is the Principle of Minimum Entropy Generation (PMEG). We establish in support of this hypothesis that if the fluctuations in the allometry variables have fractal statistics then the PMEG entails the AR between a complex host network and a subnetwork strongly coupled to it.

23.1 Introduction

Benoit Mandelbrot [1] was fond of scientific history as well as mathematics and we know he would have liked a piece discussing allometry relations. So we have attempted to blend the new with the old and produce a fresh perspective on the implications and origins of allometry. He identified a number of allometry relations (ARs) masquerading under a variety of empirical "laws" and argued that they were a consequence of complex phenomena not having characteristic scales. Subsequent explanations of ARs have involved geometric fractals, such as the fractal nutrient supply network describing metabolic networks [2], and more recently statistical fractals that couple the origin of ARs to the scaling of the probability density function (*pdf*) [3]. Herein we use the concept of entropy to tie together the various approaches to discovering the origins of the ARs.

23.1.1 A little history

At the time the German polymath Gauss [4] and the American mathematician Adrian [5] introduced the law of frequency of errors and the French physicist Laplace [6] proved the central limit theorem, the French naturalist and zoologist Curvier [7] determined that the brain mass of a mammal increases more slowly than does its total body mass (TBM). Cuvier's experiments initiated the field of allometry that relates the size of an organ to that of the host organism with mass serving as a measure of size. The statistical analysis of experiment data and fractal geometry purportedly converge in the twentieth century to explain the origin of ARs.

Allometry was given its first mathematical formulation in terms of the differential growth of body parts by Huxley [8]. In his treatment of the dynamics of differential growth the two rates were assumed proportional to one another. In this way if X is a living sub-network observable with growth rate α and Y is a living host network observable with growth rate β then

$$\frac{1}{\alpha X}\frac{dX}{dt} = \frac{1}{\beta Y}\frac{dY}{dt} \tag{23.1}$$

Equation (23.1) can be directly integrated to obtain the time-independent allometry growth law:

$$X = aY^b \tag{23.2}$$

where a and $b(=\alpha/\beta)$ have historically been empirically determined independent constants. The theoretical form of the AR considered by Huxley is the basis of subsequent theoretical discussion in such excellent books as Schmidt-Nielson's [9] and Calder's [10]. Savageau [11] provided a "first principles" derivation of Eq. (23.2) using the properties of the underlying determinants of complex dynamic systems to generalize Huxleys analysis. Thus, intraspecies AR appeared to be well established by theory, at least from a broad perspective if not in terms of biological mechanisms.

ARs also can be found in the geomorphology of rivers [12]; the range over which herbivores and carnivores roam [13]; the mass accumulation of plants [14]; information transmission in computer chips and the human brain [15]; and in many other places. The fact that ARs appear in the most unexpected complex phenomena underscores the importance of body size and has motivated the development of a number of theoretical models to explain Eq. (23.2). The reductionist models that identify the underlying mechanisms that entail the AR focus on the determination of a specific value of the allometry exponent, often leaving the allometry coefficient unexplained [3].

The statistical models focus on providing the best empirical fits to the allometry parameters [16] and often provide no mechanistic understanding of the values obtained. Virtually all of the reductionist theories rely on a strategy involving optimization in which energy is minimized subject to a number of constraints, for example, in biology such constraints are conservation of energy, maximum efficiency or the minimal use of material. Herein we take a different tack and examine the implications of assuming that the couplings of the various sub-networks within a host network are not determined by energy considerations but rather by the generation and transfer of entropy (information) and consequently through the properties of their *pdf*s.

23.1.2 Theoretical AR

Natural Science encapsulates the information contained in vast amounts of biological and botanical data through ARs, which were introduced at the beginning of the nineteenth century. The allometry laws often are discussed as if they described a single class of phenomena when, in fact, they actually characterize two distinct groups [9, 10, 17]. In Group 1 the allometry law interrelates variables within a specific species such as the TBM to the mass of a specific organ, this is intraspecies allometry. In Group 2 the allometry law interrelates variables across species such as basal metabolic rate (BMR) to TBM for mammals [9, 10]; this is interspecies allometry.

Both theoretical and observational controversies arose in interspecies metabolic ARs concerning the value of the allometry exponent b. In the middle of the nineteenth century Sarrus and Rameaux [18] reasoned that the heat loss from a warm-blooded animal must be approximately proportional to the animal's free surface. This simple dimensional analysis requires $b = 2/3$ as reviewed in a number of excellent sources [9, 10]. In 1932 Kleiber [19] expected to obtain this value using a large amount of data for metabolic ARs. Instead he established by means of linear regression of the logarithmically transformed observational data on

$$\ln(X) = b\ln(Y) + \ln(a) \tag{23.3}$$

that $b \approx 3/4$. Subsequent observational studies have suggested other values of the phenomenological exponent b [9, 10, 20, 21].

Controversy also arises in the theoretical explanation as to why the allometry exponent b should take on specific values. The quarter-power allometry scaling law is explained by West *et al.* [2] using geometric scaling arguments from Mandelbrots fractal physics to establish the value $b = 3/4$ in fractal network transport and other quarter-power scaling laws in biology. The constraint on the allometry coefficient a is that it intercept the vertical axis when the property being fit vanishes.

In Huxley's book [8] there are literally dozens of X, Y-pairs shown to satisfy Eq. (23.3). In such intraspecies allometry there is relatively little variability and Eq. (23.3) fits the single species data quite well. However the power-law index b changes markedly from species to species [22], and when fitting interspecies data the fluctuations in the linear regression to Eq. (23.3) are substantial. It has been suggested by a number of authors that in such cases the parameter a be treated as a random variable [20, 21] to account for the observed allometry variability.

23.1.3 Empirical AR

Careful reading of Huxleys book reveals that Eq. (23.3) is not being directly fit to the data because the data do not always consist of individual measurements of X and Y. The empirical data consists of averages over a collection of such measurements on individual specimens and are here denoted by $\langle X \rangle$ and $\langle Y \rangle$. This same duality permeates the literature subsequent to Huxley in which the theoretical discussions focus on (X, Y) whereas data fits are concerned with $(\langle X\rangle, \langle Y\rangle)$. Consequently the empirical form for the allometry law is

$$\langle X\rangle = \overline{a}\langle Y^{\overline{b}}\rangle \tag{23.4}$$

and the overbar on the parameters denote fitted values.

West and West [3] attempt to reconcile these very different AR descriptions by considering the average of Eq. (23.2) over an ensemble of realizations of the data

$$\langle X\rangle = a\langle Y^{b}\rangle \tag{23.5}$$

and note the inequality $\langle Y^b\rangle \neq \langle Y\rangle^b$ with $b \neq 1$. In order to derive the empirical AR from Eq. (23.2) they introduce deviations of the variables from their average values $X = \langle X\rangle + \delta X$ and $Y = \langle Y\rangle + \delta Y$ into the equation. After some algebraic manipulation the allometry coefficient can be expressed as $a = \overline{a}+\delta a$, where $\overline{a}$ is obtained from data fitting and δa is the fluctuation from the fitted value that to first order is linear in the relative fluctuations $\delta X/\langle X\rangle$ and $\delta Y/\langle Y\rangle$. Thus, Eq. (23.5) can be expressed as

$$\langle X\rangle = (\overline{a} + \delta a)\langle Y\rangle^{\overline{b}} \tag{23.6}$$

and when $\delta a/\overline{a} \ll 1$ simplifies to the empirical AR Eq. (23.4).

We emphasize that the empirical AR only follows from Eq. (23.2) when $\delta a/\overline{a} \ll 1$ implying that the distribution of fluctuations must be very narrow. This condition on the fluctuations in the allometry coefficient was directly tested using a probability density function (*pdf*) obtained from data [3] and it was determined for metabolic data that the mean-square fluctuations were of the same order as the mean and therefore the perturbation condition could not be satisfied. Mandelbrot was drawn to this kind of inverse power-law distribution and he was remarkably successful in identifying it in a myriad of complex phenomena [1].

In Sec. 23.2 we study the scaling of the *pdf* and determine the resulting form of the network entropy. The changes in the average observable are shown to be determined by the changes in the network's entropy and consequently the average network dynamics are shown to determine the empirical AR. The entropy view is discussed in Sec. 23.4 where the value of the allometry exponent is found to determine the stability of the network. This perspective is elaborated in a recent review [23]. In Sec. 23.5 we draw some conclusions.

23.2 Homogeneous scaling and entropy

Note that Eq. (23.2) was one of the first forms of scaling outside the physical sciences that had experimental justification and is the lowest-order solution to the renormalization group homogeneous equation [24]

$$Z(ct) = c^{\mu_z} Z(t) \tag{23.7}$$

that relates the dependent variable Z to the independent variable t and c is a positive constant. This homogeneous scaling relation was used by Stevens [17] to derive Eq. (23.2) with Z identified with the property X and t with the network size Y. We note however that in most, if not all, complex phenomena the variables are stochastic, either due to internal nonlinear interactions or through coupling to an infinitely uncertain external environment [25]. Under these conditions the homogeneous renormalization group relation does not pertain to the stochastic variable and Eq. (23.2) is not its solution. Instead Eq. (23.7) denotes a property of the *pdf* where $P(z,t)dz$ is the probability that the dynamic variable $Z(t)$ lies in the interval $(z, z+dz)$ at time t, given that we are discussing dynamics.

The scaling in Eq. (23.7) implies that the *pdf* scales as [26]:

$$P(z,t) = \frac{1}{t^{\mu_z}} F_z\left(\frac{z}{t^{\mu_z}}\right) \tag{23.8}$$

The form of the function $F_z(.)$ is here left unspecified. It could have the form of the normal distribution of Gauss and Adrian, in which case $\mu_z = 1/2$ would describe a classical diffusion process fully characterized by the mean and variance. On the other hand it could more generally have the form of an α-stable Lévy process in which case the scaling index is the inverse of the Lévy parameter $\mu_z = 1/\alpha$ and the variance diverges [26]. The variability in the amplitude fluctuations in this latter case is quite large and the statistics are intermittent. One way to quantify the scaling behavior of a data set is by using the associated entropy.

Demetrius *et al.* [27] introduce the concept of evolutionary entropy based on the dynamics of age-structured populations using the *pdf* of the parental age of a randomly chosen newborn individual. Herein we use the *pdf* for the stochastic variable $Z(t)$ describing various aspects of the network. The Gibbs-Boltzmann entropy associated with the distribution given by Eq. (23.8) is

$$S_z(t) = -\int P(z,t) \ln P(z,t) dz = S_z^0 + \mu_z \ln t, \tag{23.9}$$

which increases logarithmically with time and the reference entropy

$$S_z^0 \equiv -\int F_z(q) \ln F_z(q) dq > 0$$

is integrated over the full range of the scaled variable $q = z/t^{\mu_z}$. The entropy used here has the same form as that obtained by Demetrius *et al.* [27] with $\mu_z = 1$ but the parameter t had previously been interpreted as biological time and not clock time as it is here. Another useful quantity is the average of the stochastic variable

$$\langle Z(t) \rangle = \int z P(z,t) dz = \overline{Z} t^{\mu_z} \tag{23.10}$$

and the constant coefficient is $\overline{Z} \equiv \int q F_z(q) dq$. Inserting the clock time from Eq. (23.9) into (23.10) provides a relation between the average dynamic variable and the entropy change from its reference value $\Delta S_z \equiv S_z(t) - S_z^0$:

$$\langle Z \rangle = \overline{Z} e^{\Delta S_z} \tag{23.11}$$

Consequently changes in the average variable associated with the network dynamics are induced by associated network entropy changes.

If Z is a measure of the network size Eq. (23.11) asserts that changes in the average network size are produced by changes in the network entropy. Such changes are achievable by the network interacting with its environment, for example by eating. This influence is discussed in the next section.

23.3 Dynamic averages

The time derivative of the average value in Eq. (23.11) determines the relative change in the generic average variable $\langle Z(t) \rangle$ by the change in the associated entropy $\Delta S_z(t)$:

$$\frac{1}{\langle Z \rangle} \frac{d\langle Z \rangle}{dt} = \frac{d\Delta S_z}{dt} \tag{23.12}$$

It is clear from Eq. (23.12) that changes in the average observables require the generation of entropy.

We now assume that sizes of the network Y and sub-network X have the same statistical distribution with appropriately indexed parameters. This assumption is similar in spirit to that made by Huxley [8] on the proportionality of the growth rates and which enabled him to "derive"

Eq. (23.2) as the solution to his equations. By assuming an equality in the form of the *pdf*s and not in the growth rates themselves the dynamic variable $Z(t)$ can be identified with the sub-network size X resulting in the relative change in the average value

$$\frac{1}{\langle X\rangle}\frac{d\langle X\rangle}{dt}=\frac{dS_x}{dt} \tag{23.13}$$

The sub-network consequently generates entropy with the growth in its percentage average size. In the same way the dynamics of the average size of the network necessary to balance the growth of the sub-network is

$$\frac{1}{\langle Y\rangle}\frac{d\langle Y\rangle}{dt}=\frac{dS_y}{dt}=\frac{\mu_y}{\mu_x}\frac{dS_x}{dt} \tag{23.14}$$

The second equality is obtained by taking the time derivatives of the entropies for X and Y and eliminating the time between them. Combining Eqs. (23.13) and (23.14) yields

$$\frac{1}{\mu_x}\frac{1}{\langle X\rangle}\frac{d\langle X\rangle}{dt}=\frac{1}{\mu_y}\frac{1}{\langle Y\rangle}\frac{d\langle Y\rangle}{dt} \tag{23.15}$$

which looks remarkably like the equations hypothesized by Huxley with the dynamic variables replaced by their averages and the growth rates replaced with the power-law indices from the associated *pdf*s.

Equation (23.15) can be directly integrated to yield the heuristic AR

$$\langle X\rangle=\alpha\langle Y\rangle^{\beta} \tag{23.16}$$

with the allometry parameters given by

$$\alpha\equiv\frac{\overline{X}}{\overline{Y}^{b}} \quad \text{and} \quad \beta\equiv\frac{\mu_x}{\mu_y} \tag{23.17}$$

It is noteworthy that the AR given by Eq. (23.16) is not that given by Eq. (23.2) but is instead the empirical AR given by Eq. (23.8) with $\delta a/\overline{a}\ll 1$ for which no previous theory was available.

23.4 The entropy view

In the present context entropy measures the lack of information connected with the disorder in the hidden degrees of freedom in an econetwork. An econetwork has low entropy (high information) and is basically in an unstable situation that eventually follows its normal evolution towards stability and high entropy [28]. Consequently for a species embedded in an econetwork to maintain stability the econetwork must continuously supply negative entropy (information) to the species to keep its entropy low. The change in entropy of the species ΔS_x is therefore balanced by the change in entropy of the overall network $-\Delta S_y$, such that the total change in entropy is $\Delta S_T = \Delta S_x - \Delta S_y \leq 0$. It is therefore necessary for an econetwork to extract negentropy from the environment in the form of nourishment to compensate for the negentropy it transfers to the species. Negentropy was introduced by Schrodinger in his classic lectures on the nature of life [29].

The stability condition denoted by the empirical AR requires compensation for the entropy changes associated with the dynamics of $\langle X\rangle$ by the entropy changes associated with the dynamics of $\langle Y\rangle$. Using $\Delta S_T \leq 0$, the PMEG requires that the magnitude of the entropy generated by the host network be greater than that generated by the sub-network:

$$\left|\frac{dS_x}{dt}\right| \leq \left|\frac{dS_y}{dt}\right| \tag{23.18}$$

Consequently using Eqs. (23.13) and (23.14) in (23.18) establishes the stability condition

$$\frac{\mu_y}{\mu_x} \geq 1 \quad \text{or equivalently } b \leq 1 \tag{23.19}$$

Thus, the allometry exponent, independently of any geometrical considerations, is less than or equal to one. This stability condition on the allometry exponent is satisfied by the vast majority of data sets [12–16, 21].

23.5 Conclusions

ARs are often considered to be remarkable statistical patterns that appear in data over a variety of complex phenomena. Various attempts at predicting the origin of Eq. (23.2) have focused most often on deriving specific values of the allometry exponent and showing how that value explains large data sets [14]. However, the reductionist approach has been criticized by a number of investigators, including [2, 16, 21]. The difference between the theoretical and empirical forms of the AR given by Eqs. (23.2) and (23.11), respectively, is quite striking. It is Eq. (23.11) that is obtained from the interspecies data sets and Eq. (23.2) is only sometimes obtained from intraspecies data sets. Statistical data analysis yields the empirical

AR given by Eq. (23.11) as does the probability calculus [3, 23] indicating that the allometry scaling is a consequence of the *pdf* scaling.

Herein we have followed Mandelbrot's style of argument to extract maximum information from the sole fact that empirical ARs imply scaling of the random fluctuations in the variables of complex networks. This phenomenological approach suggested introducing the thermodynamic concept of entropy and the complimentary information theory concept of negentropy [28, 29]. The minimization of the total entropy generation produced by the balance between the entropy increase generated within a sub-network and the negentropy increase extracted by the host network from the environment does three things: (1) it places a stability constraint on the power-law indices of the scaling *pdf*s such that $b \leq 1$; (2) it interrelates the relative growth in the network variable averages and (3) the integrated average growths yield the empirical AR given by Eq. (23.11). Consequently, the empirical AR is entailed by the PMEG.

Bibliography

[1] B.B. Mandelbrot, *Fractals, Form, Chance and Dimension* (W.H. Freeman, San Francisco, 1977); *The Fractal Geometry of Nature* (W.H. Freeman, San Francisco, 1982).
[2] B.J. West, J.H. Brown, B.J. Enquist, *Science* **276** (1997) 122.
[3] B.J. West, D. West, *Europhys. Lett.* **97** (2012) 48002.
[4] C.F. Gauss, *Theoria motus corporum coelestrium* (Hamburg, 1809); English transl., *Theory of Motion of Heavenly Bodies Moving about the Sun in Conic Sections* (Dover, New York, 1963).
[5] R. Adrian, *The Analyst; or Mathematical Museum* **1** (1809) 93.
[6] P.S. Laplace, *Analytic Theory of Probabilities* (1810).
[7] G. Cuvier, *Recherches sur les ossemens fossils* (Paris, 1812).
[8] J.S. Huxley, *Problems of Relative Growth* (Dial Press, New York, 1931).
[9] K. Schmidt-Nielsen, *Scaling, Why is Animal Size so Important?* (Cambridge Univ. Pr., Cambridge. 1984).
[10] W.W. Calder III, *Size, Function and Life History* (Harvard Univ. Pr., Cambridge, 1984).
[11] M.A. Savageau, *Proc. Nat. Acad. Sci. USA* **76** (1979) 6023.
[12] A. Maritan, R. Rigon, J.B. Banavar, A. Rinaldo, *Geophys. Res. Lett.* **29** (2001) 1508.
[13] G. Woodward, B. Ebenman, M. Emmerson, J.M. Montoya, J.M. Olesen, A. Valido, P.H. Warren, *Trends in Ecol. and Evol.* **20** (2005) 402.
[14] K.J. Niklas, *Biol. Rev.* **79** (2004) 871.
[15] V. Beiu, W. Ibrahim, *ISCAS 2008. IEEE Int. Symp. Circuits and Systems* (2008) 640.

[16] D.S. Glazier, *Biol. Rev.* **80** (2005) 611.
[17] C.F. Stevens, *J. Biol.* **8** (2009) 14.
[18] Sarrus, Rameauz, *Bull. Acad. Roy. Med. (Paris)* **3** (1838-39) 1094.
[19] M. Kleiber, *Hilgarida* **6** (1932) 315.
[20] A.A. Heusner, *J. Exp. Biol.* **160** (1991) 25.
[21] P.S. Dodds, D.H. Rothman, J.S. Weitz, *J. Theor. Biol.* **209** (2001) 9.
[22] P.C. Withers, *Comparative Animal Physiology* (Saunders, Fort Worth, 1992).
[23] D. West, B.J. West, *Phys. of Life Rev.* **10** (2013) 210.
[24] P. Meakin, *Fractals, scaling and growth far from equilibrium* (Cambridge Univ. Pr., Cambridge, 1998).
[25] K. Lindenberg, B.J. West, *The Nonequilibrium Statistical Mechanics of Open and Closed Systems* (VCH, New York, 1990).
[26] B.J. West, M. Bologna, P. Grigolini, *Physics of Fractal Operators* (Springer, Berlin, 2003).
[27] L. Demetrius, S. Legendre, P. Harrenöes, *Bull. Math. Biol.* **71** (2009) 800.
[28] L. Brillouin, *Science and Information Theory* (Academic Pr., New York, 1962).
[29] E. Schrödinger, *What is Life?* (Cambridge Univ. Pr., Cambridge, 1944).

Chapter 24

Music and fractals

Charles Wuorinen

Any of the arts may produce exemplars that have fractal characteristics. There may be fractal painting, fractal poetry, and the like. But these will always be specific instances, not necessarily displaying intrinsic properties of the art-medium itself. Only music, I believe, of all the arts possesses an intrinsically fractal character, so that its very nature is fractally determined. Thus, it is reasonable to assert that any instance of music is fractal.

Music's fractal characteristics seem to manifest themselves on three distinguishable levels. These levels form a progression from the concrete-physical to the metaphorical-structural, and for this reason tend, even as they increase in artistic significance, to diminish in specificity.

The first of these is the acoustic signal that carries music. Here we have a one-dimensional function of time (consisting only of variation in air pressure against the eardrum) which has been shown to be a form of scaling noise, and which often possesses a $1/f$ probability spectrum. Thus, even though a given patch of music may be fully composed and thus consciously determined, from a statistical point of view it has the character of a scale-free noise. Perhaps this is because in many respects music can be conceived as a summing of periodic frequencies of varying onset times, durations, envelopes: that is, simultaneous and successive pitches and their (usually, for all intents and purposes) harmonic spectra. From a statistical point of view, often these pitches are randomly concatenated, since the principles by which they are assembled belong to the laws of Art, not Science.

The second level on which we may consider music to be fractal is that of its fundamental materials: pitch and rhythm. Of these two, pitch (noise-making instruments notwithstanding) is overwhelmingly the fundamental constituent of virtually all music as actually practiced. But musical time, in all cultures that have any conception of rhythmic organization — most notably in the musics of the West, of India, of West and Central Africa, the Middle East, and of Southeast Asia — is based upon self-similar divisions of the time-continuum.

Distinguishing self-similarity and self-affinity is important. Self-similar sets can have several different scaling factors, as shown on the left side of Fig. 24.1. Sets in at least two dimensions (time and pitch, for instance) can have different scaling factors in different directions, illustrated on the right side of Fig. 24.1.

In the West these go to make up meter and rhythm; in India they manifest as tal, vibhag, and matra. These divisions are hierarchically organized, the smaller nested within the larger by either a single ratio of reduction, or a small set of ratios (usually only 2 or 3).

The case of pitch is even more basic, for all musics that organize pitch at all (and here we exclude such pitch-primitive systems as those of much

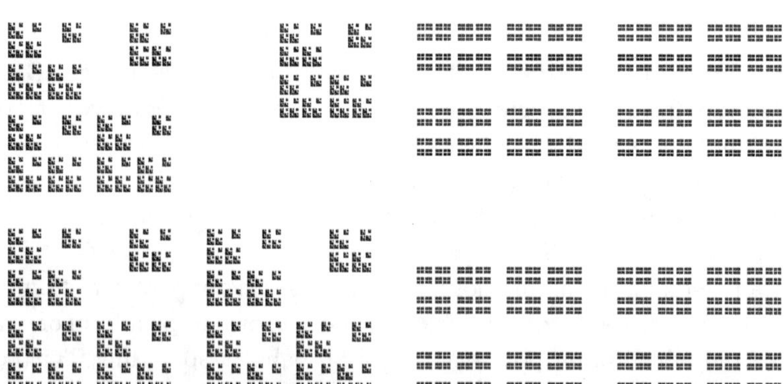

Figure 24.1 Left: a self similar fractal with two distinct scaling factors. Right: a self-affine fractal.

African Music) do so on the basis of a continuum whose fundamental steps are small and generated by successively applying a constant ratio, thus producing a self-similar structure.

In western music, this step is the semitone, which in the equal-tempered system is related to its neighbors by a multiplier which is the 12th root of 2. Western music is unique in the world in that for the past several centuries it alone has made regular use in actual compositions of the entire theoretically available repertoire of pitches from the underlying continuum. Other musics typically make a selection from their theoretical continuum for a specific piece, and organize this subset into a scale, mode, or other construct to provide the basic pitch material. Here the pitch-generation principle becomes self-similar with more than one "frequency multiplier" at work, but the principle remains the same. Almost always, there will be only 2 ratios in such schemes, with the larger an integral multiple (almost always 2 or 3) of the basic unit. Indeed the diatonic scales (and their earlier progenitors the modes) of older western music behave this way, with only 2 steps — the semitone and whole tone — generating the scale; but the case here is different from the scalar organization of non-western musics in that through modulation, the whole repertoire of twelve distinct pitches per octave is available, and has in fact been used since the end of the fourteenth century. In non-western musics, on the other hand, typically no pitches not in the pre-selected scale or mode will be introduced in the course of a composition: in a raga performance, for instance, it would be unthinkable for pitch elements (except for ornamentation) to be structurally introduced that are not elements of the originating mode.

Moreover, especially in Western music, the role of the 2:1 frequency ratio — the octave — is critical, in that it defines a set of equivalences mod 12 that produce what we call "pitch classes". This notion asserts that all pitches related by integral numbers of octaves are functionally identical; this is reflected in the traditional terminology, which (since the 17th century) indifferently calls every member of such a class by the same letter-name: If someone refers merely to "C", you will still not know which actual note is intended; the specific octave must also be given. Thus the pitch-continuum is self-similarly divided not just by the semitonal ratio, but by the octave as well.

The intrinsic "fractalness" of music may also be inferred indirectly from certain negatives. Traditionally the conceptually separable components of musical sound have been divided into pitch, rhythm (musical time, pulsation and duration), timbre or tone color, envelope, and loudness. But only the first two have since antiquity received serious theoretical analysis — for all that recently attempts have been made to deal with the other categories. While plenty of psycoacoustic investigations into musical perception have been made — for example into musical loudness, which has very little to do with physical sound pressure — musicians themselves have never been able (or very much inclined) to try to "organize" timbre and loudness in the musically and quantitatively precise ways they apply to pitch and time. Since timbre and loudness seem not to have fractal characteristics as they figure in music, their absence from any significant conceptual role beyond the vague domain of "expressiveness" seems to argue in support of a fractal character for music.

Finally, at the highest and artistically most significant level, we come to actual compositions and their structure. While some works are true examples of self-similar nesting (see below), in general the statements one can make about compositional "fractalness" are metaphorical. But, for instance, in traditional western diatonic-tonal music, there is a strong tendency for similar structures to appear on different time-scales. Thus the same harmonic progression may determine the course of a whole movement, of a sizable section of it, or of a single short phrase. Traditional terminology shows, by its indifference to scale, how deeply imbedded self-similar structures in compositions are: "C major" can refer to a single sonority, the key of a phrase, a movement-section, a whole movement, or the key of an entire work. And in post-tonal music similar structures on differing time-scales are certainly present.

Almost all of my work of the past forty years provides instances of fractalness. Well before I had become aware of the work of Mandelbrot in the mid-seventies, I had devised a compositional method which, while derived from the earlier work of Schoenberg and Babbitt, was nevertheless (as I discovered) explicitly fractal in nature, since it was based on the nesting of similar time-and-pitch-interval structures within larger replicas of themselves. This can be described as self-affine, because the scalings along the time and pitch axes can be different. Alternatively and at a later stage, I began from time to time generating compositional structure through iterative means, which of course led to a result largely equivalent to that of top-down nesting.

The "nesting method" typically begins with an ordered set of pitch-classes, whose intervals are chosen not merely for their appropriateness in the design of satisfactory harmony, but also so that, when these same intervals are temporally interpreted, they can conduce to the development of meaningful large-scale structure. The total interval span of the ordered set is flung over the duration of the whole work or movement. The set is defined by a succession of intervals of varying sizes, and these intervals may be used to divide a large span of time into sections whose lengths are directly proportional to the interval-sizes of the original succession. Within each of these large sections, then, a similar process is effected. Usually, three levels of such divisions are sufficient to produce a set of nested self-similar temporal intervals which can then be used to define the structure of the composition; the most detailed level can often be interpreted as foreground rhythm. This definition in most cases is achieved by attaching pitches to the onset points of the durational intervals — these pitches originating in the arrays generated from the same ordered set that has produced the time-interval structure I have just described. Moreover, the time-structure need not be "monophonic": there may be several strands of division unfolding simultaneously, and these will be generated as described, from transformations of the original ordered set, temporarily interpreted. Usually these transformations are combinations of the traditional transposition, retrograde, inversion operations; or multiplicative (by 5, 7, or 11, mod 12); or by [cyclic] permutation.

It should be clear that what I have sketched above is not really "composing" in the sense we might historically apply to the familiar figures of the western tradition. It is rather the preparation for composing, which these figures had provided for them by the conventions of their own day and place. But in our age, for the last century there has been no set of stable, generally acknowledged conventions. The wise composer therefore will provide them for himself; and what I have described above is an attempt to do so. Having made such preparation, then, I have found it possible to compose with a kind of intuitive freedom which still assures macrostructural coherence. Those who try for this coherence without structural underpinnings usually fail.

Since antiquity, musicians have been fascinated by their intuition that music has a special relationship to the natural world. By this I mean more than simply that music like all else is embedded in the world of nature. Rather it would seem that the structure of music somehow mirrors that of natural phenomena. In ancient times the Pythagorean-Aristotelian cosmology, in which nested planet-carrying spheres, whose intervals of separation were thought to be directly proportional to the "basic" musical intervals, constituted not only a vivid demonstration of this intuition, but also perhaps the first attempt at modeling the universe by quantitative means. (Can we therefore say that the "scientific method" owes its origin to music?) From then on until the discovery of the overtone series in the 17th century, the simple harmonic divisions of a vibrating string were cited to justify, by an appeal to nature, the pitch materials of music as actually sounded. The same continued after the partials series was identified. Trying to justify the musical practice of a specific time and place by this type of superficial appeal to nature has always caused conceptual problems. The arguments need not be rehearsed here, but suffice it to say that looking to the partials series for validation has never been successful. If, however, we acknowledge an intrinsically fractal character for music, then its intimate link with many of the world's natural shapes and processes becomes automatic — it is one such process.

This, finally, may explain another matter. I have often been struck by the fact that although music is by its very nature the most abstract of the arts and the hardest of comprehension, it is also the one about which the ignorant and indifferent have the most confident opinions. Why should this be? Relationships of rhythm to heartbeats, breathing, etc., to one side, it seems to me that the perceived visceral character of music must come from its special affinity to natural processes and shapes. The perception of music, however, is more difficult than that of the other arts. The visual and literary art-object is under the temporal control of the reader or viewer. The dance and the theater deal in controlled flows of time, to be sure, but it is time very crudely quantized when compared to music, where a tenth of a second can be critical in successional, and a fiftieth in simultaneous, relationships. In music alone, the would-be perceiver is fully controlled by the artwork. He cannot slow it down or speed it up; he has only the perceptual present to deal with, supported by memory and anticipation. Musical understanding therefore requires long experience and training. But because of music's fractal character — its "naturalness" — the casual hearer seems to assume an automatic competence to judge. That is why we see those members of the uninformed public, who would never presume set up their opinions of art or literature as normative, happy to do so with music.

Chapter 25

Stories about Benoit

The Yale University mathematics department hosted a memorial for Benoit on April 29 and 30, 2011. The first day of the meeting consisted of three technical talks on some aspects of fractals, Benoit's principal intellectual legacy. Bernard Sapoval spoke on fractals in physics, Peter Jones on fractals in mathematics, and Nassim Taleb on fractals in finance.

The second day heard personal accounts, stories about Benoit. In addition, six were sent in by people who wanted to share something, but could not attend the memorial. Some lightly edited samples of these stories are the subject of this chapter. First, those sent in, then those spoken, in the order presented. Ralph Gomory was the first speaker that morning. Taken together, these stories reveal Benoit at a personal level: how he interacted with people, how people saw him. We hope you will find something new here.

Peter Jones moderated both sessions. In particular, on the second day he added comments after some of the stories. Peter and Benoit were especially close, and Peter is a thoughtful observer, so we have included his insightful comments, along with his closing remarks.

To be sure, Benoit had what often is called a strong personality, and he sometimes clashed with other strong personalities. Stresses of this sort are common among those at the top of every field; they are not interesting, nor do they reflect anything significant in anyone's legacy. If you were looking for controversy, you've come to the wrong place. This is not because we edited out mention of conflicts; we include those recounted. Rather, it's because what we all felt in that room was affection for Benoit. That he produced such a strong reaction in such a variety of people is another dimension of his life.

25.1 Michael Shlesinger: The correct way to pronounce Cargèse

I'll just start out by saying what fun it is was to be with Benoit. At conferences, the room would perk up and feel different when Benoit was there. He would sit front row center and then stand up at attention when questioning speakers. When I would lose my concentration trying to follow a lecture, Benoit would ask a question that would get to the heart of the matter. No matter how complicated a talk would become, Benoit could point out what was known, what was suspicious, and what was new. He was sparing in praise, but that made his comments the more valuable. At dinners he was full of interesting stories and historical facts explaining how the world really works. His mind was quick and his memory was remarkably vast. Today I'm in Cargèse, one of the many conference sites I shared with Benoit. He even taught me the correct way to pronounce Cargèse and I remember how he argued with a waiter about name origins in Corsica versus Sardinia. Benoit lives on scientifically, but I miss him dearly.

25.2 Ann Robertson: First-hand answers

I am extraordinarily grateful to have had the privilege of attending the fractal geometry workshops of Michael, Benoit and Nial. To be able to ask and receive detailed answers from Benoit about Jackson Pollock, Richardson and others, helped confirm my own work. When I learned from Michael that Benoit had pancreatic cancer, I decided to email him to tell him how much he inspired me. I wrote in part:

Your life and work are truly inspirational to me. You changed the way mathematicians and scientists view the world. I discuss this in every math class I teach at Connecticut College. Your first book set the stage for so much important mathematics by you and others. We are all so grateful. The five years of summer workshops at Yale were very, very important to me. Thank you for all that you have done. It is immense.

25.3 Richard Taylor: A life-changing model

I grew up in England wanting to be both an artist and a scientist. But people told me I had to choose one or the other. I was shown C. P. Snow's book *The Two Cultures*, which declared that future artists and scientists might lack a common language with which to communicate. Psychologists even declared that artists and scientists have different brains! The idea of a Renaissance man seemed to have died with Da Vinci 500 years ago!

Then, in the mid 1980s, things changed. Suddenly both my art friends and science friends were talking with equal excitement about Benoit Mandelbrot's fractals! With remarkable relief I saw a career path that could combine my scientific and artistic desires. Today, I am a Professor in Physics, Psychology and Art at the University of Oregon and consider myself to be one of the luckiest people — I use fractals to investigate an amazing spectrum of phenomena. It's impossible to get bored! None of this would have happened without Benoit Mandelbrot. If I can help today's young scientists in the way that Benoit helped me then I will have truly achieved something special. He will always be in my thoughts.

25.4 Fereydoon Family: Amazing conversations

Benoit was an extraordinary man in all measures. I had the good luck of meeting him in 1979 and spend many memorable moments with him over all these years. I will never forget the fond memories of being with him and Aliette in so many great places and having amazing conversations that you could only have with Benoit.

Benoit will be remembered forever in the scientific world, but his memory will also live in our hearts through the many discussions that we had and the stories that he so eloquently told us.

I miss you Benoit.

25.5 Ken Musgrave: Aerial perspective

I will never forget the evening I met Benoit and Aliette. I had spoken with Benoit on the phone at the advice of Dietmar Saupe. I had called him unbidden at his home to inquire about a nine-month position as his programmer at Yale. In his no-nonsense manner he said, "Take the seven PM train on Thursday from Grand Central to White Plains. I'll meet you at the station. We'll have dinner and I will interview you." (Of course, this was Tuesday and I was in Santa Cruz, California.) Never having seen Benoit's picture, or having taken no notice if I had, as I stood on the platform I imagined a skinny, goateed Frenchman with a beret and a tweed jacket. When he drove up, I was very uncertain, because he just didn't look the part. He was too big. And his head was enormous. Nor, to my recollection, had he said anything about bringing along his wife, the diminutive and charming Aliette.

So here I was in the back seat of the car of the Great Man. I had only programmed up the Mandelbrot Set and a Peano curve; I really knew practically nothing about fractals, so I was really nervous about the coming interview. Benoit drove us to a nice restaurant. All dinner long Benoit peppered me with questions. Afterwards, dazed, I could only recall answering, "I don't know," "can't help you with that," etc. It hadn't gone very well, it seemed to me. In retrospect I had to guess he appreciated my honesty. I wasn't able to demonstrate any talent other than frank admission of ignorance.

Aliette was quiet, very, very nice, and an unguent to Benoit's astringency.

After dinner Benoit said, "Look, there's a computer graphics conference in Manhattan tomorrow. I'll pick you up at eight AM. So he and Aliette showed up the next morning and drove us down to the conference at NYU. Being in the back seat with Benoit at the wheel was often quite terrifying. I came to love Benoit, but never his driving. Anyway, at this point we'd only met hours before and had only had that atrocious interview. I was trying to take the measure of the man. I was never able to succeed at that; perhaps because I was always looking in the wrong dimension.[1]

To my surprise, despite the interview, Benoit offered me the nine month position as his programmer at Yale. The idea was for me implement some of his ideas for fractal terrains with erosion features. Fractional Brownian motion, or fBm, is a nice first approximation to rugged mountains; we were after a second approximation that included the context-sensitive erosion features of real mountains. We failed and offered the problem up as a

[1]This is a math joke, and a pretty funny one. [Ed.]

challenge in Benoit's contribution to Peitgen and Saupe's 1988 book *The Science of Fractal Images.* It was eventually solved by James Bardeen, a noted astrophysicist at the University of Washington. You can see his beautiful solution in my software product, *Pandromeda's MojoWorld.*

Benoit only came to Yale about once a week, and only when he wasn't on one of his frequent junkets. I got to cultivate a relationship with the computer science department that we might flog upon their many powerful parallel computers, and play on the expensive graphics workstation he bought for me. About six weeks into my position, he went on a three week trip to Europe. Before he left, I asked, "Is it all right if I make some pictures while youre gone?" In his inimitable style he replied, "Well, render unto Caesar Caesar's due." So I rendered. When he came back, he liked what I'd done so much that he got me a faculty office next to his and hired another programmer, Craig Kolb, to do the real work.

At this point I was really out of my depth, with my Masters degree in computer science and a faculty office in the Yale math department. But Benoit would pop in about once a week and say, "Well, what do you have to show me?" One day I whipped out a picture and Benoit exclaimed, "Oh! A multifractal!" to which I sagely replied, "Whats a multifractal?" Benoit said, "It is a delicate bed of flowers you tread on."

After three years of this, David Gelernter invited me to complete my PhD in the Yale computer science department. Even though it was right next door, I almost never went over there; I had my cozy faculty office next to Benoit in the math department, and besides, I was the only person at Yale doing computer graphics research at the time. My dissertation was titled *Methods for Realistic Landscape Imaging.* It included several chapters on atmospheric models designed to provide the aerial perspective — the loss of contrast and change in color over distance in the Earth's atmosphere — that landscape painters know is the only visual cue indicating really large scale. My first-approximation models only involved some rather simple calculus, and I was concerned that they were not at all fractal. When I voiced this concern to Benoit, he quipped, "Many things that do not appear to be fractal are, in fact, fractal." He was referring to the random walk of scattered photons that cause aerial perspective. Fractal, to be sure.

(In conversations I was always way behind Benoit. For me, listening to Benoit was a three stage process: First, decipher the accent. Second, parse the strange and elaborate syntax. Third, try to figure out, "what the hell does he mean???" I got a high compliment when one of my colleagues in the computer science department said of a mock run of my dissertation defense, "Musgrave, you have a fractal train of thought." If I do, I learned it from Benoit. Endless ramifications come naturally. This is one of them. QED.)

Benoit and I would often have lunch together, just the two of us. We never talked shop. Mostly he regaled me with tales of history: The Bourbaki school, etc., and his missed opportunity to run down the pedestrians Einstein, von Neumann, and Oppenheimer all at once when the three of them crossed the street in front of his vehicle at the Institute, all deep in conversation, without looking right and left first. Had they only known who was coming, they might have looked more carefully. He, too, might have been lost in thought.

Over time Benoit and I became quite close. Nine months became twelve; twelve months became six years. Ever irreverent, I would call him Mean Uncle Benni behind his back — because of his reputation only; he was never, ever mean to me. Even he referred to our relationship as "avuncular." Under that gruff, preoccupied exterior was a warm person who loved to laugh and share knowledge and make jokes at his own expense. Sometimes his self-deprecating humor was mistaken for arrogance. Sometimes guys like Benoit make grandiose statements so that others can laugh at them; sometimes it goes over like a Sierpinski balloon.

My favorite of his self-deprecating jokes: When asked where he was from, given that his rather bizarre polyglot accent couldn't be placed exactly, he'd say, "Well, I was born in Poland, of Lithuanian descent, and raised in France, so that makes me, on average, German." Given his experience of WWII, casually hiding out in the French countryside, that was a rich one.

Occasionally Aliette would accompany Benoit on his visits to Yale. Usually, it seemed to me, he used her as an alacritous go-fer. One day I was in his office late in the afternoon when a breathless Aliette returned and declared, "I'm afraid there's nothing more I can do for you!" To which Benoit replied without hesitation, "My dear, I could not take a breath without you."

Enough said. I really love Benoit and Aliette as family, and will always miss and remember dear Uncle Benni. I once let that moniker slip in his presence and he riposted, *voce* Benoit, "Please! Do not call me that!" But forever, to me, he will remain dear Uncle Benni.

25.6 David Fracchia: This will take 11 minutes

Benoit always insisted he never touch a computer. He wouldn't even put his hands on a keyboard. He'd joke about his IBM picture which had him near a computer with the Mandelbrot Set on the screen. When I was there he used to sit next to me while I showed him the progress of my work. I decided to play a joke on him. I excused myself to go to the washroom, and instead went to Ken Musgrave's office, remotely logged into my computer, and crashed it. Then I casually walked back and accused Benoit of doing something to the computer and losing all my work! He was very flustered until I told him what I'd done. He was a great sport about it.

Also, when I first started working with Benoit, he'd give me equations to plot so he could study them. I would hard code them into a rudimentary display program, which took some time depending on the equation and what Benoit wanted highlighted. So Benoit would look at the results, then write another equation and give it to me. And wait for me to program it while he sat beside me. I told him I'd call him when it was ready, but he said he'd wait. After about 3 minutes he said "Well, this is taking too long ... if it's going to be longer than 10 minutes, then I will go back to my office and you can call me." My standard answer each time he'd ask how long it would take was "11 minutes." He got a kick out of that until I came in on the weekend and wrote an entire system to display multiple equations, zoom in or out on any of them within a subwindow, allow for comparing two at a time, and even file driven instant input of lists of equations. Then my response to "How long" was "11 seconds" ... he said "Aha, you said seconds!" like he'd caught me. But then I showed him that it was 11 seconds, not 11 minutes. I loved his reply, "You can Zoom Infinitely! Marvelous!"

Benoit used to dismiss us with, "I tire of you ... out!" I still use that line to this day.

25.7 Martin Shubik: In fond memory of Benoit

This is a relatively brief and fond note on Benoit more as a friend than as a colleague, although we were both.

I met Benoit for the first time at IBM in the 1960s, where we were both members of IBM's Thomas Watson Research Labs. At that time (especially in part due to the largesse of ONR), corporations felt that they could risk a little bit of frivolity in the form of hiring small groups of semi-paranoid twenty and thirty year olds who neither knew nor cared about corporate reality, but who each, in their own way thought that his particular problems were the most important in the world and that they were going to solve them. It was not unlike a small fancy zoo where the wiser administrators or zoo-keepers realized that their job was to keep the cages clean, make sure that the more valuable animals were fed; try to trade the lesser ones and make sure that there was sufficient attention paid to stroking and preening. The blissful ability to ignore virtually all senses of societal responsibility in the pursuit of individual ambition was high. In a bizarre way it worked and a history of the individuals in the research group headed by Ralph Gomory would be impressive.

Looking back I must note that although Benoit and I chatted on many occasions on the subject of economics and I was aware of Benoit's work on changes in the price of cotton, I had been probably overly brainwashed by "big picture finance and economics" and did not appreciate at that time it was possible to study with great impact something as mundane as cotton prices. Some years before I had done preliminary work for John Tukey on the William Beveridge time series on wheat, but we abandoned it as too dubious a series given the problems in piecing together many fragments.

I did not appreciate that Benoit had a good and clean enough series and the right methods to carry out his work with the full standards of any physical science.

In later years, when we had become close friends the full import dawned on me and Benoit convinced me that he had clean empirical evidence that the Bachelier random walk was not what was going on and that the later work of Black Scholes and the developments in CAPM did not fit many of the facts. Nevertheless it seemed to me that although Benoit did not have a theory to go with the facts, he certainly had the facts. Benoit agreed and said that at this time in this field he felt that he had played the role of Tycho Brahe but unfortunately the work of Kepler had not been done.

He did not say so but a comparison with the fight with the church at Rome and the fight with the academic church of finance still seems apposite, however by now the evidence is overwhelming but the theory is still not there. When too many theories explain the facts there is still work to be done.

One might observe, that in the best sense of the word Benoit was a true scientist, and a seeker after the truths of nature. Yet somehow a deadly combination of the finance and economics hierarchy and Benoits personality resulted in a dismissive attitude by many of those in finance and either an ignoring or a dismissive attitude by many in the economics establishment. This rankled him deeply to the very end.

He simultaneously thrived on acclaim and yet in many ways regarded himself as an outsider. He felt that the ghost of Bourbaki and French formalism drove him out of France. He regarded himself (with accuracy) as an applied mathematician, but also as an economist. He felt a tension between himself and the pure mathematicians and a rejection by the economic establishment; or at most a damming with faint praise.

Several of us managed to bring Benoit to Yale, and while here he still lived in his somewhat baronial repository of books in Westchester. He commuted to Yale, with his wife Aliette doing the driving. My wife Julie and I and the Mandelbrots had a more or less monthly ritual of going out in the evening for a Japanese restaurant meal. When it comes to pontificating on many topics I am by no means a shrinking violet, and in certain domains come well equipped with a large store of obscure but intriguing facts. One of my pleasures in going out with Benoit was that he had an eidetic memory, hence by the time the soba or the sea-weed had been finished, if we had pressed the right button we were usually treated to a monologue for the rest of the meal, with only a minor interruption from me on occasion when I could one-up an obscure fact with an even more obscure one.

I have written this small tribute purposely with a light touch out of a deep affection for a friend who in spite of his and my many faults and foibles (as our wives can easily recount) enjoyed each others company, and in many ways provided each other with advice and insight and many enjoyable hours.

25.8 Ralph Gomory: Courage

First I want to say how much I miss Benoit, even though we saw each other only at great intervals. It was always such a pleasure to be back with him. We had very good conversations, and very good understanding, in spite of all the lapse of years between our meetings.

I want to make one comment based on the fact that I knew Benoit since 1959. And so I knew Benoit for several decades during which he was totally unrecognized. Whereas you hear a great deal about his intellectual powers, his clarity, his ability to actually see what is in front of his eyes — this is an ability which most of us do not have — I just want to stress that the most striking thing when I think of Benoit is his courage. Because during all those years, he never hesitated. He stuck with his views, even as they evolved. They were always either neglected or denied, and that over a period of two decades. But, he just kept going. Many of you have never had that kind of experience. I've had it. Not for two decades, fortunately, in my case. It's very difficult to persist alone, and to persist alone for such a span of time, and at such a variance from the popular views. That is, to me, real courage.

And so when I think of Benoit, despite all the things we say about his intellect, the thing that always comes into my mind is his courage, his persistence, and his adhering to what he thought was the right thing.

25.9 Rhonda Roland Shearer: Revolutions in geometry, art, and social values

Over the years my late husband Stephen Jay Gould and I had many dinners with Benoit and Aliette, most recently discussing his archives and his legacy, and what he would physically do with his papers. At the heart of these discussions was always a basic generosity: more about how to preserve his ideas and be able to share them when he's no longer here.

I'm an artist, but my concern is artistic innovation. I noted that the greatest revolutions in art history correlated with when artists were thinking within new geometries. The two greatest revolutions were Renaissance perspective, where flat two-dimensional representational symbolic surfaces in paintings suddenly represented perspective depth; the second was of course Cubism, where artists were concerned with Henri Poincarè and how the Euclidean axioms as well as perspective that were dominating art production were able to be foiled by artists thinking within Riemannian curved space, within higher-dimensional realities. I don't mean the fourth dimension as time, but spatial geometries, promoted in the popular literature in the famous book by Edwin Abott Abott — I'm not stuttering, that was his name — who wrote the novel *Flatland*.

I looked at those two revolutions and saw that maybe there was something potent about artists thinking within a new geometry, that maybe this is generalizable. The most recent innovation in mathematics is fractal geometry. This correlated with the best-selling book by James Gleick, *Chaos. The Making of a New Science*. These scientists became celebrities, and Benoit was within this paradigm. When I discussed my ideas with Benoit, he was very interested and very supportive.

When we talk about Benoit as the multidisciplinarian, this is not just a matter of having to face down the turf wars we all know naturally exist. Academics are not in general happy about interlopers into their fields. And then when you're integrating lots of fields that only magnifies the problem. Certainly, Benoit's personality allowed him to face this. So I think it was a psychological component. I agree with Ralph Gomory about Benoit's being heroic. And then eventually people recognized, with little dissent, that Benoit was someone worthy of being allowed in their domains, because his work was good. It was solid.

And yet, this interdisciplinary study is not an easy thing, because geometry isn't just about academics, it's also about mapping social value. When you say that there's a rationality within nature, this is a very threating statement. So much of our belief system is constructed around the dualism of rational mind, Euclidean geometry, reason, and on the other hand nature and the irrational. This duality also is engendered: nature, the irrational, is feminine, versus the masculine rational Euclidean geometry. Whole taxonomic structures in art history are built on that. Norman Bryson, who used to be at Harvard, has written that landscapes or still lifes were risky for people to paint. They were considered woman painters' subjects, not like history paintings or pure abstraction. There was a devaluation in the schema of the choices you were making.

My conversations with Benoit included the idea, similar to an earmark of scientific revolutions, that fractal geometry assaulted and changed the geometric structure of this taxonomy of values. Where we would look at nature as being irrational and feminine, a fractal geometric mountain is generated as a mathematical object. How can this mountain be devalued, as it would be in the old scheme of Alfred Barr at the Museum of Modern Art? Benoit's work gave power to artists to express themselves in nature and not be coded as decorative, romantic landscape painters. Suddenly, landscapes became something else. This was very, very important.

My final point is from the history of mathematics. In a letter of Gauss, it was clear that he knew about non-Euclidean geometry. At that time, preachers would stand in halls like this and say, "Read the bible, but also read Euclid's *Elements*." So you can imagine when it was revealed by Riemann and others that the three angles of a triangle don't necessarily add up to 180° — on a concave surface the angles sum to less than 180°, on a convex surface to greater than 180° — this was considered literally an abomination. Preachers would say that non-Euclidean geometry was the devil's geometry. This is another example of how the threat of new thinking in geometry is embedded in mathematics history. In Gauss's letter, it became clear that he chose not to reveal non-Euclidean geometry to the world because it was going to be messy socially.

In this context, when you think about fractals and what Benoit did, it's much more than just presenting new mathematics. It has such significant social coding and social impact.

Non-Euclidean geometry was viewed as messy, until Einstein predicted that gravity bends the path of starlight. Verified by 1919 observations of the displacement of the positions of stars appearing near the sun during a solar eclipse, Einstein credited Riemann's curved space as the right way to understand gravity.

We don't yet have all the fundamental science that will come from fractal geometry. Like non-Euclidean geometry and general relativity, fractals are another case where science follows mathematics and the new thinking. The path is still open for looking at the greatness of what Benoit has done. Already, the impact is certainly clear, and it may be even greater still.

I'll be forever greatful for Benoit's and Aliette's support, generosity, and openness to discuss new ideas.

25.10 Jacques Peyriére: Benoit the visionary

I want to say how much I am indebted to Benoit. I was very lucky to be his postdoc in the beginning of the 70s. I met him for the first time in 1973 when I was finishing my thesis. My advisor, Jean-Pierre Kahane, introduced me to Benoit because I had just done some work on multiplicative cascades. After that I spent one year with Benoit at Yorktown Heights. At that time I did not know that this would change and orient all my scientific life up to now. We met almost daily for long conversations that always began with scientific topics and mathematics, but very quickly moved to any topics, anything. When I say "conversation" it was more-or-less some kind of monologue because Benoit was very talkative and he had lots of things to tell about anything. Listening to Benoit was very rewarding.

But he was also very attentive to people. Indeed, he was listening. Each time I had some problem, some thing to ask him, he was very attentive.

Benoit's style of working was very different from what I had known before. He had always lots of things going on, lots of ideas. He developed completely some of his ideas, for others he only made some experiments and then some conjectures. He was glad when other people would develop these things completely. He eagerly followed these developments. When the work was not fast enough, sometimes Benoit was fairly impatient, inquiring "Is it OK? Is it going on?"

Benoit was a visionary. Ralph Gomory mentioned that some of Benoit's ideas took a long time to be accepted. Not just accepted, but understood. In particular, this was the case for his work on turbulence. Benoit was attacking some strong beliefs. His work was not accepted at that time, and he had some difficulty getting it published. His main article took three years to be accepted. But this particular work gave birth to a tremendous amount of work in mathematics. Also in physics, but later on.

Benoit did not have recognition in the scientific community, especially in France, that he deserved. And he was the target of many attacks, which I have witnessed. I know Benoit was very sorry about that. He was eager to be understood and appreciated. He did not understand why he encountered so much resistance from the French scientific community.

I've an example as late as 1998. Once Julien Barral, a student of mine, submitted an article entitled "Moments, continuité, et analyse multifractale des martingales de Mandelbrot." He received a referee's report saying the naming of these processes is a distortion of historical facts. So even at that time Benoit still was the target of attacks aiming at dispossessing him of his intellectual property. Fortunately this was of no consequence at that time because the editor-in-chief of that journal agreed with the argument we developed to say that this article was well-named.

I owe so many things to Benoit.

Peter Jones. In 1977–78 I was a visiting graduate student at the University of Paris, Orsay, a member of the harmonic analysis group. In the seminars we kept hearing about the work of the great and brilliant Jacques Peyriére. He was supposed to be at Orsay, but wasn't. So I asked where is this great man? With apologies to Ralph, I was told he was at IBM that year. I was a young and foolish boy, and said, "Why would he do that?" Later I found out exactly why he would do that: he was visiting Benoit.

25.11 Bernard Sapoval: A hall linking teaching and research centers

I met Benoit in September 1982. It was for me a very interesting and new experience, because of the variety of topics that were touched at that meeting, and the fact that the people attending such meetings were each of them interested in a variety of ideas. Then I began to work on fractals. I created a group in my lab working on the general question of why fractals exist in nature, and when they exist, what are their properties. The fractals we know now may at the beginning have appeared to be strange or peculiar objects. But now their geometric complexity makes it clear that if we understand that, then perhaps we understand the workings of any irregular geometry, fractal or not. Fractals have a very rich mathematical complexity, and simultaneously a very simple aspect.

So Benoit and I kept in touch. I visited Yorktown Heights several times, and Benoit spent a sabbatical at École Polytechnique in 95. After his death, at the École Polytechnique we organized a memorial meeting which ran one day and a half, with about fifteen talks. We were lucky to have Wendelin Werner speaking about the 4/3 conjecture and its proof. Other speakers included Jacques Peyriére, Julien Barral, Carl Evertsz, and Michel Lapidus. There was the idea that France in general has been unfair to Benoit. So we wanted to try to do something about that. Three hundred fifty people attended this meeting. I think this was very successful.

It has been mentioned that Benoit had one of the richest personalities that anyone has met. He was very talkative, so if you said nothing, he was speaking. But he could be interrupted, and then he would listen to you. It was the richness of his imagination that was working all the time.

Concerning the fact that he was not elected to the French Academy of Sciences — something he resented very deeply — the fact is that in the world of science there are many people who appreciate someone who does the same types of things that they are doing. They do not appreciate someone who does something different, because when you are a researcher you think that what you have in your mind, the subject of your interest, is the most interesting thing. So because Benoit had such wide success, his not being elected was simply jealousy. I'm not sure that in this country this is the same. Here success is applauded. People like success, and like successful people. In France, sometimes they are jealous.

Benoit was also very interested to speak with young people. When he was visiting a lab, he always wanted to speak with young scientists. My lab works on experimental solid state physics, and there once Benoit gave a talk on how to write a paper in English.

The École Polytechnique depends on the minister of defense, a general. A former student from École Polytechnique, but a general. I suggested that some location should be designated with the name Benoit Mandelbrot. The general said he wanted to have the advice of the scientific community of the École Polytechnique. I know something of the behavior toward Benoit of some French mathematicians, very important persons, that I won't tell you. (I could tell you privately.) The general told me I should go to the board of the various departments of the École Polytechnique, to make the proposal of a location designated for Benoit. Their reacion was unanimous: everybody agreed. So now there is at the École Polytechnique a Passage Benoit Mandelbrot, linking the teaching buildings to the research buildings. It opens on Benoit Mandelbrot square.

Benoit is known more outside of France than in France, because he made his career at IBM. So this was one way to give, perhaps too late, but still, a tribute to Benoit Mandelbrot's contribution and originality.

Peter Jones. The French graduate students in Orsay used to call the great professors "galaxies," because each is very large and extremely important. And if they happen to come near each other, there is a terrible collision. I'm sure this has changed now.

Also, I want to mention something about the type of impact Benoit had on mathematics. As everyone here knows, Benoit always said start by looking at the picture. It was through Benoit that I came to know about Bernard Sapoval. At this point I was very interested in some extremely classical problems in mathematics. Bernard had a laboratory where he studied fractals and electrical potentials. This was quite an experience for me, because it was my job then to compute by hand these situations. The physics made a deep impression on me. In the 1990s, Lennart Carleson and I were working on a problem from the 1920s, a descendent of the Bieberbach Conjecture. This conjecture is about conformal mappings, which on the unit disc have power series expansions of the form $z + a_1z + a_2z^2 + \cdots$. The Bieberbach Conjecture is that $|a_n| \leq n$. Bieberbach proved this for the second coefficient, and this has a physical explanation. The proof is mathematical, but it has a deep physical ramification, understood since the 1920s. There's a corresponding conjecture for conformal mappings outside the unit disc. They have a Laurent expansion at infinity: $z + b_1z^{-1} + b_2z^{-2} + \cdots$. It's completely different from the Bieberbach Conjecture. Here the only coefficient known is $b_2 = 2$. It's the same proof as in the Bieberbach conjecture. Carleson and I decided we should try to look at the order of growth. Then this work that I became aware of, and similar things by Mandelbrot, were very inspirational. The answer, it turns out, is given by a fractal. So unlike the classical Bieberbach Conjecture, which has unbelievably boring geometry — the extremal case has boundary a straight line — in this case a fractal does it. There's a huge body of research centered around this area now. It turns out to be an area of mathematics that doesn't connect with conformal field theory. But the conjectures are inspired by conformal field theory, and according to the computers, correct.

This is the type of thing where Benoit had a deep impact on mathematics. For instance, I knew immediately after starting to work on this with Lenart Carleson, that the answer had to be a fractal.

25.12 Didier Mandelbrot: A comfortable home

I met Benoit Mandelbrot sometime in August of 1961, on the day I was born. My father was a wanderer in many ways. He wandered physically in various parts of the world, and he wandered intellectually. No need to describe to this audience the various places he wandered intellectually. Maybe Peter was right that he was disappointed that there is not more conflict here at Yale, but actually I think that deep down he was very grateful at finding a home here. He had a long-time home at IBM. I know Ralph Gomory played a critical role in that, which I want to thank him for that as well. I want to thank the department here at Yale for having made such a comfortable home for him for so many years. That was very much recognized and appreciated.

25.13 Larry Gould: A Mandelbrot number of 1

I knew there would be a lot of high school teachers involved in the summer fractal geometry workshops that Michael, Benoit, and Nial ran. I wanted to learn more about them. There was this telephone number of Benoit Mandelbrot, and I couldn't believe that he would return my call. But Benoit called me at home. We had a conversation for about ten minutes. Quite frankly I don't remember the details because I was overwhelmed at the time. That was one of the reasons I attended the workshops, which was a tremendous experience. It has contributed and is contributing to a book that I have been writing on symmetry for non-science majors. It will feature prominently a chapter on fractals. Benoit gave a talk in the workshops. I would ask him a question, others would ask him a question. To get Benoit to answer a question is like asking to see a portion of the city and being told that I'm glad to show you that portion of the city, but first I want to show you how wonderful the surrounding countryside is. So he would take you on a trip around the region, which was fascinating. He would relate his experiences about various things and then he would come back finally to the question. I remember I'd asked him more than once what's the definition of randomness, and the conclusion I have from what he said was that so far there isn't really a good definition of what's meant by random.

The other way in which he strongly influenced me was that he posed a question, which is related to my own field in physics, having to do with variational calculus, a way to deal with minimization in space and time. There is an intimate connection between symmetries and conservation laws. Benoit posed a question having to do with variational calculus and fractals. It's a fascinating topic that I've started to work on and I hope I can make some progress on it. It's something that never occurred to me before, looking at how fractals and symmetry are related to conserved quantities in physics.

Like Erdös numbers,[2] I'll say a Mandelbrot number of 1 means you spoke with Benoit, a Mandelbrot number of 2 means that you spoke to someone with a Mandelbrot number of 1, and so on. So my Mandelbrot number is 1. Now some Mandelbrot numbers of 1 are larger than others. Mine is small, but it is very important to me.

25.14 Michel Lapidus: Benoit as poet, creator, and rebel

The first time I heard of Benoit's work was in the early 1980s when I read his book *The Fractal Geometry of Nature* and was completely astounded by it, so very different from everything I'd learned in Paris, where mathematics was dominated by the style of Bourbaki that allowed no pictures. Absolutely forbidden. Benoit's book had many beautiful pictures, some in color, and this resonated with me. I felt that I knew the author already. In fact, I always think in pictures. Even if I think of a formula, I always have pictures associated with it, and colors. This was pure poetry to me. It's as though we were in complete resonance without having met. But Benoit's book had no theorems. There were fascinating statements about a lot of things. It was like moving in a landscape which was not familiar to you yet, but it looked like you had been there but didn't know about it.

So I went to see my friend and colleague Mark Kac. I told him I'd read this book and it's fascinating. What do you think of it? The first thing he told me was that Benoit is a poet. He encouraged me to look into it more. A few years later someone asked me what would happen if the boundary of a drum is fractal. I learned that Michael Berry had made some conjectures about that which turned out not to be completely correct, but were very

[2]The shortest path of coauthored papers between you and Paul Erdös

interesting. Later I resolved this partially. Many other people, including Bernard Sapoval, have contributed to this topic.

The first time I met Benoit was at a conference in memory of Paul Lévy at the École Polytechnique. We had a very nice, very lively, conversation. I think he already knew about the theorem I had proved and he was very interested in it. In the early 1990s he invited me to Yale as a visiting professor. Here I met Aliette, who is one of the most wonderful people I know. Benoit and I had many, many conversations that year. He told me many times how happy he was to be here at Yale in the department. He told me about the lack of controversy. He was very happy about that, but also he was very careful to avoid controversy. Benoit was surprised that he never had a fight with Serge Lang.

Benoit did involve me in a controversy. With Bob Strichartz from Cornell I organized a special session on fractals at an AMS meeting. I invited Benoit, Bob invited another person, who I won't name here. When I mentioned this to Benoit, he turned red in the face and explained to me for hours his troubles with this person. I didn't know what to do. I remember Aliette several times saying, "Benoit, leave Michel alone." Benoit told me if that person would be there during his talk, he would put up a transparency that said "X leave." Eventually we reached a gentlemen's agreement, where neither of them would attend the other's talk.

One of the happy souvenirs from my stay here is that my daughter Julie, then one year old, started walking. So one day I brought her to Benoit's office. She entered his office and put her right foot on the floor. Then Benoit said, "Aha! She's going to be right-handed." This is just an example of how he always observed things, though he did not always seem to be close to reality.

Benoit and I shared French culture. We both love the French language. We were educated in the French mathematical world which has the tradition of Bourbaki against which he rebelled and against which I rebel less. Someone who played a big role in my life, and later I think a negative role in Benoit's life, was Gustave Choquet. Even though he was an analyst, in his lectures he drew pictures. He had a beaautiful geometric mind. Benoit also had a wonderful geometric imagination. So I thought these two people should like each other, but in some sense maybe they were too close.

Benoit told me that he was thinking about retiring and perhaps moving back to France. But he knew the problems he might encounter, especially with the French scientific and mathematical communities. Benoit asked if I could arrange a dinner at our house for him and Gustave Choquet, because he knew that Gustave was very opposed to him, especially for a particular prize. But first he asked me to talk with Choquet, to see if we could arrange some gentlemen's agreement. I went to see Choquet and we talked at length. The issue with Choquet may have been some jealousy, I don't know, but I think it was also the matter of misunderstanding some of the statements that journalists made in the press, reporting Benoit's statements about the nature of proof in mathematics. Choquet thought that Benoit was opposed to proofs, or felt there was no need for proofs. And that is absolutely wrong. I remember at Yale, for instance, having had many long conversations about that with Benoit. He always told me that proof is very important, but that it was not his own style of doing mathematics, it was not his strength or inclination. He liked to make conjectures and propose problems to others, and he was very happy when people provided proofs for his conjectures. He felt both activities are an important part of mathematics. I told this to Choquet. Eventually he said alright, maybe Benoit can come back and I'll be quiet, but it was clear that he was not completely convinced. I decided it was better not to have the dinner.

This is a sad story because neither Benoit nor Choquet were Bourbakists. In fact, both thought in pictures, and were great geometers as well as analysts.

Let close with a few words about my last conversation with Benoit. This was the first week of October (2010). More than a friend to me, Benoit was like a second father. I decided to write to him about a few of the most pleasant memories we had in common. Later I called him and we spoke as always in the past, for three hours. We talked about many things. His memoir was much on his mind. At that point he knew he could not finish it. He told me he had tried all summer to write more and improve it, but he just didn't have the strength anymore. I asked him if someone could finish it and I thought that Aliette could be the one. He said that was difficult because she did not know all the stories he knew, especially the stories from before the war. But apparently that part was already finished. Then he told me some of his stories during the war. Then I don't know why this came up, but he told me a story about preparing for the ENS. He told me that his teachers told him this: you'll be either the first or the last, because either you guess the solution to the problem, and then fine, or you find some geometric way of translating the problem and you solve it. Or you don't. I told him that was funny because my professors told me very much the same thing. Benoit said he was very happy, because he knew we had something in common.

Peter Jones. I want to say something about Benoit's appointment as Sterling Professor. To arrange this I had to meet with senior members of the administration. One of them, not a scientist, asked to meet with me privately so I could help him understand better how to evaluate Benoit. When we met, the first question was, "Could you explain to me, in English, why Yale should do this, because I must have a compelling reason." I said, "Oh, yes, it's very simple. He's the world's most famous mathematician, and you can check this." The response was, "Great," and that was the end of the meeting.

25.15 Richard Hudson: Communications

I'm not a mathematician, I'm not a physicist, I'm not a chemist, I'm not anything like that. I'm a journalist. I'll describe the aspect of Benoit that I was most familiar with: communications. Communications was, as you all know, hugely important to him. There's been a lot of discussion earlier today about his role in the world of science. Cross-disciplinary, creative, international, Benoit was all of these things. But also he placed importance he on being able to communicate the significance of his work to as many people as possible, with as many different backgrounds as possible.

In 2004 I wrote a book, *The (Mis)Behavior of Markets* with Benoit. In 1997 I was in Brussels as managing editor of the *Wall Street Journal*'s European edition. A friend at IBM called up and said, "Please, you must speak to this man." I didn't know much about Benoit. I'd heard about the Mandelbrot set by then, but that was about all. You have to imagine the typical newspaper scene. It's a bunch of journalists, usually on deadline, usually on too much coffee, definitely without enough sleep, rushing to get the work done, with visiting dignitaries usually from business. CEOs, people in suits.

Benoit was different. He came in hair awry, large as some people have said, strange accent, I caught about half of what he was trying to tell me. But what I caught was the key point: you people at the *Wall Street Journal* don't understand how the markets work. That was the message, and that really appealed to me. Of course we all know that's true. If any of us understood how the markets work, we'd all be rich, or we'd all be poor.

Benoit was very good at shaping a message to the people he was trying to reach, particularly if it was someone who he thought didn't know the mathematics or the technology that he was talking about. He did think about it a little bit in advance: what is the line, what is the message, what is the hook? There may be a lot of other things mixed in with it, but it was very evident that he was very good about thinking about who the listener was. The conclusion of this was we decided we'd publish an article by Benoit.

The whole process of that was intriguing. We were the *Wall Street Journal*: people submit articles to us and we publish them the way we like them. It was a new thing for him to be pushing back. When I sent him the edit, he said, "I don't like that. No, that's wrong." A dialogue ensued, which was wonderful. The punchline to that particular story is that he has this wonderful game, "spot the fakes," showing simulations and actual charts from the market. The game is to look at the pictures and guess which are the real market charts and which are the simulations. Inherent in this is that when you publish it, you have to get it right. I'm afraid that when we published it, we mixed up the captions. [Laughter from the audience] You can imagine what the phone call was like after that publication. [A lot of laughter from the audience] The correction ranks as one of the most unusual ever for the *Wall Street Journal.*

The book followed very quickly. Benoit had an interesting message, a two-part message, some of which you heard yesterday from Taleb, that the economic and financial establishment has it wrong. But Benoit also had a more positive message: I have an idea of how to do it right. This two-part message intrigued me. It's easy to write knocking copy. It's more difficult to suggest how to go forward.

Benoit suggested that we do this book together. I said 'No' at first, and there began what in retrospect I now see was a testing period, where he sent me a biographical essay and asked for my comments. I sent him back some notes. I didn't realize this was an exam. I guess I passed. Then we wrote this book together.

It was one of the best years of my life, fifteen months, actually. Remember, I'm a journalist, not a mathematician. I began this knowing almost nothing about fractals, and Benoit did not want to teach me all about fractals. So what did I do? I went to Michael Frame's website. It's fantastic, the most amazing self-teaching course on fractals.[3] With that basis I was able to bootstrap myself up so I could start to have serious conversations with Benoit about his research. It was an intense experience,

[3]The editor is not comfortable with this comment, but could not find a graceful way to excise it.

me visiting Scarsdale, or meeting up in Paris or Brussels. Communications, again.

Benoit had a very special English writing style. It was very poetic. Peter knows this. I would say poetic in parts. There was also a bigger picture that he had in mind. He would express parts of it really well; other parts needed a little bit more connecting. It was the poetry he brought to the subject that inspired people.

This is another lesson I draw about the science world. If you're trying to sell your research, it isn't enough just to talk about the research. You have to inspire somebody, appeal to their imagination about why it's interesting, connect it to literature, connect it to cosmology, to music, things that other people know about. That's really important.

When I was still at the *Journal*, we did a conference on risk management. This was with Steve Balmer, CFOs, suits, big suits. I thought it would be fun just to have Benoit talk. He did his spot the fakes game with these masters of the universe. It was fantastic. They really didn't know quite what to make of it. Benoit did succeed in attracting their interest, and in inspring them to think differently. It isn't what they normally hear from their bankers. The evaluations after the conference were that Benoit and Steve Balmer were the two best speakers. I was surprised it worked out that way.

My fundamental point is that in this world of science research — international, cross-disciplinary, tons of money pouring in, a growing enterprise — if you don't communicate to people it's like one hand clapping in the forest. Nobody is going to hear it. Benoit really understood the importance of making sure there are two hands, and twenty hands, and twenty million hands clapping, so that there be a noise and change going on.

In conclusion, I'll say that Benoit really inspired me. So much so that after I finished the book, I quit the *Journal*. Writing that book was too much fun. So I started my own company that has nothing to do with fractals at all. However, and this is a promise to Aliette, I will continue to do everything I can to promote fractals and continue to make a noise on the subject.

Peter Jones. I'd like to share some of my experiences on similar projects. Benoit had an article in *Scientific American* in 1999. I'm trying to be delicate here, but I helped Benoit write it. Now I contributed absolutely no scientific ideas. The editor kept sending Benoit messages and at some point Benoit asked me for help. Basically I helped him smooth out the language. It was a fascinating exercise. I took the sentences, chopped them up, moved them, and smoothed the English. The editor repeated this and it turned into what I think is a really great article. Later Benoit told me the final version was exactly what he had written before. And he was right. I think this article is very important because it was written so long ago and it talks about the dangers in financial systems. He was sending out a clear warning.

I think it's time to mention some other expository abilities of Benoit and my experiences with them. So of course he gave research lectures at the math department or other places on campus. These were sources of tension for me because I would immediately read very carefully the paper he would be talking about, because I knew what would happen. Benoit was a great man, a great scientist, and he could write really well. But sometimes — not his popular lectures — but his research lectures were tough. For example, a senior professor came to me and said he couldn't understand anything Benoit said. Could I explain it? I replied that if he had come to me before the lecture, I could have told him exactly. But now I'm not sure.

A second example: a faculty member on campus demanded that I appear in his office and give him a lecture on what had happened. After five minutes, this person said, "No, it's impossible." "What's impossible?" I asked. "But Peter, I understand every word you're saying and you just told me what Benoit said. Are you sure that's what he talked about?"

Benoit had an ability to convey very difficult mathematics in terms many people could understand, but when it came to describing his latest research, it was more exciting.

25.16 Harlan Brothers: Benoit's simple faith in me

One of the things that struck me most about Benoit, aside from his brilliance and his playful sense of humor, was his unusual strength of character. I'm sure everybody here can attest to the fact that Benoit called it like he saw it. Of course, in some circles this didn't necessarily help to boost his popularity. But that's precisely the point. Although Benoit certainly would have enjoyed swifter recognition, for him intellectual honesty was much more important than popularity.

I'm reminded of a lovely recollection that Michael had posted online. When he began collaborating with Benoit, he was working on a difficult problem and derived some results that contradicted Benoit's expectations. He felt some trepidation at approaching Benoit with these results, but was quite relieved at Benoit's reaction, "Marvelous, the problem is more interesting than I'd thought."

I first met Benoit through Miguel Garcia, my wonderful professor from Gateway Community College. Miguel, I believe, was a student of Benoit's here at Yale. Benoit had called Miguel, looking for some help in editing what was at the time planned to be the 20th anniversary edition of *The Fractal Geometry of Nature.* I came here to meet with Benoit at his office. I can still picture Benoit seated, his hands on his legs, leaning slightly forward, as he said, "So, Miguel tells me that you are not the average cookie-cutter student." We spoke for a while, I shared some of my research at the time on random walks, and Benoit shared what appeared to be the better part of the *Encyclopedia Britannica.* He talked about everything from the lesser-known work of John Venn to the socio-political history of Budapest in the 16th century. It was breath-taking. I immediately began working on the book project, and soon after, with Michael and Nial on the fractal geometry workshops. It was in fact Benoit who first suggested to me that I investigate what it means to say that a piece of music is fractal. For me, that comes back to the issue of character. Benoit had no interest in my resume. He only cared about the way that I thought. In my experience it is rare to find people with his innate confidence and clarity of judgment. It was Benoit's simple faith in me that profoundly changed the course of my life.

I live near a state park and often go for long walks in the area. Whenever I look up at the passing clouds, I am reminded of Benoit.

Peter Jones. I think this is not so well-known, but Benoit had an amazing observation about planar random walks on a lattice of size ϵ. It's known that if you measure this on any fixed scale and let $\epsilon \to 0$, that the geometry converges to the geometry of Brownian motion. This is a very simple statement. The Brownian path divides the plane into connected regions, and everything is known about the statistics of them. Then if you fix ϵ small and look at these little lakes that are surrounded by the random walk on the lattice, on scales like $\epsilon^{1/2}$, all the statistics are wrong, in the sense that they don't match up with what you know happens in the continuum limit at scale $\epsilon^{1/2}$. Nothing is known about this, yet. It's really a startling observation. I asked Benoit to show me pictures because I was so surprised by this.

25.17 Nathan Cohen: Paradigm shifts

I'm a physicist. These days I'm slumming as a businessman, but my heart is still in physics and every once in a while I get the opportunity to do some.

I'd like to begin by filling you in on a couple of interesting facts. First is the *The Fractal Geometry of Nature* is available for Kindle, and therefore for iPad. This is a marvelous version of the book, much easier to navigate because it's electronic. I think you're going to see the whole cycle of people interested in finding out about fractals happening all over again because of this new availability of the electronic edition.

I also want to point out this is a very interesting room. As you probably know, Benoit was a very visual person and obviously thought in terms of geometry. I've always had that ability myself and considered it a bit of a handicap. But I want to give you a sense of how a person who thinks geometrically sees a simple thing like a room. Obviously in this room you have circles and hexagonal columns, partial hemispheres, squares, and so on. In other words, it's a collection of Euclideana. But there's another thing you might want to notice: this room is obviously designed at least partially with respect to practical aspects such as acoustics. I do have some background in acoustics and I can tell you that there's a fix that's done here with roughness, with fractals. This is to handle the problem introduced by the Euclidean shapes. It's the foam-like structure on the ceiling, there to absorb some of the higher frequencies so the whole room doesn't sound like a whispering gallery. Subtle uses of fractals, of roughness, are there all the time in nature and are now becoming more prominent in technology.

I deal mostly in fractal antennas, which is an emerging technological aspect of fractals. Many of the issues of controversy and focus and attention I see in my life almost every day. I deal with this by removing myself from it as much as possible. When I have something new to say, I drop it out there and see what happens.

That is not how Benoit approached these issues. I've heard several appropriate adjectives, including the word courageous. I don't think that's the half of it. I can't think of another person in this last century who was able to introduce so many profound ideas under one focus and be able to push them forward the way Benoit did.

I finished my Ph. D. in 1984 at Cornell. I'm an astrophysicist and therefore I knew about fractals in the context of the large-scale structure of the universe. But I had a problem back when I got my Ph. D. which I'm sure is reminiscent of the reality of today: I came out with no job and a list of 30 publications. I was told many times that I was over-qualified to be a postdoc and an assistant professor. One of the ways I dealt with that is by focusing on *The Fractal Geometry of Nature* as a way of thinking. This had an interesting ramification because the first job that I got, after putting up the shingle of consultant, was in modeling markets, both in stock options and in commodities. Now commodities are very strange. It doesn't make a lot of sense that people would look at things like orange juice as something that you would trade and try to make money off without actually owning. But that's what really happens in the commodity markets and in the options markets. My first client was interested in seeing if I would not buy into the notion that the stock market was a random walk, and were there short-term variations that I could model in some way. We did silly things like Fourier analysis. Then I thought it might be fun to try to fit some of the data with fractal analysis. The answer is yes, we were successful; yes, money was made.

I do want to tell you there's a sense that if fractals are able to predict markets, then everyone would be rich. Let me tell you why that doesn't happen. First of all, there are several fractals which fit the data, so you have to hedge which ones you want to put your money on. Second, it's a very barren activity, and having done some of this for two or three years, I realized at some point that this was not something I felt I was making progress with in my life. So the people who really are in a position to do this kind of modeling are probably not the people who want to do this modeling. So that's why people aren't rich making money off fractals in the markets.

That was around 1984 or 1985. I had the good fortune of meeting Benoit, talking with him for the first time in 1987, in Hungary. Benoit was giving a talk on cosmology. I was there for that conference and another. This was the first time I heard Benoit give the fractal talk. One of the things he put up on the screen were the usual traces showing a Koch curve in various iterations. I remember looking at these, and because I have a background as a radio astronomer, thinking wouldn't it be amusing to make an antenna out of some of those shapes.

Six months later I'd found a job at Boston University, where I stayed for 15 years. I finally got a chance to have some fun with this, and said wouldn't it be great to try to bend a wire into some of these shapes and see if it had any differences in the antenna. I did this in my apartment. I was motivated because my landlord had a "No Antenna" clause in the lease.

I found, anecdotally at that time, that you could bend a piece of wire into some of these shapes. This made the antenna smaller and higher in performance, which was totally unexpected. That was around 1988.

By 1990 or 91, I went to my dean and said I've done all this work with fractals, trying to make them into antennas, and I'm getting some interesting results, but I can't do this in my apartment anymore. I need to be able to use the big space down the hall as a test range to make measurements. And I need a closet to store the materials afterwards. My dean said "What are you doing?" I showed him some of the antennas and he said "I did that when I was 15. I took a coat hanger and replaced the antenna on my car."

So he wouldn't let me have a closet, wouldn't let me store my stuff on campus. The result of that is Boston University had to give me a patent exemption on fractal antennas. I believe it's the first time in their history. I'm not blaming BU for that, but just pointing out that it's often hard, even now, for people to understand the uses of fractals.

By 1993–94 I'd been able to do enough of my own research with the equipment I could buy at electronics flea markets to pull together a manuscript. It turns out there are some very interesting aspects of using fractals as antennas. I reported this in 1995. I did send the manuscript off to Benoit before it was published. Being the polymath he was, he told me everything that was wrong with the manuscript. I wasn't entirely ready for that, but I thought about it overnight and called him back. I said I'm grateful for your comments and I think you're right. I'm going to throw out the manuscript and write it again. The second manuscript still had some issues based on his comments but I felt intellectually strong enough on those points that I didn't let him push them through. The manuscript was published and it's the fundamental paper on fractal antennas, which is now quite a field. These days I'm pursuing it as the head of a business instead of in academia.

By the way, I did retire from Boston University in 2002. In 2000 Benoit asked me if I wanted him to speak up for me, to try to get me tenure at BU. I thought about it for a while. He didn't realize at that time I was trying to make the transition out. So I declined his offer. I point this out because I think Benoit did a lot of things behind the scenes on people's behalf that he doesn't get credit for.

I'm going to finish up by pointing out that Benoit was responsible for several paradigm shifts. I know that's one of those obnoxious terms that's been abused, but I'm referring to the Thomas Kuhn definition of a paradigm shift invoking extraordinary working hypotheses based on new data, and the reluctance to absorb that. Fractals are probably the most important new way of looking at shapes in a number of technological fields that's come about in I don't know how many years. So it's amazing that it hasn't been adopted yet. The reason is there's resistance because a lot of stuff works just fine. It isn't just a matter of trying to make something better. It has to be something that's better, that's cheaper, any of a variety of other things on a checklist that have to work. If you go back to Thomas Kuhn, there's a misconception that people have. They think that paradigm shifts are adopted when there's compelling new data showing the old paradigm doesn't work and the new one has to be accepted. Oh my goodness, that's wrong. It's not a compelling issue, which is what most scientists would think. It's overwhelming. An analogy: compelling would be detecting extraterrestrial life based on a meteor from Mars containing DNA. That's not good enough in a paradigm shift. In a paradigm shift, you'd have aliens from Zubenelgenubi land their UFO on top of your head and then go on the Oprah show. That's overwhelming. In many circumstances that's the kind of influence needed in the future for people to really appreciate using fractals in a number of technologies.

Benoit was really an unusual person, unique in many ways. His courage is something you don't normally see in science and in mathematics. I think it's going to take a while for other people to have that kind of courage to push through these shifts.

25.18 Marc-Olivier Coppens: A fax from an undergrad

Benoit meant a great deal to me in many different ways. First as a teacher, as an advisor. He became a mentor, and finally a very good friend who I miss dearly now. I met Benoit in 1993, when I was an undergraduate student in chemical engineering. I was starting to write my final engineering thesis on the fractal geometry of surface roughness of porous catalysts, looking at the effect of roughness on diffusion and reaction in catalysis. Now I had been introduced to fractals first as a high school student in the 1980s. I won the first prize in the Belgium mathematics olympiad. The book I won was *The Beauty of Fractals* by Peitgen and Richter. I loved also computers and computing, so very soon after that I started playing around with fractals and got very excited also by the beautiful mathematics, as well as the pictures I could generate overnight. (Then it took overnight to generate a Julia set or a Mandelbrot set.) So when I had the chance to start with research, being in this field of chemical engineering, I was really thrilled to have the opportunity to work on a topic that would involve fractal geometry.

I started on research end of 92, beginning of 93. It happens that my father is a concert pianist and he was going to give a concert in New York. I would accompany my parents, so maybe this would be an opportunity to meet Benoit, who worked at the Thomas J. Watson Research Center. But he is so high up, so famous, that he probably would not have time at all. Remember this was the time before email was common, so I wrote a nice letter and faxed it to IBM. To my big surprise, the next day there was a reply from Benoit Mandelbrot, a hand-written reply, too, beautiful handwriting. I was just amazed. This probably was one of the very best days in my life. This was extremely influential to me in the longer term. I was an undergraduate, just starting to work in applying fractals to catalysis, I had zero publications, so who was I? He wrote that he was very busy at IBM, but why don't I take the train to Scarsdale on Sunday and Benoit would meet me at the train station. That, I have to say, is so very much Benoit Mandelbrot. A great scientist, but also a wonderful human being. He doesn't care about your CV, or whether you would be a Nobel Laureate or a high school kid. He takes interest in other people's stories, he listens, he responds, and he typically also influences the people he meets.

When I did meet him, taking that train in April 93, he picked me up at the station and took me to his home. It was a bit like the Wizard of Oz. I felt very special meeting him there. The half hour discussion I thought I would have with him turned out to be many, many hours. Aliette gave us some food because we were so engaged in parentheses within parentheses within parentheses of discussion. But this was just the beginning of so many more where every time there was something new to learn, something new to enjoy. Maybe not the *Encyclopedia Britannica*, but the *Encyclopedia Britannica, reinterpreted.* He would make links that other people would not see. That is what deep science and engineering, being a deep thinker, really means.

That conversation was very important to me because I learned such a great deal, even about how a scientist could work. For example, later I learned that Benoit was very interested in applications of fractals. He considered himself not only a scientist, but also an engineer. He was interested in technology, and especially in recent years, in seeing where fractals were being applied.

Also, Benoit introduced me to many other scientists. This is how later I met Bernard Sapoval. Benoit mentioned that Bernard Sapoval would organize a Gordon Conference on fractals. Gordon conferences are informal meetings, a very good way to meet and speak and focus on science. Benoit introduced me to the concept of a research poster. To me a poster was a way to advertise a concert. This gave me a lesson in how to work with students. From high school to undergraduates. Let them get involved in research directly. Let them present a poster, let them give a talk at a meeting.

When I wrote my Ph. D., 94 to 96, Benoit invited me to work with him here at Yale in the mathematics department, and as a research fellow in the chemical engineering department. Working with Benoit was a special experience, because every conversation you would have with him would not be typical of an advisor talking with a student. Instead of focusing only on the specific research topic, our conversations would go in every possible direction. I would learn a lot about the periphery, not necessarily about the theme itself, but in some way it influenced what I was going to do.

What amazed me also was his idea of visuals and of drawings. We were working on this multiplicative cascade of functions, a new way to generate multifractals, to expand multinomial multifractals. He said why don't you try just this? I think the result will be as follows, and he would draw a picture. When I did all the computations and thought deeply about the problem, it was very close to what he thought. A very important lesson was not just of his insights and his intuition, but also the process that I learned by talking to him and when I thought that maybe something is different. For example, both in some of his notes and in articles by him, I saw something I thought might not be correct. I was almost afraid of saying that, because I thought probably I'm wrong. It was a big surprise to me then that he took every one of these comments and criticisms seriously. He'd say, "Well probably it is. Let me look at it again." He would never deny something or be defensive. This attitude often is misunderstood. At conferences he would stand up and say, "I did this before. Look in my paper of 1962." First of all, this would be true. But also, just as we see in *The Fractal Geometry of Nature*, he is very careful about citing other people. This happened also when I was working on these multifractal cascades. He said, "Well, you might want to look at a theorem, I believe it's by Weyl, and there's also Hardy and Littlewood's paper. I believe it's 1916." Then I would go to the library, find this article, and in a footnote there was some mention about uniformity of random numbers. I was baffled. His memory was amazing, but also amazing was how he made connections to the things I was doing.

I'd like to allude to some of the amusing things of working with Benoit. When I started working with him, I drove up to Scarsdale and stayed the first two nights there. Pretty often we would drive together to Yale and it would be Aliette who would be driving. It was raining a lot and there was always condensation in the car. He would say, "You have to push this button," and then there would be even more condensation. Aliette would say, "No, I think it's this button. You have to have some fresh air." He could be particularly stubborn about some things. The next he would adapt and say, "You know, I think you were right. Maybe we should try that button."

After we became friends, he would call me up to discuss issues that concerned him. For the memoirs he would ask how should I write this up? When preparing his TED talk, which became very popular and exposed a new generation to fractals, he would ask if I had any ideas about how to explain this particular concept. He would always be open, and for me that was always a sense of extreme modesty. Always there was this idea of trying to explain things better, and also of self-doubt, put his own work always in perspective. Was this right? Was this wrong? Does this make sense? I think this is a great sign of someone who is extremely intelligent.

Another amusing thing was that when working here at Yale he tried to eat little at lunch. Aliette probably tried to encourage him to do so. Typically we would go to Lord's. He liked the big hall, though the food I'm not so sure. He would go for the little bowl of soup and maybe a sandwich on the side. He ate so modestly. Such a big man. Surely he's still hungry. Then on the way back to 10 Hillhouse Avenue he would say maybe we should take a detour down a side street. He'd say that it is getting hot, maybe we should have an extra scoop. And he'd have an extra scoop. He felt much better after that.

He was a great scientist, but also an engineer really interested in applications, and an artist most of all. Coming from a family of artists I think this is very important. Benoit saw the beauty in the world and opened my eyes to a lot of new beauty in this world. I will always remember that.

25.19 Nial Neger: Life in the fast lane

I want to share two stories. First about how I met Benoit. I was a high school mathematics teacher, retired in the spring of 1999. By August of 99,

I wasn't going to miss going back into the classroom, but I could see that I was going to miss mathematics. I had been doing some reading about fractals and thought why don't I take a formal course in fractal geometry. The only place I could find one was here at Yale, so I arranged to audit the course. That year the course, mathematics 290, was being taught by Michael Frame. So I took the course and got very enthusiastic. So enthusiastic that by the end of the semester I wanted to continue on. But the course stopped. I approached Michael and asked if there was something I could do to continue studying fractal geometry. He asked me to come to his office to discuss what I could do.

It was the end of the semester, and anyone who teaches won't be surprised that when I got to Michael's office there was a line of students who had just realized the semester had ended and maybe they needed help with some topics. So I stood there waiting patiently in the hall, because I figured the students came first. While I was waiting, down the hall came Benoit Mandelbrot and Aliette. He was going to his office, which was right next to Michael's. As he walked by, he looked at me, nodded, and said, "Hello." I thought Wow, he just said Hello. This is more than I can believe. A couple of minutes later he came out of his office, going to get a cup of coffee. He stopped and asked if I was one of Michael's students. I said, "Yes I am. I'm looking for something to do at the end of the semester to continue some work in fractal geometry." He said, "Oh," went off, got his coffee, came back, stopped, and said, "When you are finished talking with Michael, stop by my office. I have some projects."

So I saw Michael. We discussed what we were going to do. At the time, he and Benoit were collaborating on a book *Fractals, Graphics, and Mathematics Education*, a book about teaching fractals. We decided I would help with the editing of that book.

Then I stopped in to Benoit's office. He and Aliette were there. I thought I'd be there two or three minutes. He began telling me about his projects. And did he have projects. We talked about the teaching book and how I might help with that. During the conversation, he asked if I'd ever organized or run workshops. While I was a teacher, I had done quite a few workshops. At the time, I didn't know that Benoit was very seriously interested in teaching teachers so they could teach their students about fractals. He was very serious about this. He lit up when I said, "Oh, yes, I can put on workshops." Before I knew it, Michael, Benoit, and I were planning a workshop for teachers. We gave the first one in the summer of 2000, and it continued for five years after that.

I never figured out why he stopped and talked with me in the hall. It might be that I was a little older than the rest of Michael's students. That chance meeting certainly changed my life profoudly.

Later in the fall someone was interested in making a film about the Mandelbrot set. He had produced a trailer. Benoit called me and asked if I'd be interested in watching it with him. I said, "Sure, why not? Where is it?" He said it was in a theater on Broadway in Manhattan. So my wife Liz and I met Benoit and Aliette, the manager of the theater, the film producer, and his assistant. The seven of us went into this enormous empty theater to watch what he had done. I believe he was making an IMAX film. We made some general comments, then left. A couple of days later, Benoit called and was not very happy with the film. The producer wasn't taking the right approach to the film, he had't picked good sections of the Mandelbrot set, and so on. When we were finished discussing the film, Benoit said, "Well, how do you like life in the fast lane?" And I thought he's done it again, he's hit the nail on the head. That's Benoit Mandelbrot, that's what he's all about: Life in the fast lane. He introduced that to me mathematically, and it changed me profoundly.

25.20 Miguel Garcia: Benoit loved to proselytize

In 1987 I was fortunate to be awarded a fellowship for teachers in local Connecticut colleges that allowed me to attend seminars and go to classes at Yale. My sponsor that year was Ronald Coifman, who was then the chair of the math department. I had made a proposal of the things I wanted to do and then Ronald said "This year we're bringing Professor Mandelbrot in. Perhaps you'd be interested in working with him." I knew about Professor Mandelbrot, of course. This opportunity seemed beyond belief. Indeed, I approached him, and it was a transformative event for me. Among other things, I also should tell you that I was able to sit in the graduate course that he gave that first year. The audience had people from the geology department, from engineering departments, you name it. And there was such fervor and such interest in the fact that this great man was here at Yale. Half of the math department was sitting in the class.

In addition to that, I started working with him because he had brought Ken Musgrave to do all the fractal renderings. I started working with Ken, and Benoit asked me to bring in some students from Gateway to help. It was just amazing.

He was not just a great intellectual — his position in the history of mathematics and in the history of ideas is there in the world forever — but his humanity, his sense of humor, the generosity of his time and his counsel, were really compelling.

Not too many years later he brought in Michael, and that allowed him to get into education, an area where he had such a strong interest. Benoit loved to proselytize all the beautiful things he knew and felt that people needed to hear. Even prior to your [Neger's and Frame's] fractal workshops, I organized several workshops for teachers at my own school. These were ultimately due to Benoit, because he got me interested in fractals. I cannot tell you how many dozens and dozens of mathematics professionals have been introduced to these ideas because of that germ that Benoit laid down.

Then one day in the 90s, I got a call from Benoit, asking if I had a student that could work with him on some projects. I said, "Ah, do I have a student for you." That was Harlan Brothers.

I am going to miss Benoit tremendously, but his memory will always be with us.

25.21 Peter Jones: A combination of simplicity and magic

From the bottom of my heart we must thank Benoit for this. Benoit was a very, very speical person. First of all, I miss him very badly. We wrote only one paper together. It's a short paper, but Benoit had a very deep impact on me. I want to find a way to express this type of impact. Every schoolchild knows about the Mandelbrot set. Beautiful pictures, it has a deep connection to all sorts of things. But I think there's actually a much greater impact. It's the Mandelbrot mind-set. This is not a set in the sense of topology, but it is the mind-set. Benoit approached the world with an enormous curiosity. He didn't come in and say I am studying field X, I'm going to look at Y. We will examine the following property Z. He would go in and ask What is it? Let's take a look. Maybe it's related to this, and maybe it isn't. When you do things like this with a deep scientific background of the type that Benoit had, then magical things start happening. You start looking at things that almost by definition people have not looked at before. This is where the really great revolutions in science come from.

I was impressed over and over again because while so many things in my work are not directly descended from Benoit's work, nevertheless, his work was a guiding light. "You can see sort of where to go because the picture looks like this."

I really hope people will remember Benoit's many senses. He had a childlike curiosity; he wanted to understand how everything looks, and then bring a very deep scientific background to knowing how to look. The end result is some sort of combination of simplicity and magic.

Finally, I would like to thank Aliette and Didier for sharing Benoit with us.

Chapter 26

Some final thoughts

Michael Frame

One afternoon Benoit invited me to a master's tea in his college at Yale. He and a group of colleagues discussed opera, the distribution of galaxies, Hokusai's art, Stoppard's *Arcadia*, many other topics. Despite the conversation's swerves, Benoit contributed detailed opinions to each new direction. When he excused himself to speak with someone on the other side of the room, a diminuitive Cuban poet who joined the group late turned to me, clearly Benoit's guest, and asked, "Who is that enormous man?" Before I could reply, another said, "Oh, that's Benoit Mandelbrot. He reinvented geometry." "Ah, a scientist. Then why does he know opera and art?"

Why, indeed? Benoit was familiar with many fields, worked in many fields, did important work in many fields, but few people saw the whole picture. Some who knew Benoit's work in finance didn't know of his innovations in Kleinian groups. Few physicists investigating DLA had read Benoit's thermodynamics of discourse. Those who thought all he did was the Mandelbrot set had no notion that Benoit pioneered measuring physical roughness by an application of the arcane mathematics of Hausdorff dimension. Benoit influenced many areas in science, and as underlined by some of the chapters in this volume, he also left an imprint on music, art, architecture, engineering, medicine, and education. Yes, as the Cuban poet said, Benoit was an enormous man.

We've seen hints of Benoit's curiosity, playfulness, and sense of humor, facets less familiar than his scientific brilliance and breadth of interests. And Benoit had skills in the physical world that surprised people unfamiliar with his teenage metalworking experience. For example, he arrived grease-covered at a conference when his car broke down in transit, stranding Benoit and his passengers. Benoit removed his jacket, opened the hood, and did something or other to fix the car. (This is not a skill I have, so the mechanical details of his story found no purchase in my memory.) Of all the Benoit anecdotes I've told, this one has produced the most looks of incedulity. Yes, Benoit's story is complex.

I'll close this volume with a sketch of another dimension of Benoit's life. During one of my first visits to Yale, Benoit told the story of his family's flight from Warsaw to Paris, and when Paris became too dangerous, to the town of Tulle. With a nod to my childhood home, Benoit said Tulle was in the Appalachia of France. At some personal risk, people there kept Benoit's family safe during the occupation. After that, Benoit's life unfolded in a complicated way, evidently providing the right background for him to find the thread of fractals woven through so much of the worlds grown, built, and imagined.

Not long after sharing that story, Benoit mentioned that he had been invited to be the guest of honor at a celebration in Tulle. He and Aliette would be in France then, so a side trip would be easily arranged. Benoit said he wanted to use this visit to thank the people of Tulle, whose kindness and generosity likely saved his life, that their good hearts deserved some of the responsibility for all his work. I was charmed. This would be a beautiful gesture, recognizing the actions of these people. Benoit did appreciate the work of others, happily shared a grin and "Bon," when one of us found something new and interesting. But what he would say at Tulle went far beyond that. This would be a wonderful evening.

When we met after this trip, Benoit summarized the highlights: scientific, historical, and cultural notes woven together, his usual mode of story telling. But he said nothing of Tulle. When I asked, he replied, "It was very pleasant." "Very pleasant?" I thought, "Very pleasant! Half the audience should have been weeping at that story told by this person." What had happened?

Then a flash of insight. "You didn't tell them, did you?"

"No, Michael, I didn't."

"Why in the world not?" This was the only time in twenty years that I raised my voice to Benoit. "They should have heard that story. It would have meant so much to them. How did you not tell them?"

"Michael, I am an old man. I have not yet lost control of my emotions in public, and I knew if I told this story that I would break down. I'll tell them later. There will be other opportunities."

But there weren't. Other projects, other travels, work on his *Selecta*, work on his memoirs. Then he got sick. Then he died. He never told this story to the people of Tulle. Better than most, Benoit understood the vagaries of circumstance. We can each of us control very little of what happens to us. Things important but difficult we postpone because we think we have a long time to do them. But we may not, and eventually, we do not.

Sadly, I did not learn from this. For you see, I did not tell Benoit how impossibly lucky I was when he invited me to work with him. He knew the fact, that I resigned a tenured full professorship at Union College for an adjunct professorship at Yale, just for the chance to have many face-to-face conversations with him, to learn a bit of how his mind worked. But I never told him how dizzying was my first walk down the fourth floor hall in Dunham Lab to his office, when I was about to get to spend a year — and, it turned out, twenty years — with him. I didn't tell him, and now I can't.

So Benoit and I made similar mistakes. This hardly seems worth calling another dimension of his life, but it points the way to one, I believe. When Benoit and I talked about fractals, we spoke almost always of physical or mathematical fractals. On a few occasions we speculated about fractal aspects of how we live. Benoit's not telling the people of Tulle was a large event. My not telling Benoit was similar, on a smaller scale. Events of these sizes are noticed easily, but our lives are rife with situations where we should say something, but don't. The opportunity vanishes, we forget the circumstance, the possibility of a connection is gone, maybe we scarcely notice. I think this idea, a fractal geometry of behavior, was beginning to grow in Benoit's mind. Had he lived longer, would Benoit have pursued these directions? We cannot know, but given enough time, can you think of any place Benoit's energetic imagination would not have gone?

I cannot.

Index

www.ingramcontent.com/pod-product-compliance
Lightning Source LLC
LaVergne TN
LVHW010635300926
849161LV00005B/946